I, Too, Sing America

THE LIFE OF LANGSTON HUGHES

Volume I: 1902–1941

I, Too, Sing America

Arnold Rampersad

New York Oxford
OXFORD UNIVERSITY PRESS
1986

Oxford University Press

Oxford New York Toronto
Delhi Bombay Calcutta Madras Karachi
Petaling Jaya Singapore Hong Kong Tokyo
Nairobi Dar es Salaam Cape Town
Melbourne Auckland

and associated companies in
Beirut Berlin Ibadan Nicosia

Published by Oxford University Press, Inc.,
200 Madison Avenue, New York, New York 10016

Oxford is a registered trademark of Oxford University Press

Library of Congress Cataloging-in-Publication Data
Rampersad, Arnold.
The life of Langston Hughes.
Bibliography: v. 1, p. Includes index.
Contents: v. 1. 1902–1941, I, too, sing America.
1. Hughes, Langston, 1902–1967—Biography.
2. Poets, American—20th century—Biography. I. Title.
PS3515.U274Z698 1986 818'.5209 [B] 86-2565
ISBN 0-19-504011-2 (v. 1)

2 4 6 8 10 9 7 5 3
Printed in the United States of America
on acid-free paper

CONTENTS

Hang yourself, poet, in your own words. Otherwise, you are dead.

LANGSTON HUGHES, 1964

THE LIFE OF LANGSTON HUGHES

1

A KANSAS BOYHOOD
1902 to 1915

> This is a song for the genius child.
> Sing it softly, for the song is wild . . .
>
> "Genius Child," 1937

IN SOME RESPECTS he grew up a motherless and a fatherless child, who never forgot the hurts of his childhood. But in his life, as in his art, Langston Hughes laughed so often and so loudly that the tragedy of his earliest years, which is the way he remembered them, was finally almost always hidden. He did not fool everyone. If many people relaxed before his boyish glamour, a perceptive few also glimpsed an original unhappiness behind his chronic chuckle and ready, remarkable laugh, or intuited in his many sentimental gestures, especially to the young, the memory of early pain. In the last Christmas of his life, for example, when he sent Marianne Moore a copy of his latest book, an anthology of short stories, Hughes typically pointed to the youngest author in the collection. Moore should read the last story, written "by the youngest good writer I know, Alice Walker, just past 21." But Moore was not fooled; she sagely looked past Hughes's act of generosity and saw his most recessed motive. "Inimitable, irresistible Langston," she saluted him. "I do not know why you were not spoiled with love and care, from the cradle, on, and were not a proud boy!"

Far from being spoiled with love and care, Hughes grew up with a wrenching sense of having been a passed-around child who craved affection but received it only in episodes. This unappeased hunger left him—in spite of his gift of laughter—a divided man. When he was past fifty, a scholarly friend pointed out to Hughes his repeated use of the theme of the "tragic mulatto." Typically, a young man of mixed race is caught disastrously between the black and the white worlds, but especially between longing for acknowledgment by his white father and being disowned by him. Hughes had taken up this theme so often, the scholar mused, he was perhaps unconsciously drawn to it. For once candid about his past, instead of teasing it into vagueness, Hughes admitted that he identified with such a doomed young man even though neither of his own parents had been white. "I was pretty conscious of the derivation of my own understanding of such a boy," he conceded, "from my own parental situation."

In his autobiography *The Big Sea,* Hughes smiled and smiled and gave only hints of his ambivalence toward his mother. "My grandmother raised me until I was twelve years old," he wrote. "Sometimes I was with my mother, but not often." About his father, however, he was blunt: "I hated my father." In turn, his father did not so much hate Langston as pity him and loathe his son's absurd enterprise—to live by writing, and to do so by writing mostly about black people. With so much force on either side, their break was like the snapping of a bone. But the split, however painful, must have often seemed preferable to Hughes to the sheer untidiness of his relationship to his mother. Far from hating his mother, Hughes loved her hopelessly. In turn, she undoubtedly loved him, if in her own way. Once, self-serving, she summed up their relationship: "Some time I feel that you & I were never as close, Heart & heart as we should be, but I have loved you very dearly and if I failed in some things it was lack of knowledge."

Hughes's final reply, in a sense, was his unkempt portrait of her in *The Big Sea;* few readers would have guessed that she had died of cancer within a year of his writing. Nor was he warmer to his grandmother, with whom he had lived until he was thirteen, and who had nurtured his imagination with tales of heroism but without the loving tenderness the boy would have preferred. "When my grandmother died, I didn't cry," he remembered. "Something about my grandmother's stories (without her ever having said so) taught me the uselessness of crying about anything."

As successful as his life seemed to be by its end, with honors and awards inspired by more than forty books, and the adulation of thousands of readers, Hughes's favorite phonograph record over the years, spun in his bachelor suite late into the Harlem night, remained Billie Holiday's chilly moaning of "God Bless The Child That's Got His Own." Eventually he had gotten his own, but at a stiff price. He had paid in years of nomadic loneliness and a furtive sexuality; he would die without ever having married, and without a known lover or a child. If by the end he was also famous and even beloved, Hughes knew that he had been cheated early of a richer emotional life. Parents could be so cruel! "My theory is," he wrote not long before he died, "children should be born without parents—if born they must be."

Whatever Hughes's theory about parents, however, he was also born into a relationship with his family's past, into a relationship with history, so intimate as to be almost sensual. Much was expected by his ancestors. They demanded, from the moment his elders recognized the boy's unusual intelligence and began to talk to him about Duty and The Race, that he had a messianic obligation to the Afro-American people, and through them to America. Among these ghostly but commanding figures were a white Virginia planter, the poet's great-grandfather, who had defied the mores of the South to live with the black woman he loved and their children; two of their sons—one who risked almost everything in fighting against slavery and segregation, another who had also fought for freedom but lived to serve in the U.S. Congress and represent his country in Haiti; a black hero killed at Harpers Ferry in John Brown's band; and John Brown himself, an ancestor of Langston Hughes by virtue of the blood spilled there. For these men, their best spokesman in Hughes's youth would be his proud, tale-telling grandmother; from their demands, ironically aided by parental neglect, which made him anxious for love and a settled identity, would come much of the purpose of his life.

He was born near midnight on February 1, 1902, in the city of Joplin, Missouri. The date of his birth he would take on faith, since Missouri did not require the registration of infants, and his birth was never entered officially there. He was named James Langston Hughes. Soon the first name (after his father) was ignored, and he grew up as Langston Hughes. His earliest memory was not of Joplin but of a house of two bedrooms and a side sitting-room at 732 Alabama Street in Lawrence, Kansas. Off to one side was a shed kitchen, and behind the house stood a woodshed, an outhouse, and a pump for drawing water. The neighborhood, quiet and shaded, nestled at the foot of a hill crowned by Kansas University; from the top one could see plains of wheat rolling to the horizon, and the silty green crook of the Kansas River where it bent toward the northern edge of Lawrence before drifting east toward the Missouri. Here, in almost the exact center of the continental United States, Hughes would pass most of the first thirteen years of his life.

Mostly he lived with his maternal grandmother, "a small woman, brown, slightly bent, with very long hair almost to her waist and only slightly gray in places. . . . Her face was very wrinkled like an Indian squaw's." Mary Sampson Patterson Leary Langston was almost seventy years old when Langston was born. He remembered her spending most of her time in an old-fashioned rocking chair; except to attend the weekly meeting of her lodge, Mary Langston generally stayed at home. At night she read a chapter of the Bible, combed out her long hair, rolled it under a white nightcap, and went to bed. "She never shouted or got happy, and at night when she knelt down and prayed, she prayed silently." But she also read to her grandson from Grimm's fairy tales, or from the Bible and whatever magazines and newspapers she could afford. Or she held him in her lap and in a calm, clipped voice related tales of heroism, of slavery and freedom, and especially of brave men and women who had striven to aid the colored race. "Through my grandmother's stories always life moved, moved heroically toward an end," Hughes would recall. But of black folkways, so important later to her grandson, she said nothing. "She had been away from the South since her youth so she evidently did not remember any Negro folk-stories. At least, she never told me any."

Of Indian, French, and some African ancestry, Mary Langston had been born free in North Carolina, where she grew up as the ward of a prosperous colored mason and his wife in Fayetteville. In 1855, about nineteen years old, she survived an attempt to enslave her. With papers from two white men attesting that "she is of most respectable & moral character without reproach" she fled north in 1857 to the Abolitionist stronghold of Oberlin, Ohio, where she enrolled in the preparatory department of the college. There she was courted by another citizen of Fayetteville, also of French, Indian, and African blood, Lewis Sheridan Leary. A harness maker by profession, Leary was handsome, athletic, and musical; he was also a dedicated Abolitionist. Married in 1858, the young couple served as conductors on the underground railroad that brought slaves to freedom in the North.

One evening in 1859, when Leary was twenty-four, he rode away without telling his wife, who was pregnant, where he was going. On the night of October 16 he was in John Brown's party of twenty-one men when they attacked the federal arsenal at Harpers Ferry. The next day Leary was cut down by gunfire as he tried to cross the Shenandoah. Dragged to a carpenter's shop, he endured two days of bitter pain before he died. A friend brought his bloodstained, bullet-riddled shawl home to Mary in Oberlin. Despite her loss, she always revered the memory of John Brown. "His soul is marching on," she wrote with a shaking hand late in her life; "I am proud that he and his followers are not forgoten who braved death for Liberty to an opress race." Lewis Leary's shawl remained a symbol for her of his martyrdom; she still wore it fifty years after his death, or used it to cover her young grandchild, Langston Hughes, while he slept at night.

Not long after Harpers Ferry, Leary's daughter, Loise, was born. The Abolitionist veterans Wendell Phillips and James Redpath contributed to her upkeep, and she eventually became a teacher in the South before her death near the end of the century. In 1867, Mary Leary unsuccessfully sought a job as a teacher among the former slaves in the South; she also turned down an invitation from the government of Haiti to live there as an honored guest. Then, ten years a widow, she married another passionate lover of freedom, Charles Howard Langston, the son of Ralph Quarles, a prosperous white Virginia planter, and Lucy Langston, a former slave of Indian and African blood. Ostracized by the Louisa County gentry, Ralph Quarles and Lucy Langston had lived together for decades like man and wife; at his death, he bequeathed all his money and property to their children. Charles and his older brother Gideon left for Ohio to enter the preparatory department of Oberlin College as the first blacks to take advantage of Oberlin's radical policy on admissions; eventually they were followed by a younger brother, John Mercer Langston.

Like his friend Lewis Leary, Charles Langston was an extreme Abolitionist. In 1858 he starred in the celebrated Oberlin-Wellington rescue, when radicals forcibly prevented a fugitive slave, John Price, from being returned to the South. Found guilty after a trial of inciting to riot and violating the Fugitive Slave Law, Charles Langston passionately defended his act as one against American bigotry. "I was tried by a jury who were prejudiced," he declared, "prosecuted by an officer who was prejudiced, and defended, though ably, by counsel who were prejudiced." One listener in the courtroom was John Brown himself, who tried hard to recruit Langston for the Harpers Ferry raid. A month after the attack, Langston stood up for Brown and his followers: "Capt. Brown was engaged in no vile, base, sordid, malicious or selfish enterprise. His aims and ends were lofty, noble, generous, benevolent, humane and Godlike. His actions were in perfect harmony with, and resulted from the teaching of the Bible, of our Revolutionary fathers and of every true and faithful anti-slavery man in this country and the world."

In 1862, Charles Langston moved to Kansas. There, too old to serve in the Civil War, he recruited soldiers for the 54th and 55th (colored) Massachusetts regiments. When his first marriage ended, probably with the death of his wife, he returned to Ohio to court the widow Mary Leary, over fifteen years his junior; they were married there on January 18, 1869. With a foster son, Desalines Langston, named after the revolutionary Haitian general, they lived on a farm of 122 acres in Lakeview, just north of Lawrence, Kansas. As the years passed, Charles Langston lost little of his radical zeal. When a son was born in 1870 he baptized him Nathaniel Turner Langston, after "Gen. Nat Turner, the hero of Southampton," as he had called the most famous slave insurrectionist of Virginia. Three years later, Carolina Mercer Langston was born.

For a while, Charles Langston seemed to have a future in Republican politics in Kansas. There was talk of an appointment as minister to Liberia and even of nomination as a candidate for lieutenant governor of Kansas. But he and all other blacks were steadily elbowed out onto the margin of political life in Kansas; the *Atchison Globe* scorned him as a "saddle colored idiot" for thinking that a black man could win statewide office in Kansas. In 1888, he moved his family to Lawrence, giving up the farm for the house at 732 Alabama Street and part interest in a thriving grocery store on the main Lawrence thoroughfare, Massachusetts Street. As black settlers poured into the area around eastern Kansas, fleeing from poverty and terrorism in the South, "Colonel" Langston assumed a prime position in the community. Elected President of the Colored Benevolent Society, and Grand Master of the black Masonic Fraternity of Kansas, he held other high positions in the shadowy but powerful Masonic world. As associate editor of the local black paper, the *Historic Times,* he championed the cause of his race; he founded the "Inter-State Literary Society" in 1891, opened his home to its meetings, and repeatedly urged young people to pursue their education, especially at the state university.

Adding to Charles Langston's prestige was the fame of his youngest brother; John Mercer Langston (with Frederick Douglass and Booker T. Washington) was one of the three best-known black Americans of the nineteenth century. Earning three degrees at Oberlin College, and admission to the Ohio bar, he became the first black American to hold office—as clerk of two Ohio townships—by popular vote. Between 1866 and 1869 he worked as an inspector general of the Freedmen's Bureau; afterwards he became professor of law, dean, and acting president of Howard University. He represented the United States as minister to Haiti and chargé d'affaires to Santo Domingo, then took up the presidency of a Virginia college. Crowning his career, John Langston was elected to Congress in 1890 from a district in Virginia. Like his brother Charles, John Langston was no pacifist. He once knocked down a white man in a courtroom for insulting his race (the court exonerated Langston); in Washington, D.C., he took an axe to a barrier some whites had set up to bar his elegant daily passage to Congress. He, too, had been a proud Abolitionist.

Sharing a platform with William Lloyd Garrison, Theodore Parker, Wendell Phillips, Charles Sumner, and Antoinette Brown, he had boldly asserted the rights of his race. "As the mountain eagle hates the cage, loathes confinement and longs to be free," he had thundered, "so the colored man hates chains, loathes his enslavement and longs to shoulder the responsibilities of dignified life. He longs to stand in the Church, in the State, a man. He longs to stand up *a man* upon the great theatre of existence, every where a man; for verily he is a man."

The Langston name was famous enough; but when Charles Langston died in Lawrence in November, 1892, he left almost nothing to his family. At 732 Alabama Street, the mortgage man became a forbidding figure. Carrie Langston was still in the Lawrence High School; Desalines, who had a barber's chair in Kansas City, had his own family to support; Nat Turner was briefly associate editor of the *Historic Times,* before taking a job in the local flour mill. An accident there led to his death in 1897, but to his mother's bitter disappointment, the mill refused to compensate her. "My earliest thought of riches and bosses being synonymous with oppression and callousness," her grandson would remember, "probably came from her seldom but forcefully expressed resentment against the mill interests for whom my Uncle Nathaniel had worked and died."

Mary Langston had an even greater reason for disappointment; for her and other oldtime black settlers, Lawrence had become a lost Eden. The town had been founded as an enterprise in Abolition by a party of settlers, sponsored by the New England Emigrant Aid Society, who had swept across the Missouri River in 1854, just ahead of pro-slavery forces, and pitched their tents on the windy bluffs above the Kansas and Wakarusa valleys. Lawrence became a center of anti-slavery radicalism; John Brown first made his name when he and his sons killed five pro-slavers on the banks of the Pottawatomie River during guerilla warfare in the Kansas territory. But on August 21, 1863, William Quantrill and his Confederate "irregulars" exacted revenge for the loss of Kansas. Storming Lawrence, they killed or wounded every man they found, and destroyed seventy-five businesses and one hundred homes. Quantrill's raid became the centerpiece of official Lawrence history.

The rival history, however, told wherever oldtime black settlers gathered, was of the steady demeaning of Lawrence into a segregated town. The "exodusters"—the last, the biggest, and the poorest wave of black settlers—met an often hostile white population and an uneasy, ambivalent group of settled blacks. By the time of Langston Hughes's childhood, all blacks were barred from formerly open churches, hotels, restaurants, and other social establishments. In one restaurant, blacks ate at the end of the lunch counter furthest from public view, blocked off additionally by a wooden barrier. At the theater they were restricted to the back of the balcony—"Coon Hill," or the "Buzzards' Roost," or "Nigger Heaven."

While a small group of doctors, lawyers, schoolteachers, entrepreneurs, and other strivers lived apart as the Langstons did, among whites, the masses of blacks crowded into diseased and deteriorating neighborhoods in the east "Bottoms'' on the swampish floodplain along the river, or in separate North Lawrence. A schism grew between the elite group and the black masses. ''We are a childish race satisfied to imitate those things that appeal to a child,'' Langston Hughes's only black teacher in Lawrence later wrote to him, speaking harshly for their class about their race, who filled the courts as petty criminals, bootleggers, and prostitutes. The white *Lawrence Daily Journal* made small distinction between criminals and the respectable. When the paper sponsored a ''Children's Day Party'' at the local Woodland Park in the summer of 1910, when Langston Hughes was eight, it assured its readers that ''the colored children have no desire to attend a social event of this kind and . . . will not want to go. This is purely a social affair and of course everyone in town knows what that means.''

For Mary Langston's daughter, the death of her father and the descent of Lawrence into segregation were harsh blows. Fresh from her father's farm in 1888, Carrie Langston had become at fifteen one of the belles of black Lawrence. At eighteen, encouraged by her father, she read papers before his Inter-State Literary Society and even recited her own poems. Light olive in complexion, stylish and popular, Carrie became a star in the St. Luke's Progressive Club. When the rival Warren Street Second Baptist Church founded its own cultural society, she was elected ''Critic.'' Graduating from high school following her father's death, she worked for a while as ''Deputy District Clerk'' in the city courthouse. In April, 1894, as ''Mercer Langston,'' she finished a ten-week course in kindergarten and primary school teaching at the Kansas State Normal School in Emporia. Although she never matriculated at Kansas University, in the fall she enrolled in a course in German, then tried another in English during the spring term.

With her taste for musicals, plays, novels, and plain fun, Carrie found life at home with her mother difficult. Mary Langston considered novels ''cheap''; once she attended a movie and thereafter thought movies also ''cheap.'' She sang only hymns—never the black spirituals, much less the popular songs. She refused to set foot in a motorcar. Altogether different, her daughter practiced elocution and longed for the professional stage. As a child she had played with George Walker, who left Lawrence to become a great music hall celebrity, most famous for his partnership from 1895 with the greatest of black vaudeville stars, Bert Williams. But Carrie was not only a woman but also a black, and her talent was never great enough to overcome these handicaps. Her failure to enter the professional theater began to unhinge her; she became theatrical.

When at last she made her move, she went not to Broadway but into a sagebrush fantasy. Around 1898, she reached Guthrie in the Oklahoma territory; twelve miles away, up a muddy road, was the embryo of an all-black town, Langston, named after her famous uncle, John Mercer Langston. Carrie Langston reached what the *New York Times* had once hailed as a ''black Mecca'' to find it barely a dream, although the Colored Agricultural and Normal University, popularly (and later officially) called Langston University, opened in September, 1898. Hired to teach school in Guthrie, she soon caught the attention of probably the most ambitious young black man in the territory, a sometime schoolteacher, law clerk, farmer, grocery operator, homesteader, and surveyor's assistant originally from Charlestown, Indiana. James Nathaniel Hughes was two years older than Carrie, small but taut and tough, and bronze in color. Both his grandfathers had been prominent whites from Kentucky—a Jewish slave trader, English by birth, in Ashland, and a distiller in Henry County named Sam Clay, reputedly a relative of the statesman Henry Clay. His father, James H. Hughes, born a slave, had died in 1887 a respected Indiana farmer and a presiding elder of the Methodist church. Two of his brothers had ridden with the U.S. cavalry as "buffalo" soldiers; James Hughes himself had at least twice earned teaching certificates, once passed the civil service examination for the post office in Louisville, then languished as he waited for a position. He had read law and worked in a lawyer's office in Wichita, Kansas, then come out to the Oklahoma territory, where his homestead amounted to 160 acres. But he was thinking of moving on; he had been prevented from taking the state bar examination because of his race.

James Hughes and Carrie Langston were married on April 30, 1899, in Guthrie, Oklahoma, before Judge J. L. Foster of the Probate Court. Within days, apparently, Carrie Hughes was pregnant. Within weeks, if not sooner, she discovered that her husband's ambition had a price. Where she loved to spend money, he was tightfisted. He was unsentimental, even cold. Detesting the poor, he especially disliked the black poor. "My father hated Negroes," Langston Hughes would judge. "I think he hated himself, too, for being a Negro. He disliked all of his family because they were Negroes." Where Carrie's parents had instilled in her a sense of *noblesse oblige,* Jim Hughes seemed to look on most blacks as lazy, undeserving cowards. The marriage would not last long.

In Joplin, Missouri, across the border from the northeast corner of Oklahoma, he found a decent job. Joplin was a boom town at the turn of the century, crude very near its surface, but gilded by the revenue from mineral deposits discovered fifty years before by a black slave named Pete, near a place white people still gratefully called the "Nigger Diggins." The biggest mining company, Missouri Lead and Zinc, advertised "EXCEPTIONAL Opportunities for Prospecting and Mining." When a freshly formed outfit from West Virginia, the Lincoln Mining Company, hired Hughes as a stenographer at $25 a month, he moved with his wife into a rented cottage at 1602 Missouri Avenue. The couple lived among whites, far from the grim shacks in the Bottoms where most of Joplin's blacks lived.

Tragedy soon struck. To the U.S. census taker, who came around on June 5, 1900, the couple reported the loss of their first child, a boy; apparently he was buried earlier that year beneath the gravel at the Fairview Cemetery out on Maiden Lane. According to the cemetery records, one "Hughs Inf of J M (col)" was interred in the Potter's Field of Fairview on February 8, 1900 ("Potter's Field" suggests that the burial had been at the city's expense). Only a year later, however, Jim Hughes had just over $200 in the bank, an amount equal to about two-thirds of his annual salary. In March, 1901, he apparently sold most of his belongings, including a bicycle and a bed, and left for Buffalo, New York. In mid-July, he and Carrie were still in Buffalo, but he had begun to plan a move to Cuba, then a "protectorate" of the United States. By July, Carrie was pregnant again, and probably unable or unwilling to travel overseas. Near September 1, in the middle of her pregnancy, she headed west toward Kansas and Oklahoma. She was on an outing with Jim's youngest brother, John Hughes, at a racecourse in St. Louis, when word swept through the crowd that President William McKinley had just been shot in Buffalo.

Why Carrie returned to Joplin is not clear, nor is it certain that James Hughes was with her when his son was born on February 1, 1902. The previous December he had landed in Cuba with over $800 in cash. His venture there soon failed; perhaps he withdrew in time to return to Joplin for the birth of his child. In any event, he was quickly gone again, this time to Mexico. On October 29, 1903, he commenced work as confidential secretary to the general manager of the American-owned Pullman Company in Mexico City.

His decision to abandon Joplin was wise. In April, 1903, a white mob stormed the city jail and lynched a black man arrested for the murder of a white policeman. Attacking the Bottoms, the mob burned houses indiscriminately. The next day, hundreds of terrified blacks thronged the railway station, their possessions in rough boxes and bags. After the sheriff gathered a force and finally restored order, many families returned to their homes; but "some of them never came back to the city," a local white historian reported, "and some others, feeling that the event had been a great humiliation to their race, moved away."

From the start Langston saw little of his mother, but much of the road—on which he would spend a great part of the rest of his life. Carrie Hughes took him to Indianapolis to visit one of his father's sisters, Mrs. Sarita "Sallie" J. Garvin, and also to Buffalo and to Colorado, but mostly he lived with his grandmother while Carrie worked elsewhere. Then, when Langston was five, his parents attempted a reconciliation. Carrie, Mary Langston, and the boy set off for Mexico to join his father. The reunion ended in effect at 11:34 on the night of April 14, 1907, when the ground rumbled and shook for four and a half minutes throughout a vast region of Mexico that included Mexico City, Acapulco, Chilapa, and Chilpancingo. Langston remembered his father carrying him out bodily into the street, where people were praying feverishly on their knees. With the last aftershock, hundreds of people were dead and thousands injured and homeless. The cathedral in Mexico was cracked from top to bottom; spiders and scorpions crawled from crevices. Terrified, Carrie Hughes fled Mexico with her son and mother.

Not long after, left with his grandmother in Lawrence, Langston began a precocious discovery of loneliness. At least once, by his own admission, he ran away from home. About a year after his mother's return from Mexico she came to take him away again. She was working in nearby Topeka, the capital of Kansas, as a stenographer in the office of a black attorney named James Guy, and also for a local black newspaper, the *Plain Dealer*. Her son joined her in a room on the second floor of a building at 115 West Fifth Street in the midst of the mainly white commercial section of the city. After the solemnity of the house on Alabama Street, the child was pleased to be in downtown Topeka and to be near two men who lived on the same floor: a kindly, very old white man, by profession an architect, and a young black artist who painted wonderful pictures of tigers and lions roaming in the jungle. But Langston found life with his mother often unpleasant. Her failure to reconcile with Jim Hughes had left her bitter; he sent money for his son, but hardly enough to assuage her sense of hurt. At thirty-five, Carrie also began to face the painful likelihood that she would never have a career on the stage. Loving fine clothes, new books, plays, and travel, she indulged her tastes when she could but most of the time she barely survived. In the winter she took Langston into the streets under cover of darkness to pick up odd bits of wood for kindling, jumping on abandoned packing crates to break them up. When the sturdier pieces stuck out of the little monkey stove, she laughed and called them "long-branch kindling." In unhappier times, Carrie remembered her husband and vented her anger on their son. Langston was the spitting image of "that devil." "You're just like Jim Hughes," she would say. "You're as evil as Jim Hughes." When the child retreated into hurt silence, Carrie sometimes found his withdrawal so maddening she called him dull and stupid.

There were happier times. Even as a child, Langston was his mother's son in his passion for the theater and the road. They took the train together to Kansas City to visit Desalines Langston, who ran a barbershop on Charlotte Street in the city. Langston never forgot his first visit—the rattle of the train wheels, the bellow of the brakemen, the clanging bells at the bustling station, and at the end of the journey wonderful food at his "uncle" Des's and, at last, the music hall, where Hughes's own lifelong love of the theater was born. He owed to his mother another lasting pleasure. In Topeka, Carrie took him regularly to the public library, a small but impressive ivy-covered stone building on the grounds of the state capitol; he was entranced by the bright silence of the reading room, the big chairs, the long smooth tables, the attentive librarians who fetched books at his command. He learned to read and write; his first surviving letter concerns a book. To one of his mother's friends, Carolyn F. Battle ("Dear Carrie"), he sent a note: "I am well. I got one tooth out the chair fell on it. I want to see you. I got the book. Ma read too stories in it. I like them. when are you coming?"

His most memorable early encounter with segregation came in the late summer of 1908, when he was ready to start school. Carrie Hughes tried to place him in the first grade class of the nearby Harrison Street School in Topeka, but was rebuffed and told to take him instead to the Washington School for colored children across the railroad tracks, a considerable distance away. Carrie argued with the principal that she worked every day, and that Langston was too small a child to go through the city streets alone. When the principal, Eli S. Foster, refused to admit him, she appealed directly to the school board, argued her case, and won. (Ironically, in 1954 another victory over the Topeka Board of Education, before the U.S. Supreme Court, would mark the official end of segregation in the United States.)

Victory came at a price: Hughes began school a doubly marked child, black and a troublemaker. His teacher seemed to some people gentle and kind, but from the first day she tried to break him. Seating everyone but Langston alphabetically, she installed the baffled child deep in a corner, at the end of the last row. She peppered him with unkind remarks even after it was clear that he was a superior student. One day she took licorice sticks away from a white boy. "You don't want to eat these," she said loudly for the class to hear; "they'll make you black like Langston. You don't want to be black, do you?" At midterm she graded him "excellent" in three of his five subjects, but still taunted him. Some of his schoolmates slapped, stoned, or snowballed him on his way home; to his defense came other children, white themselves, from his neighborhood. Nevertheless, Langston did not lose his liking for school. His clothes neatly pressed each night by his mother, already in love with compliments, he was a model student.

In mid-April, 1909, before the end of his first school year, Carrie Hughes withdrew Langston from Harrison and returned him to Lawrence and his grandmother. Carrie herself was off to Colorado Springs, Colorado, where on a visit that summer the boy saw unforgettable mountains with great white pointed peaks rising above the clouds. At the end of the summer, however, he was back with his grandmother in Lawrence. By this time, Mary Langston was past seventy and seldom left her home. Once, she took him to Topeka to hear a speech by the greatest colored man in the world, Booker T. Washington, the head of Tuskegee Institute in Alabama. On another rare excursion, she and Langston went to Osawatomie, Kansas, on August 31, 1910, for the dedication of the John Brown Memorial Battlefield there. As the last surviving widow of the Harpers Ferry force, Mary Langston was given a seat of honor on the platform (as Hughes later remembered). With the spirit of John Brown at hand, former president Theodore Roosevelt delivered his almost radical, celebrated "New Nationalism" speech, in which he stressed the primacy of humanity over property rights, and called for a powerful central government to curb big business and to reform the courts. For Mary Langston, the event was the last, long-delayed honor of her life; for her grandson, only eight years old and largely unconscious of its meaning, it was an unacknowledged summing-up of the radical heritage to which he belonged by birth, and a prophecy of his life to come.

When Mary Langston stayed at home and forbade her grandson to go out after school, one reason was her hatred of segregation. Because blacks were refused membership in the church of her choice, which was probably Presbyterian, she attended no church at all. Some segregation, however, was beyond her control. When Hughes entered the second grade of the Pinckney School in the late summer of 1909, he joined the other black children of the first three grades in one classroom supervised by a black schoolteacher. (Curiously, integration began in the fourth grade.) Nevertheless, he was fortunate in his teacher. Remembering Charles Langston as a "deep thinker," Mary J. Dillard was touched by his grandson. At seven years Langston was "a dreamy little boy," but very good in class. Outside of the classroom, however, Langston and the other black children knew the harsher side of segregation. White children could swim freely at the YMCA and join the Boy Scouts; blacks could do neither. Colored boys could not take part in the grammar school track meets, play on the school teams, or form a team to play against whites. The Patee Theater on Massachusetts Street used to admit blacks; then one day, when Langston put down his coin, the woman in the box office pushed it back and pointed to a new sign: NO COLORED ADMITTED. "Your people can't come anymore," she told Langston. Until the Bowersock Opera House opened in January, 1912, with a "Coon Hill" or "Nigger Heaven" of its own, he could only listen wistfully while his white friends talked of the latest exploits of Charlie Chaplin, Dustin Farnum, and Mary Pickford.

Often he cried for his mother to come and take him to Kansas City, where she now lived. When she could, Carrie Hughes came to Lawrence to fetch him, or met him at the train station in the colored "Bottoms" section of the city. Together they saw plays, not all of them for children—*Under Two Flags, Uncle Tom's Cabin, Little Lord Fauntleroy, Buster Brown*. They attended the opera *Faust*. When the renowned Sarah Bernhardt came to town and Carrie could not afford his ticket, Langston howled in disappointment; the theater was in his blood. When he was not with his mother in Kansas City he often played in his Uncle Desalines's barbershop, or wandered the streets nearby, including the red-light district, and patronized the nickelodeons to his heart's content.

Feeling his mother's absence as rejection, Langston dropped deeper and deeper into fantasy. From his grandmother, too, came little warmth; he had outgrown her tales of heroism. Accustomed to measuring her words, she now hoarded them; he would remember her, finally, as kind but "old, old." Langston began to believe "in nothing but books and the wonderful world in books—where if people suffered, they suffered in beautiful language, not in monosyllables, as we did in Kansas." Often he passed the hours between the end of school and sleep in a kind of trance, reading and dreaming while his grandmother rocked quietly in her chair before the coal stove. Reading to escape loneliness, he felt it only more acutely. He was drifting through Florence L. Barclay's *The Mistress of Shenstone* near twilight one day when he awoke to a fearful revelation. "Catch it, Melvin. Catch it," he heard boys playing ball on a vacant lot yell. "Boys playing ball in the dusk, running and shouting across the vacant lot on the corner. For the first time loneliness strikes me, strikes me terribly, settling down in a dull ache round about, in the dusk of the twilight, in the laughter of the playing boys. . . . I am lonely."

Among the university buildings at the top of the hill he discovered the morgue of the medical school. Admitted by mischievous students, he stared in fascination as they lightheartedly cut up cadavers. He returned again and again to watch them. Once, when he ran away from home, he spent the night at the morgue. All his life Hughes would be fascinated by death—"Dear Lovely Death," as he called one of his collections of verse, "that taketh all things under wing." Hughes would also so richly assert the joy and the social purposes of life that most of his readers seldom notice this deadly undertow in his art; yet his fascination was certified throughout his career by many poems about melancholy, loneliness, and suicide.

If fascination with death was part of a strong passive tendency, the will to rebel was also strong. On at least two public occasions he struck back at his mother. Once, when she came from Kansas City for a Sunday school concert, no prompting could make Langston start his memorized speech. He went further when Carrie presented a program of dramatics at St. Luke's Church. As "The Mother of the Gracchi," Carrie wrapped "togas" around her son and another boy, who were supposed to stand pitifully on stage while Cornelia, celebrated for her devotion to her sons, lamented their fate. Whether or not he saw Carrie's choice of roles as ironic, he ruined his mother's show. While she emoted, he tried out a variety of grimaces on the audience, who first tittered, then guffawed. When Carrie redoubled her efforts, Langston stepped up his own. "Wilder and wilder I mugged, as the poem mounted, batted and rolled my eyes, until the entire assemblage burst into uncontrollable laughter." When she found out what he had done, Carrie whipped him soundly. But Langston had gained his revenge.

His life with his grandmother was not much more pleasant. Mary Langston's situation had become desperate. Langston never forgot his humiliation when another boy, passing the summer with him in Lawrence, wrote home pleading to return because Mary Langston served mostly dandelion greens; reading the letter, Hughes had burst into tears. Sometimes his grandmother rented a room in her home to black students at the university; at other times she moved out entirely in order to rent the whole house, and lived with friends. Her economies sometimes upset her grandson. "She kept me neatly and cleanly pressed, but she did not mind my wearing made over clothes . . . or even the shoes women would give her for herself and which were very embarassing to a boy to have to wear. She would always say, 'It is not what you wear, it is what you *are* that counts.' But other children were inclined to make fun of what one wore."

Mary Langston's economizing led, however, to a friendship that would be of great importance to the man Langston would become. In 1909, 1913, and 1914, according to city directories, Mary Langston lived not at her home but at 731 New York Street in Lawrence. Between 1909 and the day in 1915 he left town for good, Langston probably lived at least three times at 731 New York Street, the home of James W. Reed and his wife, Mary. Born a slave in Missouri, "Uncle" Reed was a ditchdigger and pipelayer for the local Kennedy Plumbing Company. "Auntie" Reed, some fifteen years younger, was a longtime resident of Lawrence and a member of St. Luke's Church, where she ran the Sunday school. On Sunday, while Auntie Reed passed virtually the whole day in church, Uncle Reed honored mostly himself. Southern style, he "washed his overalls every Sunday morning (a grievous sin) in a big iron pot in the back yard, or then just sat and smoked his pipe under the grape arbor in summer, in winter on a bench behind the kitchen range. But both of them were very good and kind—the one who went to church and the one who didn't." Without children of their own, they drenched Langston with love. Uncle Reed's city job and their vegetable garden and cows and chickens ensured that the Reeds set a bountiful table. Mary Reed sold eggs and milk on the side; they let Langston collect the eggs and walk the cows to pasture. Nearby stood the station of the Atchison, Topeka, and Santa Fe Railroad, and shining tracks that led to Chicago. A few footsteps beyond the house, immense and mysterious to an imaginative boy, flowed the Kansas River.

"For me," Hughes would write, "there have never been any better people in the world. I loved them very much." Tiny but energetic and indomitable, Auntie Reed made only one demand of Langston—attendance at her Sunday school. He grumbled a little, and dated his dislike of church from one translucent spring morning when he was forced to stay indoors to memorize verses from the Bible. Through Mary Reed, however, a fresh window opened on the black race and the world, one shut tight at his grandmother's house. Mary Reed was a Methodist Episcopal, but she carried no grudge against the black Baptists over on Warren Street. In fact, the focal point of Sunday, apart from church service and the Sunday school, became the forum of the Warren Street Baptist Church. There, university students and talented townsfolk gathered to sing and play classical music, recite poetry, read original essays and other compositions, and discuss the affairs of the day, especially as they affected The Race. What fascinated Langston was not the decorous forum but the drama of black religion, with its fiery sermons, inspired responses, and passionate, skilled singing. In contrast to St. Luke's, the Warren Street Baptists drew a poorer, more "down-home" Southern congregation. The sight of black-skinned worshippers captured by the Holy Spirit, their ecstatic groans and cries, fired young Langston's imagination. Here was high drama enacted by people of his race but well beyond his grandmother's world, people with skin glistening dark as jet (far darker than that of the pale Langstons), ancient women with colored rags around braided hair, old men with gnarled black hands, some of whom must have been born in slavery. The boy felt both at home and yet utterly apart, at no time more so then when he watched Auntie Reed herself transported by religious ecstasy while he, impenetrable, watched in wonder.

On a visit to Kansas City he became aware of yet another aspect of black culture on which he would draw later as an artist and an individual. At an open air theatre on Independence Avenue, from an orchestra of blind musicians, Hughes first heard the blues. The music seemed to cry, but the words somehow laughed. The effect on him was one of piercing sadness, as if his deepest loneliness had been harmonized. He would remember the refrain of one song he heard, and employ it brilliantly at a crucial point in his career as a poet:

> I got de Weary Blues
> And I can't be satisfied.
> Got de Weary Blues
> And can't be satisfied—
> I ain't happy no mo'
> And I wish that I had died . . .

Between the church and the blues singers, and in spite of his youth and his cloistered life with Mary Langston, the world of black feeling and art opened before Langston. He neither felt religion nor could sing the blues, and yet both the religious drama and the secular music soothed and diverted him from his sense of solitude. They also alerted him to a power and a privacy of language residing in the despised race to which he belonged; approaching the church and the blues as an outsider, because of his grandmother's own forbidding distance, Langston only respected them the more.

About the age of twelve, he found his first job; he gathered maple seed and sold it most likely to the Bartheldes Seed Company on Massachusetts Street. Then he started to deliver the *Saturday Evening Post* and a weekly newspaper, almost certainly the *Lawrence Democrat,* which (unlike the *Daily Journal*) occasionally published stories about black life in Lawrence. And for a short time, before he was warned off by a local editor, he distributed the best-selling socialist weekly in America, *Appeal To Reason,* published in Girard, Kansas. With money from these jobs, Lawrence began to unfold for him. For a nickel he could ride the city streetcar that circled the town, going all the way out to Woodland Park. As a partisan he attended football games in the university stadium, where Kansas played Nebraska or Missouri and the stands boomed to the ritual cheers of the Jay Hawk supporters: "Rock Chalk! / Jay Hawk! / K. U.!"

Meanwhile, he had established himself as a student. Leaving the segregated classes at Pinckney, he entered the integrated New York School (on New York Street). In at least one monthly report, his fifth grade teacher marked him "excellent" in each of his eleven subjects. To the younger blacks in a mainly white school, Langston was their star—well bred, good looking, amiable if somewhat reserved, and an excellent student. "He was a very gentle, kind of delicate boy," May Hampton remembered; "he never was in any argument. If we fussed he went on about his business, he didn't take any part in it." Luella Patterson remembered his high cheekbones and his curly black hair; Langston was "a quiet, calm and collected person, and very courteous." John Taylor knew him better than the girls did—and knew Mary Langston's tight rein: "She kept him kind of under her thumb." Langston was "a very hardworking boy," who apparently never studied much but was brilliant, always near the top of the class.

His reserve broke down once. Entering the seventh grade at the Central School in 1914, Langston passed into the care of a white teacher who decided to institute segregated seating in her class. She either compelled or induced all the black children to move to a separate row. Langston moved with the rest, but with mounting anger. His teacher had not reckoned on someone of Langston's background—the legacy of Harpers Ferry, his grandfather's fierce speeches, his grandmother's pride. Printing cards that said JIM CROW ROW, Langston defiantly propped one on each black child's desk. When she bore down on him, he flew into the schoolyard screaming that his teacher had a Jim Crow row, she had a Jim Crow row. An administrator tried to restrain him, but Langston fought off the man. He was expelled. Led by a black doctor, a delegation of parents visited the school to complain about the teacher's actions; Carrie Hughes also arrived to plead her son's case. Finally he was reinstated. But the idea of Jim Crow seating was dropped.

Fifty years or so later, at her home in Lawrence, the teacher recalled his flight from her class. "Of course," she told a young white woman with a tape recorder, "that stirred up all the nigger pupils and they went home and told their mothers about it." As she remembered him, Langston Hughes had no talent, much less promise. "Did I tell you," she mused, "he was a bad combination—part Indian, part Nigra, and part white? He was resentful, although he didn't show it in the class, as far as that goes."

Both at 732 Alabama Street and 731 New York Street, Langston had grown up in virtually all-white neighborhoods, a point of pride certainly with Mary Langston and the Reeds, but a major source of his sense of isolation and loneliness, as well as of a consciousness steadily dividing between the world of the Langstons, on one hand, and of the masses of blacks on the other. The white world represented a third force, entirely beyond his control but perhaps associated, in his mind, with the values of Mary Langston. His grandmother at one time quietly traded jellies and preserves with some of her white neighbors on Alabama, and one white lady across an alley, who worked at the university, always looked out for the old lady and her grandson, sending or even bringing soup herself if either of them was ill. Further, some of the white children played with him. He roamed the nearby woods with a boy named Paul—who innocently explained one day that he was better than Langston because Langston was a "nigger." Other neighborhood children, whose father worked downtown in a big department store, were forbidden to play with blacks. Yet somehow, Langston retained an oddly balanced view of the white world. Those who were friendly were essentially without color; those who were cold and hostile were not so much white as the enemy. But he also developed a chronic if slight fear of all whites, who could be now friendly, now needlessly cruel.

Hughes's sense of balance was maintained by at least two factors. One was his precociously developed rational sense, which kept his judgment under control. He could reason and watch where others rushed toward danger, just as he learned to govern his sexuality with apparent ease. The other factor was his sense of himself, fostered by his grandmother, his mother, the Reeds, and other striving blacks, as a little prince—even if he was almost a pauper. For Hughes, this sense of aristocracy would have little to do with money; its values were essentially those of radical Abolitionism tempered with strong racial pride. With John Brown, Lewis Leary, Charles Langston, and John Mercer Langston behind him, he nursed a sense that he was obliged within his lifetime, in some way, to match their deeds. Because of these men and, above all, his grandmother, no one could convince the boy that he was intrinsically inferior. His greatest psychological wound had been inflicted not by racism but by parental neglect; he would measure all future hurts against this primary wound and find them, on the whole, easier to endure.

How early he began to see himself as an artist is unclear. Some of his friends would remember Langston writing poems as a child. Certainly his mother had written verse; the town also boasted a publishing poet, the bohemian Harry Kemp, who once genially mocked the black Bottoms in heroic couplets for the *Daily Journal* ("THE BOTTOMS now I sing, where whiskey flows / And two per cent, make life couleur de rose"). Almost reverentially, Carrie had once pointed out Kemp to her young son. Also, there was immediate precedent for Langston to think of being a writer in Charles Langston's printed speeches and John Mercer Langston's autobiography, *From The Virginia Plantation to the National*

Capitol—even though the book was tedious and boastful. In addition, the wealthy Washington cousins it mentioned seemed to want little to do with their poor Kansas cousins. (Years later, Langston learned of two other connections to literary men: in Cleveland, his mother apparently established with the respected black novelist and short story writer Charles W. Chesnutt that they were cousins, since Mary Langston's mother, Johanna Sampson, and Chesnutt's grandmother Chloe Sampson had been sisters, growing up in North Carolina; and, according to the white Quarles family in Virginia, his great-grandfather Ralph Quarles was descended from the seventeenth century English religious poet, Francis Quarles, author of *A Feast for Wormes*.)

But Langston remembered doing no writing, only reading. "My earliest memories of written words," he would recall, were "those of W. E. B. Du Bois and the Bible"—Du Bois's magnificent essays on black life and culture in *The Souls of Black Folk* (1903), and his stern articles as editor of the *Crisis* magazine, published first in 1910. The stirring poems of Paul Laurence Dunbar, of whom Mary Langston approved, could not have been far behind. Then came novels: all of Zane Grey and Harold Bell Wright, and early work by Edna Ferber, as well as *The Rosary* and *The Mistress of Shenstone* by the inspirational British writer Florence L. Barclay. He read everything he could find on Haiti, where John Mercer Langston had lived and to which Mary Langston had been invited; he leafed through Charles Langston's main legacy to his family, a collection of his speeches. But Hughes was gripped by *Uncle Tom's Cabin* and a first reading of *Adventures of Huckleberry Finn* so thrilling that he remained a lifelong admirer of Mark Twain. Always he returned to Dunbar, the true fireside poet of the black American home; thanks to Mary Langston and Auntie Reed, the Bible was never far away; and, week after week, racial and harrowing, came the black *Chicago Defender,* with sensational details of lynchings and gallant exhortations to the race.

Hughes would probably have remained in Lawrence and entered high school there. In 1915, however, his life changed. One day late in March, Mary Langston fell seriously ill for perhaps the first time in her grandson's memory; a physician came to tend her, but she was beyond help. Near three o'clock on the morning of April 8, someone woke the boy to tell him that his grandmother was dead. As the cause of death, the doctor offered a catch-all medical term, "fecal impaction." She was 79 years old. Langston did not cry. Her own cool example did not suggest tears, and he felt none. Instead, because it was Thursday, he worried whether he would be allowed to deliver his newspapers. He wasn't. True to her independent spirit, Mary Langston was buried from her home, not from a church. Wearing a dramatically long black veil, her daughter Carrie interred her beside her husband Charles and her son Nat Turner in the grassy Oak Hill Cemetery in Lawrence.

Something else had changed in Langston's life. His mother was now Mrs. Carrie Clark, wife of Homer Clark, a sometime "chef cook" from Topeka, where his mother was a prominent black churchwoman. Where and when Carrie divorced Jim Hughes and married Homer Clark are not known. And with them was a two-year-old boy named Gwyn Shannon Clark, apparently Homer's son by a previous marriage. Langston immediately liked the boy; hungry for family, during the rest of his life he called Gwyn his brother and concealed the fact that they were not related by blood. Carrie, Homer, and Gwyn Clark lived briefly with Langston at 732 Alabama Street, before Homer left town to look for work. When Carrie departed with Gwyn to join Homer, she left Langston with the Reeds, but promised to send for him as soon as possible, once school was out. The house at 732 Alabama Street, which the 1910 U.S. census declared (perhaps erroneously) to be free of any mortgage, was lost. Hughes himself remembered "the mortgage man who always came worrying my grandmother for the interest due on her house. When my grandmother died, he got the house."

Langston was now a few months past his thirteenth birthday. These months with the Reeds following his grandmother's death were bittersweet; he loved the Reeds but wanted to be with Carrie. Given his freedom, he took a job his grandmother would have loathed. Working after school at a white hotel for fifty cents a week, he cleaned and shined brass spittoons and mirrors, scoured the toilets, scrubbed the halls, and kept the lobby neat. With his new riches, he haunted the Bowersock Opera House, once sitting high in splendid isolation on "Coon Hill" while the acclaimed dancer Ruth St. Denis performed far below, across the footlights. Loitering near the Santa Fe railroad station, he looked down the tracks toward Chicago. He thought mostly of the bright city lights, he later wrote; but his mother also lived near Chicago. And where was his father? Now, as throughout most of his childhood, Langston indulged himself with a dream of James Nathaniel Hughes in Mexico that had nothing of his mother's resentment to ruin it. Jim Hughes was "a kind of strong, bronze cowboy, in a big Mexican hat, going back and forth from his business in the city to his ranch in the mountains, free," in a landscape of "mountains and sun and cacti: Mexico!"

Apparently at some point in this interlude with the Reeds he discovered sex. In a memoir published a dozen years later, he told of Auntie Reed warning him about girls and pointing to the misfortune of young Clarence next door, who had gotten someone pregnant and been forced to marry her. Langston also mentioned a late afternoon tryst at the invitation of a slightly older girl. Much later, in a perhaps thinly fictionalized sketch, he suggested how much he found out that day: "It was dusk-dark, and outdoors, but quiet behind the shed. Right there in the open we played house. And right there for the first time (precocious) I ejaculated." If accurate as autobiography, as it probably is, this was a curious admission from someone who later would be regarded by many of his friends as asexual, without noticeable erotic feeling for either women or men.

This moment of sexual initiation seemed subsequently to Hughes to have been of little importance in itself, and especially so in comparison to another event that took place during this time when he waited to be summoned by his mother. Grateful to the Reeds for their affection and generosity, Langston was nevertheless anxious to be gone. The result apparently was a volatile confusion of feelings of gratitude and guilt, of hope and abandonment, that primed him for what he would later identify in his autobiography *The Big Sea* as the most harrowing episode of his childhood. As she had done many times before, Mary Reed had been taking him to an extended revival sponsored by her church. Every night, for what seemed like weeks, the rafters had rung with sermons, hymns, prayer, and shouting as sinner after sinner was led to Christ. To conclude the revival, a special evening near the end was set aside for the youngest sinners. At home, Auntie Reed talked to him incessantly about the coming service and its importance to his life and, because she loved him, to her own. She assured Langston that "when you were saved you saw a light, and something happened to you inside! And Jesus came into your life! And God was with you from then on! She said you could see and hear and feel Jesus in your soul."

Imaginative and susceptible, Langston believed her. To a hypersensitive boy who had spent his conscious life waiting for his mother and hoping for his father, the notion of a savior coming who thereafter would be always at his side struck and sounded deep. On the children's evening he went eagerly to church. As the preacher preached and the congregation amened, Auntie Reed hovered expectantly about him with several of the other enthusiasts. One by one, the children were saved, until only Langston and another boy remained. Finally the boy whispered to Langston that he couldn't wait any longer, he was going to be saved. Now only Langston was left. Auntie Reed knelt down beside him, praying aloud. The old folks in the amen-corner prayed aloud. "The whole congregation prayed for me alone, in a mighty wail of moans and voices. And I kept waiting serenely for Jesus, waiting, waiting—but he didn't come. I wanted to see him, but nothing happened to me. Nothing!"

Then he could wait no longer. "And the preacher said do you love God? And I said yes. And the preacher said do you accept him? And I said yes. And I was converted." The church erupted in jubilation. Throwing her arms around him, Auntie Reed led him home in triumph. But Langston was not triumphant. That night, he would remember, he cried for the last time in his life but once. "I cried, in bed alone, and couldn't stop. I buried my head under the quilts, but my aunt heard me. She woke up and told my uncle I was crying because the Holy Ghost had come into my life, and because I had seen Jesus. But I was really crying because I couldn't bear to tell her that I had lied, that I had deceived everybody in the church, that I hadn't seen Jesus, and that now I didn't believe there was a Jesus any more, since he didn't come to help me." And elsewhere: "I had waited for Him and He hadn't come."

If indeed Hughes never cried again but once in his life, his loss of faith was not only in Jesus but also in all those on whom he had waited in vain to save him from loneliness. Since he had seen Jesus come to old black men and women, even to Auntie Reed herself, then the fault, if Jesus now failed to come to Langston, must have been the boy's own. Langston could not rest there; he oscillated between self-blame and rage at his parents—and at Jesus. One act, however, was completely his own. He had repaid the love Auntie Reed and the church folk had showered on him with a hideous lie. Loving his mother, at some quite conscious level he also knew now that his love was quite hopeless. And he knew, too, that he would have to prepare himself for the future by some forceful opposition to her and to the world of the Langstons. This opposition would be the equivalent of the episode when he refused to recite in front of her, or destroyed her own recitation by mocking it.

But opposition to his mother could not at the same time offend the Reeds. Whatever course his life took, it would have to be in some way reparational to the Reeds, who had protected and cherished him through much of his childhood, and to the black congregation into which Mary Reed had brought him as a shining star. Even for her, Langston would not pretend to love Jesus; but the traditional forms of teenage rebellion—neglect of studies, open defiance of parental authority—were also impossible, because of what the Reeds stood for. His will to rebel, in fact, would take in time a subtler form; it would combine an ordinary sense of duty with a seemingly erratic bohemianism that would gradually merge into his commitment to being a writer, whose principal subject would be the black masses. In the process, Langston would turn his back, to his mother's anger, on almost all bourgeois measures of success; while she lived, Carrie and Langston would never cease to struggle over her desire for comfort and his determination to live the life he had chosen.

At thirteen, Hughes probably already viewed the black world both as an insider and, far more importantly, an outsider. The view from outside did not lead to clinical objectivity, much less alienation. Once outside, every intimate force in Hughes would drive him back toward seeking the love and approval of the race, which would become the grand obsession of his life. Already he had begun to identify not his family but the poorest and most despised blacks as the object of his ultimate desire to please. He would *need* the race, and would need to appease the race, to an extent felt by few other blacks, and by no other important black writer. This psychological craving was a quality far more rare than race pride or a merely defensive antagonism against whites; it originated in an equally rare combination of a sense of racial destiny with a keen knowledge of childhood hurt.

The sense of hurt, of having been tragically misunderstood, was there even as Hughes waited for a call from his mother in the summer of 1915. It would be there twenty years later, as he approached middle age but still faced the old tension between them. ''*Nobody loves a genius child,*'' Hughes would write:

> Can you love an eagle,
> Tame or wild?
> Wild or tame,
> Can you love a monster
> Of frightening name?
>
> *Nobody loves a genius child.*
>
> *Kill him*—and let his soul run wild!

2

OUTSETTING BARD
1915 to 1921

> For today you have the strength of youth
> And youth is tomorrow's man.
> So work! Map out your life
> And with wisdom make your plan.
>
> "To Youth," 1918

WHEN HIS MOTHER summoned him late in the summer of 1915, Langston rode a train east to Lincoln, Illinois, a prosperous little town in Logan County in the middle of the state. Like Lawrence, Kansas, Lincoln had a place in American history: it was the sole site named for Abraham Lincoln *before* he became president, "when his only reputation, as he trailed the mud circuit, was that of a good lawyer and an honest man." Tradition had it that the power of attorney to lay out the town was drawn up in his law office in nearby Springfield. One day in 1853, according to a cherished but perhaps unreliable source, the tall attorney took out his pocket knife, split open a watermelon, and christened the town-site with its juice—but not before advising the developers that he "never knew anything named Lincoln that amounted to much."

In the summer of 1915, the population of Lincoln stood at about 12,000, which included enough blacks to fill the small African Methodist Episcopal Zion Church on Sunday. As in Kansas, race relations in Illinois had soured. In Springfield, once a major stop on the underground railroad, whites had lynched a black man in 1908; out of the reaction to that incident had come eventually the National Association for the Advancement of Colored People. Late in the summer when Hughes reached Lincoln, the local *Evening Star* ("Negro Confesses He Is Guilty") reported an attempt by hundreds of whites to seize an alleged black murderer from his jail. Balanced against this reality was the legend of freedom associated with the name Abe Lincoln as the great emancipator of the slaves. Hughes knew the reality but also cherished the dream; in a pivotal poem composed five years later, he would weave Lincoln's name prominently into his text in praise of the black heritage.

For another reason, the town of Lincoln would occupy a fundamental place in his career. Here, according to his account, Hughes first wrote verse. In an election for class poet in the eighth grade at Lincoln's Central School—in which he and a girl were the only blacks—"my classmates, knowing that a poem had to have rhythm, elected me unanimously—thinking, no doubt, that I had some, being a Negro." If this election came near the end of the year, as his autobiography *The Big Sea* suggests, Hughes was wrong. By this time, the teachers in the eighth grade at the Central School, which opened that fall, had proof that Hughes wrote well. His history teacher, Laura Armstrong, who would recall "a coloured boy, yellow rather than black, with silky, curly black hair, regular features, handsome," knew him to be "highly intelligent, standing intellectually head and shoulders above the group." Frances Dyer was drawn to him not only because they shared the same birthday, February 1, but because he was oddly mature; she remembered being careful not to lecture to him over the heads of the other children. Nine years after he left her English classes, when Hughes published poetry for the first time in *Vanity Fair* magazine, Ethel Welch would consult her old grade books and report in a Lincoln newspaper that Hughes had been one of the three best students in composition in the 1915 to 1916 school year.

His election as class poet probably had less to do with rhythm and race than with his perceived merits and popularity; he made friends among his white schoolmates, and even secured a part-time job with a merchant downtown. In any event, the name of Langston Hughes as class poet appeared on the printed program when eighty graduating students assembled in the gymnasium of the Central School with their teachers and families on the afternoon of May 31, 1916. Helped by Ethel Welch and his mother, he showered lines of praise on his teachers and classmates. "In the first half of the poem, I said that our school had the finest teachers there ever were. And in the latter half, I said our class was the greatest class ever graduated." Langston was a grand success. "Naturally everybody applauded loudly," he would recall. "That was the way I began to write poetry."

The president of the school board, Dr. Robert Goebel, presented certificates of promotion to Lincoln High School. Langston, however, would not attend school there. That morning, Homer Clark had left for Cleveland in search of a job. "My life has been filled with a great envy," Hughes would write later, "for those persons who have grown up in one place, whose folks stayed put, and who have always had a home to come back to."

Near the end of the summer, Langston, Carrie, and Gwyn Clark arrived in Cleveland to join Homer, who had found work first as a conductor, then as a machinist in a steel mill. After some searching, they found a basement apartment at 11217 Ashbury Avenue, north of Euclid Avenue and not far from Wade Park, the new Cleveland Museum of Art, and Severance Hall, where the Cleveland Symphony played. For Hughes, his arrival marked the first true satisfaction of what would be his lifelong passion for cities; the older he became, the less tolerant he grew of the country. Cleveland was poised at a crucial moment in its history. After fifty years of exploiting its position on Lake Erie between the oil and coal fields of Pennsylvania to the east and the iron mines of Minnesota to the west, the city had begun to rival Detroit and Chicago in mineral industries. With Europe at war, and the United States driving toward world dominance in manufacturing, an extraordinary demand for manpower had encouraged the mass migration of blacks from the South. Between 1910 and 1920, the black population of Cleveland more than tripled. At 35,000, it was still a small percentage of a city numbering over 700,000, but the Negro influx ("a great dark tide," Hughes would call it) grew steadily as the South stagnated and industrial northern cities like Cleveland flourished.

"As always, the white neighborhoods resented Negroes moving closer and closer," Hughes observed; "but when the whites did give way, they gave way at very profitable rentals." For blacks, the street of dreams was Euclid Avenue, which passed not far from Langston's home. On it stood the mansions of "Millionaires' Row," fine shops and department stores, ornate banks, tall office and apartment buildings, glittering theaters and music halls. To black migrants, however, the city was a bemusing mixture of new freedom and old bondage; for all their proximity to great wealth, the city was not really their own. Blacks were by no means the only poor in Cleveland; on Woodlawn Avenue, migrants from Europe—Russians, Hungarians, Poles, Italians—also knew tenement and slum. But to blacks gradually fell the worst jobs, with other employment more and more difficult to find as segregation hardened. Crime followed—petty thievery, violence, bootlegging, gambling, and prostitution—on a scale unknown to Langston in his small towns in Kansas and Illinois.

Fortunately for him, Hughes found in his school a refuge from these pressures and from his clammy basement apartment. Central High School, founded in 1846, was the oldest institution of its kind in the United States west of the Alleghenies and probably the best public school in Cleveland. From its imposing, multi-storied neo-Gothic stone and brick building, capped by a bell tower and clock, had passed many future community leaders. Its most famous alumnus was John D. Rockefeller; Laura Spelman Rockefeller, class of 1855, had donated the tower clock. In 1916, when Langston enrolled, school spirit was still very strong. Several of the teachers had graduated from Central; the following year, the highly respected Edward L. Harris returned as principal after previously serving for twenty-three years and developing most of the nearly two dozen clubs and societies at the school. Central's greatest asset, however, was a diverse student body dominated by the children and grandchildren of European immigrants often fiercely interested in education. They had replaced the aristocrats of Cleveland, who were withdrawing from the center of the city to more exclusive surroundings.

Once again Hughes found himself one of the few blacks in an overwhelmingly white student body. The result was not, however, alienation from his own race. As a freshman, working after school in a little shop owned by a Mrs. Kitzmiller, he sold ice cream, watermelon, and other refreshments to people who seemed to him unusually excitable but also unusually exciting—"the gayest and the bravest people possible—these Negroes from the Southern ghettos—facing tremendous odds, working and laughing and trying to get somewhere in the world." Nevertheless, the heart of his life was school, and there he saw few black faces. He met no hostility from his fellow students or his teachers. By his second year, his best friend was a fairhaired, square-jawed, sensitive Polish boy, Sartur Andrzejewski. In the classroom, Hughes more than held his own. His best grades as a freshman came in the area in which he half-expected to make his career, graphic art, which he studied for four years at Central. Although he had some natural skill as a draughtsman, Hughes was also clearly fascinated by his applied arts teacher, Clara Dieke, a rather cool and not very popular Central alumna he would credit with teaching him "about law and order in life and art, and about sticking to a thing until it is done."

After a successful freshman year, he spent the summer of 1917 working in the huge Cleveland department store, Halle's. Running a dumbwaiter between the storerooms and the various departments made him aware as never before of the gulf between the rich and the poor, and of his grandmother's strictures against the powerful. "I was continually amazed at trays of perfume that cost fifty dollars a bottle," he remembered, "ladies' lace collars at twenty-five, and useless little gadgets like gold cigarette lighters that were worth more than six months' rent on the house where we lived. Yet some people could afford to buy such things without a thought. And did buy them."

Hughes had barely become accustomed to living with his mother when their home fell apart again. Worn down by the intense heat of the steel mill, Homer Clark gave up his lucrative job for a janitor's broom. Carrie looked for work but found only a job as a maid. When the tension between the couple became unbearable, Homer Clark boarded a train for Chicago. Soon after, hoping for a reconciliation and taking little Gwyn Clark with her, Carrie went after him.

Left behind, Langston took an attic room in a house at 2266 East 86th Street. "I couldn't afford to eat in a restaurant, and the only thing I knew how to cook myself in the kitchen of the house where I roomed was rice, which I boiled to a paste. Rice and hot dogs, rice and hot dogs, every night for dinner. Then I read myself to sleep." Sometimes he went to Sartur Andrzejewski's, where Sartur's mother and her rosy-cheeked daughters Sabrina and Regina served him sausages and cabbage cooked in sweetened vinegar. The family "had about them a quaint and kindly foreign air, bubbling with hospitality. They were devout Catholics, who lived well and were very jolly." Then, in a cottage on East 38th Street occupied by Russell and Rowena Jelliffe, a young white couple lately come from graduate school at the University of Chicago after undergraduate years at Oberlin, Hughes found another refuge. Idealistic but shrewd, the Jelliffes were starting a life of community service somewhat in the manner of Jane Addams at Hull House in Chicago but adapted finally to the needs of the local community. Hughes became one of the first teachers at what later became Karamu House. What he learned about drawing and painting at Central, he passed on to the poor neighborhood children at the "Playground House," as the Jelliffes' first community center was called. In their warm, book-filled home, he spent more and more time reading and dozing before going back to his rooming house. At Central High he also continued to do well. In the spring of 1918 he tried out as a sophomore for the track and field team, took part in quarter-mile races and the high jump, and represented Central in the relays.

That summer, Hughes visited his mother in Chicago. Homer Clark had moved on again, leaving Carrie and his son in a dismal second floor room on Wabash Avenue, with windows that opened onto the tracks of the elevated train; just beyond the walls, cars rumbled deep into the night. Carrie's employer, a milliner in the fashionable Loop district for whom she worked as a maid, hired Hughes to deliver hats. His job took him into the immaculate, tree-lined North Shore suburbs. On Sundays, however, he strolled amidst a Black Belt vaster and more exciting than anything he had seen before, even though the waves of black migrants were bitterly resented by local whites; the worst riot in Chicago history was one year away. One day, when he wandered into a Polish neighborhood, a gang of boys stopped him: "We don't 'low no niggers in this street." A blow to the jaw knocked his cap off his head; battling and scratching, he fixed on his dirty cap on the ground until they drove him out in a hail of oaths and stones. His jaw swollen and his pride bruised, Hughes was too ashamed to tell his mother about the beating.

Unhappy, he found Chicago "vast, ugly, brutal, monotonous." On the teeming streets he was sometimes propositioned by whores and at least once by a man. But he took a pleasant hike to the sand dunes near the city with a party from the colored YMCA, and for the first time he heard the music of the Holiness and Sanctified churches—the storefront tabernacles ministering to southern blacks stunned by life in the North. The gospel music reminded him of the Baptists in Lawrence, but it possessed a far greater intensity in stepped-up rhythms driven by fierce handclapping and wild tambourines; in polyrhythmic sermons, black ministers showed off all the gifts of great actors. Hughes never forgot one preacher who depicted the ascent of Calvary by climbing onto a piano stool, then up onto the keyboard, before finally planting himself on the piano top, arms extended as if crucified, preaching eloquently all the while.

At the end of the summer, Langston returned to Cleveland alone, knowing that he might never again live with his mother. His anxiety fostered what had already become an important part of his life. The previous February, as a sophomore, he had published his first pieces of verse in the Central High *Monthly,* the expertly produced school magazine founded over twenty-five years before by principal Edward Harris. Two poems were juvenile efforts at satire; the third piece, "The Red Cross Nurse" ("Angels of Mercy they call them, / Those wonderful women of earth"), was on a more serious subject, the war effort. In the following number, Langston returned with more sophomoric satire and another solemn piece about Central's graduates and the world war ("Central's heart has a memory / That will live for many a day"). In April, a Hughes sketch, a satirical portrait of a woman and a dog, adorned the front cover of the *Monthly.* Inside were two of his poems: "My Loves" ("I loves to see de big white [moon] / A shinin' in de sky"), no doubt influenced by Paul Laurence Dunbar; and the Longfellow-like "To Youth," turgid with didactic purpose ("When you have gone your way thru life / And the years have onward rolled . . . ").

In the fall of 1918, Hughes joined the editorial staff of the *Monthly.* Of four poems in the October number, three supported the war effort; the last, "Welcome, Little Freshmen," indicates how comfortable he now felt at Central. In December came his first short story, "Those Who Have No Turkey." Visiting her rich aunt in Cleveland for the first time at Thanksgiving, a country girl finds out from a newsboy that his family is too poor to afford a turkey. Just as her mother would have done on the farm, she impulsively invites them all to dinner. Of immigrant stock, ill-spoken and rough but jovial, the family enjoys the meal. Her aunt remains a snob, but the girl knows now that country ways are best. "Well, ma," she adds as the story ends, "I never knew before that there are people in the world who have no turkey on Thanksgiving." (Where Langston himself spent Thanksgiving, 1918, is not known; he was still living alone.)

The January, 1919 *Monthly* was virtually a Langston Hughes number; in addition to two poems and a lengthy short story, he became editor-in-chief of the prestigious "Belfry Owl," an autonomous section of the *Monthly* that commented satirically on life at Central. Superior to his Thanksgiving tale, the story "Seventy-Five Dollars" tells of children left on their own in a Cleveland slum after their mother's death. "Very lonesome" Joe wants desperately to finish high school. By scrimping, they save $75, enough to allow him to continue school; but when another brother robs a store owner, the money goes to pay him back. "For the first time since he was a little boy Joe cried," the story concludes. "Two big tear drops trickled down to the tip of his freckled nose and fell upon the window sill. Outside the moonlight veiled the poor houses of Primrose Street in misty radiance. Stars sparkled and glimmered in the far reaches of the night."

Of the two poems, one was conventional in form ("Come with me to Little-Boy Land / Where the houses are castles big and grand"). But the other broke decisively with Langston's poetic influences to this point. "A Song of the Soul of Central" was not only his longest published poem—just under one hundred lines—but is "in free verse," as the magazine noted. "Oh Central," the poem begins, "Great, grey Alma Mater, / You have lived for many years." Often prosaic, sometimes banal, this paean to Central High ("you, great grey mother") nevertheless clearly affirmed young Langston Hughes as a spiritual and poetic child of the good grey poet, Walt Whitman:

> A soul is composed of one's thoughts and dreams.
> Your thoughts are of children
> And your dreams always youthful. . . .
> Children of all peoples and all creeds
> Come to you and you make them welcome . . .

In large part, Hughes owed this leap in the quality of his work to his new English teacher, Ethel Weimer, a former student at Central and perhaps its most popular and progressive instructor. Unafraid of the new, she held up to him the example of poets such as Edgar Lee Masters, Edwin Arlington Robinson, Amy Lowell, Vachel Lindsay, Carl Sandburg, and, towering above them all, Walt Whitman, who had died only twenty-seven years before. His admiration for

Whitman would last the longest; Carl Sandburg, however, became his "guiding star."

If Whitman and Sandburg freed Hughes from the tyranny of traditional forms, they also led him toward a version of literary modernism not without limitations. Although Sandburg had published in Harriet Monroe's innovative *Poetry* magazine, his modernism, populist in nature, was quite unlike that of Ezra Pound, T. S. Eliot, and Wallace Stevens, whose elite standards would soon define the term. Given his already vibrant social conscience, however, Hughes had no real choice between the rival camps. And in apprenticing himself to Whitman and Sandburg, he freed himself from the most conventional kinds of imitation, especially sentimental Anglophilia; he joined a rival tradition grounded in a passion for native, democratic themes and flexible forms, especially free verse. Just back from Chicago, he was primed to appreciate the achievement of Sandburg in his *Chicago Poems,* published in 1916. Hughes was drawn by everything he learned about the Swedish-American proletarian poet—his birth and upbringing in Illinois about sixty miles from Lincoln, his drifting among lowly jobs, his troubled social conscience, his contempt for gentility, his rebellious commitment to the people.

> Carl Sandburg's poems
> Fall on the white pages of his books
> Like blood-clots of song
> From the wounds of humanity.
> I know a lover of life sings
> When Carl Sandburg sings.
> I know a lover of all the living
> Sings then.

Hughes's youthful work, on the other hand, revealed another, more private concern. In the short story "Seventy-Five Dollars," he clearly drew on his sense of a sad discontinuity between his mother and himself; surely he is the "very lonesome" boy who hungers for a higher life which his delinquent family would deny him. So, too, both the free verse poem on Central ("you, great grey mother") and the invitation to "Little-Boy Land" indicate, as did other poems, the remarkable extent to which, at seventeen, Langston still saw himself as a child, and one in need of a loving mother. Perhaps this is the outstanding fact about his writing thus far. Also worth noting, however, are the strong social conscience exhibited in his prose, and the absence of blacks. The balance between the will to be a child and the will to protest social injustice is very delicate; both seek assertion, but each also wishes to preserve the other. They merge in Hughes's idealism and the idealistic images already principal in his art. Of these, the most important would be that of the future as a dream.

At Central, Langston wrote at least one other short story, "Mary Winosky," ostensibly based on a newspaper report of a scrubwoman who left an estate of $8,000. Lonely, pathetic Mary, an immigrant from Europe deserted by her husband, waits faithfully for his return, then collapses and dies with the news that he has been killed in the war. Behind this study in sentimentality, however, is a trenchant narrator. ''Many of the metropolitan daily papers,'' Hughes began, ''crowded with news of the war, the Russian situation and heated editorials against Bolshevism,'' paid scant attention to the scrubwoman's death. At the end, he is almost sardonic: ''Many parents read [the news item] to their children as an excellent example of thrift and economy.''

Hughes's social conscience was strongly reinforced at Central High by his exposure to its cosmopolitan student body—in particular to its Jews, who were mostly the children of immigrants from Russia and eastern Europe and commanded his respect as a group. They ''were almost all interested in more than basketball and the glee club.'' In the pairings and flirtations at Central, he began to see the shadow of prejudice but felt it least with the Jewish students. He escorted a Jewish girl to his first symphony concert; according to her best friend Pauline Luke decades later, Sarah Sapir, who edited the drama section of the *Monthly,* had a crush on Langston but was too timid to show it. Sometimes he attended synagogue services, where he loved the songs of the cantors. Once, at a mass prayer meeting for victims of a Polish pogrom, he watched in awe as men and women wailed out their grief, beating their heads against the walls of the auditorium. In part because it promised the end of pogroms in Russia, the Bolshevik revolution was fervently celebrated at Central. Although all the Jewish students were not socialist, some steered Hughes straight toward socialism. They took him to hear Eugene V. Debs, who strongly opposed the war (but in being both pro-socialist and a supporter of the war effort Hughes in fact reflected the dissident leftist viewpoint of the *Cleveland Citizen,* organ of the Central Labor Council of the city). Hughes later credited Jewish students with lending him copies of Ethel Voynich's *The Gadfly,* Romain Rolland's *Jean-Christophe,* the *Liberator,* and the *Socialist Call.* Racing through John Reed's *Ten Days That Shook The World,* the youths had pined for the barricades of Petrograd.

As Hughes would recall it, his introduction to the *Liberator* was crucial. Max Eastman's magazine, which revived the banned radical journal *Masses* (edited by Eastman, Floyd Dell, and John Reed), included the work of a black writer who would soon find a place among Hughes's poetic idols. ''I learned from it the revolutionary attitude toward Negroes,'' he later stated of the *Liberator.* ''Was there not a Negro on its staff?'' The Negro was the Jamaican-born poet Claude McKay, leftist and race-proud but also a tender lyricist, who would publish *Spring in New Hampshire* in 1920 and *Harlem Shadows* in 1922. ''It was Claude McKay's example that started me on this track,'' Hughes conceded about his pro-socialist writing. In fact, McKay did not join the staff of the *Liberator* until April, 1921, after Hughes had left the school; and, unlike Hughes, McKay was very conservative in matters of poetic form (his favorite was the sonnet). However, the *Liberator* had published his verse since April, 1919, in Hughes's junior year; moreover, McKay used the sonnet to express very radical ideas. Seeking to emulate McKay, Hughes later offered several poems to the magazine without success—although Floyd Dell once wrote to say he liked one poem most of all ''but none moves us deeply.''

In spite of his interest in socialism, he was elected president of the American Civic Association, a patriotic club founded at Central High by order of an apoplectic principal Edward Harris following a pro-communist demonstration at the school—perhaps at the Armistice celebrations, when left-wing students paraded a red flag about the grounds. Here and elsewhere Hughes found himself pitched between the rival factions of Jews and Gentiles, radicals and conservatives. Since he was black, the Jews considered him not a typical Gentile; he was a radical symbol as a black, yet friendly to everyone. He was elected to the student council, served as secretary of Dr. Maurice Zeliqzon's Le Lycée Français, and was a member of the Home Garden Club, run by Latin instructor Helen Chesnutt (the only colored instructor in the school and daughter of the notable Afro-American novelist and lawyer Charles W. Chesnutt).

Attending a prestigious school did not exempt Hughes from open racism. Late in the fall of 1918, when Sarah Bernhardt appeared on stage in Cleveland, the French club attended a matinee performance. Later, in a cafeteria nearby, the cashier calmly rang up his friend Sartur Andrzejewski's bill, but became visibly enraged at seeing Langston next in line. She kept punching the cash register until the amount became ridiculous, "her face growing more and more belligerent, her skin turning red and her eyes narrowing. I could see the hatred in her face." When Langston questioned the bill, she exploded at him. The astonished boys put down their trays and left. For a long time the incident remained the most humiliating of Hughes's life. His reaction was typical of the way he would respond to personal encounters with racism. Refusing to be lured into violence or hatred of all whites or to sulk or become cynical, he subdued his rage and kept his balance. Hughes was not a coward. However, on each of the few occasions in his life when his internal pressures proved too great to bear, the cause would be private rather than racial as the cashier's insult had been. He was so self-assured about his race, and so tightly self-controlled in spite of his tenderness, that he was simply not very vulnerable to insult.

In the spring of his junior year, in 1919, he went out for track and field again; he helped the Central relay team to victory in the city high school championships. When his mother and Gwyn Clark returned to Cleveland, Langston joined them in a cramped apartment at 5709 Longfellow Avenue. (In Cleveland, he would recall ruefully, he lived only in attics or basements.)

Suddenly his father reentered his life. How long James Nathaniel Hughes had been sending money to his son, or to Carrie Clark for him, is unclear. But in the spring of 1919, he sent at least three gifts of $20 each through the Cleveland Trust Company. Then one day Langston heard directly from his father. A peremptory note announced that James N. Hughes, who expected to be in New York in June, would return home via Cleveland: "You are to accompany me to Mexico for the summer." Langston was ecstatic, Carrie outraged. She bitterly accused him of disloyalty and ingratitude when he made up his mind to go to Mexico. On June 12, unknown to Langston, his father crossed into the United States at Laredo, Texas. Later, a telegram sent by him to Cleveland went astray. One day, after waiting anxiously for word, Langston saw a little bronze man with a bushy moustache and close-cropped hair striding briskly up the street toward him. They looked at each other, then passed. They looked back.

"Are you Langston?"
"Yes. Are you my father?"
"Why weren't you at the train last night?"
"We moved, and I didn't get your wire till this morning."
"Just like niggers. Always moving!"

At a restaurant where Carrie worked as a waitress, Jim Hughes tried to start a conversation. "What's your order?" Carrie demanded. When he left a dime as a tip, she ordered someone to throw it into the street.

The journey to Mexico was a dull horror. To James Hughes, a bitter man, the United States was so much sandpaper on his nerves. In Arkansas, Langston saw a picturesque scene, some blacks at work in a cotton field. "Look at the niggers," Jim Hughes sneered. Once past the adobe border station in Laredo he spat his contempt at Mexican laborers. Baffled, Langston tried to see the workers as his father saw them. He loved the poor; his father seemed actually to hate them.

In Mexico City, after an overnight stay at a hotel, James Hughes took Langston to a house he vaguely remembered, where his father's closest friends in Mexico, the Patiño sisters, lived at 16 Santa Teresa Street, near the cathedral. Dolores, Refugio, and Rafaela Villaseñor y Patiño were pious, highly respectable, unmarried women who had befriended Jim Hughes soon after his arrival in Mexico, and who looked after his rental property in the city while he lived in Toluca, about sixty miles away. Langston found the Patiños affectionate and lovely in an old-fashioned, genteel way, with their mantillas of black lace drawn against ivory-yellow skin. His father seemed much less tense now that he was in Mexico.

The next day, Langston and James Hughes left for Toluca. The train swung slowly around the mountains, then descended into a lush, green valley past lakes on which lilies floated; rising through a brilliant blue sky to fifteen thousand feet was the snowy peak of the volcano Xinantecatl, or El Nevado de Toluca. At the station, an Indian youth named Maximiliano, silent and cowed, hoisted all their luggage onto his back, secured it with a leather thong around his neck, then trotted ahead of his master and Langston. On one side of a little park, at 3 Plaza de la Reforma, they entered a low house, blue and white with a roof of red tiles, surrounded by adobe walls. To the front was a scruffy patio garden, in the back a carefully tended corral where Langston found two horses and dozens of plump American chickens.

In addition to the Toluca house and the property in Mexico City, Jim Hughes also owned a large ranch in Temexcaltepic, in the mountains. Most of his property had been acquired in the previous ten years, but recently he had enjoyed a windfall—which partly accounted for his gesture to his son. In September, 1907, a few months after his attempted reconciliation with Carrie, he had been fired without explanation by the Pullman Company, probably by a new white manager; from the start, several white Americans had objected to dealing with a black secretary and had petitioned to have him fired. He sued the company but lost on the technical question of whether the Pullman Company in Mexico was legally American or Mexican. In 1909 he moved to Toluca to begin working for the American-owned Sultepec Electric Light and Power Company. His brown skin, knowledge of American and Mexican law, and fluency in Spanish made him especially valuable when most of the *gringos* fled before the followers of Pancho Villa and Zapata after the fall of the dictator Porfirio Diáz. Hughes was given power of attorney by his white superiors to transact any and all business, including the sale of property. A few months before his visit to Cleveland, the local chief of the secret police attempted to search the company offices at 75 Avenida Constituyentes, allegedly to look for correspondence with Zapatistas. Denouncing the move as a brazen attempt to frighten him into selling the company cheaply, Hughes resisted the police. He was thrown in jail overnight.

In March he sold the bulk of the property for $65,000 to the leading German businessman in town. On his arrival in Toluca Hughes had boarded with a German family and had many friends within the extensive German community; he spoke the language fluently, and doubtless admired German efficiency. His share of the payment had made him feel able at last to seek out his son when he visited the United States to wind up certain affairs of the company with the owners in New York.

Never intending Langston's visit to be a simple vacation, Jim Hughes began to teach his son bookkeeping. To his disgust, Langston performed weakly. "Seventeen and you can't add yet!" his father cried, bending over the ledger and showing the offended boy once again how to balance the page. "Now, hurry up and do it! Hurry up! Hurry up!"

Langston's boyhood fantasy about James Hughes quickly died. Easygoing himself, he found James's frantic pace dismaying. His father did not walk; he ran. Driving his horses hard, he rode his employees harder. He was hardest on himself. Up at five o'clock in the cold Toluca dawn, he washed quickly in icy water, then rushed to his account books and correspondence; he paused for quick meals, then worked on again. Jim Hughes's severity with Maximiliano and his housekeeper, a pleasant lady named Enriqueta de Noll, upset Langston. The Indian boy slept on a pile of sacks in the toolshed; her two children, Augusta and Roberto, were not allowed to eat in the house. Virtually addicted to food ("Langston Hughes is crazy about 'eats'," his high school yearbook would note), Langston loathed the diet of beef and beans, served in meagre portions. He ordered the larder stocked, and charged the purchases to his father's account. Furious, Jim Hughes denounced his son as a spendthrift—Langston was just like his wasteful mother, always throwing money away!

He was allowed ten pesos a week but had no place to spend them. The movies showed only on Sundays. At over seven thousand feet the town was achingly cold at night, the streets gloomy; summer brought rain and shrouds of mist across the green mountains. Befriending Maximiliano, Langston gave him cigarettes and money; in return, the Indian taught him what he could—how to ride bareback, how *serapes* were made. Maximiliano spoke very little Spanish and no English; Langston knew nothing of his Indian dialect. There were few books in his father's house, and Jim Hughes would not take him to the ranch in Temexcaltepic because Langston had not yet learned to ride well and the mountains crawled with bandits. His father also declined to take Langston to Mexico City.

Understanding now why his mother found James intolerable, Langston wrote to her. She did not reply. Carrie was herself not much better than his father, only different. "I began to wish," Langston would write, "I had never been born—not under such circumstances." He took long rides on a horse named Tito into the countryside, soothing his melancholy with visits to remote villages with decaying churches and bell towers. Depressed, bored, restless, he practiced shooting one of his father's pistols. "One day, when there was no one in the house but me, I put the pistol to my head and held it there, loaded, a long time, and wondered if I would be any happier if I were to pull the trigger. . . . I put the pistol down and went back to my bookkeeping."

Finally in August his father agreed to a return visit to Mexico City. On the day of their departure they awoke unusually early. Wretchedly cold, Langston sat down to his breakfast. His father looked up impatiently from where he sat gulping down his food. "Hurry up!" he barked. "Suddenly my stomach began to turn over and over," Langston would remember. "And I could not swallow another mouthful. Waves of heat engulfed me. My eyes burned. My body shook. I wanted more than anything on earth to hit my father, but instead I got up from the table and went back to bed. The bed went round and round and the room turned dark. Anger clotted in every vein, and my tongue tasted like dry blood." His father went on alone. Langston stayed in bed, with Maximiliano sitting on the floor outside his room. Days later, Langston still had not eaten. A doctor ordered him taken to the American hospital in Mexico City. After he found out how much his father was paying the hospital, Langston settled deep into his bed. He was indeed ill: his red blood cell count had dropped dramatically. The doctors diagnosed the presence of a virus, but Langston knew the real reason for his illness: "I hated my father."

This violent illness, in which he lost control of his body and developed certain symptoms of anemia, would not be the last of his life to be mainly psychosomatic. He would not collapse in this way more than two or three times in his life; but these episodes were only intense versions of his chronic unwillingness to vent anger. He could not explode; he found it hard even to seethe openly, which was extraordinary in someone with so hyperactive a social conscience, a writer who had shown already the great extent to which he conceived of his gift in terms of its moral energy. Faced with injustice aimed directly at him, even from his father, he revealed the moral indignation of a clam. The truth seems to be that while Hughes's anger ran deep, his fundamental urge, lodged close to his sexual instinct in its intimacy, was in fact toward passivity. The urge to restrain moral indignation is not in itself abnormal; what was unusual was Hughes's failure to cry out at some point short of physical collapse.

Or was Hughes engaged through his illness in a ritual of self-punishment and thus of self-absolution? By the time of his collapse he knew he was a child without parents. In fact, he had begun a reconstitution of the self too long delayed, while he conceived of his identity in terms of desire for his absent mother and father, and lived in a state of suspended identity. Now he knew that there was no point to his waiting—that he had allowed himself to be a child too long. So long, in fact, that in certain ways Hughes would always be, to an extraordinary though not always clearly identifiable extent, a child. He would have a child's charm and glittering sense of wonder and innocence, a child's fondness for the dream, and an empathy for children that led to several books for them; but perhaps also, in an unusually protracted form, certain adult disabilities associated with childhood—sexual reticence, an anxiety to please, a constant need for approval and reinforcement. At some point in Mexico City, however, he began to look to his adult future. His hopes would rally and then re-form according to the most fulfilling experiences of his past life. Inevitably he would consider among these the silent martyrdom of his grandmother, the love of the Reeds, and the almost sensual gratification that his words in print gave him.

In September he left Toluca for home. Passing as a Mexican in San Antonio, he bought a Pullman berth to Cleveland. On the first evening out a white man squinted at him across the dining table, then leaped up: "You're a nigger, ain't you?" At St. Louis a clerk in a soda fountain asked bluntly whether he was Mexican or American. When Langston conceded that he was indeed American, the clerk turned away. "I knew I was home in the U.S.A."

He returned to a nation inflamed by racial strife. While he had been in Mexico, black America had suffered a summer of bloodshed, notably in Chicago, that had stirred Claude McKay in the *Liberator* to write his most memorable poem, "If We Must Die," exhorting blacks to go down fighting. With expectations raised dramatically by migration and the war, racial tension had led to violence and death for almost an entire year. Between June and the end of the year, whites lynched seventy-six blacks; riots erupted in some two dozen cities. The white secretary of the NAACP, John Shillady, appearing in Texas to prevent his organization from being closed down by the state, was beaten unconscious on a street in Austin. Late in July, thirty-eight persons died in a Chicago riot begun after a black swimmer crossed an imaginary line in the water separating a "black" beach from a "white" beach. Two hundred blacks died in Arkansas in a series of riots that followed protests against exploitation in the cotton industry. These events, reported in detail in the Cleveland papers, must have weighed heavily on Langston's mind as he tried to make sense of his father's hostility to blacks. At last he began to understand fully what his youth and the democracy at Central no doubt had lulled him into ignoring—that he was a black man in a nation hostile to all people of his race.

Back in Cleveland he made peace with his mother as best he could; tension was now a permanent part of their lives. To help her, Langston worked after school in the staff dining room of a city hotel. At Central High, however, he entered his senior year as one of the most admired students. As an officer in the drill corps he strutted about in his crisp khaki uniform, or showed off his school sweater bedecked with track and field letters and club pins. Handsome and athletic but modest and amiable, he stood out even in the cosmopolitan mass of Central. A writer compiling a list of ideal features among Central students selected Langston's soulful eyes. His relay team again won the city championships, and for the second year he wore the red and blue of Central as its representative in the high jump; he cleared five feet six inches, or about two inches over his head. He was elected again to the student council. Encouraged no doubt by his friend Sartur Andrzejewski, probably Central's leading thespian, Langston acted in plays. He continued his French and as an officer in the French club, and he corresponded with a young student in Paris. In the winter he became in effect the Jewish candidate for class president, since the Christian majority at Central had no intention of electing a Jew to the post. A group of over thirty students, by no means all Jewish, signed a petition of support. In spite of the backing of Messrs. Klein, Krutchkoff, Meyer, Rabinowitz, Horowitz, Glattstein, and others, Hughes lost. However, he became editor-in-chief of the *Annual,* and class poet.

The painful summer in Mexico had strengthened him as a writer. Growing more political and more racial, he matured; by the spring of 1920 he had a whole notebook of poems—or of verses and poems, as he distinguished between his lighter, satirical pieces and his weightier compositions. He was proud of certain poems, such as his Dunbar-influenced "Just Because I Loves You" ("That's de reason why / My soul is full of color / Like de wings of a butterfly"). More solemnly, he wrote about the steel mills that had ruined Homer Clark,

The mills
That grind and grind,
That grind out steel
And grind away the lives
Of men . . .

The mills grind out "new steel, / Old men."

But Hughes made an even greater artistic leap. In his major poem as a senior, in the spring of 1920, he gave the first sign that he might possess a degree of distinction as a poet. He had shown little interest in girls thus far; he was far more often in the company of Sartur Andrzejewski—whose main ambition, as the *Annual* put it, was to remain "a bachelor." But when the late winter and spring found the seniors in an amorous mood, Langston himself went to dances at different schools. In the gymnasium of Longwood High he was smitten by a girl with dark brown skin and large, lovely eyes. Not long from the South, she was still in junior high school but almost as old as Langston. She was certainly unlike most of the girls he knew at Central—white girls like Sarah Sapir, Elsa Schreiber, and Helen Osborne, or Bess Rivkin and her twin sister Rose, or Sadie Krutchkoff, or the star scholar Helen Baldwin. But she was what he wanted. He flashed his grin, and she was charmed. Soon he was calling on her; when he saw her once in a flaming red dress, Hughes suddenly knew what she meant to him:

When Susanna Jones wears red
Her face is like an ancient cameo
Turned brown by the ages.

Come with a blast of trumpets,
Jesus!

When Susanna Jones wears red
A queen from some time-dead Egyptian night
Walks once again.

Blow trumpets, Jesus!

And the beauty of Susanna Jones in red
Wakes in my heart a love-fire sharp like pain.

Sweet silver trumpets,
Jesus!

Apparently he took the finished poem to show Rowena Jelliffe at the "Playground House," where he taught art classes to neighborhood children. "Langston was absolutely ecstatic," she would recall. "He glowed with satisfaction at what he had done. He was so tremendously happy, as if he was sure he had done something very good." In "When Sue Wears Red," Langston turned away from the delusion that he was a raceless creator, a poet who happened to be black, to face the fact that race was of absolute concern to him and thus had to be reflected in his art. In praising the beauty of a black woman, he praised the beauty and history of the African race in America. Astutely, too, he had blended the rhythm of conventional English with another, Afro-American rhythm—not the broad dialect of "Just Because I Loves You," with its caricature of black speech, but one echoing the religious cries he had heard in church. He reached this new stage assisted probably by several sources of which he might have been unconscious: Whitman's few but deft portraits of blacks in *Leaves of Grass;* Sandburg's crudely powerful evocations of black music and life, as in his "Jazz Fantasies"; the freedom of form that both Whitman and Sandburg sanctioned; Dunbar's poetic vignettes of black life; perhaps Du Bois's essays on Africa and on the black woman, soon collected in his *Darkwater: Voices from Behind the Veil;* Claude McKay's dignified zest. But doubtless behind them all was Hughes's experience of the previous summer. One result of the strife with his father was a rapid expansion of Langston's racial sense. What James Hughes hated, Langston Hughes would love; what his father loved, Langston would spurn. James Hughes loathed black women. His son would praise them.

Langston would do so even if, as the poem covertly suggests, the poet can more admire than possess this queenly woman. Susanna Jones stands black and proud on her pedestal, which is a fine place for admiration but a poor place for love. In this respect Hughes's life and writing, like Walt Whitman's, would coexist in paradox.

With this poem, in a sense, Hughes was ready to leave Central. On June 16, 1920, the school held its seventy-second commencement exercise; to the brassy strains of a "Marche Militaire" by Schubert the procession of students, teachers, and guests entered the school auditorium. A retired army general (yet another distinguished Central alumnus) delivered the commencement address. The Glee Club sang, and class valedictorian Helen Baldwin orated. Then principal Edward Harris presented the graduating class of 127 seniors to the head of the Board of Education, who handed out the diplomas. With a fine academic record, Langston received a First Grade certificate. At the dance later that evening the band played all the latest songs, including Irving Berlin's current hit, "A Pretty Girl is Like a Melody."

Hughes would always think affectionately and with gratitude of Central High. Entering the school at a fortunate moment in its history, he had found an enlightened, cosmopolitan place; a few years later, he would have met a very different Central High, almost entirely black, its creative spirit largely dissipated by severe urban pressures. At least one of his teachers wondered whether Hughes understood exactly how fortunate he was. Charles Ozanne (later credited by Langston with alerting him to world politics in history classes) offered him a candid evaluation: "You have a fine appearance, fair financial resources, if I understand your situation, good mental ability, pleasing and gentlemanly manners." Less clear was whether Langston had "the grim intense determination" to make a success of life: "I only recall how smoothly things seemed to go for you in the school, and how keen your enjoyment was of that social and club life into which you were cordially admitted." Now, when Langston should be thinking of a life of service to blacks, Ozanne begged leave to express the fear "that your path may be made too easy, so that you may think of the years as periods that you can enjoy and have a good time in, without wrestling with the grim realities of life where manhood is tested, and works its mighty work."

Although Ozanne would not be the last person to wonder if the charming, ever smiling Hughes were not too easygoing, his thoughts were very much on the future. His friends Sadie Krutchkoff, Sollie Jacobson, and the Kreinberg brothers were all admitted to Columbia University in New York. Columbia—or New York—sounded exactly right to Langston, but the expensive school was beyond his means. Unless his father would help him. Sartur Andrzejewski and other students expected to enter the leading college in Cleveland, Western Reserve University, which interested Langston much less. He wanted to move on; besides, even with a college degree he would not get very far as a black in Cleveland. He would probably find work as a teacher, but he had another goal in mind. Showing a calculation, or a practicality, often missed by observers taken in by his apparent impulsiveness and innocence, Hughes set aside his anger at his father and decided to return to Mexico to try to convince him to send his son to Columbia in New York.

A month after graduation, Hughes left Cleveland for Toluca. Again the parting from his mother was painful, with Carrie openly scornful of his talk about college. He should be working to help her! "In certain ways she had become a hard woman," Rowena Jelliffe would remember of Carrie Clark. "She was concerned only with money. She was always pounding Langston about it. Carrie hammered at him all the time with the idea that he wasn't going the way she wanted him to go. She thought he was wasting his time writing poetry; she rebuffed him on those *Monthly* poems. The odd thing is, Langston never complained.'' Guilty and depressed, he was ambivalent about his mother's anger. Life had been hard on her, but he had not caused her unhappiness; if anything, she and his father had caused his own. Now he must look out for himself. Two ideas seem to have been gathering strength in Hughes's mind. The first was that he would not return to Cleveland. The second was that his future lay as a writer. If no black American had ever lived solely by writing, Hughes was gathering the courage to try. A college education would be secondary to this goal. He would credit the vivid stories of Guy de Maupassant, which he read in French at school, with making him ''really want to be a writer and write stories about Negroes, so true that people in far-away lands would read them—even after I was dead.''

On the afternoon of his first day out he took his dinner, then returned to his seat to stare out of the train window and brood on what he had left behind and the life that awaited him now in Mexico. Cheerlessly he thought of his angry mother and his forbidding father. In particular, he brooded on his father's hatred of blacks; nothing else in James Hughes so alienated his son. In a year when W. E. B. Du Bois was predicting the coming of race war, when Marcus Garvey was preparing a grand meeting in New York with the cry ''Back to Africa,'' when the Ku Klux Klan was in resurgence and blacks were being lynched with impunity, his own father sneered at ''niggers.'' Blacks seemed to Langston, even at the distance from which he sometimes viewed them, the most wonderful people in the world. This was the main legacy of his grandmother through her heroic tales, and of the Reeds, and of the black men and women in church who had loved him as a child.

The sun was setting as the train reached St. Louis and began the long passage from Illinois across the Mississippi and into Missouri, where Hughes had been born. The beauty of the hour and the setting—the great muddy river glinting in the sun, the banked and tinted summer clouds, the rush of the train toward the dark, all touched an adolescent sensibility tender after the gloomy day. The sense of beauty and death, of hope and despair, fused in his imagination. A phrase came to him, then a sentence. Drawing an envelope from his pocket, he began to scribble. In a few minutes Langston had finished a poem:

I've known rivers:
I've known rivers ancient as the world and older
than the flow of human blood in human veins.

My soul has grown deep like the rivers.

I bathed in the Euphrates when dawns were young.
I built my hut near the Congo and it lulled me to sleep.
I looked upon the Nile and raised the pyramids above it.
I heard the singing of the Mississippi when Abe
Lincoln went down to New Orleans, and I've seen
its muddy bosom turn all golden in the sunset.

I've known rivers;
  Ancient, dusky rivers.

My soul has grown deep like the rivers.

With its allusions to deep dusky rivers, the setting sun, sleep, and the soul, "The Negro Speaks of Rivers" is suffused with the image of death and, simultaneously, the idea of deathlessness. As in Whitman's philosophy, only the knowledge of death can bring the primal spark of poetry and life. Here Langston Hughes became "the outsetting bard," in Whitman's phrase, the poet who sings of life because at last he has known death. Balanced between the knowledge of love and of death, the poetic will gathers force. From the depths of grief the poet sweeps back to life by clinging to his greatest faith, which is in his people and his sense of kinship with them. His frail, intimidated self, as well as the image of his father, are liquidated. A man-child is born, soft-spoken, almost casual, yet noble and proud, and black as Africa. The muddy river is his race, the primal source out of which he is born anew; on that "muddy bosom" of the race as black mother, or grandmother, he rests secure forever. The angle of the sun on the muddy water is like the angle of a poet's vision, which turns mud into gold. The diction of the poem is simple and unaffected either by dialect or rhetorical excess; its eloquence is like that of the best of the black spirituals.

His intense mood continued as the train swept south. One night, when Langston sat down to dinner, a white Texan rose and walked out elaborately. Otherwise there was no hostility—"nothing except stares of mild disapproval," as he noted in a journal he kept of the trip in recognition that he was now a writer. One passenger, a cosmopolite with international stickers on his luggage, talked freely with him. Another, Jewish, assured Langston "that he had known at once that I was Mexican. I did not tell him otherwise." Langston was annoyed at the man's presumption—"but Jews are warm hearted people and seldom prejudiced." At San Antonio he pulled his hat down over his curly hair and, in Spanish, secured a comfortable Pullman berth to Laredo, where word swept the train that Nuevo Laredo across the border was under attack by rebel soldiers. "Maybe there will be adventures," Langston scribbled. "Who knows? I hope so."

For six pesos he spent the night in a fleabag. "Of course it's far from being the Ritz-Carlton," he reasoned, "but then I couldn't stop there anyhow for I am Colored. But here nothing is barred from me. I am among my own people, for . . . Mexico is a brown man's country. Do you blame them for fearing a 'gringo' invasion with its attendant horror of color hatred?" Wandering the streets, he heard a peon singing of the "soldados Mexicanos." Although Langston couldn't understand all the words, "I know he is singing to the glory of Mexico." But where was the glory? Every child seemed to be a bootblack; beggars haunted the streets. "The revolutions have made so many poor," he sighed, "broken on the wheel of the freedom they are seeking." Retiring for the night, he heard music and noise from the *cantinas* nearby but gazed upward through the iron bars at his window: "Tonight the sky is filled with those lovely jewels which evening wears upon her velvet gown. High above the Rio Grande, above the two cities, above the two countries they sparkle and glow, and one big star is winking and twinkling as if he were laughing at my littleness—at the littleness of all men with their schemes of hatred and war, and their eternal bickerings."

On the train the next day, the Mexicans were uniformly "good humoured, kind and courteous." At twilight they were still rolling south. Langston looked out of the window: "It is sunset and my car window frames a Maxfield Parrish painting. Mexico's great jagged mountains are bathed in yellow wine and honey. There are some old peaks far in the background that have wrapped a purple veil about themselves, and sit huddled like Indians, silent. But there in the foreground, sunset's wonder colors of crimson and amber and gold change the dull gray mountains, these stark rugged mountains, into magic hills, the dream mountains of childhood. Only God could paint such a picture and only Mexico could be the canvas."

Next morning, cold and stiff, he was in a less romantic mood. Testily he scrutinized his fellow passengers—the Syrian next to him, reading an Arabic newspaper; "a fat and slightly stupid Chilean priest who is as greedy as a pig." As the day warmed, however, his good temper returned. At one stop he noted "a most glorious beggar. Old, but grand and tall and majestic. He wore his cloak of rags as a king might wear a robe, and begged for pennies with an ancient dignity. There are so many beggars in Mexico but I shall not forget this one soon—this old recaller of Arabian nights." Trying his Spanish on a boy dressed *muy Americano,* he heard in reply the most amusing prep-school English. Going to Toluca? "Why, that's my home!" the boy piped. He was Luis Henckel, son of the man to whom Langston's father had sold the electric company.

Sharing a large cheese, lemonade, and beer, Langston and Luis talked as far as San Luis Potosi, where soldiers, including an imposing general, stomped aboard. Declining a seat, however, the general sank democratically to the floor. In one corner of the train, flirtatious, was "a most striking blond," Hughes noted in his journal, "beautiful, but big and slightly coarse, and oh so 'Won't you make those eyes at me' style." The general bought her a beer, but Langston was not impressed. "She tries so hard to be cute, but she's too old and too big. It's like a cow trying to be a calf." Sympathizing with her shivering little dog ("with so many men around he gets little attention") Langston much preferred the company of Luis Henckel, who was rich and young but spent a long time talking earnestly with a poor Indian woman ("you would not think him wealthy. His manners are too good"). He and Langston shared with her their supper of crackers and sticky milk candy. "All the Mexicans I've met have been so kind, friendly, jolly. It seems to be a national virtue—this friendliness to strangers and their courteousness to one another."

The two boys took turns sharing a seat. At one o'clock, with Mexico City not very far away, Langston dozed fitfully in the aisle. At two he got up to stretch. At three he was wide awake. Someone had seen a pale white glow against the sky—the lights of Mexico City. Immediately the car erupted into turmoil.

Women pulled down dusty hats from the racks overhead, suitcases were opened and snapped shut, ready for the arrival. At dawn the train rolled into the capital.

In Toluca, Langston found one significant change at the house at 3 Plaza de la Reforma. James Hughes had a new housekeeper, a German woman named Bertha Schultz, who had emigrated recently to Mexico following the death of her husband in the war. With her was her eldest daughter, Lotte, about ten years old. A kindly woman, somewhat stout, with dull blue eyes and chestnut brown hair, Frau Schultz spoke no Spanish and little English, but conversed in German with James Hughes. Langston liked her at once; he liked her cooking even more. With Frau Schultz, his father was much less stingy about groceries; in fact, he was altogether a more pleasant man. Unaware of the depth of his son's disaffection, he was pleased to have Langston back. However, James Hughes made one thing clear: he had no intention of sending his son to college in the fall.

Working hard on his Spanish, Langston was soon almost comfortable reading books borrowed from the local library, especially the works by the contemporary Spanish novelist Blasco Ibáñez, author of *The Four Horsemen of the Apocalypse* and, more to Langston's liking, the realistic *Cañas y Barro* and *Cuentas Valencianas,* sketches of Valencia life. He made some friends of roughly his age in Toluca. Chief among these was the lighthearted Tomas Valero, Jr., whose father owned the largest department store, La Samaritana, on the crowded Jesus Carranza Avenue in Toluca. Valero and Hughes were often among the strolling throng of young men watching the chaperoned señoritas promenade in the cloistered walkway, Los Portales, around the main square.

James Hughes's financial position was not as rosy as it had been the previous summer. Bandits had run off all the cattle and sheep on the farm in Temexcaltepic. Twice stopped on the road and robbed of everything but his underwear, Hughes was lucky to be alive. Toward the end of the summer, after government troops swept the area, Langston and his father finally set out in a heavily armed party of German and Mexican mine owners and ranchers for the mountains. At a high pass called Las Cruces, the bodies of three Indian *bandidos* swung as a warning to their comrades. After a long ride the party reached his father's holdings, the Rancho de Cienguillas. Although the ranch covered the whole side of a mountain and more, the terrain was wild and desolate, the mines flooded, the peasants poor and dispirited.

Here, finally, James Hughes detailed his plan for his son's life (a plan "I had never dreamed of before," Langston would write). Studying abroad, preferably in Switzerland and Germany, his son would become a mining engineer and settle down in Mexico. When peace returned, the mines would be worth a fortune. Dismayed, Langston protested that he was weak in mathematics (his lowest grade at Central, although respectable, had been in algebra when he was a freshman).

"You can learn anything you put your mind to," Jim Hughes retorted. "And engineering is something that will make you some money. What do you want to do, live like a nigger all your life? Look at your mother, waiting table in a restaurant! Don't you want to get anywhere?"

Timidly, Langston told his father about his dream of being a writer.

"A writer?" Jim Hughes asked, incredulous. "A writer? Do they make any money?"

Who had ever heard of a successful colored writer? Alexandre Dumas, Langston timidly offered. That was in France, his father countered, where a colored man had a chance. Langston should learn a skill that would take him away from the United States, "where you have to live like a nigger with niggers." The discussion died.

But Langston wanted no part of exploited peasants and hanged Indian bandits. Nor was he ashamed to "live like a nigger with niggers." And he had already set his sights on New York. That summer, feeding on the tension between his father and himself, he wrote steadily. Two significant poems, "Aunt Sue's Stories" and "Mother to Son," emerged. One is an oblique tribute to his grandmother; the other might be an even more covert apology to his mother. In both, however, he celebrated not his relatives but the "niggers" his father despised. In both poems, too, the poet speaks as a child again, or to a child comforted and nurtured by a maternal figure. The poet converts the facts of his life and his childlike obsession with maternal love into an adult political and cultural plea. Writing of the self but transcending narcissism, Hughes inscribed himself into and out of his poems by fixing at last not on his personal hurt but on the only force that could heal him—the love and regard of his race.

In "Aunt Sue's Stories" he perhaps blended the images of his proud grandmother and his loving Auntie Reed:

> Aunt Sue has a head full of stories,
> Aunt Sue has a whole heart full of stories.
> Summer nights on the front porch
> Aunt Sue cuddles a brown-faced child to her bosom
> And tells him stories.

She tells of "black slaves / Singing sorrow songs on the banks of a mighty river," who "mingle themselves softly / In the dark shadows that cross and recross / Aunt Sue's stories." The child knows that these stories, and the slaves, are real. "The dark-faced child is quiet / Of a summer night / Listening to Aunt Sue's stories."

In "Mother to Son," Hughes boldly reclaimed the use of dialect by black poets after it had fallen into disrepute and disuse ("a jingle in a broken tongue," Dunbar dejectedly had called his own dialect verses) as the language of hated stereotypes, stressing either a gross humor or maudlin sentimentality. In "Mother to Son," dialect allows a humble black woman—and through her all black women—to speak nobly:

> Well, son, I'll tell you:
> Life for me ain't been no crystal stair.
> It's had tacks in it,
> And splinters,
> And boards torn up,
> And places with no carpet on the floor—
> Bare;

But all the time
I'se been a-climbin on,
And reachin' landin's,
And turnin' corners,
And sometimes goin' in the dark
Where there ain't been no light.
So boy, don't you turn back;
Don't you set down on the steps,
'Cause you find it's kinder hard;
Don't you fall now—
For I'se still goin', honey,
I'se still climbin',
And life for me ain't been no crystal stair.

Less successful, precisely because it is blunt and avoids the image of the vulnerable child and the themes of love and danger, is "Negro": "I am a Negro: / Black as the night is black, / Black like the depths of my Africa." In using the word "black" rather than Negro, Hughes was following a radical minority tradition in Afro-American letters. W. E. B. Du Bois, also a mulatto, had done the same earlier in his poem "Song of the Smoke" ("The blacker the mantle, the mightier the man! / For blackness was ancient ere whiteness began"). But perhaps Hughes was also thinking of the charismatic and anti-mulatto nationalist hero Marcus Garvey, who in August opened a month-long national convention of his Universal Negro Improvement Association in New York with news about the founding of the Black Star line to repatriate Afro-Americans to Africa. Certainly, however, the poem began in Carl Sandburg's well-intentioned yet sensational "Nigger" (from *Chicago Poems*): "I am the nigger. / Singer of Songs, / Dancer. . . ." While Sandburg emphasized sex, violence, madness, and exploitation, however, Hughes stressed black historicity and dignity, rooted in "the depths of my Africa." Although Sandburg's poem is livelier, in "Negro" Hughes showed his growing if respectful emancipation from the influence of his "guiding star."

Other poems, untouched by race, form a rival mode that reveals the extent to which race made Hughes a better poet. Less intense, almost vagrant in spirit, these raceless poems are less precise as art. Entranced by rain and night and the moon, by "the dust of dreams," "the purple and rose of old memories," the poet drifts. Love is defined by its absence or loss; death brings no hope of new life.

Flowers are happy in summer;
In autumn they die and are blown away.
Dry and withered,
Their petals dance on the wind,
Like little brown butterflies.

In general, this passivity reflects the negative aspect of Hughes's impulse to delay the onset of manhood, or the related consequences of his having been brought up without strong male example, in the care of a proud but essentially defeated woman such as his grandmother. Even if his sexuality remains ambiguous or androgynous, Hughes as a poetic persona is capable of achieving adulthood in certain moments of poetic creation, when momentarily he assumes full command of his ego. However, he reaches these moments most brilliantly (as in "The Negro Speaks of Rivers," "Aunt Sue's Stories," and "Mother to Son") only when he first becomes a child again, then transcends childhood by celebrating the unique power of the black race to nurture him.

Thus it was fitting, almost inevitable, that when he sought to enter the world as a poet (that is, to publish for the first time in a national magazine) Hughes offered material written from a child's point of view, or with deliberately childlike technique, to a magazine for black children. The journal was the monthly *Brownies' Book,* founded in New York in October, 1919, by W. E. B. Du Bois and run by Du Bois, Jessie Redmon Fauset, and Augustus Granville Dill (respectively the editor, the literary editor, and the business manager of the *Crisis*). On September 20, 1920, Hughes sent three poems to the *Brownies' Book.*

His verse reached a sympathetic but also very capable editor in Jessie Fauset. From an old black Philadelphia family, Fauset had graduated Phi Beta Kappa from Cornell in 1905; she earned a master's degree at the University of Pennsylvania, and later did more work in French, her favorite language, at the Sorbonne. Between 1906 and 1919 she had taught at the highly respected M Street School (renamed Dunbar High in 1916). In 1919, Fauset resigned to join the staff of the *Crisis,* edited by her longtime friend and mentor Du Bois, who had been predicting in its pages an upsurge of literary activity in the black world; with the *Crisis* at the peak of its circulation, Fauset had been hired to stimulate the younger authors. A writer herself, she would publish four novels in the coming years.

From New York early in October, Jessie Fauset replied that she found one poem "very charming" and would print it sometime soon; did Mr. Hughes have original Mexican stories for children, or know of Mexican games for children? At once Langston forwarded an article on Mexican games, as well as other poems. The January, 1921 number of the *Brownies' Book* carried two poems by Langston Hughes, "Fairies" and "Winter Sweetness":

> The little house is sugar,
> Its roof with snow is piled,
> And from its tiny window,
> Peeps a maple-sugar child . . .

Delicately, Langston moved towards his first adult statement. Near New Year's Day, 1921, he sent Fauset "The Negro Speaks of Rivers," written the previous July. On January 18, she accepted it: "I think it has a very fine touch." She would use it "probably very soon"—not in the children's magazine but in the adult *Crisis.* With the timidity of a child but the guile of an ancient, Langston had begun his seduction of Harlem.

Also, he had made himself independent of his father. The previous September Langston had started teaching morning classes in English to young women at a private school run by a Señorita Padilla; in the afternoon, he instructed students at Luis Tovar's business college. In addition, he gave private lessons to the teenaged children of the mayor of Toluca. Steadily his savings mounted. Popular as a teacher, Langston had every chance to savor the pleasures of exile. With his brown skin, slightly curly hair, Indian features, and his father's position, he could have courted a señorita and slipped smoothly into a Mexican destiny—or, remaining single, melted into the population as a *muy simpatico* American expatriate. An older female student fell in love with him apparently, but he kept his professional distance. Invited to parties in the "best" homes, he developed a circle of friends. When the older boys at the leading high school planned a two-day walk up the volcano Xinantecatl, he went along in the party of forty, taking notes carefully for a small article soon accepted by Jessie Fauset in New York ("the moon had gone down and the forest was in inky blackness. The low burning camp-fires gave a little light. A long way off and deep down in the night-covered valley, we saw the white lights of Toluca, shining like a cluster of sunken stars in the darkness").

At least once, by his own admission, Langston went to the bordello in Toluca that catered to army officers and other gentlemen (another house was less exclusive). This visit probably marked his adult sexual initiation. "In Cleveland," he would admit, "the boys I knew hadn't yet worked up to that." Perhaps Hughes's exposure in Mexico to an attitude to sex very different from that at home prepared him for the way he would cope with sexual desire for much of the rest of his life. In Toluca, the young men knew and respected the madam of the bordello—even if no one spoke to her when she walked in the Portales, a stately figure dressed all in black. Only three times in his autobiographical writings would Hughes speak of being in love with women. More often, however—describing incidents in Toluca, Havana, Madrid, Paris, Lagos, Japan, and the Soviet Union—he would write casually about patronizing prostitutes or their near equivalent. Perhaps Hughes was telling the truth, though not necessarily the whole truth, about his sexual history. This is an important consideration, given the later speculation, without convincing evidence, that he was a homosexual. Although delayed maturation can be consistent with homosexual feeling, the key place of childhood not only in Hughes's poetry but in his founding of a career suggests psycho-sexual complication of a kind far more rare than homosexuality. What Langston appears to have sought and felt for much of his early life was a quality of ageless, sexless, inspired innocence, Peter Pan-like, which race and even sex brought down to earth without sullying. "Langston," a close friend in some of his more bohemian years would say, "was like the legendary Virgin who walks through mud without soiling even the hem of her robe."

However innocent his temperament was, he had no fear of blood. He loved bullfighting, even if he claimed that he couldn't write about it—bullfighting had to be seen, heard, and smelt, "dust and tobacco and animals and leather, sweat and blood and the scent of death." On weekends in Mexico City, Langston saw Rudolfo Gaona, Mexico's greatest bullfighter, and the famed Spanish matador Ignacio Sánchez Mejías, whose death in the bullring would be lamented famously by the poet García Lorca. At the annual festival bullfight for the charities of La Cavadonga, when Gaona fought, Langston joined the charge into the arena for souvenirs. Leaping the *barreras* and tearing his trousers from ankle to knee, he plucked and carried off the burnt orange banderillas, tipped with tiny fruit made of silk and tinselled gold ("they were beautiful things to torture an animal with," he would recall). Bullfighting "is not a game or a sport. It's life playing deliberately with death. Except that death is alive, too, taking an active part."

On these weekends in Mexico City he usually stayed with the three Patiño sisters in their house behind the cathedral near the Plaza of the Constitution. Attending vespers and Sunday mass with them, he liked the gloomy interiors of the churches, the dolorous Virgins and crucifixes dripping with what looked in the dark, after the corrida, like real blood. As in the United States, the drama of religion appealed to him, not its dogma: Jesus was dead, ritual was alive. Latin culture, like black culture, seemed richly elemental compared to white Protestantism; Hughes felt very much at home in Mexico.

Perhaps he fell in with certain young writers in the capital. The important Mexican poet Carlos Pellicer, three years older than Langston, and involved between 1920 and 1923 with the most gifted young literary progressives in Mexico City in the journal *México Moderno,* would later suggest that he knew Hughes from his days in Toluca. Certainly at some point Langston presented a copy of "The Negro Speaks of Rivers" to Pellicer, who would cherish it as a "precious" poem from his "very good friend." Whether they met then or later, when Hughes lived in Mexico in 1935, the two poets had much in common. Hughes and Pellicer were members of a tiny international advance guard that would eventually include Pablo Neruda of Chile, Jorge Luis Borges of Argentina, Léopold Sédar Senghor of Senegal, Jacques Roumain of Haiti, Aimé Césaire of Martinique, and Nicolás Guillén of Cuba. Their aim was to develop, even as they composed in the languages of Europe, an aesthetic tied to a sense of myth, geography, history, and culture that was truly indigenous to their countries rather than merely reflective of European trends, whether conservative or avant garde. In 1921, the artistic acknowledgement of the Indian roots of Spanish-American culture awaited the return from Europe of the Mexican painters Diego Rivera and David Alfaro Siquieros; with José Clemente Orozco, they would form the heart of the modern Mexican revolution in art, based on veneration of the Indian past. In Spain, García Lorca's fascination with the gypsy and other folk cultures showed a similar concern—as did his interest in black Harlem in *The Poet in New York.*

A similar recognition by Anglo-Saxon, North American culture of its powerful African influence would never come. But Hughes's poems such as "The Negro Speaks of Rivers," "When Sue Wears Red," "Aunt Sue's Stories," and "The Negro" were linked organically to the historic shift by Mexican, Cuban, and other Latin American poets away from Europe in the depiction of their cultures. Hughes's poems would appear in *Contemporáneos,* the important journal of cultural change in Mexico published between 1928 and 1931; elsewhere, in 1931, as part of the international movement, Borges translated "The Negro

Speaks of Rivers.'' Langston himself would translate and publish, or attempt to publish, works by Mexican, Cuban, Chilean, Haitian, and Spanish writers, including books by Guillén, Roumain, Gabriela Mistral, and García Lorca. Hughes, however, had begun to shift away from imitation while virtually all the other writers, although often nationalistic, still hesitated to grapple with the legacy of racism and exploitation in their history. His achievement, when he was barely nineteen years old, reflected not only the force of his unusual personal circumstances but also the extent to which black American intellectuals, notably W. E. B. Du Bois, had been driven by American racism to develop a sense of ethnic pride, not chauvinist but international, in opposition to North American and European chauvinism. From the start, the subtitle of Du Bois's *Crisis* was ''A Record of the Darker Races.''

As alluring as Mexico was, Langston set as a goal his return to the United States. ''I began to wish for some Negro friends to pal around with,'' he later wrote; but his aim was more serious, as his continuing courtship of the *Crisis* showed. Late in February, Jessie Fauset received an account of daily life in Toluca and was ''completely charmed'' by it. The essay and a poem would appear in the April *Crisis.* A handwritten postscript at last opened the door on which Langston had tapped gently over the months. ''May I be a bit curious?'' she asked. ''You write so well and sympathetically—who are you, and whence and why do you live in Mexico?'' Langston knew exactly how to reply: very modestly. In May he sent another poem and a play of three or four pages—the first he had written—for children, ''The Gold Piece.'' A peasant couple, Pablo and Rosa, are elated to find a valuable gold piece. But a poor old woman appears; she has a blind son, whom the doctors will cure if she brings enough money. ''Ah!'' she sighs, ''I would sell all that I have if my boy could only see again!'' Pablo and Rosa name all the things they want to buy with the gold piece; would she give them up? ''I would give up all my dreams,'' the old woman vows, ''if my son were to see again.'' Giving her the gold piece, Pablo and Rosa settle happily by the fire. Pleasant and moralistic for the audience of children, the little play is the optimistic, reverse side of the three short stories he had written in Cleveland. Money brings false happiness; true happiness comes with sacrifice and a sense of brotherhood.

With ''The Gold Piece,'' Hughes now had published in at least four genres (poetry, fiction, the essay, and drama), a variety he would only augment during the rest of his life. In June, 1921, ''The Negro Speaks of Rivers'' appeared in the *Crisis;* the next month it was reprinted in the influential white *Literary Digest.* When he sent the *Crisis* a new group of poems, Fauset declared herself ''very much pleased'' with almost all; she also asked Langston to think about preparing an entire page of poems for either the *Brownies' Book* or the *Crisis*—even Dr. Du Bois himself was impressed with his work. In a few short months Hughes had become virtually the house poet of the most important journal in black America.

Sniffing at the magazine containing his son's work, Jim Hughes asked only two questions. ''How long did it take you to write that?'' And, ''Did they pay you anything?'' They hadn't. Although the *Crisis* enjoyed a circulation of almost 100,000 copies, Langston never dreamed of asking for pay. Nevertheless, his father was impressed. In a compromise with Langston, he agreed to pay for a year at Columbia University. If Langston did well, he would consider further support.

Langston applied to the university and was quickly admitted. He almost never made it there. His father's housekeeper, Frau Schultz, had a young German friend, shy, plain, awkward, who worked as a housekeeper for an old, recently widowed German brewery master. The girl, Herta, often visited Frau Schultz at the Hughes home. One day, when Herta was visiting and Langston happened to be away, the old man came unannounced. Apparently he suspected that Herta and Langston were lovers. Pushing his way into the room where the women sat, he shot Herta in the head, jaw, and shoulder. Then he shot Frau Schultz in her right arm, breaking it. He looked for Langston. Unable to find him, he went to the police station and turned himself in. Miraculously, both women survived the shooting.

Langston's year in Mexico came to an end. In mid-August, he turned his position at Luis Tovar's business institute over to a white woman from Arkansas who nervously confessed to have never met an educated Negro; while he explained her duties she watched him closely out of the corners of her eyes as if she expected him to attack. She was an excellent augury for his return to the United States—but perhaps she was only the white South, "Beast-strong, / Idiot-brained," as he called it in a poem about this time. But nothing could dull his expectation of New York, where an extraordinary black company had opened on Broadway the past June in the acclaimed *Shuffle Along,* a musical partly written by an alumnus of Central High in Cleveland, Noble Sissle, and his friend Eubie Blake. Hungry for the company of his own people, hungry for fame, Langston was eager to leave for New York.

His father, who went with him to Mexico City for his departure, promised to wire money to Langston once he reached Columbia University. At the Buena Vista station father and son parted. If James Hughes felt regret, Langston had none. *"Gracias a dios!"* he remembered murmuring to himself as the train pulled out for Vera Cruz on the Caribbean. He never saw his father again.

The rich brine smell of the Caribbean, his first scent of the sea, reached him as he rode the train eastward through dense fog. The next day he saw the wide flashing ocean, green and white in the tremendous sunlight. Vera Cruz was hot, humid, and malarial; even the mosquitoes whimpered.

Sailing across the Bahía de Campeche to the port near Mérida in the Yucatan, Hughes generously allowed a young Mexican, Daniel García, on his way to Detroit and automotive school, to share his stateroom. Langston also gave him lessons in English.

At Mérida, so many passengers were sick that the ship was quarantined there and at Havana, the next port of call. At Havana, Daniel García gave way to a very old Cuban assigned to Hughes's cabin. With him was a large crate of young chickens, which the old man insisted on keeping in the cabin. Langston was very understanding, but he was happy when the ship at last rounded the Straits of Florida and steamed north for New York.

3

MY PEOPLE
1921 to 1923

Hold fast to dreams, my son,
  For if dreams die
Life is a broken-winged bird
  That cannot fly . . .

"Dreams," 1923

"THERE is no thrill in all the world like entering, for the first time, New York harbor,—coming in from the flat monotony of the sea to this rise of dreams and beauty. New York is truly the dream city,—city of the towers near God, city of hopes and visions, of spires seeking in the windy air loveliness and perfection." Thus Hughes would remember his first sight of Manhattan early on the evening of Sunday, September 4, 1921. By the time his friends Daniel García and the old Cuban with his crate of "chickens" (young gamecocks, Hughes found out) had cleared customs, it was past nine o'clock. A taxi sped them to Times Square and the Hotel America, a favorite with Spanish-speaking visitors, where they paid the exorbitant sum of $18 for a suite for the night. Early next morning, Labor Day, Hughes put the old man on a train going upstate, then escorted Daniel García to Long Island to try to locate a relative. Finally, after returning to the hotel to collect his bags, he boarded the Bronx Park Subway uptown to Harlem.

After more than a year in Mexico, Langston was hungry to see his own people. Outside the station at 135th Street, "I stood there, dropped my bags, took a deep breath and felt happy again." In the heart of Harlem, Hughes walked to the single most important address for an unconnected young black man arriving in New York after the war—181 West 135th Street, the Harlem branch of the YMCA. For $7 a week he secured a room on the fourth floor, set down his bags, and was off to explore the community that the sophisticated secretary of the NAACP, James Weldon Johnson, himself a poet and a novelist as well as a former consular representative of the United States, had confidently predicted would soon be "the greatest Negro city in the world."

Harlem in 1921 was only a portion of its future expanse. Blacks had not yet swept down to Central Park North, nor captured the heights of "Sugar Hill" along St. Nicholas Avenue, nor laid siege below Morningside Heights to the walls of Columbia University. Pockets of whites still lived, albeit nervously, among the newcomers. The "Negro city," however, was expanding steadily as the same pressures that had multiplied the black population of Cleveland and Chicago also drove migrants from the South to the various boroughs of New York, which doubled its number of blacks from 1910 to 1920. Starting cautiously in 1905 with the stealthy admission of colored tenants to one apartment house on 134th Street near Fifth Avenue, and aided mightily by unscrupulous real-estate interests within both races, the district of Harlem had absorbed the great majority of the migrants. What they found, in spite of rising rents and shrinking apartments, was housing stock of generally superb quality. The area between the thoroughfares of 125th Street to the south and 145th Street to the north, roughly bounded by Fifth Avenue and Eighth Avenue, comprised such fine urban real estate that no other black community could match its air of settled prosperity. Although Afro-Americans had lived in New York for three centuries, Harlem at last allowed them, in Johnson's words, "better, cleaner, more modern, more airy, more sunny houses than they ever lived in before."

The key to Harlem's lucent promise was its strategic location ("holding the handle of Manhattan," as Claude McKay put it) within the most influential city in the nation. Here were the headquarters of the most important black magazines—Du Bois's *Crisis,* long confident of a coming renaissance in black culture, and the socialist *Messenger,* edited by young A. Philip Randolph and Chandler Owen, who sneered radically now and then about the limitations of bourgeois literature but liberally published such poetry and fiction. Within two years would come the aptly named *Opportunity,* the voice of the National Urban League, to be edited by the farsighted sociologist and cultural entrepreneur Charles Spurgeon Johnson. Manhattan was also Broadway, where in 1917 the white writer Ridgely Torrence's three one-act plays, with all-Negro casts, had marked an epoch in American drama by depicting black themes and characters seriously. In 1920 in Greenwich Village had come Eugene O'Neill's *The Emperor Jones,* with the black actor Charles Gilpin acclaimed in the title role. And on May 23, 1921, only months before Hughes arrived, *Shuffle Along* had opened at the Sixty-third Street Theatre to be hailed soon by David Belasco as the greatest Broadway musical in years, with its star performers and a major hit "I'm Just Wild About Harry." "I was in love with Harlem long before I got there," Hughes later recalled, thinking no doubt about the years that followed his arrival in 1921. "Had I been a rich young man, I would have bought a house in Harlem and built musical steps up to the front door, and installed chimes that at the press of a button played Ellington tunes."

But first there was the matter of Columbia University. Crowning an imposing hill in Manhattan and designed to stun, the campus only confirmed Hughes's worst fears about the life his father was forcing on him ("I had come to New York to attend Columbia," he would admit, "but *really* why I had come was to see Harlem"). With its monumental granite and limestone library dominating an austere quadrangle of red brick halls, the university exuded an air of impersonal grandeur. On the western flank of the quadrangle was Nemesis itself, the School of Mines building, where—if his father had his way—Langston would spend much of the next four years of his life. When he sought a place to live, however, he ran into trouble. At the housing office he was told first that all rooms were taken, then (after offering proof of his reservation) grudgingly given the right to live for one year in Hartley Hall. "There were maybe a dozen of us that year at Columbia," George N. Redd '25 recalled of his fellow black students. "And Langston Hughes was definitely the only one allowed to live in a dormitory. I don't know how he managed that, but he did." Hughes's application from Mexico and the absence of any indication of race on his letters and transcripts had tricked the system of exclusion. (At Harvard the following year President A. Lawrence Lowell threatened to bar black students altogether rather than allow them to live in the dormitories; in 1924 Columbia students burned a cross in front of Furnald Hall to smoke out a black resident.) In ten-story Hartley Hall, which housed three hundred students, Hughes was assigned perhaps the worst room, the one nearest the noisy entrance way. For this accommodation, he paid $105 for the year.

Or was supposed to pay. Tuition and fees were due September 21, but the date passed without money coming from Mexico. Given an extension, he cabled his father about the matter; James Hughes replied that he had sent the money by registered mail ten days before. On Monday, September 26, Langston moved into Hartley Hall. One day before the deadline, on October 4, he again wired that nothing had come. Two telegrams brought no response. Deeply embarrassed, he secured one postponement of the tuition payment, then another; appealing finally to one of the New York principals of the Sultepec Light and Power Company of Toluca, Langston secured a loan of $100. He dipped into his own funds to pay his dormitory room costs in advance for the year, deposited $37.50 on the tuition bill, and borrowed $10 more to buy some of his books. At this point, the wife of one of the company officials came from Brooklyn with a personal loan of $250.

Lodged in one of the richest colleges in the world, Hughes lived anomalously, like a pauper. Instead of buying his textbooks, he borrowed most from the library; unable to pay a laundry, he washed his socks and handkerchiefs by hand at night. November came before he finally received the sum of $532 from his father, whose drafts to the Cleveland Trust Company had apparently been lost in the mail. The delay cast a pall over his first few weeks at Columbia, where Hughes enrolled in courses in written and oral French, English composition, contemporary civilization, physics, and physical education. Except for his English instructor, himself only two years out of the college and killing time at Columbia, Hughes's professors bored him. Although his English teacher entertained the class by reading aloud from the writings of the iconoclastic journalist H. L. Mencken, the French instructor drilled an enormous class like a pompous little corporal, and the lectures on physics were incomprehensible from the start. In spite of Langston's track letters at Central, he found that he loathed physical education.

His mind was elsewhere. "It hurt me to the soul to leave Harlem," he would confess. "I had always been slightly afraid of American white people, and I knew that Columbia University was full of them." In Hartley Hall, where he made few friends, Hughes was closest to a Chinese fellow from Honolulu, Yee Sing Chun, who took him to Chinatown and passed on a smattering of Mandarin. Among the whites, a student named Alfred Henry Best, whose family lived on then-fashionable Riverside Drive along the Hudson, was most agreeable. But another wealthy white student would press Hughes for help with his class assignments, then dash off with other students—but without Langston—in the company of chorus girls: "See you tomorrow, Lang, third hour, and we'll get on them French verbs. I don't need no verbs tonight." The class of 1925 would be distinguished, but Hughes never met two of its future stars—Lionel Trilling and Lou Gehrig; nevertheless, he joined his classmates in a group picture for the yearbook. He came no closer to most of them, so tightly was the color line drawn at Columbia. If Yee Sing Chun was hurt when white girls refused to dance with him, Hughes was no longer so innocent. Still, he was wounded when, trying for a position on the Columbia *Spectator,* "they assigned me to gather frat house and society news, an assignment impossible for a colored boy to fill, as they knew."

Drawn by his presence and the dazzle of New York, his mother arrived to drain what was left of his savings from Mexico. Their most memorable outing was downtown to the NAACP offices on Fifth Avenue in Greenwich Village. The invitation had come from Jessie Fauset of the *Crisis* after she discovered—not from the bashful Hughes—that her mysterious Mexican correspondent was not only in town but at Columbia University. But she quickly invited him, writing "you have been so generous to us and our work." The idea of meeting W. E. B. Du Bois, the greatest of Afro-American intellectuals and the fiery heart of the NAACP as editor of the *Crisis,* terrified Hughes: "What would I say? What should I do? How could I act—not to appear as dumb as I felt myself to be?" He decided to take along Carrie Clark, "for I knew she would do the talking, since my mother loved to talk and meet people."

Nevertheless, he made a strong impression at the *Crisis* office and later over a lunch of filet of sole at the liberal Civic Club on East 34th Street. The fifty-three-year-old Du Bois, with patrician features and an infamously glacial manner with strangers—"a cold, acid hauteur of spirit," wrote Claude McKay, "which is not lessened when he vouchsafes a smile"—melted a little before Langston; his only son, whose early death he had painfully recorded in *The Souls of Black Folk,* would not have been much older. Fauset turned out to be "a gracious, tan-brown lady, a little plump, with a fine smile and gentle eyes." Some people found her affected ("too prim school-marmish and stilted for me," said McKay), but she was sufficiently charmed to ask Hughes to call on her at her home at 130 West 142nd. "I mean this," she assured him, "because I feel that we are genuinely indebted to you at this office." Another conquest was Augustus Granville Dill, business manager of the *Crisis* and a graduate of Harvard and Atlanta, where he had been taught by Du Bois (who would fire him within a few years, after Dill was arrested on a charge of public—homosexual—indecency). A short, brown, gushing man with a very big head, Dill was a skilled lover of the *"pianoforte,"* as he called it. Soon Langston made friends of both Fauset and Dill. With Fauset he discussed plays and movies they might write together; Dill was so entranced by Hughes's first visit that "from that day until this—*and even to the end,*" he wrote years later, Ruth to Langston's Naomi, " 'Where *I* am there may *ye* be also'."

To his father, Hughes sent news of his social success. But the tension between them did not slacken, because money was the weight on the line. On November 27, when the first snow of the season fell on Manhattan, he asked his father for money to buy winter underwear and an overcoat. There was no reply. In his mid-term examinations he acquitted himself reasonably well for a freshman—two B's in English and Contemporary Civilization, a B and a C in French. But the day after Christmas, his father lashed out. He saw no reason why his son should not have an A in each subject; the C in French was intolerable, especially since he had urged Langston to go to Europe, where he could learn the language easily and study without any "prescribed limits to your capabilities." He sent $300 to New York, and a warning. "From the way things look now," he wrote, "I will have no money to put you through another term there or anywhere else, so you had better make your plans accordingly."

For the Christmas vacation, Langston moved into Dill's small but comfortable apartment (complete with a *pianoforte*) in the Phipps houses on West 64th Street. They lunched and dined several times at the Civic Club, and attended concerts and lectures; Hughes heard Dill play the pipe organ, as he did regularly, at John Hayes Holmes's famous Community Church. For two cheerful weeks Langston was submerged in cosmopolitan Manhattan life—all because of his handful of published pieces, his prestige as a Columbia student, his personal charm, and the generosity of his new friends. That fall, Hughes had only two pieces (small essays on Mexican life) in the *Crisis* and *Brownies' Book,* but the new year came in auspiciously with a late supper at the Civic Club in a party that included Dill, the editor of *Success* magazine, and Edwin Markham, author of the famous humanitarian ode "The Man with the Hoe."

All of which made James Hughes's threats and warnings seem cheap. Langston's grades were fine; his father had little idea what a university was all about. With the resumption of classes on January 3 the students opened their books (some perhaps for the first time) and stormed through the texts. When examinations started on January 19, certainly Langston was ready. With his French final the best in the large class, his grade jumped to A-minus; snaring an A in the final of Contemporary Civilization, he settled for a B-plus overall. He won only a B in English, but was chosen to be one of "a picked group (supposed to be the best)" to do advanced work next term. The physical education courses he survived—but not physics, which he dropped.

Finally he responded to his father's warning. No, $300 would *not* do for the rest of the year! If his father did not intend to support him, he should petition the Dean of Columbia College to have Langston withdraw. As a student, he could not go on if he had to worry about finances, work after school, or "stay up half the night washing my socks." He hoped, however, that at the very least "I shall be able to remain here until June." In turn, James Hughes criticized his son for not writing more often and not sending an itemized account of his expenses: "*I want this statement.*" Langston's testy remark about the dean led his father to write Dean Herbert E. Hawkes at Hamilton Hall about his son's performance and promise, then to demand an itemized account of his payments to the university.

Hawkes replied that Langston Hughes was performing "admirably"; they should not rush him. But when Langston found out about his father's inquiries, his tone became icy. Nagged about lapsed subscriptions to the *Crisis* and the *Nation,* Langston suggested that it would be best "if, in the future, you attend to these things yourself, thereby avoiding any possible confusion." Disdainfully, he sent the itemized statement of expenses. "Please don't think that I am trying to cheat you," he added, "or am spending your money foolishly. If you think that, I had rather you give me nothing. . . . I know you care very little about my going to college here and that you are not interested in what I want to study. You probably think that I am a worthless investment. . . . I cannot study and worry at the same time about every cent of money I must spend."

Relenting a little, James Hughes sent $25 more. He had already spent over $800 that year on his son, but his remarks gave Langston no chance to warm to him. A weak grade in mathematics didn't surprise him, given the fact that Langston didn't know even arithmetic in Toluca. The failure of the *Brownies' Book* in December was "sufficient proof" that the people for whom it was published "are not worth fooling away your time with." Langston sputtered that many *white* magazines had gone under in the current recession; even the *Saturday Evening Post* had reduced its size. But he argued in vain; his father's contempt for blacks, and for his son's hopes to write about and for them, persisted.

Increasingly, Langston stayed away from Morningside Heights. Downtown, at the Rand School of Social Science, founded by the American Socialist Society as a venture in educating workers, he attended stimulating lectures by Heywood Broun and Ludwig Lewisohn. More often he went to the theater. Night after night he sped downtown to pay 50 cents for the pleasure of sitting high in the gallery at the Sixty-third Street Theatre and watching Florence Mills, Caterina Jarboro, and other gifted black entertainers dazzle audiences in *Shuffle Along*. But he loved all the theaters. That season he saw W. Somerset Maugham's *The Circle* with Estelle Winwood at the Selwyn; O'Neill's new *Anna Christie* at the Vanderbilt, and also his *The Hairy Ape;* Clemence Dane's *A Bill of Divorcement* at the Times Square; Balieff's Chauve-Souris troupe from Moscow; Andreyev's *He Who Gets Slapped;* Shaw's *Mrs. Warren's Profession;* Georg Kaiser's *From Morn Till Midnight* at the Theatre Guild; Katharine Cornell in Granville-Barker's *The Madras House*. Actually preferring the cheaper seats, Langston liked few things better than sitting high in the balcony, dark and lonesome, with the stage across the footlights far below; he became a child in Kansas once again. He followed the comings and goings of the playwrights and their stars; when Bert Williams, the greatest black comedian and once the partner of Lawrence's George Walker, died on March 4, Langston passed up an important examination to join thousands of grieving Harlemites at the funeral.

Harlem became more and more his home. "That winter I spent as little time as possible on the campus. Instead, I spent as much time as I could in Harlem. . . . Everybody seemed to make me welcome. The sheer dark size of Harlem intrigued me." At lectures and readings at the Harlem Branch Library on 135th Street, Hughes met the black intelligentsia; but his main interest was the people, of whom his vision was both intensely romantic and cold. With a chronic recession, sometimes extortionate rents, poor education, and rising crime, the deterioration of Harlem had begun. Already Claude McKay, gone away to Europe, had mourned the "little dark girls" who walked Harlem streets "to bend and barter at desire's call." Soon Eric Walrond would call Harlem itself "a white man's house of assignation" in the city. For another black writer, gambling was becoming "the chief past-time of Harlem." Very early, as the Jazz Age ripened in New York, Harlem was accepting the role forced on it—that of bookie, bootlegger, and bordello to white downtown. Fastidious and yet bohemian, moral but determined never to judge his people, Hughes instead celebrated his kinship with these

> Dream-singers,
> Story-tellers,
> Dancers,
> Loud laughers in the hands of Fate—
> My People . . .

Dishwashers, elevator boys, maids, crapshooters, cooks, waiters, hairdressers, and porters—he sang the ordinary and the low. In this way he met his father's contempt for black folk, and for the poor. His father received the *Crisis*—let him read the poem there! Thus, too, he answered Columbia's snobbishness and racism toward the dark culture beneath it.

With the new year, 1922, and an early end to his Columbia career a possibility, he stepped up his publishing. In January the *Crisis* published his declarative "The Negro," written in Mexico; soon *Current Opinion* reprinted the poem in its "Poetry of Distinction" section. In their range of theme, three poems in the March *Crisis* suggested the poles of his sensibility, as well as its equator. The blunt, prosaic "Question" (rejected by Floyd Dell at the *Liberator*) is rhetorical: when "the old junk man Death" comes to toss our bodies "into the sack of oblivion," will a white "multi-millionaire" be worth more than "a Negro cotton-picker?" In sharp contrast is "The New Moon," where the poet passively observes "a half-shy young moon veiling her face like a virgin, / waiting for her lover." The best of the three is "Mexican Market Woman"; neither prosaic nor fey, it is also Hughes's only poem on a Mexican theme.

> This ancient hag
> Who sits upon the ground
> Selling her scanty wares
> Day in, day round,
> Has known high wind-swept mountains;

And the sun has made
Her skin so brown.

But if Hughes was buoyed in April to learn that a Berlin paper (no doubt socialist) had reprinted one of his poems, many of his publications while at Columbia were weak. Under the name "LANG-HU" or "LANGHU" he published four poor poems in April and May in the Columbia *Spectator,* including a whitewashed, non-dialect version of "Just Because I Loves You" from his high school days. In May, the *Crisis* carried the jejune "My Loves," also first printed in the Central High *Monthly,* as well as the only somewhat more mature dirge "To a Dead Friend" ("The moon still sends its mellow light / Through the purple blackness of the night . . ."). In June, he was in better form with two jagged but forceful free-verse pieces. "The South" was "Beautiful, like a woman, / Seductive as a dark-eyed whore,"

Passionate, cruel,
Honey-lipped, syphilitic—
That is the South.
And I, who am black, would love her
But she spits in my face . . .

His other piece, printed in a long column of lines parallel to "The South," was the defiantly jubilant "My People," about the black masses as "Loud-mouthed laughers in the hands / of Fate."

Ironically, "My People" appeared precisely at the moment he said goodbye not only to Columbia University but also to his father, and made ready to live in Harlem. In early May he received news from R. J. M. Danley, an American colleague of his father's in Mexico, that James Hughes was critically ill. While in Mexico City on business he had collapsed with a stroke; still partially paralyzed, he had been taken home to Toluca. Langston sent a message of sympathy at once. Then Danley's son visited him at Columbia, bringing with him $100 as a loan on behalf of James Hughes. Finally, Danley himself wrote to say that his father was improving but still very ill. Langston might wish to come to Mexico this summer; and what were his plans for Columbia University in the fall?

A momentous choice faced Langston, but he gave no sign of wavering on May 14 when he drafted a reply to Danley. If the money brought by young Danley was intended for a trip to Mexico, he would refund it at once if his father so desired. Definitely he would not come to Mexico in the summer; "nor do I wish to return to college this fall." Although he was truly sorry to hear that his father was ill, "we were so uncongenial when together that I am afraid I could be of no help to him now in his illness. I am sorry, but I do not see what I can do for him."

On May 22, final examinations started at the university. In spite of his knowledge that he would not be back in the fall, Hughes applied himself to the tests. Having floundered in trigonometry, he ignored the final and received no grade. He also failed physical education. But Langston finished with B-plus in English and written French, B in oral French, and a C-plus in Contemporary Civilization. Apart from his showing in math and science, he had done fairly well as a university student. But he wanted no further part of Columbia.

Haunting the wharves about South Street downtown, Hughes looked for a ship to take him far away. He was even prepared to pay to leave the United States: on May 17, applying for a passport, he declared his intention to visit Great Britain as a tourist, then France, where he would study the language. Scheduled to leave New York on May 27 on the *Orduna,* at the last minute he cancelled his plan to travel, took a room in a boardinghouse in Harlem, and began searching for a job.

He quickly discovered that his color checkmated his year at Columbia. Nine times out of ten, white employers would shake their heads in amazement: "But I didn't advertise for a colored boy." And yet only one job in a thousand would be marked "colored." Still, there was something exhilarating about his life now that he was immersed in the day-to-day culture of the people. He began to feel, as never before, the daily humiliations of black men and women; protected until now by his youth, or his white high school or university, or Mexico, Hughes was at that point in black lives where bitterness or cynicism begins to seem inevitable, even natural. In at least one poem he mourned his loss of imaginative, childlike innocence:

> Now,
> In June,
> When the night is a vast softness
> Filled with blue stars,
> And broken shafts of moon-glimmer
> Fall upon the earth,
> Am I too old to see the fairies dance?
> I cannot find them anymore.

Down to his last dollar, as he later recalled, he came across a printed call for workers on a small truck-farm on rural Staten Island. Although he had never farmed, the former secretary of the Home Garden Club at Central High quickly made his way out to 2289 Richmond Avenue. There, two Greek brothers who ran a forty-three-acre vegetable farm with their wives hired him on the spot at $50 a month plus room and board, to join their other workers—a Jewish youth from the East Side of Manhattan, a Greek sailor who had jumped ship, and a seventeen-year-old boy from Atlantic Avenue in Brooklyn who was laying low after stealing $300 from a subway station. A summer job under the blazing sun was exactly what Hughes needed to burn away the dead skin of his year at Columbia. At five in the morning, as the first light glimmered over the Atlantic, he rose to a breakfast of coffee, white cheese, and olives. Then he and the others set to work ploughing, hoeing, fertilizing with manure, weeding, watering, and—best of all—harvesting beets, carrots, lettuce, onions. The work was hard and lasted all day but "I liked it. A mid-day dinner of vegetables and watermelons, wine and cheese and olives in the fields in the afternoon, and supper after dark under an oil light with the moths and mosquitoes circling round. Then, perhaps, a walk to the cross-roads store, perhaps a little talk, and bed on the hay too dead tired to think. I liked it."

Hughes set foot in Manhattan only once that summer, to see the bullfights in Rudolph Valentino's *Blood and Sand*. He was disappointed; the movie showed too much sand, too little blood to suit his taste. Avoiding Broadway altogether, he also went nowhere near the *Crisis*, having no desire to explain why he had dropped out of Columbia. In August, however, the journal published his little poem about June, "After Many Springs," as well as the accomplished "Danse Africaine" ("the low beating of the tom-toms, / Slow. . . . slow / Low . . . . slow . . ."). Jessie Fauset wrote that she was reading some of his poems in lectures on modern Negro poetry at New York high schools. "They seem to take them seriously," Langston chuckled nervously in a letter to his father. "Ha! Ha!"

Under his housekeeper Frau Schultz's care (James Hughes would marry her in January, 1924) his father's health had improved, although his right arm remained paralyzed. But whatever his condition, Langston remained firm about not returning. He would use his father's $100 for extension classes at Columbia in the fall; as for regular attendance at the university, "I don't want to go back." Nor would he return to Mexico: "I don't like Toluca." His father should not send any more of his money: "I have had enough; and do not worry about things—rest, get well." Then he decided to send back the $100, which obviously weighed on his conscience, as did his general response to his father in a time of need. His guilt deepening, for the first time in their exchanges he ended his letters with "Love, Langston." But although he kept writing to James Hughes, no reply came. At Christmas, he would renew his father's subscription to the *Crisis* as a gift; no acknowledgment, much less thanks, came from Toluca. By his last letter without reply, the son's signature had congealed into "Sincerely yours, Langston Hughes."

Near Labor Day, the first anniversary of his arrival in New York, the Greeks released their summer help. With money saved, and deeply tanned and muscled after the long hot days of bending and lifting, Hughes visited his mother, Homer Clark, and Gwyn Clark at their latest home, in the town of McKeesport near Pittsburgh, Pennsylvania. Within days, he returned to a comfortable roominghouse at 156 West 141st Street in Harlem ("everywhere I roomed I had the good fortune to have lovely landladies. If I did not like a landlady's looks, I would not move in with her, maybe that is why"). Although Hughes's name was listed in the directory of the Columbia University Extension School, he did not attend classes. Instead, he took a job delivering orders for the fashionable florist Charles Thorley of 605 Fifth Avenue. The flowers were beautiful but extremely expensive, each box usually costing more than his month's pay. Many of the clients were very well known—Baroness d'Erlanger at the Ritz, the actress Marion Davies on her yacht, the Roosevelts at Oyster Bay, Vivienne Segal at the Empire. Mostly Hughes surrendered his blossoms to a servant; "sometimes you would catch a glimpse of the great one, though, and then you would feel a little more cheerful, having laid eyes on some famous and successful person.'' Ironically, this job was both the closest he had come physically to fame, and the ''lowest'' he had sunk in the world. As a delivery boy, Hughes was now truly one of ''My People.''

The job at Thorley's lasted about a month. Arriving late one morning after a busy night on the job, he met Thorley himself on the front steps. When the florist rudely brushed aside all explanations of his late arrival, Hughes resigned on the spot and demanded his wages. Thorley, however, had the last word: Langston had just received his salary for the previous week, and was therefore entitled to nothing. His days now became a collage of clipped newspapers and application forms. But this time he was determined to try the sea: ''It seemed to me now that if I had to work for low wages at dull jobs, I might just as well see the world.'' In mid-October he was at the United States Shipping Board office on Broadway when the call came for a messboy, one of the few jobs sometimes open to colored seamen. Rushing forward, he found suprisingly little competition—the job was his for the asking.

At the docks Hughes boarded a launch with no idea where he would be going. He didn't care: the seven seas had called, and he was responding. Near Staten Island, when the launch drew up under a rusty freighter, he scrambled aboard and addressed a sailor.

> ''Where are we going?''
> The sailor looked at me a long time. Finally he said:
> ''Nowhere.''
> I couldn't believe my ears.
> ''This boat ain't going nowhere.''

Instead of heading seaward, the freighter bore the miserable Langston up the Hudson under the majestic Palisades on the New Jersey shore, past Irvington, Tarrytown, Nyack on the Tappan Zee, and Ossining. At Jones Point, a few miles south of West Point, it finally dropped anchor among a great fleet of vessels, obsolete after the war, owned by the U.S. Shipping Board. In ghostly rows, side by side, 109 ships anchored at Jones Point that fall. Five were ''mother'' ships, housing the men who tended the others; most of the seamen oiled machinery in the remote chance the vessels would be needed again.

Once over his disappointment, Langston liked his new life. Jones Point, he wrote, was ''a most picturesque place . . . a forest of masts, smoke-stacks, and cables; and autumn-tinted hills on both sides.'' The ships rode where the joint Harriman and Bear Mountain State Parks first touched the water heading north. Heavy with chestnut, pine, larch, and cedar, the Ramapo and Hudson mountains rose above a thousand feet, then fell swiftly to the river to form part of the magnificent Palisades Interstate Park. Although this park was the most popular in the eastern United States, by mid-autumn the sailors passed the time in virtual seclusion. Langston would have ample time to think and to write.

Signing on as a ''saloon messman'' on the *Oronoke,* he also served on the *Bellbuckle,* the ship furthest out on the water, and the *West Hassayampa.* Some of the crew had special reasons for working on the surplus fleet. Hughes's immediate superior, a redheaded, redfaced fellow, was a drunk; entrusting all his keys to the messman, he sometimes withdrew for days into his cabin with a large supply of liquor. But everyone was friendly. A Spanish sailor from Seville prattled about gypsies and their folksongs. "Scotty," a stocky albino Scotsman still in his teens, entertained them with long stories in a wonderful brogue. Erik Olsson, a Swede twenty years old, drew beautifully (once he sketched Hughes as a Jewish or Biblical type) and talked of going to design school some day. In addition, the ships' libraries were well stocked—Langston remembered reading Butler's *The Way of All Flesh,* Conrad's *Heart of Darkness,* D'Annunzio's *The Flame of Life.*

These sailors, he decided, were "the finest fellows I've ever met—fellows you can touch and know and be friends with"; after Columbia, life at Jones Point was "like fresh air and night stars after three hours in a dull movie show." The nights could be rare, Hughes would recall, with endless games of poker and blackjack and tales of women and fights and faraway places like Bombay and Cape Town. Best of all was the singing—in French, Spanish, Italian, Swedish, English—that made him happy to be with these men but also long for the open seas. Meanwhile, he rested in a kind of suspended animation. When the chill December winds swept over the water and the old ships crunched against the gathering ice, Jones Point became "a little white village almost pushed into the water by the snow-covered hills."

Isolated, Hughes wrote as never before. Columbia had been poor for his poetry. There he had written nothing to compare with "The Negro Speaks of Rivers," "When Sue Wears Red," or "Mother to Son." By this time, Langston understood that he needed to be unhappy to write good verse; but there had been too much business attending his depression at Columbia. As a composing poet, he flourished in a moving train rushing through the dark. A rocking ship going nowhere, motion without progress, was also sweetly amniotic. In the womb of a ship, he created steadily.

Finishing a triptych, "Three Poems of Harlem," he found pathos in Harlem's music ("Does a jazz-band ever sob? / They say a jazz-band's gay . . .") and in the life of a Harlem prostitute ("Her dark brown face / Is like a withered flower / On a broken stem"). Before Harlem's religion he stood in awe: "Glory! Hallelujah! / The dawn's a-comin'! / A black old woman croons in the amen-corner of the Ebecaneezer Baptist Church." In "Gods," about ebony and ivory deities and those "of diamond and jade," he struck at the "silly puppet-gods / that the people themselves have made." "Our Land: Poem for a Decorative Panel" decries the pallor of white "civilization":

> We should have a land of sun,
> Of gorgeous sun,
> And a land of fragrant water
> Where the twilight is a soft bandanna handkerchief
> Of rose and gold,
> And not this land
> Where life is cold . . .

Aggressively Hughes celebrated "Joy," whom he found "driving the butcher's cart / In the arms of the butcher boy!" The image of Pierrot, the white-faced clown of French pantomime, entered three Hughes poems, influenced probably by Jules Laforgue, the French symbolist poet who had pioneered free verse in France, translated Whitman, and written ironical, sentimental complaints in the clown's image. Perhaps Hughes saw himself as dreamy, star-crossed Pierrot, who turns his back on "Simple John," his house, his wife, and a life of being merely "good":

> For Pierrot loved the long, white road,
> And Pierrot loved the moon,
> And Pierrot loved a star-filled sky
> And the breath of a rose in June . . .

But almost all of Hughes's poems that autumn and winter were shot through with a sense of deep sadness. One tender expression of grief (sometimes taken insensitively as proof of his homosexual feeling) was dedicated to someone Langston called only "F. S."

> I loved my friend.
> He went away from me.
> There's nothing more to say.
> The poem ends,
> Soft as it began,—
> I loved my friend.

Never far away was the nurturing thought of death, as in "Poem":

> I am waiting for my mother.
> She is Death.
>   Say it very softly.
>   Say it very slowly
>   If you choose,
> I am waiting for my mother,
>   Death.

Although he wrote steadily, Hughes worried about the quality of his work; almost never in his life would he be sure about his own estimate of a poem. But the loyal Jessie Fauset urged him on: "I certainly do not think you are standing still." Without approving of his decision to leave Columbia, she stuck by him—although he had broken a promise to teach her Spanish in order to be ready to sail away. Fauset talked Hughes up to Donald Duff of the *Liberator,* and included him in the September *Crisis* with a ditty, "The Song of a Banjo Dance." In December, the magazine carried his moving "Mother to Son," which quickly became one of the most beloved of Hughes poems.

But as pleased as he was with aspects of life at Jones Point, almost certainly Hughes also found it embarrassing to be a sailor on a ship going nowhere. When his letters came less frequently, Fauset gave him room to hide. His friendship with Augustus Granville Dill also faded. Dill was no snob, but messmen and delivery boys did not lunch at the Civic Club. In the late autumn and winter, however, one friendship certainly grew. Among the younger writers who frequented the Branch Library in Harlem, Hughes had met by far the most promising—Countee Porter Cullen, then twenty years old. The adopted son of Rev. Frederick Asbury Cullen, pastor of the large, influential Salem Methodist Episcopal Church on Seventh Avenue, Countee had topped off an excellent record in the almost all-white De Witt Clinton High School by leading the senior class in scholarship. To the further delight of Harlem, he had also won a citywide competition (sponsored by the white Federation of Women's Clubs) for the best poem by a high school student, with his "I Have a Rendezvous with Life," first published in his school's literary magazine. In certain ways Cullen was Hughes's exact opposite. Plain where Hughes was very handsome, he was often a little stiff with strangers where Langston was shy but congenial. Cullen was religious, with a Bible at hand when he composed; Hughes was secular to the bone. As a poet Hughes believed in the power of inspiration and improvisation; Cullen practiced sonnets and villanelles, honed his rhymes, and searched mightily for the exact word. Frowning on free verse and the wild men Whitman and Sandburg, he instead adored John Keats and A. E. Housman. And although he wrote touchingly of race, Cullen found no particular beauty in the black masses, as Hughes did; Africa more or less embarrassed him. While he deeply respected tradition, he saw it as being mainly the European classics. Above all, Cullen wished to be seen not as a Negro poet, but as a poet who happened to be Negro—an attitude Hughes found more than a little perverse.

For all their differences, however, Hughes and Cullen were young men of ability and integrity who liked and admired one another. When Langston made his rare excursions south to Manhattan that winter, he usually went straight to the comfortable Cullen home at 234 West 131st Street. Together they went to movies and plays, of which Hughes was far more fond than Cullen; once, Langston convinced Countee to cut three classes at New York University, where he was an undergraduate, to see the Moscow Art Players in *The Lower Depths*. On the other hand, Cullen balked at following Hughes into certain places; an exceptionally fine ballroom dancer, while Hughes was inept, Countee flashed his feet only on respectable floors. Nor did he have time to browse among the little French bookshops on Sixth Avenue, another of Hughes's urban pleasures. But he cheerfully tried to help Langston by keeping him abreast of events of interest in Manhattan. When the *Amsterdam News* accepted some of his friend's poems, Countee watched for their appearance and sent clippings from the newspaper and the *Crisis* to Jones Point.

On February 12, a grateful Hughes sent Cullen the main fruits of the past weeks, a twenty-page manuscript of fifteen poems (one poem was the Harlem triptych) inscribed "To a fellow poet, these unpublished poems." In return came Countee's "To a Brown Boy" ("That brown girl's swagger gives a twitch / To beauty like a queeen") dedicated to Hughes. "I don't know what to say about the 'For L. H.'," the slightly embarrassed Langston admitted, "but I appreciate it, and I like the poem." When Ernestine Rose of the 135th Street Branch Library organized an evening of poetry reading, Cullen offered to perform for

Hughes. Unlike Langston, Countee loved to dress up and perform in public. Hughes, impatient at the pretentiousness of most poetry readings, sent along a "divertingly original" piece called "Syllabic Poem":

> Ay ya!
> Ay ya!
> Ky ya na mina,
> Ky ya na mina.
> So lee,
> So lee nakyna.
> Ky ya na mina,
> Ky ya na mina.

"Tell them," he instructed Cullen, "that it is the poetry of sound, and that it marks the beginning of a new era, an era of revolt against the trite and outworn language of the understandable. Wouldn't that be amusing? Then they could discuss the old question as to whether artists and poets are ever sane. I doubt if we are."

With Cullen putting up with Hughes's desultory correspondence ("very beastly of me not to have written a single line," Langston apologized breezily, "I'm an absolutely shameless person, am I not?"), they selflessly shared their triumphs and their contacts. Cullen was the first friend to know that the black violinist Hall Johnson wanted to set Hughes's "Mother to Son," and that the white dramatist Ridgeley Torrence, preparing a special number on blacks for the magazine *The World Tomorrow,* had rejected "Three Poems of Harlem" but had taken three others—"Our Land," "Dreams," and "Poem," which would appear the following May. Thus Hughes was about to appear for the first time (apart from reprintings) in a white journal. From Jones Point Langston sent Countee what he could—his translation of an Andalusian song, and articles clipped from magazines, including an essay on "your beloved" A. E. Housman.

While the poems in Hughes's handwritten anthology showed a waning of the influence of Sandburg, they gave little evidence of fresh inspiration. Jones Point could not in itself provide the impetus he needed truly to grow as a poet; for that spurt, Hughes needed the encouragement that only his race could supply. Perhaps he himself did not yet understand fully what the culture meant to him as a poet. However, in his willingness to stand back and record, with minimal intervention as a craftsman, aspects of the drama of black religion or black music, Hughes had clearly shown already that he saw his own art as inferior to that of either black musicians or religionists—inferior to the sweet trumpeters and moaners in cabarets, inferior to the old black woman in the amen corner who cries to Jesus, "Glory! Hallelujah!" At the heart of his sense of inferiority—which definitely empowered rather than debilitated Hughes—was the knowledge that he stood to a great extent *outside* the culture he worshipped. Unable to believe in God or to play an instrument or to dance or sing (his singing voice was painful to hear), Hughes stood outside because so much of his life had been spent away from consistent, normal involvement with the black masses whose affection and regard he craved.

And then one night in March, in a little cabaret in Harlem, he finally wrote himself and his awkward position accurately into a poem. Letting his sense of isolation from the culture merge with his profound love and admiration for it, Hughes exposed not only his isolation and his love but also his knowledge that he perhaps could never understand fully what he so deeply admired. Within a poem based in loosely conventional form he set the earliest blues he had ever known, lyrics he had heard as a child in Kansas, so that in one and the same work he honored both the tradition of Europe and the tradition of black America. The technical virtuosity of the opening lines is seen only when one measures them against the cadences of urban black speech, derived from the South, with its glissandos, arpeggios, and sudden, unconventional stops. The result was his most powerful poem since "The Negro Speaks of Rivers" in the summer of 1920, "The Weary Blues":

Droning a drowsy syncopated tune,
Rocking back and forth to a mellow croon,
  I heard a Negro play.
Down on Lenox Avenue the other night
By the pale dull pallor of an old gas light
  He did a lazy sway. . . .
  He did a lazy sway. . . .
To the tune o' those Weary Blues.
With his ebony hands on each ivory key
He made that poor piano moan with melody.
  O Blues!
Swaying to and fro on his rickety stool
He played that sad raggy tune like a musical fool.
  Sweet Blues!
Coming from a black man's soul.
  O Blues!
In a deep song voice with a melancholy tone
I heard that Negro sing, that old piano moan—
  "Ain't got nobody in all this world,
  Ain't got nobody but ma self.
  I's gwine to quit ma frownin'
  And put ma troubles on the shelf."
Thump, thump, thump, went his foot on the floor.
He played a few chords then he sang some more—
  "I got the Weary Blues
  And I can't be satisfied.
  Got the Weary Blues
  And can't be satisfied—
  I ain't happy no mo'
  And I wish that I had died."

And far into the night he crooned that tune.
The stars went out and so did the moon.
The singer stopped playing and went to bed
While the Weary Blues echoed through his head.
He slept like a rock or a man that's dead.

Just as the classically trained black musican Scott Joplin had labored to notate ragtime in order to enshrine its beauty as art, so Hughes worked to link the lowly blues to formal poetry in order that its brilliance might be recognized by the world. He knew immediately that in so honoring the blues, he had done something unprecedented in literature. Rather than share this ''beauty of a cabaret poem'' with anyone, even Cullen, Hughes kept it from sight as, unusually, he struggled to shape its ending—''I could not achieve an ending I liked, although I worked and worked on it.'' More than two years would pass before he offered it publicly; then, ''The Weary Blues'' would open the door to his entire adult career as a published writer.

Since Langston's arrival in New York from Mexico, he had passed significant personal milestones. Repudiating his father and Columbia, he had joined the ranks of the black masses, then boldly tried for the sea. Rusticated instead at Jones Point, he had consolidated himself as a poet. With the new year, 1923, came not so much another milestone but a challenge equal to any other he had met in the past two years, in the form of a campaign of homosexual seduction aimed directly at him. Psychologically, however, the contest was for Hughes's right to retain the privileged balance he had maintained up to this point between his childlike sense of wonder, on one hand, and his will toward adult effectiveness, on the other. The sexual challenge was inevitable, though not its homosexual cast. Sooner or later, the world would have asked Hughes to assert some basic conjugal ability—to be a man, in other words, and complete his growth either by taking a wife and starting a family of his own or by giving clear evidence, in the form of recognized liaisons, of the desire to do so. Because he had so far given not the slightest indication of sexual interest in women, at least one of his closest friends, Countee Cullen, deduced that Hughes must be a homosexual. (Whether or not Hughes was a homosexual did not affect the nature of the challenge; a male lover would have posed the same threat to his desire to remain a child and to benefit from his innocence as a poet and in relationship to the race.)

Although Cullen himself expected to marry (and did so twice), by the time of his friendship with Langston he understood fully that he himself had strong homosexual feelings. This conviction neither completely relieved his sense of guilt nor made him unduly aggressive with others; Countee took his failure gracefully when Langston, with whom he was clearly interested in having an affair, seemed utterly unconscious of such an interest. Then Cullen decided to help another friend, one more worldly, who might succeed with Hughes where he had failed.

The man was Alain Leroy Locke, thirty-seven years old, whom Cullen had first met because of his interest in applying for a Rhodes scholarship. Locke was then the only black American to have won this honor (he would remain the only one until 1967). The award had followed a sterling record of scholarship in Philadelphia schools and at Harvard, from which he had graduated *magna cum laude* (Locke would also earn a doctorate in philosophy there in 1918). After graduate work at Oxford between 1907 and 1910, and further study at the University of Berlin and the Collège de France, in 1912 he had joined the faculty at Howard University in Washington, D.C., where he currently taught a variety of subjects including literature and philosophy. For all his training in philosopy and the classics, and his passion for German culture, Locke was both race-conscious and forward looking; although the reactionary Howard trustees had rejected his proposals to teach courses on race, he still hoped, if rather vaguely, to encourage a cultural movement among young black men (young black women interested him little) that would bring credit to Afro-America. For the moment, however, he seemed almost to be drifting, liked and admired but somehow isolated not only by the breadth of his education and his genuine intellectual brilliance but also because he had not yet found, in a sense, his subject, let alone his mission. So far, he had published very little—a handful of articles. Instead of boldly carving out a field in Afro-American studies, as the obsessive W. E. B. Du Bois had done by the time he was Locke's age, with weighty books on history and sociology, Locke sometimes seemed more interested in influence through the power of personality—in teaching, university politics and intrigue, and the cultivation of friendships. Certainly his prestige as a university scholar and his conspicuous sophistication, as well as his sometimes malicious wit, made his dainty, fluttering figure a welcome sight in all important black social and artistic circles.

In January, following a visit to Washington, Cullen sent Hughes's address on the *West Hassayampa* to Locke. "Write to him," Cullen urged, "and arrange to meet him. You will like him; I love him; his is such a charming childishness that I feel years older in his presence." Clearly they had previously discussed Hughes. A few days after Langston's twenty-first birthday Locke finally wrote to him. Every particular friend of Hughes, he said—mentioning Cullen and Jessie Fauset—"insists on my knowing you. Some instinct, roused not so much by the reading of your verse as from a mental picture of your state of mind, reenforces their insistence." But where the devil was the *Hassayampa?* In Patagonia? He wanted full details about Langston's life, and an assurance that he could visit him.

Flattered, and eager to impress the learned Locke, Hughes replied at once. He missed his friends in the city, but Jones Point had its compensations: "I am chasing dreams up here . . . and that's an infinitely more delightful occupation even than living in New York, where all my old dreams had been realized; college, (horrible place, but I wanted to go), Broadway and the theatres—delightful memories, Riverside Drive in the mist and Harlem. A whirling year in New York! Now I want to go to Europe." He was thinking of France for a while, then of living with the gypsies in Spain ("wild dream, isn't it?"); no sport was as "lovely" as the *corrida!* He chattered on about seeing and hearing

Jeritza, and also Chaliapin in *Boris Godunov* (''It's the experience of a lifetime''), seeing the Chauve-Souris three times, and Barrymore in *Hamlet*—the hit of the season, as far as he was concerned. He wrote also about the admirable crew at Jones Point, which he located precisely for Locke.

Immediately Locke pressed for a visit to Jones Point on Washington's Birthday later in February. He was eager to meet Hughes: ''First impressions count with me, and I rather suspect with you also and I am quite nervous, either through anticipation or some deeper instinctive feeling about our first encounter.'' Spain was wonderful indeed, but Germany was ''psychologically my native land.'' Germans had a gift for friendship, ''which cult I confess is my only religion, and has been ever since my early infatuation with Greek ideals of life.'' ''You see,'' he went on, ''I was caught up early in the coils of classicism.'' He hoped to be friends with Hughes. And he would love to meet the whole crew; he *adored* sailors—they were ''of all men most human.''

All this was a little too spiced for Hughes. Pretending not to understand what Locke meant by ''Greek ideals of life'' and ''the coils of classicism,'' he begged him, though coyly, not to come to Jones Point. He would have to cross a half-mile of slippery gangplanks and icy decks; ''and then, at the end of your journey, you would find a very stupid person, because I am always dumb in the presence of those whom I want to be friends with.''

Langston's refusal annoyed the highly sensitive Locke, who was not without a streak of petulance in his character. Cullen should forget about Hughes: ''He doesn't need any more humoring than he has already had, and I advise you to discontinue pampering his psychology.'' Over a month passed before Hughes wrote Locke again. By that time Cullen had gone to Washington to visit Locke, sparking curiosity in Langston: ''Is Mr. Locke married?'' To Locke now he sent a long, teasing letter, in which he picked up Locke's reference more than a month before to the ''coils of classicism.'' Did Locke teach Greek or Classics? He himself had read only the *Odyssey* and a few other works, ''but I love them. They were the only things college gave me, (other than a supreme dislike for college). But I am glad I have known Telemachus and the beauty of Homer's 'wine dark sea'.'' He excitedly detailed all the wonderful plays he had seen recently, but returned soon to a more personal note:

> You must be a charming friend for poets. . . . I do want your help, and friendship, and criticism, and how good you are to offer them to me. . . . Of course, I'm stupid and only a young 'kid' fascinated by his first glimpse of life, but then after so many years in a book-world and so much of striving to be a 'bright boy' and an 'intelligent young man,' it is rather nice to come here and be simple and stupid and to touch a life that is at least a living thing with no touch of books.
>
> But I am boring you. . . . I shall perhaps be here only a month or so more, until Spring comes over the high hills and down to the river's edge. There are fruit trees all along the west bank waiting for flowers . . .

This letter was so much more to Locke's liking that he got over what he called his pique at not being invited to Jones Point. He would love to teach the classics, since he was truly ''pagan to the core.'' He leavened their exchange with talk about forming a coterie of young writers—on Cullen's last visit, he had met the commandingly handsome Jean Toomer, who that year published *Cane,* a brilliant modernist collage of fiction, poetry, and drama that evoked the drama and pathos of much black life, especially in the South. The coterie could begin with the three friends; Locke raised the possibility of Hughes joining him in Europe and the middle East (he was a Bahai follower) that summer. Now Hughes, in turn, threw off a little more of his reserve. He was all for a coterie, the founding of "some little Greenwich Village of our own. But would our artists have the pose of so many of the Villagers? I hate pose or pretension of any sort. And especially sham intellectuality. I prefer simple, stupid people to half-wise pretenders. (But perhaps it's because I'm stupid myself and half ashamed of my stupidity.)" But he too was a lover of paganism, as in D'Annunzio's *Flame of Life,* about the actress Eleonora Duse. And did Locke like Whitman? "His 'Song of the Open Road' and the poems in the Calamus section? I do, very much. And have you read, or tried to read, Joyce's much discussed 'Ulysses'?"

With Hughes's references to paganism, Whitman's infamously homosexual poems, and the scandalous Joyce novel, Locke felt free to declare himself. Perhaps Cullen should be there when they met for the first time, as "a sort of shock-absorber—in case of—well in case of disillusionment on your part or fatal loss of poise on mine." He had made sure that the way to Langston was clear: "Countee already means so much to me—but he generously insists on deeding over a certain part of me to you." Indeed, Cullen wanted them all to go to Europe in the summer. But Locke also sounded a weary note, a plea to Hughes to be gentle with him: "I am too battered and disappointed in human relationships at present . . . to be anything but quiescently fatalistic even about you."

His fatalism was well placed. Under such pressure, Hughes's sexual desire, such as it was, became not so much sublimated as vaporized. He governed his sexual desires to an extent rare in a normal adult male; whether his appetite was normal and adult is impossible to say. He understood, however, that Cullen and Locke offered nothing that he wanted, or nothing that promised much for him or his poetry. If certain of his responses to Locke seemed like teasing (a habit Hughes would never quite lose with women or, perhaps, men) they were not therefore necessarily signs of sexual desire; more likely, they showed the lack of it. Nor should one infer quickly that Hughes was held back by a greater fear of public exposure as a homosexual than his friends had; of the three men, he was the only one ready, indeed eager, to be perceived as disreputable. But one thing is relatively certain because of Hughes's repeated references to himself, in letters to Locke and Cullen, as "stupid." Understanding what Locke was after, and its gravity as a challenge, in his own main defense Hughes had summoned up a show of childlike innocence so complete that even Cullen was convinced by it.

As the days grew warmer Hughes yearned to be on the road, but alone. Quoting Edna St. Vincent Millay's "Travel" to Cullen ("there isn't a train I wouldn't take / No matter where it's going") he sighed his desire to be gone. "Don't you ever get that feeling sometimes?'' he asked. ''To go somewhere, anywhere, just to be going?'' But the decorous Countee did not indulge wanderlust, and Locke's homosexual passion was finally commonplace. At last Hughes said no, with a feeble excuse: ''I can't go with you this summer. My 'I wish I could go with you' did really mean so, but for more than one reason, I must work the entire summer.'' He would try for a freighter to the Mediterranean—''And how delightful it would be to come surprisingly upon one another in some old world street! Delightful and too romantic! But, maybe, who knows?'' He would send a long-promised portrait of himself to Locke. ''The hills are at their loveliest now,'' Hughes ended maddeningly, ''and I do not like to go, but then there are other rivers in the world to see besides the Hudson. And oh! so many dreams to chase!''

He was indeed ready to return to the world. Warm days had come to Jones Point. With his sailor pal ''Scotty,'' Hughes took long hikes in the blossoming hills above the Hudson; down below they saw steamships returning after winter to ply the river between Albany and New York. To Jessie Fauset he sent a letter, his first in some time, and a dozen poems. She hastened to inform him that Robert T. Kerlin, a white professor of English and history, who was preparing an anthology (*Negro Poets and Their Poems,* published that year) including verse by Hughes, wanted a photograph. ''Be sure to get this to him,'' she ordered, ''as it is a good opening for you.'' The second week of May he was in Manhattan to be photographed. After he sent Fauset a copy gratefully inscribed, she responded with a prediction. She was not crazy about his prose, but ''you assuredly have the true poetic touch, the divine afflatus, which will some day carry you far.'' Something in his face touched her—or was it something she had sensed, or heard in the very small world of blacks as cultivated as Locke and herself? ''I am a little worried about you,'' she confessed. ''I like your picture, it is very striking, not to say handsome. But it makes me wonder.'' She flicked one of his own lines at him: ''Hold fast to your dreams my boy.''

On June 4, Langston left Jones Point. He took with him two letters of recommendation from his superiors—James Langston Hughes was ''thoroughly efficient, honest, and reliable, and at all times obedient and polite,'' and he was ''a willing, sober, reliable and very efficient worker.'' Thus armed, he returned to Manhattan to seek a ship that would actually sail away. Visiting the docks and shipping office, he did odd jobs on some of the smaller vessels. His prospects seemed good. Offered a berth to Canada and another to Cuba, he held out for something more ambitious. Still another was going to Mexico; Hughes probably would have preferred to go back to Jones Point. Finally, ecstatic, he signed on board a ship leaving for storied Constantinople.

He spent as much time as he could with Cullen, who was himself about to leave town not for Europe but for his usual summer work—as a busboy at a hotel in Atlantic City, New Jersey. At first Cullen gave Hughes no clue about the pain he had inflicted on Locke. Or on Countee himself. ''No more skeins have been tangled,'' Cullen assured the unhappy Locke; ''indeed I am doing my utmost to unravel those which are twisted. My attitude now (I don't know how long I shall have endurance to maintain it) is one of indifference—that is, as far as I myself am concerned.'' Then Cullen pulled the curtain aside for Hughes, or tried to. He showed Hughes a letter Locke had written him about the effects of Langston's behavior. Still Hughes appeared not to know exactly what was beneath the triangle of friendship. ''He was sympathetic,'' Cullen wrote Locke, ''but I do not believe that he fully understood the situation.'' Locke's disappointment turned, as it easily turned, into petty malice. But Cullen honorably pleaded reason: ''I think you are probably doing Langston an injustice. . . . I am sure you both have not fully understood one another.''

Suddenly Hughes lost his position on the ship to Constantinople. Downcast, he searched the wharves again. As he thought of the Hudson and the stark docks of New York, a sad little poem, very much like one by Cullen's beloved A. E. Housman in its murmuring about dreamy ''lads,'' came to him:

> The spring is not so beautiful there,—
> But dream ships sail away
> To where the spring is wondrous rare
> And life is gay.
> The spring is not so beautiful there,—
> But lads put out to sea
> Who carry beauties in their hearts
> And dreams, like me.

Then, on June 11, he found his ship. The weatherbeaten freighter *West Hesseltine* was set to leave for the worst summer voyage in the world, down the steamy, malarial west coast of Africa. Captains could not be fastidious in selecting their crew. Besides, the steward was not white but Filipino. He glanced at Hughes's excellent letters and hired him on the spot.

Africa! The continent was a mystery known to few black Americans. As a child Hughes had dreamed of exploring its forbidding jungles and mighty kingdoms. Over the winter he had struggled with the Africa of Conrad's *Heart of Darkness,* consoling himself with Jessie Fauset's confirmation that Conrad was very difficult reading indeed. He knew of Dr. Du Bois's gallant if futile efforts, through the meetings of his Pan African Congress in 1919 and 1921, and scheduled again for that year, 1923, to alter the destructive patterns of European imperialism which had despoiled Africa; certainly Hughes also knew about Marcus Garvey's hope to return blacks from America to Africa. But Langston had probably never met anyone who actually had been there—not even Du Bois or Garvey had set foot on the soil of the Motherland. But he, not yet twenty-two, was going to Africa!

Late on the afternoon of Wednesday, June 13, the *West Hesseltine* steamed slowly past the fog-shrouded Battery at the tip of Manhattan. Soon the Statue of Liberty came up to starboard. The ship gathered speed to make the open sea beyond the Verrazano Narrows between Staten Island and Brooklyn. By the last glimmers of the sun Hughes and the other sailors saw the lighthouse on the point at Sandy Hook on the New Jersey shore.

In the lighted fo'c'sle the sailors unpacked their boxes and bags for the months ahead. One item in Langston's luggage set him apart from the rest of the crew. He was taking with him a box of books, mostly the detritus of his year at Columbia and the mark of his devotion from childhood to the lonely world of the written word. Suddenly, as he later recalled, he felt a powerful sense of revulsion at all that books had meant in his life. Leaning over the rail, he pitched them one by one into the darkness; with each he felt a burden fly away.

> It was like throwing a million bricks out of my heart—for it wasn't only the books that I wanted to throw away, but every thing unpleasant and miserable out of my past: the memory of my father, the poverty and uncertainties of my mother's life, the stupidities of color-prejudice, black in a white world, the fear of not finding a job, the bewilderment of no one to talk to about things that trouble you, the feeling of always being controlled by others—by parents, by employers, by some outer necessity not your own. All those things I wanted to throw away.

Hughes stood there awhile in the darkness, salt spray blowing in his face. All was quiet except for the voluptuous heave of the ocean and the pounding of the engines as the *West Hesseltine* surged into the dark. Toward Africa!

One book only did Hughes save. He had flung overboard the symbols of his hurt. But he had also kept the symbol of his best self, and of what he hoped to be. He saved his copy of Walt Whitman's *Leaves of Grass:* "I had no intention of throwing that one away."

4

ON THE BIG SEA
1923 to 1924

Old Walt Whitman
Went finding and seeking,
Finding less than sought
Seeking more than found,
Every detail minding
Of the seeking or the finding . . .

"Old Walt," 1954

ON JUNE 23, 1923, after crossing the Atlantic in calm, sunny weather, the *West Hesseltine* dropped anchor in fifteen fathoms off Horta in the Azores. To Hughes, thrilled to be across the ocean for the first time, Horta seemed like "a picture-book town" of houses painted "like toy Noah's arks, palm trees, nuns in flaring bonnets, oxen pulling wooden-wheeled carts, scores of brown-white children begging for cigarettes and pennies." Four days later, the Canaries too seemed like "fairy islands, all sharp peaks of red rock and bright sandy beaches and little green fields dropped like patchwork between the beaches and the rocks, with the sea making a blue-white fringe around." His anticipation of Africa began to build. Finally, just after dawn on July 3, a long, low line of stunted hills drifted out of the ocean. As the sun rose, Hughes saw sandy beaches and thatched-roof huts, white houses high on wooded slopes and, at a fort guarding the port of Dakar in Senegal, the tricolor of France. First a pilot, then a doctor and some customs officials, all white Frenchmen, boarded the vessel offshore. Then, at 8:16 in the morning, the *West Hesseltine* docked at Dakar.

"My Africa, Motherland of the Negro peoples! And me a Negro! Africa! The real thing, to be touched and seen, not merely read about in a book." In fact, Hughes's first impression was of crudeness and absurdity. Wandering through the town in ninety degree heat, his head spinning from glasses of cheap white wine, Langston that day saw Africa as ridiculous—black men dressed in billowy white gowns, sweating market women with bare breasts, children stark naked to the world. Giddy, he sat down to describe the scene to his mother. "You should see the clothes they wear here," he wrote Carrie, "everything from overcoats to nothing. I have laughed until I can't. No two people dress alike. Some have on capes, some shawls, some pants, some wear blue clothes fastened around their necks and feet blowing out like sails behind. Some have on preachers' coats, others knee pants like bloomers, with halfhose and garters. It's a scream!''

At least for the next eighteen days, he would see Africa almost as insensitively, from a distance, in a blur of exotic images. Shore leave was denied the crew until the *West Hesseltine* reached Lagos, Nigeria; on July 4, after only one night in Dakar, the ship steamed south. Five days later, it reached Portugese Bissao; here the pilot was a wizened black man who threaded the vessel through a green maze of islands off the estuary of the Rio Geba to a brief anchorage in seven fathoms off Bissao. Suddenly a boat approached, rowed by fierce-looking oarsmen; but they had only come for sacks of mail, which the men bore away as if in victory. Southward, past the Bijagós archipelago, shaggy with forests once haunted by African pirates, the *West Hesseltine* pushed on. Near Conakry in French Guinea the first sudden gusts of heavy rain washed the decks. At Freetown, in the British colony of Sierra Leone, the ship took on four rugged surf-boats for use further down the coast, where safe harbors were rare, as well as a full complement of African help. Men and boys from the Kru tribe would load and unload cargo; some of the boys were also assigned to mess duty, relieving Hughes and his messmates of most of their chores.

Continuing down the coast to Sekondi and Takoradi in the Gold Coast territory, the *West Hesseltine* steamed on to brief stops at the tiny ports of Saltpond and Winneba, before reaching the major harbor of Accra on the Gulf of Guinea. Here, at the ancient capital of the Ga kingdom, Hughes watched as the Krus off-loaded over eighty tons of cargo. Next came Ada at the wide mouth of the Volta River, and Keta, as the *West Hesseltine* passed out of Gold Coast waters into the seas off Dahomey. Now the weather turned bad. Late on the afternoon of July 19, dark thundershowers swept in from the south to meet the vessel, driving smaller boats back to the shore. Steaming slowly through heavy rain and a violent sea, the ship reached Cotonou near sunset the following day. With the waves subsiding, the surf-boats made their first trip to land with passengers and cargo. On July 21, the *West Hesseltine* at last dropped anchor off the port of Lagos in Nigeria.

Cosmopolitan to its core, Lagos was home to Moslem and Christian merchants, to Hausa, Ibo, Fulani, and Yoruba. A major slave center in the nineteenth century, it bustled now as the mercantile heart of the colony carved out imperiously by the British only a few years before. While the Krus labored to unload over 400 tons of cargo, Hughes and his shipmates enjoyed a week of shore leave. Then, just after dawn on July 29, the *West Hesseltine* left Lagos; the next day, it reached the port of Bonny in the great delta of the Niger. Steaming slowly through heavy rain and between walls of dense jungle, the ship dropped anchor at Dawes Island to take on a seasoned pilot for the tricky run to the wharf at Port Harcourt. There, Langston found an Africa more remote than anything he had seen thus far. In Port Harcourt, ostriches stalked the streets; a white man strolled by, followed by a completely naked black manservant; the barefooted black women were curiously short and dyed their nails with henna. The visit was also memorable for a boy to whom Hughes's heart went out, the golden-skinned son of an African woman and an Englishman who had deserted them, leaving mother and son to be shunned by her people. "Is it true," the boy asked Hughes, "that in America the black people are friendly to the mulatto people?" The youth ("Poor kid!") reminded Hughes more than a little of himself—except that instead of indulging his sense of apartness, Langston had plunged into the life of the race. "He looked very lonely, as he stood on the dock the day our ship hauled anchor. He had taken my address to write me in America, and once, a year later, I had a letter from him, but only one, because I have a way of not answering letters when I don't know what to say."

The first of August, clear and calm, found the *West Hesseltine* in the deep, long bay at the mouth of the Cross River, near the equator. As the ship eased through mangrove swamps, Hughes saw snakes and monkeys, yellow leaves and crimson blossoms that shone like stars in the gloom. At Calabar among the hills they left 173 pieces of lumber, then returned to the sea to push on to Douala in the Cameroons. Rain was falling as the ship passed between the 13,400 foot Cameroon Mountain and the island of Fernando Po. On August 8, when they crossed the Equator, Langston's head was shaved, according to custom, with oil from the ship's tanks. Baptized by Father Neptune (Chips the carpenter, with a wooden crown and a false beard), he was presented with a certificate of membership in the Amalgamated Society, The Sons of Neptune, formally signed by first officer E. S. Jansen. As if in celebration, the water that night flamed in a phosphorescent glow—"a million fallen stars foam in the wake of the ship and streaks of light move where fish swim near the surface."

The next day, when they reached the wide mouth of the Congo, the *West Hesseltine* was once again in Belgian territory. Nowhere was the journey more surreal than on this passage ninety miles up the Congo through thin forests, then semi-arid plains of parched palms and blasted yellow grass. At Boma, which they reached in mid-afternoon, the *West Hesseltine* was moored to a great tree, like a dinghy. Instead of going ashore in a surf-boat with everyone else, Langston tried to disembark via the rope to the tree; he fell headlong into the Congo—a symbolic African baptism perhaps, except that he could not swim. Beyond Boma the next day, as rain fell lightly, the river narrowed and ran swiftly between sheer, forbidding cliffs until, at the end of a broad curve, the white houses of Matadi came into view. Here, nearly 100 miles upriver in the infamous Belgian Congo, where Hughes made his deepest penetration of Africa, the signs of colonial repression were open. On the docks, black soldiers paced with guns and fixed bayonets; nowhere did Hughes see the examples of native enterprise found elsewhere in Africa. The various tribal villages, each on a separate highland, seemed joyless, the town shabby and nondescript, its streets lined by untidy mango trees. For Hughes, Matadi meant the ninth circle of European colonialism, "the dirtiest, saddest lot of Negro workers seen in

Africa.'' Even the twilight, blue-green, depressed him. The night was hot, the sky heavy with remote stars.

On August 14, the vessel entered the spectacular blue bay at San Paolo de Loanda in Portugese Angola. Two days later, it dropped anchor in Lobito Bay, more than halfway down the vast Angola coast. Here the journey south ended.

By this time, Hughes's sense of Africa and his voyage into the tropics had changed drastically. He had left New York with a tremendous sense of excitement. Old sailors called the route down West Africa ''the worst trip in the world,'' he would write. But ''for me it wasn't. For me it was the 'Great Adventure'.'' Still, nothing at Jones Point had prepared him for life on the *West Hesseltine*. Almost from the first day, the crew of some forty men seemed bent on mutiny. The skipper, a fat and ineffectual Dutchman named Breckwoeldt (he looked to Hughes ''like the father of the Katzenjammer Kids''), had blustered and threatened but barely controlled them. At Horta, after much carousing on shore, a seaman was left behind. The chief engineer, backed by his all-Irish engine room group, feuded openly with the captain, especially after clogged fuel lines repeatedly spilled oil onto the deck. The crew despised the passengers, who were almost all white missionaries except for one dedicated follower of Marcus Garvey, a West Indian tailor going to Africa to preach the gospel of racial uplift according to Savile Row. Sapped by bottles of warm beer and quarts of Johnny Walker whiskey, discipline fell apart. At least once the crew fought a pitched battle with a rival ship; once, the enraged captain threatened to declare them in mutiny after they refused to stop a baseball game and return to ship. ''The poor missionaries . . . were in a state of continual distress,'' Hughes would recall. ''They wrote irate letters back to the New York office. . . . They said they would never sail on that line again. Some even threatened to sue the company. Others wrote their congressmen.''

For ''James L. Hughes,'' as Langston craftily entered his name in the ship's register, disruption began in the cabin he shared with other messmen, including a Filipino serving boy, a young, dark brown Puerto Rican named Raphael who claimed to dislike women but slept with a pair of silk stockings tucked under his pillow, and an irrepressible, black-skinned, 19-year-old native of Kentucky, George Green, formerly a valet to a female impersonator. Green joked and sang and lazed almost all day, often sprawled stark naked in the cabin. Langston liked Green; at Horta, after meandering unsteadily through the town drinking cheap brandy, the two had been among the last four to stagger back to the *West Hesseltine*. As at Jones Point, Hughes made friends easily. He was a dutiful worker, and his knowledge of Spanish boosted his popularity among the Latins on board; his only trouble came from an officer from the South with whom Langston almost came to blows because of the man's insistence on segregation in the mess room.

''He is really squeezing life like a lemon,'' Cullen marvelled about Hughes to their mutual friend Harold Jackman. ''Think of being in Africa!'' On the steamy *West Hesseltine,* and in ports of call from the Azores to Angola, Hughes had every opportunity to learn things Columbia University could never teach him. At Las Palmas, Langston, a Puerto Rican seaman named Ernesto Rivera, and two or three others (but not the disreputable George Green) passed the night at a brothel that inspired a poem, "To The Dark Mercedes Of 'El Palacio De Amor' ":

> Mercedes is a jungle-lily in a death house.
> Mercedes is a doomed star.
> Mercedes is a charnel rose.
> Go where gold
> Will fall at the feet of your beauty,
> Mercedes.
> Go where they will pay you well
> For your loveliness.

At Lagos, with their first pay since leaving New York, most of the crew rushed to the bars and bordellos. Like François Villon's francs, Hughes would note approvingly, their dollars went *"tous aux tavernes et aux filles."* He was evidently among them at least once, if not more often, for he would remember distinctly the "vile houses of rotting women" in Lagos. Nor were all his experiences heterosexual. Decades later, Hughes would confide to one of his secretaries that the first homosexual episode of his life, a swift exchange initiated by an aggressive crewman, with Hughes as the "male" partner, came on this voyage. In some hastily compiled notes on sex, also composed late in life, his visit to Santa Cruz in Tenerife ("Tenerife. All those houses on the hills") would evoke a similar association.

> "Won't it hurt you," I said.
> "Not unless it's square," he said. "Are you square?"
> "Could be," I said.
> "Let's see," he said . . .

Not all of Hughes's learning was about sensation. By the southernmost point of the journey, off the coast of Angola, he saw Africa differently. Grown accustomed to tropic exoticism—to the roar and diminuendo of surf, the detonation of waves on distant beaches, the African profusion of flowers and iridescent sunsets—he was no longer teased by "the bare, pointed breasts of women in the market places," or the Kru men bathing on deck, their sex tucked between their legs in deference to the missionaries ("evidently not realizing it then stuck out behind"), or the black oarsmen magnificently naked, as Langston wrote Cullen, "save for a whisp of loin-cloth." With mounting anger he turned his attention to the main results of Europe's greedy feeding on Africa in the name of progress and civilization. The lessons began on the *West Hesseltine* itself. From Tom Pey, one of the Kru workers, Hughes discovered that the African men were paid only two shillings a day, plus board, while the little boys earned nothing at all; Hughes himself, among the lowest paid of the crew, earned $35 a month. Uncomplaining, the Kru endured the white man's orders and the backbreaking work. Africa had become "ten-year-old wharf rats offering nightly to take sailors to see 'my sister, two shillings,' elephantiasis amd swollen bodies under palm trees, white men with guns at their belts, inns and taverns with signs up, EUROPEANS ONLY, missionary churches with the Negroes in the back seats and the whites who teach Jesus in the front rows." "The white man dominates Africa," Hughes would write. "He takes produce, and lives, very much as he chooses. . . . And the Africans are baffled and humble. They listen to the missionaries and bow down before the Lord, but they bow much lower before the traders, who carry whips and guns and are protected by white laws, made in Europe for the black colonies." In at least one poem, "The White Ones," Hughes identified with the baffled hurt of black Africans forced to endure white domination:

> I do not hate you,
> For your faces are beautiful, too;
> I do not hate you,
> Your faces are whirling lights of loveliness and
> splendor, too;
> Yet why do you torture me,
> O, white strong ones,
> Why do you torture me?

In such work, Hughes affirmed his kinship with Africans. To his dismay, however, he quickly discovered that no African thought of him as one of them: "The Africans looked at me and would not believe I was a Negro." Nappy-headed, black George Green (to Green's mortification) was a native son; Hughes was a white man. "They looked at my copper-brown skin and straight black hair—like my grandmother's Indian hair, except a little curly—and they said: 'You—white man'." That he would want to be considered black struck the Africans as perverse, perhaps even subtle mockery. In vain he protested that he was not white. "You are not black either," one Kru told him flatly. Their rejection only stirred Hughes further to assert the unity of blacks everywhere, as in his little poem "Brothers": "We are related—you and I. / You from the West Indies, / I from Kentucky." And both were related to Africa. (Hughes was surely thinking of George Green of Kentucky, their Puerto Rican cabin mate Raphael—or the seaman Rivera—and a Kru such as Tom Pey.) His anxiety over Africa also inspired "Poem":

> The night is beautiful
> So the faces of my people.
>
> The stars are beautiful,
> So the eyes of my people.
>
> Beautiful, also, is the sun
> Beautiful, also, are the souls of my
> people.

And more contrived, though with a typically childlike, perhaps androgynous persona, was his "Dream Variation":

> To fling my arms wide,
> In some place of the sun,

To whirl and to dance
Till the white day is done.
Then rest at cool evening
Beneath a tall tree
While night comes on gently
Dark like me,—
That is my dream!

To fling my arms wide
In the face of the sun,
Dance! whirl! whirl!
Till the quick day is done.
Rest at pale evening . . .
A tall, slim tree, . . .
Night coming tenderly,
Black like me.

In these poems and others written in Africa, Hughes responded emotionally to the most dangerous lies of European colonialism.

Heading north from Angola, the *West Hesseltine* took on cargo for the United States. The first stop was in the Cameroons, this time at Kribi. Then, steaming across the great western crook of the continent, the ship anchored briefly off the mouth of the Forcados River in Nigeria. Upstream they reached Burutu, a major port for the *West Hesseltine*. Working every day for a week, the Krus loaded 1770 casks of palm oil deep into the hold while the crew revelled in nine days of shore leave—"Oh, the little black girls of Burutu!" Langston later recalled. If he enjoyed the women of Burutu, he also talked about race and politics in the United States with at least one African, to whom he later sent copies of the *Crisis* from New York. And one evening he and Tom Pey, his Kru friend, visited an old Moslem trader who received Langston as an honored guest and proudly opened boxes of his African treasures—beaten brass from Benin, exquisitely woven cloth, dried skins of wild animals, soft white feathers of jungle birds. But when Hughes heard ritual drumming and begged Tom Pey to take him to the ceremony, the Kru refused. The god Omali would not tolerate the presence of a white man. Disconsolate, Langston now saw the *West Hesseltine* and the other foreign vessels with deeply alienated eyes—"tall, black, sinister ships, high above the water."

On the last day of August the freighter reached Accra again. In one hard day's work the Krus loaded 4800 bags of cocoa beans for the chocolate factories of the United States, with still more beans loaded at the next stop, Sekondi. Although the marketplaces flashed with colors and the black girls were striking with their bandannas and strips of bright cloth twined about their bodies, Hughes remained troubled. Losing track of the month—it was already September—he heard black oarsmen chanting as they rowed to the ship, and he felt the wall between his world and theirs:

Singing black boatmen,
An August morning

In the thick white fog at Sekondi,
Coming out to take cargo
From anchored alien ships—
You do not know the fog
We strange so-civilized ones
Sail in always.

Assinie in the Ivory Coast was the start of the most dangerous phase of the voyage, when the ship loaded mahogany logs destined to make furniture for the parlors of the United States. Bobbing and spinning treacherously, the massive logs were floated to the side of the ship and fastened with chains by the diving Kru men. One mistake could mean death. When the water at Assinie proved too rough even for the hardy Kru, the *West Hesseltine* pushed on to Grand Bassam, looking in vain for calm water. Then in three days at Grand Lahou they raised 145 logs. Twice the winch cables snapped under the strain. Although the ship was almost a month behind schedule and supplies were running low, on September 10 they returned to Grand Bassam, which Hughes found delightful; its streets were shaded with palm and almond trees, and the Africans seemed clean and assured. But the ship stayed on too long. Growing bored and restless, the crew drank heavily—Hughes later marvelled at "the millions of whiskey bottles that must be buried in the sea along the West Coast." One night, when two young African women rowed out to the ship, they were stealthily admitted. Then, after one was snatched away by an officer, the other was set upon by the men, one after another, while Hughes watched dispassionately. "Each time a man would rise, the little African girl on the floor would say: 'Mon-nee! Mon-nee!' But nobody had a cent, yet they wouldn't let her get up. Finally I couldn't bear listening to her crying: 'Mon-nee!' any more, so I went to bed. But the festival went on all night." To Langston's greater disgust, one of the firemen openly brought young black boys to the ship and sodomized them for a shilling or so. Hughes longed for the ship to leave.

By September 26, when the last log was hoisted aboard, the hold of the *West Hesseltine* groaned with copra, palm oil, cocoa beans, and logs. In return for machinery, tools, canned goods, and Hollywood films, "we took away riches out of the earth, loaded by human hands." On October 1, at Freetown in Sierra Leone, the Kru workers were discharged. The next morning, in the middle of a rain squall, two passengers boarded at Conakry for New York. A few days later the vessel reached the main port in the Cape Verde Islands, where the fuel tanks were refilled. Finally, at 10:30 on the evening of October 6, the *West Hesseltine* weighed anchor for the crossing. "We are homeward bound," Hughes had alerted Countee Cullen:

> Only one more port and then across to Manhattan—and Harlem. We are a month late now but then we have almost a full cargo—palm oil, cocoa beans, and mahogany logs, and here we load ginger and pepper. . . . We've been every where,—up rivers, branch rivers and creeks, surf ports, and harbor ports, and visited about every colony of the West Coast. . . . I have . . . seen more of Africa than I ever expected to see. And tonight the sunset! Gleaming copper and gold and then the

tropical soft green after-glow of twilight, and now stars in the water and luminous phosforescent foam on the little waves about the ship and ahead the light of Freetown toward which we are steering through the soft darkness.

I'll be up to see you soon after this letter. I am anxious to be back again. Just a little bit homesick for New York.

At a quarter past nine on the morning of October 21, the bright beginning of a spectacular Indian summer day in New York, the *West Hesseltine* docked in Brooklyn. When the paymaster arrived, Hughes and the rest of the crew were handed their money, then fired. Their behavior had sunk to new depths on the return trip, as bad weather pushed the ship far off course and, without fresh supplies, their meals turned from indifferent to very bad. But Hughes left Brooklyn in high spirits. Clinging to him as he sauntered away from the ship was a companion from Africa—Jocko, a large, red monkey purchased in the Congo for an old shirt, a pair of old shoes, and three shillings, and intended as a present for Langston's "brother," Gwyn Clark. With Jocko on his arm and money in his pocket, Hughes made for Harlem.

He was a magnificent sight. "Langston is back from his African trip," Cullen alerted Alain Locke, "looking like a virile brown god." After buying a suit for a visit to his mother in McKeesport, Hughes then impulsively delayed his leave to attend the premiere of Eleanora Duse in her first tour of the United States in two decades. At Jones Point he had read *The Flame of Life,* and was determined now to see the actress D'Annunzio had evoked so movingly. At the Metropolitan Opera House on the evening of October 29, he was in the throng that took every seat and stood three and four deep as Duse opened in Ibsen's *La Donna del Mare*. For him the performance was a flop: "She seemed just a tiny little old woman, on an enormous stage, speaking in a foreign language, before an audience that didn't understand." Between the expensive delay and the need to board Jocko, then secure him a warm parlor car on the train, Hughes reached McKeesport broke and fearful of Carrie Clark's tongue ("I could hear my mother now: 'Coming home with only a monkey, heh?'"). At 1037 Rose Street in McKeesport, her home, he hid the animal in a bag until the right moment; then Jocko leaped out, "big as the jungle." Gwyn Clark fled; Carrie sprang back in horror; Homer's jaw dropped. But Jocko was soon a neighborhood celebrity, scampering in the trees along the river nearby, going with Homer to amuse his poolhall friends. Only Carrie remained hostile. When an overexcited Jocko defecated on a pooltable and Homer paid to clean it, the monkey's days on Rose Street were numbered. Hughes had barely left McKeesport before Carrie sold the beast to a pet shop.

On November 8, while in McKeesport, Hughes earned a little money with a reading at the local colored YMCA. A week later he was back in Harlem, staying at a boarding house at 233 West 136th Street. Fresh from Africa, Langston glowed now in a special light; also, the August *Crisis* had devoted an entire page to his poems, including "My People," "Three Poems of Harlem," and the dazzling "Jazzonia" ("Oh, singing tree! / Oh, shining rivers of the soul!"). Hughes's success was only part of the rising excitement in educated Harlem. Jessie Fauset had started a novel; Jean Toomer of Washington, whose avant-garde

Cane had appeared earlier that year, had begun to be seen about Harlem; at the 135th Street library, the energetic, brownskinned Regina Anderson had joined Ernestine Rose's excellent staff. With evening volunteers such as Jessie Fauset and Ethel Ray Nance of the Urban League, the library hosted an increasing number of readings, lectures, and soirees that brought Harlem booklovers together. Best of all, a cosy little circle of promising young men and women, encouraged mostly by Jessie Fauset (still the literary editor of the *Crisis*), was now writing or working seriously in the arts in Harlem. Among them were Gwendolyn Bennett of Texas, a writer and artist of definite talent; the short story writer Eric Walrond of British Guiana, who had come to Harlem from Panama some years before; and Cullen's best friend, the handsome, urbane schoolteacher Harold Jackman, who neither wrote nor painted but patronized all the arts. In this circle, Hughes was welcomed back from Africa as a hero.

Yet he was uncertain about his future, unsure whether he should go back to sea or return to school—not to Columbia University, but to a black college. Quickly, Cullen assured him that he was still admired in Washington. Although Alain Locke had written him off definitively ("As for Langston—he is a fool—never again—I swear it"), the generous Cullen had insisted that in spite of Hughes's "irresponsible streak" he was worth pursuing. In fact, Locke's loneliness was so sharp that he was eager to forgive Langston. Pragmatically, in turn, Langston himself had begun to think afresh of Locke—or, more accurately, of Howard University in Washington, where Locke was influential. The professor had appealed to Hughes to come and help him with his work: "I am humbly and desperately in need of a companion."

Instead, Hughes took a job on an ancient tub sailing between Hoboken and the West Indies. The cook, however, a tyrannical old black Barbadian with a savage scar on his face, was so offensive that Hughes quit before the ship sailed. On December 8, he found something much better—a freighter going to Europe. The *McKeesport* was about to leave for Rotterdam with a cargo of grain and flour. Just after "James L. Hughes" signed on as a messboy at $42 a month, and as the first snow of the season was falling on New York, the *McKeesport* put out to sea. A storm broke that evening with terrific force and persisted across the Atlantic. With the ship rolling and plunging, her stern sometimes pointing to the sky, propellers flailing clear of the water, the voyage seemed doomed. A messman stepped into a scalding pot of boiled cabbage set down on the deck. On the fourth day out, the chief engineer, Franklin Robert Adams, going to meet his fiancée in Europe, died of double pneumonia. Hughes sadly watched his burial at sea.

Battered, the *McKeesport* reached Europe on the cold, drizzly dawn of December 22. Rotterdam was covered in snow, the canals frozen, but soon Hughes and his friends were drinking Hulskamps gin in a cozy waterfront tavern. He liked everything about Rotterdam—Chinatown, the white canals, the red-cheeked children clopping in wooden shoes, the quaint houses he had read about as a child. As a Christmas treat, a watchman at the dock took home all the black fellows on the *McKeesport*. The man's son-in-law was from Paris; Hughes tried out his rusty high school and college French, and thought a great deal about Paris.

On January 22, after a return trip in weather even worse than in December, the *McKeesport* was back in Hoboken. Again Hughes faced the choice of returning to sea in a few days or going to Washington. Putting off the decision, he attended the new black musical *Runnin' Wild,* which sensationally introduced the Charleston dance to Broadway, and *Salome,* controversially produced by the black Ethiopian Art Players. Hughes also saw a fair amount of Countee Cullen. On February 1, as Langston celebrated his twenty-second birthday quietly at 234 West 131st Street, Cullen's home, the moment seemed right for a decision. Reading over Alain Locke's letters to him, he decided to act. The next morning he sent a telegram: "MAY I COME NOW PLEASE LET ME KNOW TONIGHT BY WIRE."

Two days later, without receiving a telegram (Locke sent a letter, which apparently never reached Hughes) Langston apologized to Locke. "Forgive me for the sudden and unexpected message I sent you," he wrote; "I'm sorry." He should have known that one couldn't begin school in the middle of the term, but "a sudden desire came over me to come to you then, right then, to stay with you and know you. I need to know you. But I am so stupid sometimes." He still intended to come to Washington, but not now. He was leaving the next day for Holland.

On February 5, two days later, the *McKeesport* left Hoboken for Holland. Once again the ship ran into miserable weather. Langston's mind, however, was not on the storm. He was tense about what to do next—whether to return and go off to Washington and Alain Locke or stay with the ship while he saved money. Or jump ship, and go to Paris. Looking for a sign from fate, he found one when he clashed with the steward one Sunday morning. Although a large pile of left-over chicken sat in the galley, the steward refused to let Hughes have a single piece. Chicken was reserved for the officers—the crew ate beef stew. Outraged, Langston stormed off to his cabin.

On the evening of February 20 the *McKeesport* reached Holland. The next day, Thursday, when the captain advanced money to the men, a smiling Langston took as much as he could get, about $20. Friday was a ship's holiday—George Washington's birthday; on Saturday, when heads were next counted, eight men were missing. Two never showed up. One was James L. Hughes, who was reported officially as having deserted his ship.

At the frontier, where the French border inspectors threw open the doors of his third-class coach and let in the swirling snow, Hughes purchased a visa to enter France. When the train reached the Gare du Nord in Paris early next morning, he changed his few remaining dollars into about 150 francs and bought the cheapest breakfast possible, a cup of coffee and a skimpy croissant. Leaving his bag at the station, he boarded a bus going to the Opéra. At first, falling snow obscured his view as the bus drove down rue Lafayette. But when he stepped off the vehicle across the street from the Café de la Paix, his heart almost stopped as postcards of Paris came alive before his eyes. The Place Vendôme was ahead. Walking down rue de la Paix into the Place Vendôme, he turned and found himself unmistakably in the Place de la Concorde. The Seine was at hand, into which snowflakes were delicately falling. The Champs Elysées rolled before him, with the Arc de Triomphe in the distance. Dizzy from the magic of the moment—but also from hunger and the cold—he trotted along the river to the warm Louvre.

Directed to Montmartre by a black doorman at the American Express office on rue Scribe, Hughes stumbled on a covey of American blacks at a little café near the corner of rue Pigalle and rue de la Pruyère. Musicians afraid of competition, they were rude to him. What instrument did he play? Did he sing? Did he dance? Then why was he in Paris? Taking a room for one night, Hughes had his first French dinner, a very humble affair of *boeuf au gros sel* and cream cheese with sugar. The next morning, after coffee and another croissant, he looked for work. Trying all the English-speaking places—the American Express office, the American Library, the Embassy—he returned, tired and dispirited, to the café in Montmartre. He was inquiring aloud about a cheap hotel when a trim young white woman offered to help him. She knew just the place, if he would follow her. She led him through the narrow streets off the Place Clichy to a run-down six-story hotel at 15 rue Nollet, where an attic room cost only 35 francs per week. Thanking the young woman, who spoke French and a little English with an East European accent, Hughes left to recover his bag.

When he returned, he found that his luck had improved even further: the young woman had moved in with him. Sonya was a Russian emigré dancer, perhaps two years older than Langston, who had slipped out of the Soviet Union via Constantinople and Vienna with her five-year-old child. When her dance troupe broke up somewhere in the south of France, she had headed for Paris. Her child was staying elsewhere. Pretty and experienced, she soon found work as a bar girl in the famous Zelli's nightclub on the rue Fontaine; she and Hughes began to eat better.

Searching for a job for himself in depressed Paris, where foreign laborers were very unpopular, he was chased *(''Sale étranger!'')* from one construction site by irate Frenchmen. Hughes knew no one in France but had been given two addresses. The first was for Claude McKay, who unfortunately had just left for the south. The other led him to one of Jessie Fauset's former students from her days teaching at the M Street School in Washington. An extremely light-skinned young man, and a scholarly graduate of Williams College, Rayford W. Logan had been one of the few Negro officers in the war. When Hughes approached him at a bar, Logan was hobbling on crutches with a leg broken in a recent traffic accident. He would recall meeting ''this rather disreputable young Negro, not a bohemian at all, just poor,'' who introduced himself as a friend of Miss Fauset.

Logan promised to help Hughes find a job, but Langston found one first on his own. In a little alley off the Boulevard Clichy, he spotted a club without a *chasseur* standing guard outside. The owner, Madame Moffat, a brownskinned woman from Martinique, liked his youthful looks and offered him the job at five francs a night, plus dinner. Hughes took it at once. Sporting an old blue and gold military hat lately purchased at the Paris flea market, he arrived each night as Madame Moffat was dining with her lover, a striking, very thin Roumanian young woman with purple circles about her eyes and heavily rouged lips. He ate his dinner, then guarded the door of the Cozy Corner. Unfortunately, the little *boîte de nuit* attracted a fighting crowd—gentlemen who warred with canes, prostitutes who snapped champagne glasses to slash their rivals. Supposed to stop fights, Hughes vanished instead at the sound of breaking glass.

When Rayford Logan heard of an opening at the well-known Le Grand Duc nightclub at 52 rue Pigalle in Montmartre, Hughes quit his dangerous job and was hired by the part-owner and manager, Eugene Bullard, a much-decorated American war veteran of the Lafayette Flying Corps, in which he had been the only black pilot, and a retired prizefighter who had won forty-two bouts. The entertainers were also American, as was the remarkable cook Bruce, an enormous, one-eyed, heavy drinking, supremely neurotic Negro who cooked fried chicken and other southern-style food, and for whom Hughes was hired as a *sous-chef* (a *plongeur,* really, or dishwasher). The two waiters, Romeo and Luigi, were friendly young Italians. By the end of his third week in Paris Langston was settled into life at Le Grand Duc. Arriving at work at eleven in the evening, he helped Bruce the cook, but kept an ear cocked to the music of the jazz band and the singing of the star attraction of the club, the regal Florence Embry Jones. The clientele was of a superior class; a prince now and then was not unusual, and American celebrities such as the novelist Anita Loos came for Bruce's home cooking. Around three in the morning, when Florence's husband, the pianist Palmer Jones, showed up from his job at Les Ambassadeurs with other American musicians such as trumpeter Cricket Smith, clarinetist Frank Withers, jazz violinist Louis Jones, and the drummer Buddy Gilmore, they jammed freely until dawn. These sessions did not stop after the replacement of Florence Jones by an unknown but vivacious brownskinned redhead from New York, Ada Beatrice Queen Victoria Louisa Virginia Smith, who called herself "Bricktop" and made up for an only adequate voice with sensational energy (she put sin back into syncopation, they said). After the jam sessions, the musicians and the staff sat around talking and drinking coffee and left-over champagne until the morning flowered. Some time after seven, Langston made his way home, past the shrouded white dome of Sacre Coeur, to go to sleep.

From his garret room with slanting ceiling and double windows under the eaves on rue Nollet he looked out over the chimneys of Paris. Lonely, he found Paris hard to take at first, as he wrote to Countee: "Kid, stay in Harlem! The French are the most franc-loving, sou-clutching, hard-faced, hard-worked, cold and half-starved set of people I've ever seen in life. Heat—unknown. Hot water—what is it? You even pay for a smile here. Nothing, absolutely nothing is given away. You even pay for water in a restaurant or the use of the toilette. And do they like Americans *of any color?* They do not!!! Paris—old and ugly and dirty. Style, class? You see more well dressed people in a New York subway station in five seconds than I've seen in all my three weeks in Paris. Little old New York for me! But the colored people here are fine, there are lots of us."

For a brief while, Sonya was good company. Then one day she lifted several hundred francs from a drunken Swede. She and Hughes were at a café celebrating her emancipation from Zelli's when the Swede, now sober, passed by. Sonya slid under the table and made plans to leave town. Recovering her costumes from the pawnshop, she left Hughes and Paris to return to her child and the dance. But he made some new friends in March when Rayford Logan asked him to deliver a package to a young woman living on the Boulevard St. Michel across from the Luxembourg Gardens in the Latin Quarter. There he found not one but three lovely young educated colored women, all British. Two were sisters from Jamaica, Amy and Gwen Sinclair, studying French in Paris; the third, Anne Marie Coussey, Logan's friend, was more purely English in her speech and manner. Attractive, intelligent, and clever with words, brownskinned Anne Coussey was pleased to meet Hughes; she had read some of his poems in the *Crisis*. He found her mature and refined; her upper-class British enunciation was fascinating, and she also seemed to have lots of money.

Returning to see the young women, he met another Jamaican, the gifted and somewhat driven daughter of a wealthy businessman. Violette Hope-Pantin had studied at the University of London, the Sorbonne, and in Berlin, and was also a very good pianist. Calling regularly on the four friends at the rooms near the Luxembourg, Hughes enjoyed talking with them about politics and art, offering the black American perspective. They formed easily the most polished quartet of colored young women he had yet known; to his dismay, however, he soon saw that the Jamaican girls—typical West Indians—were snobbish about skin color. Claude McKay of Jamaica, Hughes's favorite black writer, was a fine poet, yes, but oh so *dark!* Anne Coussey was different. Growing up among whites, she knew how little the various shades of color meant when they saw Negroes. Hughes liked her for this astuteness; then he liked her for herself.

After the cramped, cold room on rue Nollet and the greasy kitchen of the Grand Duc, he found it very pleasant to cross Paris in the late winter afternoon and sink into an armchair near the fireplace while Anne brewed hot English tea and served chocolate eclairs or sweet biscuits from London. Although his appetite astonished her, he often brought cakes and sweets himself; she liked that about him. She didn't care at all for his Montmartre nightclub—"that beast of a place," she would call it—but her young poet was certainly intelligent and handsome, and made no demands. Soon he called her "Nan," and he was her "Teddy." In the early days of spring they strolled among the children and the shrubs in the shadow of the Luxembourg Palace, or travelled out to Versailles to walk in the cool green forest. Nan and Teddy loved to recite Claude McKay's "Spring in New Hampshire" ("Too green the springing April grass, / Too blue the silver-speckled sky / For me to linger here, alas . . ."). Together they heard Raquel Meller and Damia sing, and the young star Maurice Chevalier. Nan grew very fond of Teddy. By May, Teddy was in love.

He told her a little about himself—about his mother and his little "brother" Gwyn Clark, and about Jessie Fauset and Dr. Du Bois at the NAACP, and about Columbia University. Wanting to be liked for what he was—*sous-chef,* rover, troubador, poet—he told her little more. In turn, she did not boast about her family, as she might have done. Her mother was Ambah Orbah Coussey, a full-blooded member of the Aboradizo clan of the Fanti of the Gold Coast. Ambah Orbah's granduncle, Philip Quaque, had been the first African ordained as an Anglican priest in the Gold Coast. Anne's father was Charles Louis Romain Pierre Coussey, a man of mixed blood, whose Scots grandfather had migrated to Africa and built his fortune on the slave trade. Charles Coussey's father had studied at St. Andrew's College in Scotland, and married a woman of mixed Scots and Senegalese blood; they came to Africa in the middle of the nineteenth century. Educated in England, Charles Coussey returned to Africa well prepared for a career in business; eventually he traveled up and down the west coast of Africa working for Swanzy Miller, the big English firm that eventually became the United Africa Company. All his children went to England to be educated. Leaving Africa when she was five, Anne had attended boarding school in Brighton, then trained as a business secretary. She had been about to go home to Africa when her father fell ill and moved to London for treatment. She lived with him and a sister on Adelaide Road in Hampstead, but was taking time off in Paris to learn French and also weaving at a school run by Isadora Duncan's brother Raymond.

There was also a suitor in England. In fact, Nan was practically engaged but did not want to marry this Englishman. Longing to have a baby, as she confided to Hughes, she was nevertheless strong enough to wait for someone closer to her ideal. Teddy was definitely closer to her ideal, but was he serious? Younger than she was by some years, he sometimes seemed even younger. His job at Le Grand Duc was a lark, but surely he intended to go back to university? "I wonder how you have been able to stick to it all this time," she would ask him; "don't you think it will be possible for you to get home soon and complete your education? After all what you are doing is leading nowhere, and you are wasting some of the best years of your life." When she raised this point, Teddy didn't like Nan nearly as much as when she took him as he was; she fed his suspicion that marriage was hostile to his great goal. Writing to tell Cullen about Anne Coussey, Hughes had put the matter in cool perspective. "So you are in love at last?" wrote a bemused Countee. "I question its depths, however, when you tell me so resignedly that you are determined that nothing shall come of it, basing your decision upon chimeric demerits."

Far from wasting his time, Hughes was growing as a poet. Before sailing on the *McKeesport* for Europe he had left Cullen a number of his poems to be placed. In February the *Crisis* had published "Brothers," about the bond between blacks everywhere. In March, Hughes made his first appearance in an adult black magazine other than the *Crisis* (although the previous year, the *Amsterdam News* had published two poems, including "Justice," about the "blind goddess to which we black are wise. / Her bandage hides two festering sores / That once, perhaps, were eyes"). In March, the socialist *Messenger,* edited by A. Philip Randolph and Chandler Owen, published two pieces, "Gods" and "Grant Park"; and "The White Ones" appeared the same month in *Opportunity: A Journal of Negro Life,* founded the previous year by the Urban League. "Gods," from his hand-written anthology sent to Cullen from Jones Point, was strongly anti-religious; "Grant Park" is also bitter:

> The haunting face of poverty,
> The hands of pain,
> The rough, gargantuan feet of fate,

The nails of conscience in a soul
That didn't want to do wrong—
You can see what they've done
To brothers of mine
In one back-yard of Fifth Avenue.
You can see what they've done
To brothers of mine—
Sleepers on iron benches
Behind the Library in Grant Park.

Even harsher, in the May *Messenger,* was the sardonic "Prayer for a Winter Night" ("O, Great God of Cold and Winter, / Wrap the earth about in an icy blanket / And freeze the poor in their beds"). And in the June *Crisis* came "Lament for Dark Peoples," in which Langston mourned the coming of the white man and the caging of the red and the black into "the circus of civilization."

Hughes had sought outlets for his verse other than the *Crisis* in part because Jessie Fauset now disapproved of several of his poems. Hesitant to publish his most aggressive verse, in June she returned a poem about black miners in South Africa, "in the hope that you will revise and smooth it. You know it is possible to be rugged without being crude." Yet she freely admitted that Du Bois "was satisfied with the form and said the thought permitted it." The *Messenger* would publish either that poem or one defiant of her response:

In the Johannesburg Mines
There are 240,000
Native Africans working.
What kind of poem
Would you
Make out of that?
240,000 natives
Working in the
Johannesburg mines.

Fauset continued to press him to write more frequently in rhyme, although she accepted his liberal way of mixing rhyme and free verse. Grateful to her for giving him his start, Hughes tolerated her habit of "improving" his lines with changes in punctuation and even of words; Fauset remained "my brown goddess."

In spite of Fauset's—and Countee Cullen's—certain disapproval, however, Hughes struck out boldly in another direction that spring. Following the lead of the jazz musicians, whose riffs filled his head every night at Le Grand Duc, he worked to get their complex rhythms, more modern than those of the blues, into his verse. But instead of lyric tribute, as in "Jazzonia," or irony and obliquity, as in "The Weary Blues," he allowed the masses to speak for themselves in "Negro Dancers":

"Me an' ma baby's
Got two mo' ways,

Two mo' ways to do de buck!
Da, da,
Da, da da!
Two mo' ways to do de buck!''

''White folks, laugh! / White folks, pray!''—but the dancers and the dance go on. ''Do you like it? Do you get it?'' Hughes asked Cullen excitedly. ''We'll dance! Let the white world tear itself to pieces. Let the white folks worry. We know two more joyous steps—two more ways to do de buck! C'est vrai?''

Both Countee and Fauset, however, preferred rhymed, sentimental verse, marinated in pathos. Meeting their standards, Hughes's ''Song For A Suicide'' appeared in the May *Crisis*.

Oh, the sea is deep
And a knife is sharp
And a poison acid burns,
But they all bring rest
In a deep, long sleep
For which the tired soul yearns—
They all bring rest in a nothingness
From where no road returns.

When he heard that some blacks took the poem to be about race, Hughes impatiently protested that ''people are taking it all wrong. It's purely personal, not racial. If I choose to kill myself, I'm not asking anybody else to die, or to mourn either. Least of all the whole Negro race.'' The speculation showed how little some readers understood Hughes. The poems about race and politics, in fact, routed his innate tendencies toward melancholy; in a real sense, they kept him alive.

Like Cullen and Jessie Fauset, Anne Coussey much preferred the sweet lyrics to the wilder jazz verse. She did not like ''Fascination'' in the June *Crisis,* an awkward tribute to feminine beauty (''Her teeth are as white as the meat of an apple'') written at Jones Point. She expressed her opinions freely; perhaps it was her way of prodding Hughes. Her sister Christine came over from England for a brief vacation and to have a look at the poet. Still, as the spring ended and Anne prepared to leave Paris, Langston had still made no serious move. When Anne proposed a trip together to Italy, Hughes excused himself. He said nothing about going back to college, nothing about their future. Finally the day came for her to leave. A doctor friend of her father's would escort her home across the English channel.

Years later, Hughes would depict her departure movingly. The young lovers were betrayed by Christine and possibly a jealous Violette Hope-Pantin. Anne's father would cut her off without a penny if she did not return with the formidable Dr. Alcindor to London. The lovers, in Hughes's account, went to a little restaurant in the Place du Tertre just behind Notre Dame de Sacre Coeur. After a last supper together, they walked slowly down the hill. Buying some wild strawberries and cream, they went to rue Nollet. ''And we sat in my room on the wide stone window seat,'' he reminisced, ''dipping each berry into the cream and feeding each other, and sadly watching the sun set over Paris. And we felt very *tristes* and very young and helpless, because we could not do what we wanted to do—be happy together with no money and no fathers to worry us." Then they returned to the Boulevard St. Michel and the grim doctor, who whisked Nan away in a carriage.

Touching—but perhaps not quite accurate. In her first letter from Hampstead, Anne invited Langston to come over for a visit. Everyone was going to the seaside except Daddy and her little sister, so there would be lots of room for him at the flat. Dr. Alcindor wished to be remembered. She herself would be back in Paris late in the summer—but only passing through, and she did not expect to see him.

If Anne Marie Coussey ever resented anyone, the person would not have been her father but Langston himself. Too well bred, too controlled to task him directly about an insincere flirtation, she nevertheless had her say. After reading some of his poems in the *Crisis,* she did not bother to stifle a yawn. "I am afraid you are getting into a groove, there is a sameness about your things," she wrote; "I wonder if Paris is responsible, or is it perhaps that you are not really original?"

Two years later, almost engaged to her future husband, a bright young black lawyer from Trinidad named Hugh Wooding, she came closer to showing Hughes her true feelings:

> I wonder if you really did love me when we were in Paris. I think not, you would not even kiss me goodbye when I was coming away, do you remember? . . . I think sometimes too of that Spring time, I think I have always loved you without knowing it, but you know I am always rather cynical. I have just finished reading a book about a girl who wanted to love a man Soul to Soul, then one day she discovered she had a body, then concluded there was no such thing as love, it was all Sex. Maybe she is right. What do you think?

In the days after Anne Coussey's departure Hughes was indeed sad. Although he did not regret his choice of freedom and poetry over a "good" job and domestication, he knew nevertheless what he had sacrificed. In a beautiful lyric, "The Breath of a Rose," he wrote plaintively about fleeting love. Then, or somewhat later, he penned "A Letter to Anne" ("Since I left you, Anne, / I have seen nothing but you . . ."). And, "In the Mist of the Moon" ("In the mist of the moon I saw you, / O, Nanette, / And you were lovelier than the moon . . .").

Swept by "a sudden notion that I wanted to see the world again," he almost took a job on a boat late in June, then decided to stay in Paris. Alain Locke was coming—which, with Anne's departure, was perhaps why Hughes felt an urge to flee. This time he would wait. In April, Langston had written to Locke about his chances of getting into Howard University. Locke, having failed previously with displays of fire, had responded with ice. He indeed would be in Paris in the summer; meanwhile, he had three wishes for Langston: to learn French well, to write much verse, and "that you may have whatever happiness is possible." But Locke had also given Hughes's address to the rich art collector from Merion, Pennsylvania, Albert C. Barnes, who invited Langston in mid-June to lunch at the Café Royale near the Louvre. The meeting went very poorly; Albert C. Barnes on art reminded Hughes of his father on mining. Through Barnes, however, he was able to visit the home of the most famous French collector of African art, Paul Guillaume, where he spent some moving hours viewing expatriated treasures from the Congo, Benin, and the Sudan.

Near the end of the spring, after acquiring a black American roommate who turned out to be a drug addict, Hughes gave up the room on rue Nollet for one in an attic on the rue des Trois Frères. With his French much improved, he stepped out further into Paris and met some of the people who, within ten years, would lay the foundation for the concept of *Négritude* with the founding of *La Revue du Monde Noir* in 1930 and the more militant journal *Légitime Défense* in 1932; in *Négritude,* various French-speaking writers of Africa and the Caribbean would attempt to blend the primary racial insights of black Americans like Du Bois and Hughes with their own complex experience of European colonialism. René Maran, the author of the novel *Batouala,* winner of the Prix Goncourt in 1921, was "a darned nice fellow," so pleasant and earnest that he reminded Hughes of Countee Cullen. Langston also met Prince Kojo Touvalou Houénou, nephew of the deposed king of Dahomey, and perhaps the main founder of a new black newspaper in Paris. Hughes was invited by the staff of the race-conscious *Les Continents* magazine to submit poems (they published "A Black Pierrot" that year) and to solicit pieces from blacks in the United States; he asked Cullen for material. But though he seemed to be settling down in Paris, Hughes also thought more and more of home. Certainly Cullen wanted him to return. "Do come home and go to Howard," Countee advised. "You are missed and needed and wanted."

Near noon around the last day of July, after a long night at Le Grand Duc, he was awakened by a gentle rapping on his door.

"Qui-est-il?" he mumbled.

"Alain Locke."

When the instantly awake Hughes opened the door he saw a mild-looking, gently smiling, light-brown-skinned man; the lilting elegance of his voice was reflected in the gentility of his clothing, which contrasted starkly with Langston's barely decent dishevellment. At lunch in a bistro near the Place Clichy, Locke talked effortlessly of many things, but in particular of editing a special issue, on the Negro question, of Paul Kellogg's influential magazine *Survey Graphic.* Of one thing Locke was certain—his special issue would not be "as deadly sociological as most special issues, if I have anything to do with it." (He was perhaps thinking not only of other issues of the magazine, which emphasized social reform, but also the special black issue of *The World Tomorrow* in 1923, in which Langston had published three poems.) Locke wanted poetry from Hughes and he wanted his best, especially "Dream Variation," which Cullen had shown him. When Langston explained sheepishly that he had sent the poem to Augustus Dill in a private letter, and that Dill had promptly given it to the *Crisis,* Locke promised to use it anyway.

Eighteen months had passed since his first letter had reached Hughes at Jones

Point. Given the height to which Locke's expectation had soared, and Langston's own cry from the heart just before he fled to France, this meeting could have ended badly. Locke, however, was not disappointed by what he saw and heard, and Hughes himself was charmed. "Locke is here," he reported to Cullen. "We're having a glorious time. I like him a great deal." Knowing much about painting and sculpture, Locke steered Langston away from the dishpans of Le Grand Duc and the fleshpots of Montmartre ("I do not love your Montmartre") toward the treasures of Paris. They strolled together in the Parc Monceau, Locke's favorite place in all the city ("Mother and I have spent such happy afternoons there"); at the Opéra Comique they heard *Manon* and *Samson and Delilah,* and were mortified to miss a performance of *Die Walküre*. Locke was happy. "See Paris and die," he wrote with a flourish to Cullen. "Meet Langston and be damned." He had just done both, "but it looks as if I might live and be blessed."

"As to the joy of having met you and of being able to be with you," Locke wrote Hughes a few days later, "I can only say that it is intense enough to be sad and premonitory of change or disillusionment." Not as pagan as he had once claimed to Hughes, he could not seize the moment now without trembling at its brevity. Clearly, also, Hughes did not respond with passion. Locke felt himself falling into an "almost suicidal depth of despair and discouragement." When Langston made some annealing gesture that Locke described only as the giving of a "brotherly hand of help and hope," the older man declared his love for "sublimated things." Clearly Hughes had drawn a line, well short of a sexual relationship, beyond which he would not go.

Both men knew that Locke still held one card—his possible usefulness to Hughes in entering Howard University. Barely hidden in the behavior of both men was an element of calculation and strategy. Hughes wanted to get into Howard; Locke wanted Hughes in bed. Evidently they agreed (but without Langston giving in to Locke's proposition) that Hughes would enter the school and live at Locke's home. To some extent, this plan reflected Locke's caring for Hughes; to some extent, it was only bait. At which Langston, innocent and yet wily and self-protective, sniffed with great caution. He agreed to meet Locke in Italy later in the summer. One of the waiters at Le Grand Duc, Romeo Luppi, had invited Hughes to spend August (when the club was closed) with him at his mother's home at Desenzano on the beautiful Lago di Garda in northern Italy. After the damp spring, Langston agreed to go at once.

On August 9, he, Romeo Luppi, and the other waiter, Luigi, pushed their way into a third class carriage on a midnight train. In Turin, they stayed overnight with Luigi's parents. When Langston and Luppi reached Desenzano, the family was at the cinema. Langston's entrance between reels, when the main lights went on, was sensational. The audience had been waiting to welcome the Negro coming home with Romeo—the first Negro ever seen in Desenzano. Rising en masse to salute Romeo and Langston, they were soon pressing close to congratulate Hughes on his exceptional color; "many of them clung to me and shook my hands, while a crowd of young boys and men pulled and pushed until they had me in the midst of them in a wine shop, with a dozen big glasses of wine in front of me." He went to bed very late but very happy.

"And the sun!—Oh! What sun! I'm in love with Italy," he enthused. "I'd like to stay here forever,—almost." The lake was a richer blue than he ever imagined water could be blue, but shifted to gold and purple and grey under the sun. The village sloped down to the lake, where fishing boats unfurled sails of red, orange, and brown. Once again Hughes understood how much, as a child of the sun, he loved people of the sun. The Italians were the essence of friendliness. A ten-year-old named Giacomo, who attached himself to Hughes, reminded him so much of his brother that he sent Gwyn a postcard telling him about the crystal-clear water of Lake Garda, where he bathed each morning, and the sweet watermelons of Desenzano. "The men would take me with them to the village inns near the port and ply me with wine, or to the village dances on the sandy lake shore under the moonlight, or on picnics to the ancient olive groves, where Virgil used to walk." On August 15, under a full moon, the villagers celebrated with a dance until midnight beside the water, where a tipsy Hughes was made to dance with one girl after another. Under the great moon, he went out on the lake for a boat ride before returning to the village with the happy, singing crowd.

In the midst of these innocent pleasures, a challenge arrived from Locke. Two magnificent days, a "trance-like walk" up the Champs Elysées, and "a considerable ramble" in the Bois de Boulogne had stirred in him the urgent need "to tell you how I love you." He hoped for some last, positive intimacy with Langston "before America with her inhibitions closes down on us." Give him that moment, Locke pleaded, and he would concede the future—"and then perhaps through prosaic hours we can keep the gleam of the transcendental thing I believe our friendship was meant to be." Given their plan to have Hughes attend Howard and live with him, however, Locke's letter was a lacquered ultimatum: when next they met, Hughes would have to give in—or risk losing his mentor's support. While Locke may have respected the idea of a "transcendental" friendship, he certainly had no intention of subsidizing one. Previously, Hughes had tried to lower Locke's flame gently: "I like you immensely and certainly we are good 'pals', aren't we? And we shall work together well and produce beautiful things." He suggested a casual tour of nearby towns, but Locke's mind was set on Venice and the consummation of their affair. A somewhat peremptory card, sweeteened a little by a "Lovingly yours, Alain," announced his intention to go to Venice. "I would love to see Venice," Hughes smoothly agreed, "especially with you."

While he waited for Locke to arrive, he went out to the famous Solferino battlefield, where the French and Sardinians had fought the Austrians in 1859. He read *Madame Bovary* and thought that Emma was not only correct to kill herself, but should have done it before. Then, on the last day of August, he left Desenzano after one of the most carefree months of his life.

Keeping his rendezvous with Locke, he spent a night at Verona, where they visited the tomb of Juliet and sent Cullen a postcard. In the late summer mist, Venice was a spectacular vision; the Rialto, the Doge's Palace, and the Bridge of Sighs were far more vivid than he had ever imagined they would be. They rode in gondolas on the canals, and visited the house where Wagner lived when he wrote *Tristan,* and where he died. Joining the throng of summer tourists, they scattered pigeons in the Piazza San Marco; they listened to the municipal band and watched fireworks burst over the water. For a while Hughes also enjoyed the learned murmurings of Locke about Titian and Tintoretto, Caravaggio and Canaletto, this *palazzo* and that *ponte*. Then he grew a little restive; Venice with Locke talking, always talking, was like a surfeit of spumoni. "I began to wonder if there were no back alleys in Venice and no poor people and no slums. . . . So I went off by myself a couple of times and wandered around in sections not stressed in the guide books. And I found that there were plenty of poor people in Venice and plenty of back alleys off canals too dirty to be picturesque."

He also knew now what it might be like to live with Locke. Although Hughes would soon tell him that the five days in Venice were "certainly as pleasant as anything I can remember," Langston's interest in their plan sharply declined. Fate came to his rescue. They took the train together across Northern Italy towards Genoa on the Mediterranean, en route back to France. Travelling separately in third-class, Hughes awoke in the warm, crowded carriage to find that he had been robbed. In spite of having pinned the inside pocket of his jacket, as his grandmother had taught him as a child in Lawrence, his passport and wallet were gone.

At Genoa, they left the train; without a passport Hughes could not enter France. He would wait in Genoa, one of the busiest ports of the world, for a ship going to the United States. Leaving Langston a few lire, Locke hurried on to be sure to catch his own boat back to the United States. Perhaps his ardor, either satisfied or frustrated, had cooled; certainly he made no further attempt to help Hughes. Finding the American consul deaf to his appeals, Langston sought help from The Sailor's Rest, a seamen's aid organization at 15 via Milano, not far from the train station, then moved to the Albergo Popolare, probably part of the Albergo Poveri, the old, huge municipal home for the sick and poor high on a hill with a grand view of the bay. For only three lire a night he had a tiny room to himself, with a restaurant nearby that charged five lire for its best meal. At these prices he could hold out for about two weeks.

Joining the wharf rats on the docks, he hunted a ship going home. Right away he found a little work; painting the side of an American ship called *The City of Eureka,* he earned two dollars and eight cents. Then there was no more work. And he discovered that passage home was by no means assured. The captain of a ship, probably the *West Mahomet,* sent him to the American consul for a note of recommendation ("If you have need of a man for stand-by work the bearer, James Hughes, would like to have something to do"), but the crew, hired in New Orleans, would not accept a black fellow worker. On the beach, Hughes fell in with an English-speaking gang—two white Americans, a Scotsman, an Englishman, another Britisher of some sort, and a black West Indian. Using their wits to keep alive, they tricked a tourist here, stole a loaf there, while watching out for the Fascist police, who loved to crack foreign heads.

When his money was almost gone, Hughes gave up his single room for a bed at two lire a night and ate pasta and then more pasta, innocent of meat, with ripe figs for dessert and cheap red wine. Desperate for money, he sat down in Columbus Park in the city and wrote a little impressionistic prose piece called "Burutu Moon," about a night with the Kru tribesman Tom Pey in Africa. Mailing it to the *Crisis,* he begged to be paid for it, and at once. He sold his alarm clock, but also gave away a shirt to a Puerto Rican who had none. He read a little—some essays by Emerson and Greek plays. One ship, heading for India and China, made room for him. When he would not say that he was a cook or a baker, it left without him. The autumn chill came to Genoa, and he still had heard nothing from the *Crisis.* In dismay, he watched as one white sailor after another found passage out while he was left behind. Subdued, Hughes wrote a poem that, departing from Walt Whitman's celebrated chant, caught his unhappy mood:

> I, too, sing America.
>
> I am the darker brother.
> They send me to eat in the kitchen
> When company comes,
> But I laugh,
> And eat well,
> And grow strong.
>
> Tomorrow,
> I'll sit at the table
> When company comes.
> Nobody'll dare
> Say to me,
> "Eat in the kitchen,"
> Then.
>
> Besides,
> They'll see how beautiful I am
> And be ashamed,—
>
> I, too, am America.

Finally a ship with an all-colored crew came in. The *West Cawthon* had sailed from Philadelphia a month before; for the return voyage, Langston was signed on as a consular passenger to work without pay, chipping and painting his way to New York. At six o'clock on the evening of Sunday, October 5, after almost exactly a month in Genoa, he left port.

In marvelous fall weather the *West Cawthon* visited Livorno and then Naples, then passed achingly close to Capri. Sailing around Sicily, Langston saw Catania, Messina, Palermo, and the little barren pumice peaks of the Lipari Islands. In Spain, when they put in at Valencia, land of the writer Blasco Ibáñez, whose work he had read in Mexico, Hughes was almost stranded again. He caught the last boat back to the *West Cawthon,* but with the captain himself, who promptly denied him all further shore leave. At Alicante, he stole ashore anyway. What could they possibly do to him? Or was Hughes afraid of what awaited him at home?

On the bright, clear, cool morning of November 10, he saw the lighthouse at Sandy Hook, New Jersey, again. Polishing brass all the way to the docks at Brooklyn, he meditated no doubt on his next move; his misadventure on the train had cost him admission to Howard University, at least for the fall term.

Landing with only one coin to his name, earned by ironing a shirt for an officer, he made at once for Countee Cullen's home; over the summer the family had moved from 234 West 131st street to a fourteen-room townhouse at 2190 Seventh Avenue. Welcoming Hughes like a long-lost brother, Countee gave him a letter from his mother, Carrie, whom Cullen had just seen. She was now living in Washington with relatives from the prosperous John Mercer Langston side of the family, who wanted Langston to join them. Another letter, sent via Harold Jackman, was from Jessie Fauset, who had just arrived in Paris for a course of study. Worried about Hughes, she implored him to "fix your eye on some professional goal . . . and reach it."

He had returned to find Harlem apparently on the brink of a new day. In his absence that year, Boni and Liveright, publishers of Toomer's *Cane* (the first black American novel from a white firm in a decade), had brought out Jessie Fauset's first novel, *There Is Confusion,* a more than creditable performance even if she had too self-consciously advertised the exclusive black class to which she proudly belonged. (With mounting fury, Hughes had read a disdainful review of her book by Eric Walrond in *The New Republic*—an "ugly, childish, little review," Hughes fumed to Cullen; in a letter to Harold Jackman, he threatened to punch Walrond in the eye.) Also in 1924, but more critically acclaimed and a commercial success, had come a striking novel of racial violence in the South, Walter White's *The Fire in the Flint,* published by Alfred A. Knopf. The Atlanta-born White, assistant secretary of the NAACP and only thirty-one years old, was blond and blue-eyed but considered "black" according to Southern law and American custom. Both aggressively interested in personal influence and a keen worker for the race, White was already hard at work on another novel, as was Fauset. By 1924, too, Charles S. Johnson, the farsighted, manipulative editor of *Opportunity* magazine, founded by the Urban League the previous year, had made the journal more than a match for the venerable *Crisis* in its service to younger writers. Johnson, thirty-one years old and trained as a social scientist at the University of Chicago, but sensitive to the power of the arts, already had begun a series of brilliant maneuvers (mainly well-advertised literary contests and gala dinners that fed prominent whites and hungry young blacks) designed to bring writers and artists face to face with white publishers and patrons for the first time in American history. In May, while Hughes was washing dishes at Le Grand Duc, Johnson had helped to organize a gala party ostensibly in honor of Jessie Fauset on the appearance of her first novel—but where, to Fauset's barely suppressed but justified fury, given her vastly superior record, Johnson had hailed as the "virtual dean of the movement" his own key adviser in cultural matters: the distinguished Howard University professor, Alain Leroy Locke.

Something else had happened in the past year. With his stream of brilliant poems in the journals, Hughes's reputation among educated Harlemites had soared; only Cullen among the younger poets now rivalled him in acclaim—and mainly because Countee's work was more palatable to the middle class. On the evening of his return, when he attended a benefit cabaret party sponsored by the NAACP at Arthur "Happy" Rhone's nightclub at 143rd Street and Lenox Avenue in Harlem, Hughes received a laureate's welcome. Walter White greeted him effusively at the door, then seated him at his own table across from his wife Gladys, one of the most beautiful women in New York; her striking face and stunning Moorish costume knotted Langston's tongue. Augustus Dill rushed over to embrace him and explain that the *Crisis* had indeed sent the check for "Burutu Moon" to Genoa but had just missed Langston, who could collect the money the next day. Mary White Ovington, one of the white leaders prominent in the founding and running of the NAACP, also came over to welcome him back. Dr. Du Bois was there, and James Weldon Johnson, whose position as secretary of the NAACP only complemented his many other brilliant accomplishments in poetry, fiction, music, and diplomacy. Later, in the middle of the dance floor, Langston also met a tall white man with an oddly shambling physique, who looked down at him with brown eyes set in very delicate pink skin, and a face too solid at the jaw—from which protruded remarkably ugly teeth. Carl Van Vechten was once one of the leading music critics of New York, but more recently a popular novelist in the sophisticated, deliberately facile manner then in vogue, as in his *Peter Whiffle* and *The Tattooed Countess*. With the energetic Walter White (a fellow Knopf author) as his contact, Van Vechten was eagerly meeting many of these blacks for the first time. In the next three years, no one would be more important to Hughes—even if Van Vechten would record in his diary the next morning that he had just met "Kingston Hughes."

The evening was a great success; certainly the entertainment was wonderful. Florence Mills and Alberta Hunter, Noble Sissle, Bill "Bojangles" Robinson, and Fletcher Henderson and his orchestra. Harlem never seemed more stylish and alive to Langston; how could he think of leaving it? A few days later, when Regina Anderson of the library gave a little party for him at an apartment on Edgecombe Avenue she shared with Louella Tucker and Ethel Ray Nance, Charles S. Johnson's secretary, almost all the younger literary crowd and many of the veterans turned out to honor him. "The atmosphere was already heady," one guest remembered, "but the appearance of Langston Hughes, wearing a plaid mackinaw, smiling shyly, galvanized it." Hughes did not seem to understand the powerful impact he had made with certain poems in the *Crisis*. Asked about his latest work, he modestly pulled scraps from his pocket and opened a tattered notebook, then riveted the audience with a masterful reading. The months in Africa and Europe had been to some purpose.

Among those who congratulated him was a young Louisiana-born man, recently arrived from Los Angeles, California, who looked so like Langston that he had created a commotion in the Cullen home by appearing unannounced one day. "Countee! Countee!" Rev. Cullen had called out, "Mr. Hughes is here." A teacher at the Harlem Academy, a small private school run by the Seventh Day Adventists, of which he was a member, Arna Bontemps was determined to become a writer; his dream and his decision to move to New York had been spurred by the appearance of his first published poem in the August *Crisis* that year. Not nine months younger than Langston, Bontemps nevertheless looked up to the much-published, much-travelled Hughes. Bontemps, Cullen, Hughes, and one or two others left the apartment together. Although Langston was shivering in his thin mackinaw, as Bontemps would remember, "we forgot the cold when first one and then another recalled the poem by him that they liked best." Cullen chose, not surprisingly, "Song for a Suicide." Bontemps had missed it, so Hughes promised to send it to him soon.

Within a few years, Bontemps would become Hughes's closest friend for life and his most devoted correspondent. But November also brought losses. Mysteriously, his friendship with Countee Cullen ended during his few days in the city—"one week in New York," he wrote privately a few months later (without naming Cullen), "during which I lose my boyish faith in friendship and learn one of the peculiar prices a friend can ask for favors." Although the exact cause of the break is unknown, certainly Hughes took the initiative in ending the friendship. A baffled and regretful Countee wrote to Locke on November 23 asking about his former friend, now in Washington. "Any news of Langston? If your analysis of his speech is correct 'tis a pity that a sincere and devoted friendship should be knifed merely to dissemble."

Cullen's "dissemble" and Hughes's reference to "peculiar prices" suggest that Countee expected certain responses from Langston that the latter emphatically refused to give, and that Hughes had then denied assertions that Cullen took to be facts. Perhaps Cullen, stirred by reports from Locke of his happy days with Langston in Europe, had pushed for a degree of intimacy with Hughes from which he had recoiled. Would Langston have broken with Cullen over a sexual proposition? Perhaps, but more likely so if Cullen had countered Hughes's refusal by revealing that he knew all about Locke and Langston. Until that moment, Langston would have been unaware of the extent to which Cullen was privy to his business with Locke. About his sexuality Hughes was highly secretive, and would become only more so throughout his life. He would have viewed Cullen's knowledge, and Locke's telling, as a grave violation of his confidence.

Amiable to his soul, Cullen prided himself on quarreling with no one. Langston, too, found it hard all his life even to strain a friendship. However, he snapped this one in two. Then he left Harlem for Washington, D.C.

5

WE HAVE TOMORROW
1924 to 1926

And dawn today
Broad arch above the road we came.
We march!

"Youth," 1924

IN WASHINGTON, Hughes's plan to live with his prosperous relatives in the elite LeDroit Park section of the capital soon collapsed. Having heard for years about the glories of Washington black society, he had expected much: "It must be rich and amusing and fine." By the time he arrived at his cousins' home, however, his mother and Gwyn Clark (Homer Clark was gone again) were already quite uneasy about her decision to accept an offer of hospitality from the descendants of her famous uncle, John Mercer Langston. Soon, Hughes himself was uncomfortable. His shabby clothing, including the plaid mackinaw he wore instead of a winter coat, and his free admissions about jumping ship, dishwashing in a night club, and beachcombing in Genoa, did not sit well with his hosts. "My cousins introduced me as just back from Europe, but they didn't say I came by chipping decks on a freight ship—which seemed to me an essential explanation."

Nevertheless, he met the best people. "The people themselves," he would explain two years later, "assured me that they were the best people,—and they seemed to know. Never before, anywhere, had I seen persons of influence,—men with some money, women with some beauty, teachers with some education,—quite so audibly sure of their own importance and their high places in the community. So many pompous gentlemen never before did I meet. Nor so many ladies with their chests swelled like pouter-pigeons whose mouths uttered formal sentences in frightfully correct English. I admit I was awed by these best people." Soon it was clear that Hughes would never fit comfortably into life on a city block that included a Methodist bishop, a judge, the U.S. Recorder of Deeds, the secretary-treasurer of Howard University, two doctors, four lawyers, and an architect. Introduced to the world of the black bourgeoisie—nowhere more exclusive than in Washington—Langston found it, on the whole, insufferable. The younger blacks were obsessed by money and position, fur coats and flashy cars: "their ideals seemed most Nordic and un-Negro." Lightskinned women coolly snubbed their darker acquaintances; college men boasted of attending "pink" teas graced by only "blue-veined" belles almost indistinguishable from whites. To such people, Hughes's bohemian poverty must have seemed silly; hypersensitive, he was sure he heard them snicker. "They had all the manners and airs of reactionary, ill-bred *nouveaux riches*—except that they were not really rich. Just middle class."

At least at first, Langston tried to live up to the expectations of LeDroit Park. With the educator and civil rights leader Mary Church Terrell, perhaps the most influential black woman in Washington, he discussed a possible appointment as a page at the Library of Congress. When he found himself shunted from one pompous race "leader" to another, however, Hughes gave up the idea and took a job with a black weekly, the Washington *Sentinel*. He cared little for its prestige, he wrote to Harold Jackman—"just so it yields the bucks." The job yielded him few. Assigned not to write stories but to hawk advertising space on commission, Hughes did poorly and soon quit the paper.

His next job, working in a wet-wash laundry, was a slap in the face to the black bourgeoisie. But if the job was unsavory, it also paid $12 a week, which was more than he had ever earned. Ironically, he took it following the appearance, emblazoned over an entire page of the January, 1925 *Crisis*, of his long, Sandburg-like poem, "Song to a Negro Wash-woman":

> Oh, wash-woman,
> Arms elbow-deep in white suds,
> Soul washed clean,
> Clothes washed clean,—
> I have many songs to sing to you
> Could I but find the words . . .

Neither the job nor the poem went over well with his relatives and their friends, most of whom took the *Crisis* and sometimes read it. One day, Hughes looked up from his bags of soiled laundry to see his mother hurrying towards him in tears after some fresh humiliation at home. With young Gwyn, they moved into a shabby little apartment near the laundry, two rooms on the second floor at 1749 S Street.

Coming down with a bad cold, Hughes gave up the steamy laundry, then took a job downtown as a pantryman in an oyster house; the first week he ate so many oysters that he broke out in a painful rash. Finally he found a position that offered better pay, far greater prestige, and—at first—the promise of intellectual stimulation. Langston went to work for the extraordinary Dr. Carter G. Woodson, one of the founders in 1916 of the Association for the Study of Negro Life and History, and editor of its influential *Journal of Negro History*. Woodson, a child of former slaves, had graduated from high school at twenty-two but eventually earned a master's degree from the University of Chicago and, in 1912, a doctorate from Harvard. Operating for many years with virtually no help and at great personal sacrifice, he had edited the journal while also writing and publishing highly respected books on black history and education. In 1921, in an effort to build a first-rate black publishing house, Woodson established the Associated Publishers; grants of $25,000 each from the Carnegie Corporation and the Rockefeller Foundation the following year had allowed him to expand his staff and step up his already remarkable efforts. In 1925, when Hughes joined his staff, Woodson's efforts as author, editor, and publisher were hailed by W. E. B. Du Bois—more a rival than a friend—as "the most striking piece of scientific work for the Negro race" done in the past decade, and "a marvellous accomplishment."

Woodson worked Hughes hard. Langston fired the furnace early in the morning, dusted the furniture, sorted mail, answered some pieces himself, wrapped and mailed books, banked the furnace at night, and, when his employer was away, supervised the entire office. These were his secondary duties. His main job was to help in the preparation of Woodson's gargantuan current project, *Free Negro Heads of Familes in the United States in 1830,* a list of some thirty thousand persons that was scheduled for publication that year. Hughes's task was to arrange all the names in alphabetical order, then to check the list through all the various stages of publication. Fortunately, the 50-year-old Woodson was a fatherly employer who found no task too menial for his own hands; often prickly with other scholars, he was forgiving with his young staff. Once, when he returned suddenly from a trip to find Langston and some other workers crouched in the shipping room playing cards, he fired no one. Instead, he quietly delivered a very effective talk on their responsibilities to the race and to themselves.

Woodson's work was clearly valuable, and the pay good, but Hughes was not happy. "Although I realized what a fine contribution Dr. Woodson was making to the Negro people and to America," he would remember, "I personally did not like the work I had to do. Besides, it hurt my eyes." Clearly he was not ready to settle down; after the months in Paris and Africa, and at Jones Point and on the Staten Island farm, Woodson's musty office and demanding schedule must have loomed like a prison. And the city of Washington, beautiful and heavily populated with blacks but grimly segregated, was actually an affront. Unlike in Manhattan, legitimate theatres refused to sell tickets to Negroes; no restaurant downtown would serve them even a cup of coffee. Against these indignities the high-toned black middle class rarely protested. Visitors were more likely to be offended, as when in May of that year the musicologist Nathaniel Dett pulled his choir off the stage of the Washington Auditorium rather than perform there with several hundred blacks trapped in the balconies at an "All-American Festival of Music."

For a while, too, Hughes had few friends. Of Alain Locke he saw little; at Howard University, Locke was deeply involved in a bitter student strike and other wide-ranging disputes with the administration that would result in his suddenly being fired in June. Langston knew that he would have to get into Howard on his own. In any case, he kept his distance; Locke believed (so he wrote to Claude McKay) that Hughes had been warned against him by "busybodies." But, cannily, Langston made sure that Locke had almost a complete file of his poems for the special Negro number of the *Survey Graphic* his friend was editing. And, occasionally stopping by Locke's apartment at 1326 R Street, NW, at least twice he left teasing notes: "10 p.m.—You are out on such a snowy night! Or do you think I'm a robber and won't let me in?" And: "Not even a rainy night keeps you home. / Fair or foul though the weather be—you roam."

Isolated, unable to stomach Washington's high life, Langston sought out its low. Lowest was black Seventh Street, what Jean Toomer in *Cane* had boldly called a "crude-boned, soft-skinned wedge of nigger life." There the plainest, blackest folk in the city lived, sang the blues, shot pool and one another, guffawed and hollered out tall tales, and devoured watermelon and greasy barbecue without apology. On grimy Seventh Street, Hughes would write, poor blacks "looked at the dome of the Capitol and laughed out loud." This was a major step in Langston's knowledge of the masses. Older and more experienced as a writer than in his questing excursions into Harlem as a student from Columbia, he quickly recognized that these blacks, closer to the South, were far more the real thing than their striving cousins in New York. At the store-front churches Hughes listened raptly to sermons and hymns; elsewhere he hunted low-down blues and jazz. Trying to capture their spirit, Hughes himself sang openly until one day, as he crossed Rock Creek Bridge, a man rushed over.

> "Son, what's the matter? Are you ill?"
> "No," I said. "Just singing."
> "I thought you were groaning," he commented. "Sorry!"

"After that," Hughes admitted, "I never sang my verses aloud in the street any more." However, his respect for the power of black music deepened among the Washington poor. "Like the waves of the sea coming one after another," he would write, thinking of Seventh Street, "like the earth moving around the sun, night, day—night, day—night, day—forever, so is the undertow of black music with its rhythm that never betrays you, its strength like the beat of the human heart, its humor, and its rooted power."

"I tried to write poems like the songs they sang on Seventh Street—gay songs, because you had to be gay or die; sad songs, because you couldn't help being sad sometimes. But gay or sad, you kept on living and you kept on going." Without having become much more confident about his ability to judge the quality of his new poems, Hughes nevertheless persisted at his craft. "I always put them away new for several weeks in a bottom drawer," he would recall. "Then I would take them out and re-read them. If they seemed bad, I would throw them away." In a year in which he consolidated his position as a poet rather than broke important new ground, he also confirmed his basic criterion as an artist. At the center was a vigilance about the need to find new ways, based on a steadfast loyalty to the forms of black culture, to express black consciousness—and, in so doing, to assist at its passage into the hostile modern world. But poetic modernism as defined by elitism, hyper-intellectualism, and a privacy of language was not for Hughes. Even as he wrote in the monumental shadow of *The Waste Land,* published only three years before by another Missouri-born poet, but one in flight from his native mud toward the higher ground of self-proclaimed classical, Anglo-Catholic, and royalist values, Hughes was unable to identify his poetical self with the exigencies of post-war European civilization, or to internalize its tragic sense of impotence and decay, of a fatally fragmented rather than a still unified consciousness. Liberal and generous in his spirit, and well aware of the special demands of the modern world, Hughes nevertheless could only be highly selective in appealing to the therapy of tradition when tradition was defined to exclude the only audience—the masses of blacks—that finally mattered to him. Out of his imagination he had fashioned an aesthetic to suit, above all, their needs, not to amuse their masters. For all his air of ignorance and innocence, Hughes knew more or less exactly what he was doing, and why he was doing it. If T. S. Eliot wrote in tongues, so occasionally did Langston Hughes—but to a radically different purpose. In "To A Negro Jazz Band in a Parisian Cabaret," he urges the black musicians to "Play that thing" for the white lords and ladies, "whores and gigolos," and "the school teachers out on a spree. Play it!":

> May I?
> *Mais oui,*
> *Mein Gott!*
> *Parece una rumba.*
> *¡Que rumba!*
> Play it, jazz band!
> You've got seven languages to speak in
> And then some.
> Can I?
> Sure.

In his fourteen months in Washington, Langston published more poems than he had ever done before, and in his now familiar, often paradoxical range from leftist radicalism (sharpened by his months in colonial Africa) to the old nihilistic gloom where his muse languished until reprieved by conscience. The February number of the socialist *Messenger* carried four of his pieces, including "To Certain Intellectuals" ("You are no friend of mine / For I am poor, / Black . . .") and "Steel Mills," from his high school days. In March, in April (on the same page with an article by Joseph Stalin), and in July, he published six equally biting poems in Earl Browder's communist *Workers Monthly* (into which the *Liberator* had been subsumed). One of Hughes's offerings was "Poem to a Dead Soldier":

> Ice-cold passion
> And a bitter breath
> Adorned the bed
> Of Youth and Death—
> Youth, the young soldier
> Who went to the wars
> And embraced white Death,
> The vilest of whores . . .

On the other hand, three poems in the May *Buccaneer,* a Dallas, Texas, poetry magazine, reflected a far more private, almost existential anguish. One was on suicide (''Death comes like a mother / To hold you in her arms . . .''). Another was ''Prayer'':

> I ask you this:
> Which way to go?
> I ask you this:
> Which sin to bear?
> Which crown to put
> Upon my hair?
> I do not know,
> Lord God,
> I do not know.

Most of his poems, however, fell between the poles of radical aggression and existential angst. These pieces found a home most easily in the race-conscious but essentially moderate *Opportunity* and *Crisis*. In the first he published ''Troubled Woman'':

> She stands
> In the quiet darkness,
> This troubled woman,
> Bowed by weariness and pain,
> Like an
> Autumn flower
> In the frozen rain.
> Like a
> Wind-blown autumn flower
> That never lifts its head
> Again.

Still his major outlet, with Jessie Fauset as literary editor, *Crisis* published six of his poems in December. In at least one, ''Summer Night,'' Hughes sounded a gloomy tone, lamenting the loneliness of

> My soul
> Empty as the silence,
> Empty with a vague,
> Aching emptiness
> Desiring,
> Needing someone,
> Something . . .

But more significant—more political—was the ballad ''Cross'' (''My old man's a white old man / And my old mother's black . . .''), in which Hughes first explored a theme he would exploit for the rest of his life. Spoken by a once bitter, now repentant but still confused young mulatto (''I wonder where I'm gonna die, / Being neither white nor black?'') the poem would reduce audiences, both black and white, to an uneasy silence for decades to come. From this seed of a poem would come in later years a major short story, a long-running Broadway play, and an acclaimed opera.

Writing steadily, Hughes was nevertheless relieved when March brought both the first spring blossoms to Washington and a wave of personal good fortune that began with the appearance of Locke's special number of the *Survey Graphic,* "Harlem: Mecca of the New Negro," with several superb portraits of Harlem blacks by the Bavaria-born artist Winold Reiss. The special number proved to be such a sensation that the *Survey Graphic* sold twice its usual number of copies that month. In two eloquent essays, Locke compared the cultural significance of Harlem to the role of Dublin and Prague in modern Irish and Czech culture. Until now, he argued, blacks were linked more by "a common condition rather than a common consciousness; a problem in common rather than a life in common." But in Harlem, "Negro life is seizing upon its first chances for group expression and self-determination." Reviving a term first used at the turn of the century, Locke heralded the arrival on the American scene of "the New Negro," a generation of young men and women "vibrant with a new psychology" brilliantly expressed by one of their best poets, Langston Hughes:

> We have tomorrow
> Bright before us
> Like a flame.
>
> Yesterday, a night-gone thing
> A sun-down name.
>
> And dawn today
> Broad arch above the road we came.
> We march!

Deft prose pieces by Dr. Du Bois, Albert C. Barnes, the anthropologist Melville Herskovits, Charles S. Johnson, Kelly Miller of Howard University, Walter White, James Weldon Johnson, the black bibliophile Arthur Schomburg, and the inventive race historian J. A. Rogers were set off by a rich selection of poetry. Cullen and McKay were both handsomely represented, but Hughes also shone with ten poems on five different pages. The special number was a triumph, although McKay and Toomer would be enraged by Locke's deceptions as an editor (he "tricked and misused me," Toomer fumed; for McKay, Locke destroyed "every vestige of intellectual and fraternal understanding" between them). Its appearance later that year—in much revised form—as a book, *The New Negro,* from Albert and Charles Boni, seemed not only to certify the existence of a great awakening in black America but also to endow it with a Bible.

The stir created by the special number pleased Hughes, but it also led to an incident that further embittered him against bourgeois blacks when "The Literature Lovers of Washington, D.C." (Locke was chairman, but the group comprised mainly women) organized a dinner for March 3 at the Phillis Wheatley YWCA in honor of the younger writers. When Hughes let the group know that he did not own formal dinner wear, an exception was made for him. Then one of the organizers informed his mother, also invited, that while Langston would be acceptable out of formal attire, she would not be excused; since presumably she had nothing suitable to wear, she should stay home. Carrie was devastated; the insult so enraged Hughes that he wrote Jessie Fauset and Claude

McKay, both in France, breathing fire against the black middle class. When he refused to set foot in the hall, Countee Cullen, who loved to dress up, was brought from New York to eat chicken *à la king* and to star at the banquet, with Locke presiding.

Fortunately, not all of the Washington middle class was as benighted as the "Literature Lovers." His *Survey Graphic* showing led Hughes to 1461 S Street, NW, the home of the eccentric but lovable and maternal Georgia Douglas Johnson. Well known in the community as the author of two books of verse, Mrs. Johnson (she had two grown sons) served cake and wine every Saturday evening to a growing circle of talented young writers and older visitors. At her home Hughes met Rudolph "Bud" Fisher, a Phi Beta Kappa graduate of Brown University and an intern in medicine at Howard who by May had two stories accepted by the *Atlantic Monthly;* the playwright Willis Richardson, toiling in a clerkship in the Civil Service; Lewis Alexander, a promising actor and writer; the young poet Clarissa Scott, whom Langston had met in New York; and playwright Marita Bonner, Dutton Ferguson, Hallie Queen, and John P. Davis, who would soon enter the Harvard Law School. Among the older visitors were Alain Locke himself, whose verbal jousts with the quick-witted Fisher dazzled Hughes, and the seasoned poet Angelina Grimké, educated at the famous Boston Latin High School, and for many years an English teacher at Dunbar High School in Washington, where Jessie Fauset had also taught.

But Hughes's favorite among the regulars was probably Richard Bruce Nugent, a willowy, intelligent, but rebellious nineteen-year-old of Italianate coloring and good looks, who hoped vaguely to make a career of painting and acting. Meanwhile, Nugent aimed mainly to shock. From the black upper class (one distant relative may have been Blanche K. Bruce, a United States senator during Reconstruction and later register of the treasury), Nugent nevertheless sometimes dispensed with wearing socks in public, sometimes with shoes; only a few years later, when he had more fully developed his personality, he would be remembered as "a soft young fellow with a purr, like a cat's, and a little gold bead in one ear." In a culture that shrank from the mention of homosexuality, Nugent at nineteen was already openly gay, a fact that cost him many friendships—but not one with Hughes. Definitely not sexual, according to Nugent, their relationship was yet warm and spontaneous. "We struck up a really extraordinary friendship," he recalled. "We couldn't leave each other the evening I first met him. I mean, he would walk me home and then I would walk him home and then he would walk me home and I'd walk him home and it went on all night." Although only four years older, Hughes quickly became Nugent's idol. "He was a made-to-order Hero for me. . . . He had done everything—all the things young men dream of but never quite get done—worked on ships, gone to exotic places . . . written poetry that had appeared in print—everything. I suppose his looks contributed to the glamorous ideal, too . . . as did his voice and gentle manner."

By April, the main topic in Mrs. Johnson's salon was almost certainly the literary contest, replete with prestigious judges and prizes of money in a variety of categories, shrewdly organized by Charles S. Johnson of *Opportunity* magazine, with the results to be announced at a banquet even more impressive than his "coming-out" party for the young writers at the Civic Club a year before, while Langston had been in Paris. Just about every one of Georgia Douglas Johnson's "children" had submitted entries, and several intended to travel to New York for the dinner. At Charles Johnson's personal behest, Hughes had sent in several pieces—including his major unpublished poem, "The Weary Blues," begun in March, 1923, at Jones Point but studiously withheld since then. Auspiciously, Johnson had accepted it as "a rather unique contrast" to the usual Hughes style.

Anxious for relief from Washington, Langston could hardly wait to leave. Then he was knocked flat on his back by an attack of malaria; Washington, he wailed, was "worse than the coasts of Africa." With his "brother" Gwyn also ill, and his mother in need of money, Hughes could hardly justify spending money to travel to New York on the chance that he might win a prize. But Jessie Fauset, just back from Europe, insisted that he attend the banquet. She would lend him the train fare, save him a seat at her table with Regina Anderson, and, as lagniappe, take him to Winold Reiss's studio. Fauset was certain that Hughes's hour was at hand. "Your ship," she assured him, "will be coming in soon."

His ship was so long overdue, Hughes made sure he met it. On May 1, still wan from his illness but spruced up in a dark vested suit, he presented himself at the Fifth Avenue restaurant near 24th Street in downtown Manhattan to join the greatest gathering of black and white literati ever assembled in one room. Representatives had come from every publishing house sympathetic to the cause, including Alfred A. Knopf, Boni and Liveright, Macmillan, Harper's, and Harcourt, Brace. Fannie Hurst was there, and Dorothy Canfield Fisher, Zona Gale, Van Wyck Brooks, and various Van Dorens. Among the judges of the more than 730 entries were Eugene O'Neill, John Farrar, Witter Bynner, Alexander Woollcott, and Robert Benchley. Virtually all the younger black writers, shiny-faced with anticipation, were present, including Hughes's former friend and his main poetical arch-rival Countee Cullen, who had been publishing with such dazzling authority that in the previous November alone he had placed poems in four different white journals. The third prize in poetry went jointly to Cullen and Hughes. The second prize went to Cullen, for the ineptly titled "To One Who Said Me Nay." Then, to Hughes's profound satisfaction, the first prize was his—for "The Weary Blues," which James Weldon Johnson orated magnificently to the banquet. Crowned in glory as the best poet, Langston mingled afterwards with the other prize winners, including the young sociologist E. Franklin Frazier, the promising poet Frank Horne, Eric Walrond, and an earthy, conspicuously intelligent young woman from Florida and Howard University in Washington, drawn to New York by the growing cultural movement and the chance to attend Barnard College, where she would enroll in the fall. Witty and flamboyant, Zora Neale Hurston impressed Hughes at once. She "is a clever girl, isn't she?" he soon wrote a friend; "I would like to know her." By the following year, he would know her very well, as he would another recent migrant present at the dinner, who had just been driven to New York by his ambition as an artist—Aaron Douglas, a painter from Topeka, Kansas, where Hughes had lived as a child.

Also in the crush to congratulate Hughes was Carl Van Vechten, the tall, shambling white man with ugly buck teeth whom he had met on the night of his return from Europe. Eric Walrond, Van Vechten, and a white friend, Rita Romilly, were going out for a night on the town—as soon as Carl's wife, the popular, Russian-born actress Fania Marinoff, who didn't care much for Van Vechten's hard drinking and Harlem club-crawling, was safely home. The Harlem YMCA was sponsoring a benefit dance at the Manhattan Casino; then some of the performers from the glamorous Cotton Club, which admitted very few blacks, would be performing at the more egalitarian Bamville. Would Hughes care to join them? Accepting the invitation, Hughes met Van Vechten, Walrond, and Romilly that evening at the Manhattan Casino. Before the last round of drinks, Van Vechten had invited Langston to visit him at home the next day—and to bring along his manuscript of poems.

Since their first meeting on November 10 the previous year, Van Vechten had plunged more deeply into black society than probably any white man had ever done in New York. If some blacks, including Du Bois, Jessie Fauset, and Cullen, found his easy manner subtly patronizing, to others Van Vechten was a welcome pioneer bravely flinging open doors of opportunity (Claude McKay, for example, was eager to meet a white man who bothered to be subtle in his patronizing). Already his friendship with James Weldon Johnson and with the gregarious Walter White had deepened; he had also become a confidant of the stirring young actor and singer Paul Robeson, a Rutgers University and Columbia Law School graduate and the star of Eugene O'Neill's *All God's Chillun Got Wings* and O'Neill's revival of *The Emperor Jones* ("The Negro Speaks of Rivers" had graced the program cover). Van Vechten had also captivated the wealthiest young black woman in Harlem and its boisterous but sensitive joy-goddess, A'Lelia Walker, the free-spending heiress to a fortune accumulated through a hair straightening process developed by her mother, Madame C. J. Walker. Shattering white custom, Van Vechten invited these and other blacks to his home at parties often so integrated that Harlem columnists had begun to report them routinely, as if the Van Vechtens were nothing but blue-veined Negroes with lots of money, who happened to live downtown.

An odd mixture of the frivolous and the farsighted, of decadence and intellectual high seriousness, Van Vechten was an iconoclast who had long championed the new in literature, music, and dance; he was a close friend of Gertrude Stein's (and would become her literary executor), had attended the second performance in Paris of Stravinsky's *Le Sacre du Printemps,* and had been the first dance critic in the United States. In fashion, too, he spurned tradition with his soft ruffled shirts and elegant but increasingly casual wear; he was apparently the first man in America daring enough to appear in public with a watch on his wrist. In his private life, he concealed little and denied himself less. His intaglio ring, set in gold, showed Leda entertaining her swan; Van Vechten was unashamed of his nature, which was largely homosexual. To some people he seemed obsessively frivolous. "With him," a close friend would explain, " 'amusing' things were essential things; whimsicality was the note they must sound to have significance. Life was perceived to be a fastidious circus, and strange conjunctions were more prized than the ordinary relationships rooted in eternity." The aesthetic butterfly, however, had also been an early supporter of the NAACP; hardly a Bolshevik, he had nevertheless once vacationed in Italy with John Reed. His father, who had contributed to the founding of a school for blacks in Mississippi, had taught Van Vechten to respect Negroes (two, apparently, actually lived in Cedar Rapids, Iowa, while he was growing up there). Now he had discovered that blacks deserved not so much his respect as his admiration. To some extent, the revelation had come home early the previous year, 1924, at Paul Whiteman's celebrated concert, "An Experiment in Modern Music," at Aeolian Hall on West 43rd Street in Manhattan; as a leading music critic, Van Vechten had been prominent in an audience that included Stravinsky, Rachmaninoff, Virgil Thomson, and Jascha Heifetz for the premiere of young George Gershwin's *Rhapsody in Blue*. Proclaiming Whiteman's band to be greater than any American symphony orchestra, Van Vechten had called jazz the only indigenously American music of true distinction. Then he had set about learning as much as he could of its black roots, as opposed to its white branches. With Walter White, a fellow Knopf author, as his guide, he had begun his digging in Harlem.

At five o'clock on May 2, the day after the *Opportunity* dinner, Hughes arrived at Van Vechten's home at 150 West 55th Street. The apartment was impressive—rich drapes and antique Persian rugs, gorgeous malachite tables and statuettes and silver, fine china and paintings and other pictures, including some of Van Vechten's own amateur photographs (an early picture, taken in 1896, had been of two black children in front of Harriet Beecher Stowe's home in Cincinnati). At his host's invitation, Hughes read several of his poems; then they talked about his passion for the blues, and what he hoped to do with this music in his poems. An urban sophisticate, Van Vechten nevertheless believed in the power of the folk tradition (he himself had written at least one essay on Iowa folk songs, such as they were). Vehemently against black imitation of hackneyed white forms, he wanted to learn as much as possible about the blues. Above all, Van Vechten claimed, he wanted to help Hughes. Specifically, he asked Langston (they were on a first name basis from the start) to leave his manuscript overnight.

Hughes did so apparently without hesitation. Unlike Cullen, who had politely shunned Van Vechten's overtures, Langston harbored only a slight suspicion of friendly whites like Van Vechten or of admirers in general. In spite of his loneliness and poverty and his occasional claims of stupidity, Hughes fully expected to be admired; because he had lived more among whites than blacks all his life, he was sure that he could read them with reasonable accuracy. (He had also grown generally in confidence; a photograph taken that weekend on the roof at 580 St. Nicholas Avenue, where Regina Anderson, Ethel Ray Nance, and Louella Tucker shared an apartment, caught Langston to the front of a line of gifted men—he was also the shortest—his hands easy in his pockets, smiling at posterity with elegant assurance.) He left the poems behind. Then, at three the next day, Sunday, he dropped in again at Van Vechten's home for another hour before leaving to catch a train to Washington. Returning the manuscript with profuse compliments, "Carlo" suggested several changes; when Langston had made his corrections, Van Vechten promised, he would try to find him a publisher. Hughes should call his book "The Weary Blues."

Barely half a week passed before Langston returned home from his chores at Dr. Woodson's office to find a heavy vellum envelope awaiting him at 1749 S Street. That Van Vechten had written first so touched Hughes that he actually mentioned it in his reply. Within two or three days another thick envelope was in his mailbox. Where was the collection of poems? "I shall do my best to get it published," Van Vechten vouched, "and that should be easy because it is a beautiful book." Throwing himself now into revising the manuscript, Hughes cut thirty poems at one sitting; on May 10, when he mailed it to New York, he gave Van Vechten permission to cut even more. The new version reached Van Vechten on the morning of May 13; before noon he sent Hughes a note of encouragement: "Your work has such a subtle sensitiveness that it improves with every reading." As it happened, he was to lunch with Alfred A. Knopf himself that day. "I shall ask him to publish them and if he doesn't some one else will." Would Langston permit him to write an introduction?

Clearly moved, and returning grace for grace, Hughes danced almost to the brink of flattery. He was reading Van Vechten's novel *The Blind Bow-boy:* "It is amusing! And beautiful, very beautiful! And not a dull page in it. Yesterday and last night the rain poured down, but with your book and cigarettes and bed I spent the most delightful evening I've had for a long time. What next of yours shall I read?" Hughes made it clear that he understood the sterling compliment Van Vechten was paying in pressing the manuscript on his own house, Knopf ("Such a fine publisher!"), and in offering to write an introduction. "I would be very, very much pleased if you would do an introduction to my poems."

At that time, Van Vechten was one of the three or four most important authors at the young firm; in less than a week he had a decision in a telephone call from Knopf. With a typically dramatic flourish, he immediately sent Hughes a telegram: "LITTLE DAVID PLAY ON YOUR HARP." To which Hughes, who himself loved a touch of drama, replied: "THANKS IMMENSELY THE SILVER TRUMPETS ARE BLOWING." On May 18, Blanche Knopf, as active in the firm as her husband Alfred, wrote to their latest author. "Mr. Van Vechten has sent us your manuscript and we like it very much," she advised. "It is very delightful verse and I am glad to tell you that we want to publish it." Stuffed in the envelope were copies of a contract. Only eighteen days had passed since the *Opportunity* dinner.

"How quickly it's all been done! Bravo to you!" Hughes marvelled to Van Vechten. "You're my good angel! How shall I thank you?" But there was more—at Van Vechten's behest, the acclaimed Mexican caricaturist Miguel Covarrubias would design the cover; Carl himself would supervise the manuscript through the press. "As the old folks say," Hughes exulted, "I'll have to walk sideways to keep from flying! All this is very fine of you." He walked sideways all through the rest of May as Van Vechten, a master of publicity, began to pave the way for *The Weary Blues*. At his request, Margaret Case of the glamorous *Vanity Fair* promised to read twenty to thirty poems by Hughes in order to make a selection. "Never in my wildest dreams had I imagined a chance at Vanity Fair," Langston admitted.

What could he possibly do for Van Vechten in return? Well, there *was* something; Langston could help Carl prepare a series of articles for *Vanity Fair* on black subjects, including one about the blues, the form that had supplanted ragtime and now undergirded modern jazz, but was yet virtually unknown to Northern whites. Hurrying to oblige, Hughes brushed aside the fact that Van Vechten, a white man, most likely would make a great deal of money from essays on a black subject about which, by his own admission, he knew almost nothing. Typing up some verses he remembered from his childhood—"Freight Train Blues," "Reckless Blues," and such—he also sent a blues his messmate George Green of Kentucky had sung on the trip to Africa, about going to a gypsy to have his fortune told ("Gypsy done told me / Goddam your un-hard-lucky soul!"). Instinctively, Hughes also seized the chance to give Van Vechten a further glimpse of his own stylistic and intellectual range:

> I know very little to tell you about the Blues. They always impressed me as being very sad, sadder even than the spirituals because their sadness is not softened with tears but hardened with laughter, the absurd, incongruous laughter of a sadness without even a god to appeal to. In the Gulf Coast Blues one can feel the cold northern snows, the memory of the melancholy mists of the Louisiana low-lands, the shack that is home, the worthless lovers with hands full of gimme, mouths full of much oblige, the eternal unsatisfied longings.
>
> There seems to be a monotonous melancholy, an animal sadness running through all Negro jazz that is almost terrible at times . . .

With this swift stroke, Hughes altered the relationship between Van Vechten and himself. "What you write me about the Blues," came the response, "is so extraordinary that I want to incorporate it into my article—with due credit to Langston Hughes." "Your sense of character is extremely picturesque," Van Vechten went on; "you describe accurately—and *with your own eyes*." There was enough material in the back of Hughes's head for Langston "to write forty books—once you get started." Having shown his force to Van Vechten, however, Hughes deliberately allowed it to subside when he sent him seven poems for *Vanity Fair*. "In six of them I've tried to catch the jazz-rhythm and something of the jazz spirit," he explained. "Have I been at all successful?"

When news of Langston's success with Knopf swept through the tight little Harlem literary circle no one was more pleased than his discoverer and champion, Jessie Fauset: "Undoubtedly, the greatest kind of a future is stretching before you." Walter White, always eager to be involved, wired congratulations and an offer of help. Countee Cullen, whose own first book of poems was expected from Harper's shortly, generously but tersely offered ("Please don't think this an attempt to foist an unwelcome correspondence upon you") to write a review of Hughes's book. Taking publicity photographs, Hughes made sure to send one to his old Auntie Reed in Kansas, now Mrs. Mary Campbell following the death of her husband James. She loved the picture: "You have grown to be such a fine looking young man." Another went to the violinist Hall Johnson, who was busy setting "Mother to Son" to music (Langston was "so *sincere!*"). With the world singing his praises, Hughes moved to turn the hosannas into a scholarship, preferably to Howard University. Here, however, he was disappointed. Dean Kelly Miller reminisced touchingly about John Mercer Langston's historic contributions to the university before telling Hughes, incredibly, that there was no scholarship for him. Still Langston hoped that some money would soon come his way. On June 30, Howard formally granted him permission to register in the fall, with credit allowed for his year at Columbia.

In New York, honoring his promises of help, Van Vechten found just the right binding paper for *The Weary Blues:* "We want to make it as beautiful a book as possible." To help Van Vechten with his introduction, Hughes sent a brief essay, "L'histoire de ma vie," so dazzling that it set off a brainstorm in Van Vechten. Yes, Blanche Knopf and he completely agreed: Langston should do an autobiography! "Treat it romantically if you will, be as formless as you please, disregard chronology if you desire . . . but however you do it, I am certain not only that you can write a beautiful book, but also one that will *sell.*" He pressed Langston: "What I want you to do, therefore, is to *write this book.* . . . I know it is hard to write a book with all the other things you have to do, but *I am sure you can do it.*"

Hughes did not like the idea. He did not look forward to writing a long book of prose; he also did not look forward to stirring the demons of parental rejection and racial ambivalence lurking in his past. "I hate to think backwards," he explained to Van Vechten. "It isn't amusing. . . . I am still too much enmeshed in the affects of my young life to write clearly about it. I haven't escaped into serenity and grown old yet. I wish I could. What moron ever wrote those lines about 'carry me back to the scenes of my childhood'?" Still, mainly to please Van Vechten, he promised to try to write an autobiography. If Van Vechten was trying to shape his sense of professionalism as a writer, Langston was an almost totally compliant pupil. The two men had something in common. Neither saw any conflict between popularity and the quest for literary greatness; the idea of deliberately limiting one's audience, or defying it, was equally foreign to both. Their reasons, however, were different. Van Vechten wished to remain wealthy and to amuse many people; Hughes wanted, above all, to speak to the black masses. Because of this populist and yet morally esoteric goal, he probably saw himself from the start as the greater artist. Only obliquely would he ever signal this idea, as when he ventured to say that their taste in blues differed—"you like best the lighter ones . . . and I prefer the moanin' ones." Flinching from this veiled accusation of shallowness, Van Vechten stoutly denied the charge.

As the hot, humid Washington summer came on, and the first excitement about the Knopf decision passed, Hughes became bored and restless. Fortunately, August also brought another chance to go to New York. To match *Opportunity,* the *Crisis* had announced its own literary competition, also to be capped by a banquet; long before the *Opportunity* dinner, in fact, Amy Spingarn, the wealthy wife of Joel Elias Spingarn of the NAACP leadership, had donated a gift of $300 for this purpose. As a lure to Hughes, Du Bois himself informed him in advance that he had won two prizes, totalling $40—a third place in poetry, and a second place among essayists for an autobiographical piece called "The Fascination of Cities" (no doubt some loose leaves from his autobiography-in-progress). Just before quitting Washington, he received a draft of Van Vechten's proposed introduction to his book. "Humming-birds and bright flowers in a marsh—that's the way your introduction strikes me," Hughes gushed. "Full of color and different, I've seen no other introduction like it." Then he listed several corrections and pressed for other changes to underscore his social concern. In addition, Van Vechten should mention Georgia Douglas Johnson, who had "some really good poems and she is our best known woman poet."

On the evening of Friday, August 14, he joined Bruce Nugent, Georgia Douglas Johnson, and several others from Washington for the *Crisis* awards dinner at the Renaissance Casino at 138th Street and Seventh Avenue in Harlem. (Seeing Countee Cullen, decked out in formal wear, for the first time, Nugent irrepressibly nudged Hughes: "Cullen looks like one of the three pigs in swallowtail"; the oddly shaped Van Vechten looked "exactly like a white monkey.") Countee, however, took the first prize in poetry. Langston's relatively weak showing was offset handsomely by his prominent place in the September *Vanity Fair,* already selling briskly in Harlem; he autographed at least half a dozen copies before leaving New York. Beneath a two-paragraph introduction by Van Vechten were four poems—"Cabaret," reprinted from the *Crisis;* "To Midnight Nan at Leroy's" ("Strut and wiggle. / Shameless gal . . ."); "Fantasy in Purple" ("Beat the drums of tragedy for me. / Beat the drums of tragedy and death . . ."); and "Suicide's Note":

> The calm,
> Cool face of the river
> Asked me for a kiss.

Cullen's *Crisis* triumph and Hughes's *Vanity Fair* coup brought the men no closer to a reconciliation. Disapproving of Langston's close link to Van Vechten ("I know Carl is coining money out of the niggers") and detesting the title "The Weary Blues" as catering to whites "who want us to do only Negro things," Cullen also found Hughes's search for a black college disconcerting; he himself had just been admitted to graduate school—at Harvard. But Hughes, undeterred by what people might be saying, spent the evening after the banquet at Van Vechten's, arriving for dinner at six o'clock, welcoming Bruce Nugent there at nine. Van Vechten, for his part, had much to tell. Excited by Harlem as a possible source of fictional material, he had plunged into research for a novel of black life. The day before, a sensational title had flashed on him—"Nigger Heaven."

Cramming as much as he could into this visit to Manhattan, Hughes dined with the fine black singer and actress Abbie Mitchell, wife of the musician Will Marion Cook and mother of Mercer Cook, just graduated from Amherst College. With the tall, homely, but very amusing Hall Johnson, Hughes dropped in at Leroy's and then at Small's Paradise, two of the liveliest Harlem nightclubs; he also heard Marguerite Avery ("a real African type," he noted approvingly to Van Vechten) sing Johnson's setting of his "Mother to Son"—a thrilling moment, the first time he had listened to one of his texts being sung. After calling on Jessie Fauset at home, he attended a performance of Stravinsky's *The Fire-Bird.* On Sugar Hill in Harlem, in the apartment of Walter and Gladys White at 409 Edgecombe Avenue, the most prestigious address in Harlem, their lovely little daughter Jane sang songs for him. Very pleased with his visit, he rode the train back to Washington reading Thomas Mann's homosexual tragedy, *Death in Venice,* thinking no doubt now and then of his own drama there the previous year with Alain Locke.

He returned with a determination, revived by Van Vechten, to work on his autobiography, which he now called "Scarlet Flowers: The Autobiography of a Young Poet," because the years of his life, "like the petals of a bright flower, ha[ve] been scattered a hundred places by the wind." ("Scarlet Flowers" sounded like Louisa May Alcott, Walter White sniffed.) From the outset, however, the project had not gone well. A few gorgeous pages about episodes in his life were easy to write; the exposure of his psyche was quite another thing. With a poetic persona far bolder, more sensual and passionate and politically radical than he himself ever felt consistently, Hughes could not risk cracking a mask he had so rigorously constructed. The story of his original melancholy and sense of rejection, of his deep ambivalences and his transcendence of emptiness through art, was the last tale he wished to tell the world; he had not yet acquired the nerve (which he would master) to try to cross the perilous deep of autobiography by swimming nonchalantly on its surface. Although Van Vechten had urged him to lie or steal material (in other words, to be an artist in writing the book), Hughes's main problem lay not in the quantity of his raw material but in his closeness to it. Van Vechten had urged him to try again, but Langston's heart was elsewhere. "Because you want me to write my autobiography, I am going to enjoy doing it," he promised. "Otherwise, I wouldn't." He would resign his eye-straining job with Dr. Woodson, take a room somewhere, buy a typewriter, "and see what devoting one's life to one's art is like."

The typewriter came; he also gave up his job and took a room at the YMCA on 12th Street. Hughes was at the Y when Arna Bontemps came to Washington from Manhattan for a visit, further consolidating their friendship. Hughes continued to see a great deal of Bruce Nugent, and one day on a bus, after a young man recognized him from his picture in a black newpaper, he found another good friend—Waring Cuney, a poet and musician enrolled at the oldest black university in the country, Lincoln University in Pennsylvania. Cuney, Nugent, and Hughes quickly became almost inseparable. Speaking Spanish to pass as foreigners (or rather Hughes speaking Spanish while Nugent and Waring babbled and waved their hands) they attended the segregated theatres downtown; on hot afternoons, they appalled onlookers by strolling barefoot on 14th Street.

When Jessie Fauset came to Washington for a brief visit Langston shaped up but Nugent only increased what Miss Fauset called his "rather too deliberate eccentricities." She nagged Hughes: "Can't you straighten him out in this?"

The autobiography languished. Then the YMCA became too much for him after the police came one day and took away his roommate and the man's superb wardrobe, stolen from guests at a hotel where he worked ("but he wore their clothes beautifully," Hughes sighed). A reading at the downtown Penguin Club, his first "white" reading, was amusing; he met a former member of the cabinet, and also someone who insisted that he knew the poet Sara Teasdale.

Returning home aimless and jobless, he lounged under his mother Carrie's hostile eye and sharp tongue. The *Vanity Fair* splash brought flattering mail, including a warm letter from Ethel Welch, his old English teacher from Lincoln, Illinois, where he probably had written his first poem. From the acerbic black journalist George Schuyler came the kind of compliment that Langston craved above all others: "All your work has the flavor of the soil and the people, and for that I much admire you." In this sudden glow of success, Hughes basked indolently: "You'd think I was famous already!" But the failure either to enter college or to progress with his autobiography, his mother's nagging about money, and his undeniable nervousness about the fate of his book, all made him irritable and anxious. He clashed with his mother, who refused to feed him if he would not work. A young Howard student named Edward Lovette came by one day to introduce himself and show Hughes his own writing. For several days, Lovette brought Langston his lunch, delighted to be allowed to feed a young lion.

In October, however, when Waring Cuney sent an application form from Lincoln University, Hughes began a serious search for at least $300 for the first year there. Jessie Fauset sent him the addresses of several educational funds, but begged him to apply to Harvard, not to a black school: "You'd get just the mere formal 'book-larnin' and no contacts to speak of." Hughes knew, however, that he had been schooled away from his own people for too long; they were the true "contacts" he needed—let Cullen take his Phi Beta Kappa key and go dry up at Harvard! Coyly, Hughes approached Van Vechten: "Do you happen to know some philanthropic soul who might like to take a sporting chance on the development of genius and advance me said loan?" No, Van Vechten snapped (he thought college a waste of time); Langston should work on the autobiography: "there may be money in *that* for you." "Some big hearted person," Hughes argued plaintively to Walter White about a loan, "ought to be interested enough in the development of talent to grant me that."

Giving up his experiment in the life of art, he took a job as a busboy at the prominent Wardman Park Hotel in Washington, at a salary of $55 a month. He had afternoons off, excellent for writing; better still, as a busboy he was always close to food. On October 20, hoping against hope, he applied for admission to the dean of the college at Lincoln University. Hughes detailed his high school and Columbia honors and involvements, his travels, and also something of his success as a published poet.

> I want to come to Lincoln because I believe it to be a school of high ideals and a place where one can study and live simply. I hope I shall be admitted. Since I shall have to depend largely on my own efforts to put myself through college, if it be possible for me to procure any work at the school, I would be deeply grateful to you. And because I have lost so much time, I would like to enter in February if students are admitted then. I have had no Latin but I would be willing to remove the condition as soon as possible. I *must* go to college in order to be of more use to my race and America. I hope to teach in the South, and to widen my literary activities to the field of the short story and the novel.

His remark about teaching in the South was not insincere; he had recently urged Harold Jackman to do the same. Two days after he wrote his letter, the dean of the college informed Hughes that "we shall be very glad to admit you."

With the appearance of Cullen's book of poems, *Color,* Langston finally responded to Countee's efforts to revive their friendship. In a note mailed to Harvard, he called the volume "almost as lovely as the poems therein. It is really very beautiful." The move was not entirely disinterested; Cullen (who had won the yearly Witter Bynner prize for the best poem by an undergraduate in the United States) was editing a special number of *Palms* magazine, for which Hughes sent him fifteen pieces. Cullen's *Color* upset Jessie Fauset in one way. Individual poems were dedicated to Walter White, Eric Walrond, Harold Jackman, even to Carl Van Vechten ("To John Keats, Poet. At Spring Time"). But there was no mention of W. E. B. Du Bois, in whose *Crisis* so many Cullen poems had first appeared, although one poem was dedicated to his daughter, Yolande, whom Countee was courting and would marry. (Not surprisingly, Cullen had revoked the private dedication in February, 1923, of "To a Brown Boy" to Hughes.) Fauset begged Hughes to consider dedicating a poem in his own book to the man who "has made things possible for very many of us." At once, Hughes suggested "The Negro Speaks of Rivers," which would be the only dedication (other than "To F. S.") of any single poem in *The Weary Blues.* Du Bois, unaware of Fauset's move, was touched: "I should be honored to have a poem of yours dedicated to me."

In mid-November, when Hughes received the complete proofs of *The Weary Blues,* his nervousness about the volume began to build again. Covarrubias's strength was in caricature, and blacks hated to be caricatured. And what of Van Vechten's introduction, which he had sanctioned? Would blacks find it patronizing? Obviously, Hughes himself wondered how much of his integrity, if any, he had surrendered in his closeness with Van Vechten. His conscience, in fact, was clear on this score; publicly and privately, he stuck by Van Vechten for the rest of his life. But in a letter to Gwendolyn Bennett, now studying art in Paris, he worried about these matters. Toughly, she reminded him that he wasn't writing for colored people only. "And if they who have a chance to have a kinship of race with you dont like your things," she advised, "well, let them go hang!"

Thus buttressed, he concentrated on preparing the way for the book. Although he had planned readings in Manhattan and Washington, Hughes was eager for a real publicity break. Earlier in the fall, working through Charles S.

Johnson, he had tried to snare Carl Sandburg with greetings and a copy of "To Midnight Nan at Leroy's" dedicated to him. Sandburg had brushed them off politely: "It is fine to have such letters as yours, and the words of greeting from Langston Hughes." Then, near the last day of November, Vachel Lindsay came to the Wardman Park Hotel to give a reading. One of the best-known writers in America, Lindsay was a troubadour poet and an orator of sensational theatricality, especially when he launched into pieces from his bombastic *General William Booth Enters into Heaven* (1913) and *The Congo* (1914). Like Sandburg, he was an Illinois poet, born in Springfield, who had responded to the humanitarian legacy of Abraham Lincoln and the poetic mastery of Walt Whitman; thus Hughes felt a strong kinship with Lindsay—which he immediately set out to exploit. Finding Lindsay's picture in the morning paper, he spotted him and his wife seated at a table against a wall in the dining room. Hughes copied out three of his poems, then waited for a chance. When Lindsay came in alone for dinner, Langston pounced; dropping the poems on the table, he mumbled something about admiring Lindsay, then fled to the kitchen. Peering out from behind the pantry door he saw Lindsay finish his dinner, then leave the room.

That evening, Lindsay startled his large audience by announcing that he had discovered a poet, a bona fide poet, a *Negro* poet no less, working as a busboy in their very midst. As proof, he read all three poems to the audience. The next morning Hughes found several white reporters waiting to pepper him with questions about his poetic gift (curious in a Negro) and how he had come by it. Milking the moment, he answered all the questions with his practiced humility. First the story made the local newspapers; then, picked up by the Associated Press, the item about the Negro busboy poet and Vachel Lindsay made its way into newspapers from Maine to Florida. As for Lindsay, Hughes hoped to meet him, but had to leave on November 30 for New York, where he had appointments.

With Van Vechten to bolster his confidence, Langston visited the Knopf offices to meet the staff; at Winold Reiss's studio, he sat for his pastel portrait posed in the manner of the most famous picture of John Keats (by Joseph Severn), an open book before him. But the highlight of the visit came on his second evening in Manhattan when, after dining with the Van Vechtens, he joined them in welcoming guests to a party in his honor. They included Joel Spingarn's brother Arthur, a lawyer, an accomplished bibliophile, and an NAACP leader in his own right; Miguel Covarrubias, Hughes's illustrator, exuberant and satirical as one hoped he would be after seeing his drawings; literary editor Irita Van Doren of the New York *Herald-Tribune;* Harry Block, a senior reader at Knopf; the lordly Alfred A. Knopf himself and his wife, Blanche Knopf, cool and aristocratic in a rather deliberately Gallic way; the poets Louis Untermeyer and Jean Starr Untermeyer; and—Langston could hardly believe his eyes—the actress Marie Doro ("I never in life thought I would meet Marie Doro"), whom he had seen many times on the movie screen. Among the black guests were the jazz singer Nora Holt, the composer and conductor Rosamond Johnson, brother of James Weldon Johnson, and the irreverent entertainer Taylor Gordon. That these people had come out for *him* moved Hughes deeply. In provincial Washington, people kept up "social barriers and a sort of conceited snobbishness"; here, he was received as a man to be honored. "Your party was like a dream,—not come true but still a dream," he would write Van Vechten. "Not even my first night ashore in my first foreign port was any more happy." A seal had been set on their young friendship.

Others saw Van Vechten differently; *Time* magazine would sneer (with Hughes in mind) that he "has been playing with Negroes lately . . . writing prefaces for their poems, having them around the house." But the party probably hastened Langston's next move in looking for a loan. At 9 West 73rd Street, a handsome Manhattan townhouse off Central Park, Hughes met Arthur Spingarn's sister-in-law and the benefactor of the *Crisis* literary contest, Amy Spingarn. A private elevator whisked him to her atelier where she painted and sculpted and sometimes wrote poetry (she had even published a poem in the September, 1924 *Crisis* under her maiden name, Amy Einstein). Hughes found her a soft, gracious woman who showed her pictures and poems almost timidly; while he munched on cinnamon toast and sipped tea, she recited Wordsworth and Shelley in a husky voice. Then she sketched his head for a portrait in oils. In another way, his head was in her hands—they both knew that she could easily send him to school.

The daughter of a very rich mill owner and landlord in New Jersey, Amy Einstein had passed a lonely if expensive childhood. In Europe she had learned French, German, and Italian and studied medieval art, then attended Barnard College fitfully before leaving without a degree. But after falling in love with a relatively poor but brilliant Columbia professor and poet, Joel Elias Spingarn, and successfully proposing marriage to him, her life had begun to change for the better. Her husband, famously fired from Columbia by Nicholas Murray Butler, joined the NAACP leadership in 1911 and encouraged her to take an interest in the suffragette movement and also to help talented young blacks. She had returned from Europe after the August *Crisis* banquet to find a note from Hughes thanking her for the prize money and telling her briefly of his hopes. In October she wrote from Troutbeck, her farm at Amenia, to invite him to visit her in Manhattan. As for college, she wrote, "perhaps I might be able to be of some help to you." Carefully, Hughes explained his attitude towards Columbia and other white universities; many persons, perhaps most, he knew, would prefer to invest in a Harvard education rather than one at an obscure black college. He left with only her promise to think about him.

Hughes's entire visit to New York was not spent with the rich. He met the latest recruits to the Harlem arts movement—the promising twenty-five-year-old sculptress Augusta Savage, already drawn to African themes, and Wallace Thurman, a young writer from Salt Lake City who had arrived on Labor Day from Los Angeles, where he had worked with Arna Bontemps in the post office. Thurman was a fascinating but troubled man. Able to read several lines at once, he had already raced through more books than most of his friends would read in a lifetime; very opinionated, his gift allowed him to find fault, in detail, with so many great writers that the lesser talents, including his own, shrivelled under his scrutiny. Gregarious but unhappy, he was both guilty about his active bisexuality and eager to satisfy it; within days of his arrival in New York he had been arrested on a morals charge in a public lavatory (a minister piously secured his release, according to Thurman, then propositioned him). Touchy about his jet-black skin, he nevertheless craved the company of whites and mixed-bloods. "I love to give pleasure," he once would write desperately to Langston. "Perhaps I am the incarnation of the cosmic clown." If Thurman was a clown, however, he shrewdly saw Hughes as a permanent enigma, something he could almost embrace but never quite understand. In a portrait barely disguised as fiction, Thurman would depict Hughes as "the most close-mouthed and cagey individual [Thurman] had ever known when it came to personal matters. He fended off every attempt to probe into his inner self and did this with such an unconscious and naive air that the prober soon came to one of two conclusions: Either [Hughes] had no depth whatsoever, or else he was too deep for plumbing by ordinary mortals."

Back in drab Washington, Langston returned at once to squeezing what further publicity he could from the Vachel Lindsay episode. Contacting Underwood and Underwood, the photographic news agency, he hoisted a tray of dishes onto his shoulder for their cameraman; he also secured the promise of a Sunday news story from a writer, Josephine Tighe Williams, on the *Washington Star* newspaper. Then Lindsay, who had since discovered that his "discovery" had already published widely, with a book from Knopf almost off the press, gamely met with Hughes for a few minutes. "He seems to be a kind, shy sort of fellow," Langston wrote Knopf. Later, Lindsay left a gift for Hughes at the hotel office—a handsome two-volume set of Amy Lowell's biography of Keats. Flowingly inscribed across the first six leaves was a letter to "My dear Langston Hughes," urging him to read more of Amy Lowell, especially her "New Poetry" criticism, and praising Edwin Arlington Robinson and Alfred Kreymborg. But, he added wisely, "you may know all of this better than I do." "Do not let any lionizers stampede you," he urged. "Hide and write and study and think. I know what factions do. Beware of them. I know what lionizers do. Beware of them."

Touched, Hughes at once drafted a reply. "Something of what you tell me in your letter on the fly I already knew," he conceded. As for the personal advice:

> I am wary of lionizers. I do avoid invitations to receptions and teas. I'm afraid of public dinners. And people who tell me my verse is good I seldom believe. I see no factions to which I could belong and if I did I think I should prefer to walk alone,—to be myself. Don't be afraid. I will not become conceited. If anything is important, it is my poetry not me. I do not want folks to know me but if they know and like some of my poems I am glad. Perhaps the mission of an artist is to interpret beauty to the people,—the beauty within themselves. That is what I want to do, if I consciously want to do anything with poetry. I think I write it most because I like it. With it I seek nothing except not to make work out of it or so much [as] a line or propaganda for this cause or that. I want to keep it for the beautiful thing it is.

(Lindsay was not finished with trying to help Hughes. The following March, he would lecture on *The Weary Blues* at a bookstore on East 54th Street, off Fifth Avenue.)

Lindsay's warning must have had something to do with Hughes's next major poem. In Manhattan, he had attended a party at 17 Gay Street in Greenwich Village, where the black artist Jimmy Harris lived with his wife, Dorothy Hunt Harris. There, much of the talk had been of Taos, New Mexico, especially about the literati associated with its artists' colony, including D. H. Lawrence (long gone), Mabel Dodge Luhan, and her Indian husband, Tony Luhan. Also involved was Jean Toomer, author of *Cane,* the brilliant modernist pastiche published in 1923. Intimately involved with Mabel Dodge Luhan, he was even now trying to help her petition the mystic Gurdjieff, to whom they were both devoted, to found a branch of his French-based movement in Taos. The Gurdjieff movement had already begun to bleach a talent that would never again even approach the achievement of *Cane;* in addition, the lightskinned Toomer had started to deny his race. After Horace Liveright called him a promising Negro writer, Toomer was furious: "I must insist that you never use such a word, such a thought again." Perhaps Langston heard of Toomer's attitude; certainly he eventually would write of him only with amused contempt. In 1925, however, Hughes took Toomer more solemnly. Brooding on his own future, all suddenly vivid to him with the coming of his book and perhaps the onset of fame, and with Vachel Lindsay's warning before him, written on the pages of a biography of the greatest of Romantic poets, dead at twenty-six, Hughes wondered how far fate and fame might take him away from his old dreams. The result was "A House in Taos," a poem of unusually stately architecture for Hughes, a study of barren, pretentious lives mocked by nature:

RAIN

Thunder of the Rain God
  And we three
  Smitten by beauty.

Thunder of the Rain God:
  And we three
  Weary, weary.

Thunder of the Rain God:
  And you, she and I
  Waiting for nothingness.

Do you understand the stillness
  Of this house in Taos
  Under the thunder of the Rain God?

SUN

That there should be a barren garden
About this house in Taos
Is not so strange,

But that there should be three barren hearts
In this one house in Taos—
Who carries ugly things to show the sun?

MOON

Did you ask for the beaten brass of the moon?
We can buy lovely things with money,
You, she and I,
Yet you seek,
As though you could keep,
This unbought loveliness of moon.

WIND

Touch our bodies, wind,
Our bodies are separate, individual things.
Touch our bodies, wind,
But blow quickly
Through the red, white, yellow skins
Of our bodies
To the terrible snarl,
Not mine,
Not yours,
Not hers,
But all one snarl of souls.
Blow quickly, wind,
Before we run back into the windlessness—
With our bodies—
Into the windlessness
Of our house in Taos.

Complex in comparison to his usual work, "A House in Taos" was Hughes's most ambitious venture in 1925 and the most "modern" poem he had yet written, according to the prevailing definition of modernism. If the work is not deliberately obscure, it is inherently so, in that its central concern is the loss of meaning that comes with the failure to hold to life-affirming values. Three persons, one red, one white, and one yellow (Tony Luhan, Mabel Luhan, and Jean Toomer?) pass their lives remote from society but "Smitten by beauty" and, at the same time, enamored of the money that allows their evasion of the world. Mistaking languor for meditation, they wait and wait under the laconic scrutiny of the elements "for nothingness." Instead of the unity of mind and body promised by hermetic isolation, there are only "separate, individual" bodies and "one snarl of souls." The most powerful effect in the poem comes from the contrast between the whimpering, unrhythmic human voice of the persona and the silent but implacable elements of nature. The values of these three unfortunates represent the contradiction of all that Hughes as a poet upholds. To him, mind and body unite and flourish only with a plunge into the world—not for material rewards, but for the affirmation of ideals of joy, brotherhood, and justice, on which nature itself smiles. In "A House in Taos," Hughes turned his philosophy on its head and rendered it by its obverse image; the poem is conceived out of the most radical play of ironic imagination he had yet permitted himself in his poetry. Almost at once he withdrew to safer ground. Irony was a mode Hughes employed only sparingly; he knew how easily it is misunderstood and resented by the oppressed.

Although Mabel Luhan and Van Vechten were close friends, Langston's main informant about Taos was probably the Brooklyn-born poet Witter Bynner; once friends, Bynner and Luhan now despised one another. Bynner had lived for some time in Taos, but was in New York during Langston's visit and within a few weeks had a copy of the poem.

The end of 1925 loomed without any word from Amy Spingarn. Meanwhile, James Weldon Johnson had found another potential patron for Hughes, a woman who offered not a loan but a full scholarship, but was so much for Harvard and Columbia that she had already written the deans about him. Hughes would have preferred a loan to go to Lincoln rather than a full scholarship to Harvard or Columbia. He informed Mrs. Spingarn about the offer; but, fed up with waiting for patronage, he thought about shipping out once again. Perhaps he would rove to the West Indies, then head east to India and China: "One must do something amusing." Meanwhile, as he would always do, he tried to help other artists, writing Aaron Douglas a note of praise, sending poems by Waring Cuney to *Opportunity*. "You are delightfully generous in your encouragement of other young poets," Charles S. Johnson noted. At "Scarlet Flowers," his life story, Hughes scratched without feeling in spite of Van Vechten's urging that "indubitably now is the psychological moment when everything chic is Negro."

Hughes had no idea what he would do next; he was certain only that he had to get away from Washington. Christmas drew near, but he felt no seasonal thrill. With only a portable heater, the two rooms he shared with his mother and Gwyn Clark were cold. Then, the day before Christmas, a letter arrived at last from Amy Spingarn. She had decided to help him return to college. As to the relative merits of Harvard and Lincoln, it seemed to her that "although culturally Harvard would offer more it might have the same chilling effect that Columbia had." So why not go to Lincoln as he had planned? She had set aside $300 for him for this purpose; "if you still feel like going in February let me know." Fairly tingling with gratitude, Hughes carefully drafted his reply:

> I have been thinking a long time about what to say to you and I don't know yet what it should be. But I believe this: That you do not want me to write to you the sort of things I would have to write to the scholarship people. I think you understand better than they the kind of person I am or surely you would not offer, in the quiet way you do, the wonderful thing you offer me. . . . I hate pretending and I hate untruths. And it is so hard in other ways to pay the various little prices people attach to most of the things they offer or give. And so I am happier now than I have ever been for a long time more because you offer freely and with understanding, than because of the realization of the dream which you make come true for me. Words are clumsy things. I hope you understand what I mean. . . . Surely you are the very spirit of Christmas.

Cautiously he waited for money to arrive before doing anything else. On January 7, a check for $50 came to cover his room deposit of $15 and other small expenses. Now he thanked the scholarship person, and dismissed her. To the annoyance of his mother, who was also hurt to know that she faced three years or more in which, most likely, no financial help would come from him, he gave up his job at the hotel. Again she refused to feed him if he refused to work. One evening, when he dined with Alain Locke and others after a concert by Roland Hayes, Hughes ate for the first time that day. Luckily, Ridgely Torrence took three poems for *The New Republic* and sent a check; and, at Van Vechten's urging, Irita Van Doren bought four for the New York *Herald-Tribune*.

In the middle of January the first copies of *The Weary Blues* arrived, Covarrubias's angular black bluesman playing the piano astride the red, black, and yellow jacket. To Knopf, Hughes wrote simply: "I like my book." With Alain Locke in the chair, and an admission charge of one dollar, Hughes read on January 15 at the Playhouse, 1814 N Street, NW, in Washington. In a daring move, he had selected a Seventh Street bluesman, funky and unfettered, to howl during the intermission. Locke, however, had thought of all the respectable black folk and the whites who would come, then sneaked in—to Hughes's intense annoyance—a housebroken Negro. The man played, according to Langston, "nice music, but nothing grotesque and sad at the same time, nothing primitive, and nothing very 'different'."

When he went to Baltimore, thumping his drum for the book, he discovered that Bessie Smith, greatest of blues vocalists, was in town. Backstage, the little brown poet of the blues stood before the real thing—the Empress of the Blues in her volcanic prime in 1925, big-boned, black, majestic, and with overwhelming emotional needs that moved her to shattering performances on stage. He did not impress her. Did she know Van Vechten? Yeah, and his wife Fanny (Marinoff once tried to kiss Bessie Smith, who slugged her); but she didn't like that stuff on the blues in *Vanity Fair*. What exactly were her plans? Well, she had a big tent down in 'Bam and, come summer, she was goin' to make herself "a few more thousands." Did she have a theory about the blues as Art? Naw, she didn't know nothing about no art. All she knew was that the blues had put her " 'in de money'."

Langston had an easier time with the bourgeois members of the Du Bois Circle of Baltimore. His way prepared by darkly handsome photographs, all plashy eyes, soft lips, and firm jaw, he gazed on his admirers like a little brown Byron. Every young lady he met, Hughes noted, seemed to know his suicide poems by heart, "or else, 'their souls have grown deep like the rivers'." One Sunday, when he visited the women's dormitory at Howard University to see a friend, word of his presence brought a rush of adoring students with magazines and books to be signed. One woman missed her step, fell down the stairs, and had to be taken to the hospital. Hughes loved the attention. "My own poems are about to bore me to death," he conceitedly wrote Van Vechten. "I've heard them so much in the mouths of others recently." When a request came for poems for special publication for schools in the South, he threw up his hands in elaborate despair: "If I ever get in the school books then I know I'm ruined."

At the end of January he went to New York. As he would for the rest of his life, Hughes tried to ensure that his book would reach shops patronized by blacks and that advertisements would appear in black papers and magazines. Blanche Knopf refused the latter; there was enough publicity in the white papers, which would sell books. When he raised a gentle protest, her coolness turned to ice. He visited Amy Spingarn and had a joyful reunion with Bruce Nugent, who had finally quit Washington for New York. By chance they ran into Wallace Thurman, about whose brilliance Hughes had often talked. Nugent took one look at Thurman and snubbed him. "He was so black and flat-nosed," Nugent would remember; fresh from Washington, he was sure "you couldn't be that bright if you were that black." Soon he was ashamed of himself; he and Thurman became close friends.

On Sunday, January 31, about two hundred persons gathered at the Shipwreck Inn at 107 Claremont Avenue, between West 120th and 121st Street near Columbia University, to help Hughes launch his first book. Among the crowd were the woman who had brought him through, Jessie Fauset, the one who promised him the future, Amy Spingarn, and—in his mother's place—Ethel Dudley Brown Harper, or "Aunt" Toy, a friend from Carrie's Kansas days, who now worked as a *modiste* in New York. (He was probably staying at her home at 281 Edgecombe Avenue in Harlem, where she lived with her musician husband, Emerson Harper.) The reading was a thorough success. Later, he signed books until his wrist ached; Wallace Thurman took away a copy signed for one of Hughes's greatest admirers, the heiress A'Lelia Walker.

Then, alone, Langston boarded a night train for Washington. On his way home, he reached his twenty-fourth birthday.

The city was white under a rare snowfall when he arrived. In the chill rooms at 1749 S Street he prepared to leave for his new life at Lincoln University, which he had not yet seen. As for the book, all the portents were favorable. Locke had done a review for *Palms,* and Cullen another for *Opportunity*. When V. F. Calverton of the *Modern Quarterly* circle in Baltimore wrote that his group was "extraordinarily anxious" to have him read, Hughes accepted at once. The Civic Club in Manhattan wanted a date in March. From Oberlin College, near Cleveland, came a request from a black student leader for an appearance in April. And W. C. Handy, author of the world-famous "St. Louis Blues" and easily the most famous composer associated with the form, had just commended Hughes's "entirely original" treatment of the form.

He read his poems at a testimonial dinner for Alain Locke and in various high schools, but was nervous about the fate of the book, tired, and anxious to be gone. Remote Lincoln University, deep in the countryside miles and miles away from Philadelphia, seemed just the right place for him now.

"You see," he wrote Carl Van Vechten, "I'm going into seclusion, weary of the world, like Pearl White when she retired to her convent. And I hope nobody there reads poetry."

6

A LION AT LINCOLN
1926 to 1927

Listen to yo' prophets,
Little Jesus!
Listen to yo' saints!

"Shout," 1927

"LINCOLN is wonderful," Hughes wrote to Countee Cullen after a week at the little university of just over 300 students, located amidst "trees and rolling hills and plenty of country" about forty-five miles southwest of Philadelphia. Life was "crude," but comfortable, the food "plain and solid": there was "nothing out here but the school and therefore the place has a spirit of its own, and it makes you feel as though you 'belonged', a feeling new to me because I never seemed to belong anywhere." For the first time since the segregated third grade in Kansas, Hughes was in school among his own people. His first impressions were all favorable. "Out here with the trees and rolling hills and open sky, in old clothes, and this do-as-you-please atmosphere, I rest content." After three weeks he was still happy: "I like Lincoln so well that I expect to be about six years in graduating."

He had reached the university on the cold late afternoon of Sunday, February 14, after a train ride from Philadelphia past sleepy towns that yielded to forest and brown farms stretching to the winter woodline. (The evening before, Hughes had read his poems in Baltimore before enthusiastic members of V. F. Calverton's *Modern Quarterly* set, well known for their Saturday evenings at Calverton's home, where Hughes stayed overnight.) At Lincoln, a farewell party for a departing senior turned into a welcome for Lincoln's most glamorous newcomer. Outside the night was cold and the ground was white with snow, but the party grew raucous as the students, fueled by liquor in spite of Prohibition, celebrated their last evening before the start of a new term. That night, in his room in Cresson Hall, Hughes went to bed elated by the new life before him.

In spite of the objections to his choice by Jessie Fauset and Countee Cullen, the school was prestigious. Since its founding as Ashmun Institute in 1854 by John Miller Dickey, a white Abolitionist minister from the nearby town of Oxford, the all-male school, comprising undergraduates and a small seminary, had helped to educate more doctors, lawyers, and ministers than any other black college. Lincoln's religious ties were firm but not oppressive. Seven of the nine tenured professors were ordained ministers, most with some connection to the white Presbyterian stronghold of Princeton University; Lincoln men liked to call their school "the black Princeton." Curiously, but only increasing its prestige among many blacks, the tenured faculty was exclusively white; in spite of its many professional alumni, no black had ever served on the board of trustees. And yet, in an era of unrest on major black campuses, Lincoln was virtually exempt from trouble, although there was strong alumni protest when the presidency, still vacant when Langston arrived, had been offered to a Philadelphia minister who turned out to be a sympathizer of the Ku Klux Klan. So important was Lincoln that the head of the Phelps-Stokes Fund, an authoritative voice on black schools, warned that a mistake in the appointment would have "a most unfortunate effect on the college education for Negroes, not only in America, but throughout the world."

For all its prestige, however, Lincoln was falling on hard times. A substantial endowment had been depleted by weak management and by other factors beyond its control. The students were mostly poor, the alumni hardly rich. On the charming 145-acre campus, most of the buildings were "in very very poor condition." A new science facility served the students' strong interest in pre-medical training (about half hoped to be doctors), but the three dormitories and the dining hall were overcrowded and Vail Memorial Library definitely antiquated. The faculty was uneven in quality and too small. The curriculum still stressed the classics; no courses were offered in black American or African culture, although Professor Robert Labaree discussed race and culture liberally in his popular lectures on sociology. Professor of biology Harold Fetter Grim was worshipped by the pre-medical students, and Professor Walter Livingston Wright, who astonished alumni by remembering their names decades after he had last seen them, was thought by the students to be one of America's greatest mathematicians. Other instructors were less inspiring. The mandatory Bible courses were scandalously taught by the ultra-conservative Prof. Edwin Reinke, given to haranguing the students as "adulterers and fornicators," slick dudes from the cities, or dumb "bozos" from the sticks. On the whole, however, Lincoln men held their teachers in high regard.

"To set foot on dozens of Negro campuses," Hughes wrote later about student freedom, "is like going back to mid-Victorian England," or Massachusetts in the days of the witch-hunting Puritans. This was not true of Lincoln. Student freedom was part of the gentlemen's agreement between the white faculty and its black students, out of which had grown a robust school spirit. With little supervision and no counselling, and with women absent except on gala weekends, the dining hall sometimes resembled a bear pit; theatrical ventures (with male "actresses") dissolved easily into pandemonium. The students disciplined themselves in other ways, however; compared to later generations, one of them would reflect, "their morals were higher, their purpose in living was higher, and they took their work seriously." Neither crime (except for genteel bootlegging) nor bad grammar was tolerated; studying was fashionable, and the threat of flunking out real. At mid-year, more than one unfortunate Lincolnite was summoned by the dean, quietly handed a Wanamaker Bible and train fare to Philadelphia, and sent on his sorrowful way. The next morning only his tracks in the snow told his brothers that he was gone. But the administrators were almost always forgiving. What Professor Wright wrote swiftly on the blackboard with his right hand he erased as swiftly with the left, but almost all the students passed mathematics. Lincoln policy was to let the men do as they chose, Wright declared as acting president, "because then one is always likely to do right."

The generosity of the men touched Hughes. "One can use, wear, or borrow anything anyone has," he noted happily; "there's a fine spirit of comradeship and helpfulness here." With his meager wardrobe, he was also glad that dress was far more casual than at Columbia; except on Sunday, most of the men wore thick boots and comfortable old jackets and ties. The fraternity system, all-powerful in student life, claimed him quickly. Passing up Alpha Phi Alpha, which favored light-skinned students, he pledged Omega Psi Phi, somewhat less fashionable but more democratic and the current leader in scholarship among the fraternities. (From Harvard, an alarmed Cullen begged him to reconsider in favor of the Alphas.) In arriving at mid-year, Hughes had escaped much of the terror unleashed upon freshmen, or "dogs," by the sophomores, or "gods," led by the ugliest runt among them, who was known as "Zeus," but he did not escape, or wish to escape, fraternity hazing. The Omegas insisted that he show them his "two mo' ways to do de Charleston"; as Hughes bucked and heaved to unheard music, a circle of his superiors intoned "The Weary Blues." Mocked as a silly "New Negro" and a pretentious "Boy Poet," Langston was soundly paddled by one and all—"beaten till I couldn't sit down," he noted blissfully. He would be inducted into the Beta chapter of Omega Psi Phi in the spring.

"Lincoln is more like what home ought to be than any place I've ever seen," he judged. To Leroy D. Johnson, later dean of students at the school, Hughes was "a very quiet, very nice guy who never made an issue of anything with other students." He was by no means the only gifted person there. While he attended Lincoln, the school also trained respected future academic humanists such as William V. Fontaine, Therman B. O'Daniel, Fannin Belcher, and W. Edward Farrison; Toye Davis would earn a doctorate in science at Harvard in addition to becoming a physician; the loquacious, high-spirited Thurgood Marshall would become the first black justice of the U.S. Supreme Court; eloquent Benjamin Nnamdi Azikiwe would be inaugurated as the first African governor-general of Nigeria; not many years after Azikiwe, the future leader of the African independence movement Kwame Nkrumah arrived as a student. Still, Hughes made his mark; in his senior year he would be acclaimed the "most popular" student at Lincoln. Although he would admit privately that about two years passed before he was fully comfortable among the young black men, few saw his effort. "Langston was a remarkable person," Therman O'Daniel remembered. "He fitted in very easily—he could fit in anywhere. He had the personality to get along with other people, without the slightest pretentiousness.

No doubt he had his private opinions and his private thoughts as a poet needing to create, but he was certainly one of us.''

In this first term he enrolled in mandatory Bible, took two courses in English and American literature from Prof. William Parker Finney, two others in French and Spanish, and one in algebra. But Hughes had not come to Lincoln to be a scholar—he had chosen a wilderness from which he could commute to New York. On his second weekend he was in Manhattan, reading at the Civic Club on Friday, February 26, with Jessie Fauset in the chair and Bruce Nugent and Wallace Thurman grinning encouragement from the floor. So far, Nugent had evaded all offers of steady employment, but Thurman had just become editor of the formerly radical, now bourgeois opportunist *Messenger*. After dinner on Saturday night, Hughes dropped in at Van Vechten's home. The men talked until one o'clock in the morning, with one topic being almost certainly Carl's progress on his Harlem novel, ''Nigger Heaven.'' On both Friday and Saturday night, Hughes visited the cabarets. But, perhaps seeing them now from the perspective of black Lincoln, he found them dishearteningly full of white people and ''rather dull.''

With two of his Lincoln schoolmates as secretary and manager, he sought engagements for performances with the university's excellent vocal quartet led by James ''Lord Jim'' Dorsey, later an instructor at Lincoln. Hughes read with them in Trenton, New Jersey, before members of a black teachers' association, but mostly in black churches in Cheney, Bordentown, Avondale, and other towns nearby. At Easter, however, he went alone to Manhattan for a Friday, April 2, reading at Martin's Bookshop, a little place run by one Martin Kamin at 97 Fourth Avenue. The shop could pay Hughes nothing, not even his train fare, but it was in Manhattan, and ''besides, I like Jewish people.'' Two weeks later, with *The Weary Blues* now in its second printing after selling 1200 copies (a respectable figure, although Cullen's *Color* had done far better), Hughes and the quartet went out to Cleveland, which he had not seen since 1920. His program was a triumph. Teachers and former students from Central High, including Ethel Weimer, Clara Dieke, and his best friend as a student, Sartur Andrzejewski, packed the hall. Most saw the same smiling, boyish Langston Hughes of old—until he began to read. Then Andrzejewski, for one, was amazed by his friend's poise. ''I liked that you on the platform,'' he wrote Langston, ''that you which you had never shown us before. You wear a mask so that you can keep that you for work. Not quite that either. It's terrifying to weaklings and weakness—that you, not the mask. It perhaps should make men weep some day.''

In Oberlin, he was welcomed somewhat nervously by members of the Paul Laurence Dunbar forum led by Susie Elvie Bailey, the black student leader who had invited him. The Oberlin students had vigorously debated the wisdom of bringing a blues poet to the mainly white college. But all objections were swept aside as Hughes read confidently to an audience mostly of whites, including professors. His visit was an ''immense inspiration to us here,'' one student wrote shortly after; ''you brought so much that was new and refreshing.'' Also successful was a reading in Indianapolis, with his father's sister

Mrs. Sallie Garvin in the overflow audience; Langston sold 45 copies of *The Weary Blues.* Since Indianapolis was Ku Klux Klan territory, the integrated audience at the colored YMCA was visibly tense; when Hughes read "The White Ones" ("why do you torture me, / O, white strong ones?") one white woman broke down and wept. Despite an automobile accident that left Langston bruised, the men returned home very pleased. At the end of the month Hughes made his Lincoln debut, performing with the quartet and the Glee Club directed by James Dorsey in the Mary Dod Brown Memorial Chapel. Although most of the music was either classical or some dignified version of the spirituals, Langston boldly inserted the blues into the proceedings. Royster Tate thumped the piano and saxophonist S. Leon Jackson blared out notes and tones almost certainly never before heard in the spiritual center of Lincoln University.

In his various appearances, Hughes capitalized on the flow of generally favorable reviews of *The Weary Blues,* with the degree of praise depending mostly on the attitude of the critic to blues and jazz. The *New York Times* regretted that the earthier poems were placed before the more traditional, but judged Hughes a poet of promise, as did *The New Republic.* In the New York *Herald-Tribune,* however, Du Bose Heyward, the white South Carolina-born poet and author of the popular novel of black life *Porgy,* lauded Hughes's musical sense and predicted a career "well worth watching." To Langston's surprise, the white Southern papers were almost all enthusiastic; the *New Orleans Times-Picayune* called him "sensitive and intelligent," and judged his book unquestionably one of the most interesting of the year. The radical *New Masses* naturally found Hughes not radical enough, but James Rorty placed him nevertheless at the forefront of black poets. Among black reviewers, Locke and Jessie Fauset both wrote strongly favorable pieces; in *Opportunity,* however, Cullen wondered exasperatingly whether the jazz and blues pieces were really poems at all. To Claude McKay, writing from Nice, Van Vechten's introduction was "just the right thing. Nothing overdone"; but "Caribbean Sunset," with an image of God hemorrhaging, was "intensely nauseating." (Van Vechten had implored Langston to cut this bad poem, which Cullen had singled out for praise.) "Cross," however, on the mulatto theme, was "exquisitely done. It shows that while others are vainly prating about artistic freedom among Negro writers, you have won out over all obstacles. You have opened up new vistas by touching a subject that thousands of Afro-Americans feel and yet would be afraid to touch."

In March, when the poet and editor Witter Bynner sent some Hughes poems to *Poetry* magazine, the major modernist journal, Harriet Monroe took four. From abroad came translations and flattering notices in France, Belgium, and Germany; after a piece on him came in from Central Europe, someone at Knopf teased Hughes that soon he would be second only to Kahlil Gibran as the world's most translated poet. There were setbacks—V. F. Calverton at *Modern Quarterly* balked before a fresh batch of blues, and first Ridgely Torrence, then the *Nation,* refused the caustic "Mulatto" (*"I am your son, white man!"*). But Hughes was pleased when his poem "Youth" ("We have tomorrow") appeared on the cover of the *Orange Jewel,* a magazine published by whites in

Chapel Hill, North Carolina. And at home at Lincoln, where he found at least one competent poet in Edward Silvera, Jr., his work appeared in the Lincoln *News*.

By springtime, Hughes knew for sure that he had done the right thing in choosing Lincoln. Limbering up around the college track, he looked for the old speed, then settled for long walks across the fields. He welcomed visitors to the campus—first Alain Locke, then V. F. Calverton, who both lectured. And on the warm afternoon of May 27, Carl Van Vechten swept in with two ladies in a limousine to meet Langston at the home of Robert and Mary Labaree, who had befriended Hughes. While the pious Labarees waited for pearls of wisdom from their celebrated guest, Van Vechten amused Langston by languidly sipping tea and going on about a "simply marvellous" new black male dancer Hughes *absolutely* had to see. As much as he liked Lincoln, however, Langston was restless as the spring ripened. "I've been home too long," he fretted to Locke. Offered a cabin in the woods near Indianapolis, he declined it partly out of fear of the Klan. He thought of the deep South, which he had never seen, and Haiti. As for his first examinations, beginning on June 2, he studied hard and earned Group One—the best of five groups—in four courses. In the Bible he was Group Two; in algebra, from acting president Wright, he almost fell on his face.

Hughes might have done better, but in the middle of these tests came a greater challenge. Freda Kirchwey, managing editor of the *Nation,* sent him proofs of an essay by George Schuyler called "The Negro-Art Hokum," in which the journalist, who prided himself on being the black H. L. Mencken, ridiculed the idea of a separate black American culture and aesthetic. Calling his essay "rather flippant in tone and provocative in its point of view," Kirchwey obviously hoped to provoke Hughes, who had been recommended to her by James Weldon Johnson. She wanted not so much a rebuttal of Schuyler as "rather an independent positive statement of the case for a true Negro racial art."

Within a week, the *Nation* had the finest essay of Hughes's life, "The Negro Artist and the Racial Mountain." With its tone quite personal, the main argument was against blacks who would surrender racial pride in the name of a false integration. A talented young Negro poet (undoubtedly Countee Cullen) wanted to be " 'a poet—not a Negro poet'." Which meant, Hughes went on, " 'I want to write like a white poet'; meaning subconsciously, 'I would like to be a white poet'; meaning behind that, 'I would like to be white'." But the young man would never be a great poet, "for no great poet has ever been afraid of being himself." To Hughes, "the mountain standing in the way of any true Negro art in America" was precisely "this urge within the race toward whiteness, the desire to pour racial individuality into the mold of American standardization, and to be as little Negro and as much American as possible." After hearing his jazz poems, middle-class blacks inevitably asked the same questions: "Do you think Negroes should always write about Negroes? . . . How do you find anything interesting in a place like a cabaret? Why do you write about black people? You aren't black." Only the "so-called common element'' among blacks live natural lives: ''Their joy runs, bang! into ecstasy, their religion soars to a shout. Work maybe a little today, rest a little tomorrow. Play awhile. Sing awhile. O, let's dance!'' The younger artist must cast off racial shame and ask, ''Why should I want to be white? I am a Negro—and beautiful.'' Hughes ended with a hosanna:

> Let the blare of Negro jazz bands and the bellowing voice of Bessie Smith singing Blues penetrate the closed ears of the colored near-intellectuals until they listen and perhaps understand. Let Paul Robeson singing ''Water Boy'' and Rudolph Fisher writing about the streets of Harlem, and Jean Toomer holding the heart of Georgia in his hands, and Aaron Douglas drawing strange black fantasies cause the smug Negro middle class to turn from their white, respectable, ordinary books and papers to catch a glimmer of their own beauty. We younger Negro artists who create now intend to express our individual dark-skinned selves without fear or shame. If white people are pleased we are glad. If they are not, it doesn't matter. We know we are beautiful. And ugly too. The tom-tom cries and the tom-tom laughs. If colored people are pleased we are glad. If they are not, their displeasure doesn't matter either. We build our temples for tomorrow, strong as we know how, and we stand on top of the mountain, free within ourselves.

These brave words were soon tested. Leaving Lincoln for his summer vacation, Hughes visited his mother, who was now living with Gwyn Clark in Atlantic City, New Jersey, and working as a servant for a white family. Securing a reading at a local black church, he was chanting a section of blues verse when a deacon approached the rostrum with a note from the minister: ''Do not read any more blues in my pulpit.'' This was his first encounter—but not his last—with censorship.

After a few days, Langston headed for Manhattan to spend the summer there. First, however, one piece of business had to be settled. The second weekend in July, he visited Amy Spingarn at her country home, Troutbeck, two and a half hours away by train in upstate New York. Here, in 1916, Du Bois and Joel Spingarn had hosted the famous Amenia conference aimed at uniting rival civil rights groups after the death of Booker T. Washington. Across a green lawn edged by a row of splendid sycamores sat a fairy-tale house of stone and oak, with a heavy slate roof, leaded windows, and extruding wooden beams in a quaint English country house style. The farm comprised several hundred acres, including a five-acre lake and a sparkling brook. Inside was comfortable, not fussy or formal. Staying in one of twenty-two rooms, Hughes passed a relaxed weekend with Amy and Joel Spingarn and their four teenaged children, Hope, Stephen, Honor, and Edward. Satisfied with his progress, Amy Spingarn agreed to lend him $400 for another year, renewable until he finished his bachelor's degree. The entire amount was in the form of an interest-free loan, for which Hughes never signed a note.

The *Nation* check arrived, followed by his royalties from *The Weary Blues*. On sales of 1,628 copies he had earned $319.20; after deductions for money advanced and the many books personally ordered, his check amounted to $124.17. A windfall came on July 22 when Witter Bynner, who funded the competition, informed Hughes that he had won the first prize of $150 in the Poetry Society of America contest for undergraduate poets (Countee had won the previous year) for "A House in Taos." The poem would appear in Idella Purnell's *Palms* magazine in November. Settling down to enjoy a summer in "the nicest town in the world," Hughes lodged in a rooming house at 314 West 138 Street, Wallace Thurman's home. Wherever Thurman lived was the center of a whirlwind of activity, of laughter and carousing but also of intense, rebellious creativity among the "Niggerati," Zora Neale Hurston's term for the hustling young writers who now seemed to be everywhere in Harlem. Hurston was the intelligent but earthy young woman who had caught Langston's eye at the *Opportunity* prize dinner the previous year. She, Thurman, and Bruce Nugent were only the chief "Niggerati." Aaron Douglas was painting and sketching with authority; John P. Davis from Washington and the Harvard Law School had just replaced Jessie Fauset as literary editor of the *Crisis* (apparently she wanted to write more); Cullen, a member of the "Niggerati" in spite of his Harvard degree and his propriety, had joined the editorial board of *Opportunity*. The poet and artist Gwendolyn Bennett was back from Europe (where she had met both Claude McKay and Hughes's old flame, Anne Coussey); from Boston had come the ingenuous but talented young writer Dorothy West and her good friend, the equally promising poet Helene Johnson, whose work was beginning to appear in the journals. Beautiful Dorothy Peterson, fluent in Spanish, eager for the stage, and with money of her own, was less often in Harlem but was still counted as one of them.

Although he could be "very precious, autocratic, commanding, and demoralizing if you allowed yourself to be demoralized," as Nugent would put it, Carl Van Vechten was definitely an honorary "Niggeratus." The young blacks sought his company, his advice, his introductions, and occasionally the loan of his money. They were seldom disappointed. Van Vechten had arranged for Paul Robeson's first Town Hall concert, listened to his most intimate confidences (about his first extramarital affair, for example, and with a white woman), and loaned him money. Van Vechten had renamed Hall Johnson's choir the "Harlem Jubilee," and commissioned a mural from Aaron Douglas. Dorothy Peterson, Zora Hurston, and Bruce Nugent were all devoted to him, as was W. C. Handy, the enterprising "father" of the blues, who was brought to Carl's home for cocktails by Hughes. When Handy's *Blues: An Anthology* appeared that year (with sketches by Miguel Covarrubias), Handy credited Van Vechten with wielding "the pen that set tongues to wagging, ears listening and feet dancing to the blues." And near the end of August, Langston and James Rosamond Johnson proposed to Carl (who declined the invitation) that he should collaborate with them on a revue.

Certainly Van Vechten and Hughes were closer than ever. On July 2, along with Carl's close friend, Donald Angus, they attended Bill "Bojangles" Robinson's farewell party (he was en route to London) where they celebrated with Noble Sissle's son Andrew, Avis (Mrs. Eubie) Blake, and various Harlem musicians. Three nights later Van Vechten and a group of friends kept a rendezvous with Hughes, Hurston, and John Davis at yet another Harlem club. On

July 7, Carl and Langston ate black beans and rice at a party given by the maid of a prominent actress. The following evening, Paul Robeson and Hughes dropped in at Van Vechten's in the late afternoon. Then, on July 13, just back from Amy Spingarn's, Langston accompanied Van Vechten to dinner at the home of Dorothy Harvey, the author of a fine study of Theodore Dreiser, where her sister Caroline Dudley Reagan, a theatrical producer who had taken a black revue to Paris with much success, was also present. Van Vechten was in a festive mood; that morning he had received the first bound copy of *Nigger Heaven*.

In spite of all the parties and nightclubbing, and the promiscuity and alcoholism they provoked (Van Vechten would speak of these years as passing in a haze) Hughes remained a person finally unknown; he heartily joined the carousing but his facade remained, in the end, almost impenetrable. Two years later, Thurman would vent his frustration at the impossibility of knowing Langston: "You are in the final analysis the most consarned and diabolical creature, to say nothing of being either the most egregiously simple or excessively complex person I know." Certainly Hughes's sexual life remained a mystery. As close as they were to him, neither Nugent nor Van Vechten knew much about even his basic sexual taste. Late in his life, after a friendship of thirty years with Langston, Van Vechten would confess to his own biographer that he knew two men "who seemed to thrive without having sex in their lives." Of the two, "never had he any indication that either was homosexual or heterosexual. One of them was Langston Hughes."

In the summer of 1926, Langston was hard at work. By early August he was writing lyrics and sketches for a show first proposed at least a year before to Van Vechten by Caroline Dudley Reagan, who hoped to stage a black revue starring Paul Robeson. Since the enormous success of *Shuffle Along* in 1921, Broadway had welcomed a number of black musicals, including *Strut Miss Lizzie, Liza, Plantation Days, From Dixie To Broadway,* and *Runnin' Wild;* after a lag, black music and dance had returned triumphantly with David Belasco's *Lulu Belle.* With his successes in plays by Eugene O'Neill and his recently launched singing career, Paul Robeson was the most commanding black male performer, as well as the one perhaps most sensitive to the depiction of blacks on the stage. Hughes, who deeply admired Robeson, was eager to work with him on the revue, as well as with the musician Ford Dabney, an associate of the famed black bandleader James Reese Europe in the glory days of the white dancers Vernon and Irene Castle. Soon Langston was travelling every day downtown to Mrs. Reagan's apartment in a cool old house with a courtyard garden in Greenwich Village. Slipping behind the jaunty name "Jimmy" Hughes, as he had done in his days at sea, Langston soon finished a "dramatico-musical composition" called "Leaves: A novelty song number for singer and chorus." His first effort at songwriting was a typical piece of show business, with slick lyrics and cues for scantily clad (in fig and other "leaves") chorus girls and "naked" men. Hughes was such a novice that he tried to register the piece with the Library of Congress; it was declined, since he had not sent music. "Leaves" was either a trial run or an eventually discarded number from the Reagan revue, called *O Blues!* (from Hughes's poem "The Weary Blues"), which included some broad satire and even low comedy but also revealed elements of fantasy, pathos, and social protest, all of which Langston saw as significant in the often despised blues form.

"Jimmy" Hughes's lyrics did not entirely please Langston Hughes. By midsummer, he and Zora Hurston had begun to plan a more ambitious venture—a black opera based on blues and jazz. Although Scott Joplin, Paul Laurence Dunbar, Robert Cole, Will Marion Cook, and others had helped to compose dignified black operas, the form had deteriorated into the strut-and-leer that marred many of the Broadway revues. Hughes wanted to reverse the trend; at least one year previously he had discussed with Charles S. Johnson the need to lift the general level of the black musical. Hurston would be an invaluable collaborator; while their enthusiasm for folklore and the black masses was almost equal, she had both a bookish and a native knowledge of the rural South. Hurston also had a particular musician in mind—forty-six-year-old Clarence Cameron White, who had trained as a concert violinist, studied composition with the black English musician Samuel Coleridge-Taylor, and was currently director of music at West Virginia State College. In August, when he visited New York, White met Hughes for the first time and was also taken by Hurston to meet their main supporter in the project, Van Vechten.

In steamy August, however, none of the musical projects flourished. The enthusiasm for *O Blues!* wilted. Neither it nor the folk opera was ever finished, but Hughes's association with musicians such as Robeson, White, Hall Johnson, and Handy was not wasted. All through the summer he worked hard on new poems that brought his art to a remarkable new level of proficiency. For inspiration he depended not on romantic melancholy and traditional forms but on a now classical respect for the art of the blues; unquestionably, Hughes had been toughened in his resolve to celebrate the lives of the black masses both by his months at Lincoln and by the heady confidence of the Harlem group. Out of that confidence came another idea that captivated the "Niggerati." Late one night, in the small apartment of the recently married Aaron and Alta Douglas (like Aaron, Alta was from the Topeka, Kansas, area) on fashionable Edgecombe Avenue on Sugar Hill, the talk turned to the need for a new magazine that would go beyond the staid *Crisis, Opportunity,* and *Messenger* in capturing the spirit of the younger writers. According to Nugent, the idea was Hughes's: "He suggested that maybe someone should start a magazine by, for, and about the Negro to show what we could do." Two events helped to define its proposed spirit. The first was Langston's *Nation* manifesto, "The Negro Artist and the Racial Mountain." The second was the savage reaction of the black press and other sections of Harlem to Van Vechten's *Nigger Heaven.*

Hughes had found the novel, if anything, too pro-Negro. "Colored people can't help but like it," he predicted to Locke; *Nigger Heaven* read as if it were written by "an N.A.A.C.P. official or Jessie Fauset. But it's good." Alain Locke thought the book a big step toward objectivity, "in spite of its pro-African propaganda." Langston's mother, Carrie, thought it was "our people to a 'T'." But Du Bois had delivered perhaps the most crushing review in his sixteen years of aggression at the *Crisis;* the book was "an affront to the hospitality of black folk and the intelligence of white,'' who should burn the book and read the *Police Gazette*. Hubert Harrison of the *Age* also savaged the book, and other Harlem newspapers refused to advertise it. For a while, to his great distress, Van Vechten was barred from certain clubs, until one night, flanked by the redoubtable Zora Hurston, he successfully braved the ban. To almost all the young black writers, Van Vechten's troubles were their own. The attack on him was an attack on what they themselves stood for—artistic and sexual freedom, a love of the black masses, a refusal to idealize black life, and a revolt against bourgeois hypocrisy. In July, Hughes received a frightening reminder of black snobbishness when the old poet George M. McClellan wrote to tell him that while he liked the poems in *The Weary Blues,* he had scissored from the dust-jacket ''that hideous black 'nigger' playing the piano.''

The ''Niggerati'' decided to attack. The magazine would be called *Fire!!* after Hughes's poem (set by Hall Johnson) in which an inveterate sinner confesses loudly and happily that ''Fire gonna burn ma soul!'' Thurman would be editor; he, Hughes, Hurston, John P. Davis, Gwen Bennett, Bruce Nugent, and Aaron Douglas each would contribute $50 to produce the first number. Later, nine persons would be listed as patrons, including Dorothy Peterson, Arthur Huff Fauset (Jessie's half-brother) and Van Vechten. In mid-August, when Hughes wrote Locke in Europe to break the news, a crucial stage had been reached in the New Negro movement. At the ''coming out'' *Opportunity* party in 1924, Locke had been introduced as the virtual dean of the younger writers. Two years later, the students had dispensed with the dean.

As Hughes prepared to return to Lincoln, he agreed to continue work on *O Blues!* for Caroline Reagan. On his last Saturday in Manhattan, Zora Hurston cooked a dinner in his honor; Mrs. Reagan was there, Van Vechten and Fania Marinoff, Zora's brother Everette Hurston, and a cousin. Monday, September 20, registration day at Lincoln, found him still in Manhattan. That evening, he and three Lincolnites hosted a cabaret benefit to aid Lincoln's scholarship fund (perhaps by raising their own tuition) at Small's Paradise, one of the few Harlem nightclubs owned by blacks. Patrons included Van Vechten, Walter White, and Paul Robeson; Ethel Waters (lauded by Van Vechten in *Vanity Fair* as the finest of all blues singer) and the teenaged singing star Adelaide Hall were among the entertainers.

Back at Lincoln, Hughes enjoyed his first autumn there. For a while, summer weather clung to the Pennsylvania woods but soon the fields were harvested and lay stubbled in red and brown. He signed up for a course in ''The Art of Poetry'' with a visiting instructor, Dr. Robert T. Kerlin, the white editor of *Negro Poets and Their Poems,* to which Langston had contributed in 1923 from Jones Point. In general, Hughes was more relaxed at school than ever before. A ''god'' now as a sophomore, he watched with amusement as his class tormented the freshmen. When they rashly revolted, ''Zeus'' and the lesser gods fell upon them one night and shaved the heads of almost all the class. The administration brought charges against twenty-six sophomores. In response the sophomore class chose Hughes to draw up a statement pleading collective guilt, which each member (including Thurgood Marshall) signed. The class was fined $125 and warned to desist. Such jejune activity was part of the fraternal bonding Hughes clearly craved. As the Lincoln Lions waged their football campaign against Howard and Morgan and other black colleges, he joined "the rabble" around bonfires on chill autumn nights as the dark grounds echoed with school cheers and the rumbling baritone of fight songs. In a concession to his career, Langston lived without a roommate now, the only undergraduate to do so. No one grumbled; he was Lincoln's literary lion. Lincoln was proud that he had won the Witter Bynner prize for the best poem by an American undergraduate, that one of his poems graced the cover of the October *Opportunity* and several covered an entire page of the December *New Masses.* And he continued to attract important visitors to the campus. A Princeton professor (a foreigner, of course, from Seville in Spain) once stayed late into the night at Cresson Hall listening to blues records. And Adelaide Casely-Hayford, the African poet and anti-colonialist, came out to see him; Hughes, for all his youth, was becoming a black eminence, like Du Bois or Marcus Garvey.

Once settled back at Lincoln, Langston gathered his new poems into what he hoped would be his second book, which would be named (by whom it is unclear) "Fine Clothes to the Jew," after a line from "Hard Luck":

> When hard luck overtakes you
> Nothin' for you to do
> Gather up yo' fine clothes
> An' sell 'em to de Jew . . .

On Sunday, October 3, he took the manuscript to Van Vechten, to whom the collection was dedicated. As with Hughes's first collection, they went over each of the poems; exactly what part Van Vechten played now is unclear. Three weeks later, on Saturday, October 23, Langston presented the revised "Fine Clothes to the Jew" to him to take to Knopf. Hughes was still in town on Monday to attend, along with Rudolph Fisher, a small gathering at Van Vechten's for the English novelist W. Somerset Maugham, at which Taylor Gordon and Rosamond Johnson sang. Later that evening, Langston attended a concert at Aeolian Hall on 43rd Street to hear mezzo-soprano Mina Hager end a recital of classical music with four of his poems which had recently been set by the white American composer John Alden Carpenter.

Meanwhile, having thrown himself with abandon into the first number of *Fire!!: A Quarterly Devoted to the Younger Negro Artists,* Wallace Thurman was in trouble. The New Negroes were no more dependable than the old; only Hughes and two others had sent their promised $50. Nothing—not even art work—had come from Gwen Bennett, and the aristocratic John Preston Davis ("Sir John") declined even to return telephone calls. And yet Thurman refused to cut costs. Only the very best paper would do for Aaron Douglas's drawings; the cover had to be just so. Bruce Nugent, hungry and in charge of collecting subscriptions, apparently spent most of the meager receipts on food. The *Fire!!* headquarters shifted from 17 Gay Street, the toney Greenwich Village address of Jim and Dorothy Hunt Harris, to Thurman's rooming-house. Thurman was behind in his rent, growing thin without regular meals (''ho hum''), and becoming even more manic than usual. ''Ye Gods!'' he cried to Hughes, ''my hair is already gray.'' He appealed to Langston for $100 to lure a printer; Langston turned to Van Vechten, who thought *Fire!!* rather pointless but sent some money.

Soon Hughes had a chance to help Van Vechten. Early on October 30, a frantic call awakened Langston at Lincoln. Carl had borrowed some lyrics from the popular ditty ''Shake That Thing'' (by a white composer) and published them without ascription, much less permission, in *Nigger Heaven.* Now the threat of a lawsuit promised to kill the best-selling novel as it raced to the end of its fifth large printing. Deciding to settle out of court and replace all of the many excerpts from popular songs in his novel, Van Vechten asked Hughes to write them. At seven that evening, Langston reached New York. He worked deep into the night, then slept at Van Vechten's. The next day, while a steady rain fell on the city, he finished new lines for seventeen places in the novel. Later, Langston and Carl went to a party on Park Avenue with Hugo and Susie Seligman, the debonair Eddie Wasserman, Florine and Ettie Stettheimer, and other friends. After listening to Fania Marinoff on the radio in a reading of *Twelfth Night,* they went with her and the English writer Rebecca West, Arthur Spingarn, and Walter and Gladys White to James Weldon Johnson's home to meet the famous attorney Clarence Darrow, who had taken part the previous year in the celebrated Scopes evolution trial in Tennessee. Hughes, Darrow, and Johnson each took turns reading poetry; Darrow was particularly moved by Langston's astringent ''Mulatto.''

For his work on *Nigger Heaven* Hughes received $100 from Van Vechten. Some of this money went to *Fire!!,* still a source of distress to Wally Thurman. The incorrigible Bruce Nugent refused to read proofs even of his own brilliant but difficult stream-of-consciousness story. Forced to rely on his current lover, a tall, blond Canadian fellow, Thurman deplored the fact that *Fire!!* apparently would not burn ''without Nordic fuel.'' Still, he forged ahead. Advance material pointedly linked the journal, striking with its thick cream sheets and an overlapping red and black cover adorned by a monumental Aaron Douglas design, to the notorious *Yellow Book* of Aubrey Beardsley and other Decadents. The ''Foreword'' was inflammatory: *''FIRE* . . . weaving vivid, hot designs upon an ebon bordered loom and satisfying pagan thirst for beauty unadorned . . . the flesh is sweet and real . . . the soul an inward flush of fire.''

Represented by only two poems, Hughes nevertheless watched the reception of *Fire!!* with avid interest. The white magazines *Bookman* and the *Saturday Review of Literature* barely, but politely, noticed its appearance. Dr. Du Bois, who perhaps had only glanced at *Fire!!,* called it ''a beautiful piece of printing'' and urged its support in the *Crisis.* However, other reviewers were incensed by Thurman's ''Cordelia the Crude,'' about a promiscuous Harlem teenager (*''Physically,* if not mentally,'' the story starts, ''Cordelia was a potential prostitute'') and the highly suggestive ''Smoke, Lilies and Jade, A Novel, Part I'' by ''Richard Bruce'' (Nugent) about its young hero's opiate hallucinations and androgynous fantasies. ''I have just tossed the first issue of *Fire* into the fire,'' the reviewer in the *Baltimore Afro-American* fumed; Zora Hurston's little play and her short story generally escaped criticism, but Aaron Douglas had ruined "three perfectly good pages and a cover" with his drawings, while Langston Hughes displayed "his usual ability to say nothing in many words." On the other hand, *Fire!!* moved some people. An interracial group of young Chicago writers, including Dewey Jones of the *Defender* and the excellent black poet Fenton Johnson, sought permission to bring out the third number. Incorrectly, they assumed (as Thurman did for a while) that the second was on its way. Instead the editor's personal earnings were soon legally attached to pay the printer. *Fire!!* was finished after one number.

Meanwhile, Knopf accepted "Fine Clothes to the Jew," but not without balking at the title (the firm had published *Nigger Heaven* apparently without difficulty). After Van Vechten personally defended the name, as he recorded in his journal, it was allowed to stand. Van Vechten perhaps had also chosen it, as he had chosen "The Weary Blues." Certainly, Hughes had been thinking of using "Brass Spitoons," from one of his poems. In any event, both Hughes and Van Vechten agreed on the choice, which was unfortunate. Apparently no one alerted Hughes to the effect his title would have on sales, which proved to be opposite to the result of Van Vechten's own crudeness.

In addition to news of the acceptance came other signs of success. Louis Untermeyer, revising his classic *Modern British and American Poetry,* asked permission to include at least two Hughes poems. Edwin Markham, with whom he had rung in the new year during his first winter in New York, took three poems for an anthology of his favorite verse from Chaucer to the present. The controversial "Mulatto" was taken by Henry Seidel Canby at the *Saturday Review of Literature.* And, as further proof of his eminence, Hughes was escorted upstairs to help judge the second Amy Spingarn *Crisis* competition, joining James Weldon Johnson and Babette Deutsch on the poetry panel (the first prize went to Arna Bontemps for "Nocturne at Bethesda," a stately meditation on the modern loss of spiritual values). The second weekend of November found Hughes in Manhattan to see the first talking film, Al Jolson's *The Jazz Singer,* to dine with the Van Vechtens, and to attend a Paul Robeson concert with Carl. And on Friday, November 19, marking his first reading in the South and also the first of innumerable times his fraternity, Omega Psi Phi, would sponsor him, Hughes read in the chapel at black Virginia Union in Richmond, Virginia. Making the visit even more remarkable was a small party in his honor given by a group of whites, including a young woman, led by Hunter Stagg, literary editor of the *Richmond Times-Dispatch* and a friend of Van Vechten's. Stagg reduced the tension with "Hard Daddy" cocktails (named after one of Langston's blues) of whiskey, lemon juice, maple syrup, and ice. Far less enjoyable was Hughes's appearance two days later at another inter-racial affair, this time sponsored by the Urban League in Columbus, Ohio, at which Countee Cullen was also present. Hughes detested few things more than formal meetings of blacks and whites to discuss racial harmony. The meeting, held in the Chamber of Commerce auditorium, dragged on for hours as Langston sat on the stage and listened to a succession of tedious, often hypocritical speeches. "Under such boredom," he scribbled, "no wonder the races never come together."

At Christmas he was so anxious to be in Manhattan he left before the official school break, staying in New York as he almost always did at Thurman and Bruce Nugent's rooming house at 314 West 138th Street. On Thursday, December 16, he took part in a midnight concert at the Fifth Avenue Playhouse with Hall Johnson's choir, which sang four of his poems set by Johnson. Langston dined at the James Weldon Johnsons' with the Knopfs, the Van Vechtens, the Charles S. Johnsons, young Clarissa Scott from Washington and her new husband, Hubert Delany, and other friends. After dinner, Langston read some of his poems. But the laurel went to the host, James Weldon Johnson, who read from his moving collection of poetic sermons published that year as *God's Trombones*. Johnson read the best: an old black preacher's inspired account of Genesis called "The Creation." Visiting Amy Spingarn, just back from Zurich, Hughes presented her with a set of prints of Aaron Douglas drawings illustrating some of his poems from *Opportunity,* specially prepared as his Christmas present to close friends. At the Civic Club, he picked up a little money reading poetry with the well-known poets Charles Erskine Scott Wood from California and Margaret Larkin. Then, with presents for his mother and fourteen-year-old Gwyn Clark, he went down to Atlantic City to spend Christmas. Hughes welcomed the new year, 1927, at a party thrown by Alfred and Blanche Knopf and attended by a host of celebrities, including Jascha Heifetz, Ethel Barrymore, and Fannie Hurst. To his embarrassment, he was the only man not in tails. Unlike black Washington society, however, no one seemed to mind.

Running out of money, Hughes returned to Lincoln early. He wrote to Locke for ten dollars, and got five. Then, instead of preparing for his examinations, he took a bold new step in his career. Hughes started a projected series of six short stories, the first he had written since high school, to be set against an African background and involving crewmen on a steamer called the *West Illana* but obviously modelled on the *West Hesseltine,* on which he had sailed to Africa. Almost certainly, Hughes was getting the jump on a new instructor, Charles Boothby, who that year had replaced the aged senior professor of English at Lincoln and brought new life and a revised English curriculum to the school. In the coming term Langston intended to take Boothby's course "The Short Story," which included the "Writing of original narratives."

Almost as daring as anything Hughes had ever written, these stories, like many of the poems in *Fine Clothes to the Jew,* were ripe fruit of the extraordinarily fertile summer and fall of 1926. The tight, humid parties in Harlem, the heady autumn flights from the Lincoln woods to New York, the sense of dangerous excess on which certain members of his circle thrived, all combined with the energizing effects of Langston's first months among the young black men at Lincoln to inspire him with a confidence that resulted, in both his poems and these stories, in a franker revelation of aspects of the libido than previously seen in his work. Under cover of the African dark, his ship of fiction plowed warmer, more sensual waters than he had ever sailed before; whatever their limitations as art, the four stories that resulted steam suggestively of miscegenation, adultery, promiscuity, and the turmoil of sexual repression. "The Young Glory of Him" tells of the unfulfilled love of a plain young woman passenger, the daughter of missionaries, for the handsomest, most rakish sailor of the *West Illana*—a love that ends in her suicide. "The Little Virgin" concerns a bashful young sailor who fights an older, rougher mate who had slapped an African woman, then succumbs to tropical fever and delirium. In "Bodies in the Moonlight" the rough, strong, hard seaman "Porto Rico" slashes the neck of the gentler, more passive, Hughes-like narrator in a squabble over a local beauty, Nunuma, with whom they both wish to go to bed. "Luani of the Jungles" is about the unhappy marriage of a white man, formerly a Sorbonne student, to a wealthy, educated Nigerian woman who promptly takes to the jungle once she returns to Africa with him. He must endure her hunting and fishing without him, and then her taking the chief's virile son as a lover ("A woman can have two lovers and love them both," she states simply). Tortured ("she drives me mad"), the husband tries four times to leave her, but always returns to Luani, even after she has a child by the chief's son.

The stories went to Wally Thurman, who took "Luani of the Jungles" for his new magazine, *Harlem,* and persuaded George Schuyler, his replacement at the *Messenger,* to publish the others (Thurman had just become circulation manager of the white religious magazine *The World Tomorrow*). By mid-January Hughes also had copies of *Fine Clothes to the Jew.* The first reports were very encouraging. Far from objecting to the title, Amy Spingarn liked the book even more than *The Weary Blues,* because it seemed "more out of the core of life." Her brother-in-law Arthur Spingarn, who was also Jewish, noted the title but found the book a "splendid" work, in which "Jacob and the Negro come into their own." George Schuyler praised Hughes as unquestionably "the poet of the modern Negro proletariat." After the attacks on *Nigger Heaven* and *Fire!!,* however, Hughes was nervous. "It's harder and more cynical," he explained defensively to Dewey Jones of the Chicago *Defender,* and "limited to an interpretation of the 'lower classes,' the ones to whom life is least kind. I try to catch the hurt of their lives, the monotony of their 'jobs,' and the veiled weariness of their songs. They are the people I know best."

On February 5, just as he prepared to set out on a tour for Negro History Week (founded the previous year by his Washington employer Carter G. Woodson), the black critics opened fire. Under a headline proclaiming Hughes a "SEWER DWELLER," William M. Kelley of the New York *Amsterdam News,* who once had sought his work, denounced *Fine Clothes to the Jew* as "about 100 pages of trash. . . . It reeks of the gutter and sewer." The regular reviewer of the *Philadelphia Tribune* adamantly refused to publicize it; Eustace Gay confessed that *Fine Clothes to the Jew* "disgusts me." In the *Pittsburgh Courier,* historian J. A. Rogers called it "piffling trash" that left him "positively sick." The *Chicago Whip* sneered at the dedication to Van Vechten, "a literary gutter-rat" who perhaps alone "will revel in the lecherous, lust-reeking characters that Hughes finds time to poeticize about. . . . These poems are unsanitary, insipid and repulsing." Hughes was the "poet 'low-rate' of Harlem." The following week, refining its position, the *Tribune* lamented Langston's "obsession for the more degenerate elements" of black life; the book was "a study in the perversions of the Negro."

To these and other black critics, Hughes had allowed the "secret" shame of their culture, especially its apparently unspeakable or unprintable sexual mores, to be bruited abroad by thick-lipped black whores and roustabouts. How could he have dared to publish "Red Silk Stockings"?

> Put on yo' red silk stockings,
> Black gal.
> Go out an' let de white boys
> Look at yo' legs.
>
> Ain't nothin' to do for you, nohow,
> Round this town,—
> You's too pretty.
> Put on yo' red silk stockings, gal,
> An' tomorrow's chile'll
> Be a high yaller.
>
> Go out an' let de white boys
> Look at yo' legs.

Or "Beale Street Love"?

> Love
> Is a brown man's fist
> With hard knuckles
> Crushing the lips,
> Blackening the eyes,—
> Hit me again,
> Says Clorinda.

By pandering to the taste of whites for the sensational (the critics ignored their own sensationalism, demonstrable in the scandal-ridden sheets of most black weeklies), Hughes had betrayed his race. In spite of this hostility, however, and perhaps the poorest sales of any book Hughes would ever publish (a fact he attributed to its title), *Fine Clothes to the Jew* was also his most brilliant book of poems, and one of the more astonishing books of verse ever published in the United States—comparable in the black world to *Leaves of Grass* in the white. Marking his maturity as a poet after a decade of writing, *Fine Clothes to the Jew* represents Hughes's most radical achievement in language. While *The Weary Blues* had opened with blues and dialect poems before presenting the sweeter, more traditional lyrics, a prefatory note ("the mood of the *Blues* is almost always despondency, but when they are sung people laugh") indicates the far greater extent to which *Fine Clothes to the Jew* falls deliberately within the range of authentic blues emotion and blues culture. Gone are the conventional lyrics about nature and loneliness, or poems in which the experience of the common black folk is framed by conventional poetic language and a superior, sometimes ironic poetic diction. Here, on purpose, few poems are beyond the range of utterance of common black folk, except in so far as any formal poetry by definition belongs to a more privileged world. *Fine Clothes to the Jew* was the perfect companion piece to Hughes's manifesto "The Negro Artist and the Racial Mountain."

As a measure of his deeper penetration of the culture and his increased confidence as a poet, three kinds of poems are barely present in *Fine Clothes to the Jew*—those that directly praise black people and culture, those that directly protest their condition, and those that reflect his own personal sense of desolation. For example: "Laughers," which celebrates blacks as "Loud laughers in the hands of Fate," is also probably the earliest piece in the book, having been published first as "My People" in June, 1922. "Mulatto" lodges perhaps the strongest protest, but is staged dramatically:

> . . . The Southern night is full of stars,
> Great big yellow stars.
> O, sweet as earth,
> Dusk dark bodies
> Give sweet birth
> To little yellow bastard boys.
>
> *Git on back there in the night,*
> *You ain't white.*
>
> The bright stars scatter everywhere.
> Pine wood scent in the evening air.
> A nigger night,
> A nigger joy.
>
> *I am your son, white man!*
>
> A little yellow
> Bastard boy.

Only one poem, "Sport," proposes life as an empty nothingness—as "the shivering of a great drum / Beaten with swift sticks." Sorrow and despair dominate *Fine Clothes to the Jew,* but mainly through the expressive medium of the blues and its place in the lives of poor black men and women. In "Hey!" the blues is mysterious: "I feels de blues a comin', / Wonder what de blues'll bring?" It is also, as in "Misery," soothing, or even cathartic:

> Play de blues for me.
> Play de blues for me.
> No other music
> 'Ll ease ma misery . . .

Although the blues drifts in most often on the heels of lost love, the feeling can come for other reasons and still have poetic power. "Homesick Blues":

> De railroad bridge's
> A sad song in de air.
> De railroad bridge's
> A sad song in de air.
> Ever time de trains pass
> I wants to go somewhere . . .

In *Fine Clothes to the Jew,* the singers and sorrowers are mainly women. By comparison, men are almost shallow; one man (''Bad Man'') beats his wife and ''ma side gal too'': ''Don't know why I do it but / It keeps me from feelin' blue.'' Men may be hurt in love, like the fellow in ''Po' Boy Blues'' who met ''a gal I thought was kind. / She made me lose ma money / An' almost lose ma mind.'' But the blues are sung most often, and most brilliantly, by black women. Sometimes they sing to warn their sisters (''Listen Here Blues''):

> Sweet girls, sweet girls,
> Listen here to me.
> All you sweet girls,
> Listen here to me:
> Gin an' whiskey
> Kin make you lose yo' 'ginity . . .

Or, as in ''Lament Over Love,'' their daughters:

> I hope ma chile'll
> Never love a man.
> I say I hope ma chile'll
> Never love a man.
> Cause love can hurt you
> Mo'n anything else can.

Women sing for being cheated, for having been done wrong by ''a yellow papa,'' who ''took ma last thin dime'' (''Gypsy Man''); or, as in ''Hard Daddy,'' they grieve over male coldness:

> I cried on his shoulder but
> He turned his back on me.
> Cried on his shoulder but
> He turned his back on me.
> He said a woman's cryin's
> Never gonna bother me.

Sometimes the sorrow goes deep to the bone, when loss or the prospect of loss is mixed with profound self-abnegation and despair. ''Gal's Cry For A Dying Lover'':

> . . . Hound dawg's barkin'
> Means he's gonna leave this world.
> Hound dawg's barkin'
> Means he's gonna leave this world.
> O, Lawd have mercy
> On a po' black girl.
>
> Black an' ugly
> But he sho do treat me kind.
> I'm black an' ugly
> But he sho do treat me kind.

High-in-heaben Jesus,
Please don't take this man o' mine.

But the blues can reflect great joy as well as sorrow, as in "Ma Man," where a black woman's emotional and sexual ecstasy is so overpowering it drives her into song:

When ma man looks at me
He knocks me off ma feet.
When ma man looks at me
He knocks me off ma feet.
He's got those 'lectric-shockin' eyes an'
De way he shocks me sho is sweet.

He kin play a banjo.
Lordy, he kin plunk, plunk, plunk.
He kin play a banjo.
I mean plunk, plunk . . . plunk, plunk.
He plays good when he's sober
An' better, better, better when he's drunk.

Eagle-rockin',
Daddy, eagle-rock with me.
Eagle rockin',
Come and eagle-rock with me.
Honey baby,
Eagle-rockish as I kin be!

The last stanza of this poem, the second to last in the book (as if Hughes tried to hide it) was among the most sexually teasing in American poetry—to those who understood that "eagle-rocking" was possibly more than a popular dance step. His critics had not howled without cause, but Hughes did not retreat. First at a Baptist church, then before an African Methodist Episcopal congregation in Philadelphia, he fulfilled engagements to read his poems. Then, arriving in New York for a recital by friends at Carnegie Hall, he coolly faced Floyd Calvin of the *Pittsburgh Courier* at the Knopf office on Fifth Avenue. In spite of the reviews, Hughes said, he declined to write about Vanderbilts and Goulds. At least two-thirds of all blacks were lower-class—"even I myself, belong to that class." In any event, "I have a right to portray any side of Negro life I wish to." He defended the blues singers Bessie Smith and Clara Smith as equal to the best of European folk singers, who were honored in Europe and America; and Carl Van Vechten had done more than anyone else for black artists. To the white Cleveland *Plain Dealer,* curious about the hubbub in the black press over poetry, of all things, he explained that the black reviewers still thought that "we should display our 'higher selves'—whatever they are," missing the point "that every 'ugly' poem I write is a protest against the ugliness it pictures."

When the *Pittsburgh Courier* invited Hughes to defend himself against his critics, he did not hesitate. In "These Bad New Negroes: A Critique on Critics," he identified four reasons for the attacks: the low self-esteem of the "best" blacks; their obsession with white opinion; their *nouveau riche* snobbery; and the lack of artistic and cultural training "from which to view either their own or the white man's books or pictures." As for the "ill-mannered onslaught" on Van Vechten: the man's "sincere, friendly, and helpful interest in things Negro" should have brought "serious, rather than vulgar, reviews of his book." A nine-point defense of his own views and practices ended in praise of the young writers, including Toomer, Fisher, Thurman, Cullen, Hurston, and the Lincoln poet Edward Silvera. And Hughes himself. "My poems are indelicate. But so is life," he pointed out. He wrote about "harlots and gin-bibers. But they are human. Solomon, Homer, Shakespeare, and Walt Whitman were not afraid or ashamed to include them." (Van Vechten thought the situation easy to explain; "you and I," he joked to Hughes while making an important distinction, "are the only colored people who really love *niggers.*")

Hughes was not without friends in the black press. The *New York Age* found the book evocative of the joy and pathos, beauty and ugliness of black Americans, if of the more primitive type. The poet Alice Dunbar-Nelson, once married to the master himself, Paul Laurence Dunbar, compared the book to Wordsworth and Coleridge's celebrated venture, also once maligned, into the poetical use of common speech and common lives, *Lyrical Ballads;* Hughes was "a rare poet." Theophilus Lewis praised the book in the *Messenger,* and in the *Saturday Review of Literature* Alain Locke was deft about *Fine Clothes to the Jew:* "Its open frankness will be a shock and a snare for the critic and moralist who cannot distinguish clay from mire." And Claude McKay wrote privately to congratulate him on having written a book superior to his first in all important respects.

Among white reviewers, perhaps the most perceptive evaluation came from the young cultural historian Howard Mumford Jones. Using black dialect austerely, Hughes had scraped the blues form down to the bone, and raised the folk form to literary art. "In a sense," Jones concluded, "he has contributed a really new verse form to the English language." Although, like Wordsworth, he sometimes lapsed into "vapid simplicity," if Hughes continued to grow, he was "dangerously near becoming a major American poet." V. F. Calverton, Margaret Larkin, Arthur Davison Ficke, Hunter Stagg, Abbe Niles, Babette Deutsch, Julia Peterkin, and a wide range of reviewers praised the stark lyrical simplicity and beauty of most of the verse. More than once the comparison was made to Coleridge and Wordsworth's *Lyrical Ballads;* the critics understood that Hughes was trying to effect a historic change in poetry by compelling both blacks and whites to admit the power of black language. Other critics were not so sympathetic. The *Boston Transcript* flatly preferred Countee Cullen and called some of the Hughes verse "tawdry"; the *Nation* reviewer thought that Hughes was merely transcribing folklore, not writing poetry; the *New York Times* judged the volume "uneven and flawed."

The ignorant blasts in the black press were nicely offset on March 1 when Hughes accepted an invitation ("a great honor for me") from the Walt Whitman Foundation to speak at the poet's home on Mickle Street in Camden, New Jersey. Stressing Whitman's humane depiction of blacks in his poetry, Hughes went on to describe modern free verse, and his own work, as descending from Whitman's great example. "I believe," Langston told the little gathering, "that poetry should be direct, comprehensible and the epitome of simplicity." Suspicious of theory, Hughes had nevertheless pointed to one of the main ideas behind his theory of composition—the notion of an aesthetic of simplicity, sanctioned finally by democratic culture but having a discipline and standards just as the baroque or the rococo, for example, had their own. The fact that simplicity had its dangers both extended its challenge and increased its rewards. The visit to Whitman's home left Hughes elated; to Van Vechten he mailed a postcard imprinted with an excerpt from Whitman's "Song of the Open Road": "All seems beautiful to me."

Continuing the fight, in the spring Hughes bravely sent to *Opportunity* a stinging essay about the black middle class in Washington, D.C. When the sedate Charles S. Johnson received "Our Wonderful Society: Washington," he was flabbergasted—"It's a— . . . you will pardon me if I say—Wow! . . . You may be sure that you cannot live there again after it is published." The essay would appear in the August *Opportunity*. Hughes was equally fearless, against a different enemy, when he represented Lincoln at a conference of college chapters of the YMCA on the campus of Franklin and Marshall College in Lancaster, Pennsylvania, a religiously based school where blacks were barred as students (having admitted a Negro once, the college had exhausted its sense of Christian duty). When Hughes learned further that Johns Hopkins University in Baltimore refused to teach Negroes even in correspondence courses, he started a discussion of segregation that grew so warm the alarmed conference leaders appealed to God for guidance and declared the session closed. Still, Langston was satisfied: "They really stopped discussing religion for a while."

For most of the winter, however, with Caroline Reagan's revue in hibernation and his funds low, Hughes kept close to Lincoln. In his mid-year examinations, he dropped to comfortable mediocrity in his grades, which was exactly where he wished to be. His twenty-fifth birthday passed quietly. On Friday, February 11, however, he went to New York to help welcome Van Vechten back from Hollywood, where he had gone to meet various moguls and entertain—or perhaps seek—offers to write film scripts. The big welcoming party was at Eddie Wasserman's on Saturday night, with many of Carl's admirers there, including A'lelia Walker, the James Weldon Johnsons, Harry Block, Covarrubias, Dorothy Harris, Florine Stettheimer, Nora Holt, Alain Locke, and the Metropolitan opera star Marguerite D'Alvarez (once, at Van Vechten's, Bessie Smith listened to D'Alvarez deliver an aria, then consoled her: "Honey, don't let nobody tell you you can't sing!") Two days later, Hughes dined privately with the Van Vechtens, then joined Carl again the next evening to go through steady rain to Carnegie Hall for a recital of spirituals by Rosamond Johnson and Taylor Gordon, sponsored by the Urban League.

This would be one of the most important evenings of Hughes's life. Alain

Locke was in the audience; Langston met him, probably by arrangement, during an intermission. Earlier that day, Locke had been to tea at the apartment at 399 Park Avenue of a very old and very rich woman, Charlotte van der Veer Quick Mason. The widow for over twenty years of a prominent physician and psychologist, Dr. Rufus Osgood Mason, Charlotte Mason was then a few months short of her seventy-third birthday. Some days earlier, present at a lecture by Locke on African art, she had become so excited by his ideas that Mrs. Mason had introduced herself afterwards. Immediately, she felt "tremendous rapport" between them. Finding her not only knowledgeable but also evidently quite wealthy, Locke wasted no time in asking permission to call on her. He telephoned her a few days later, and was invited to join Mrs. Mason at Carnegie Hall on February 16. First, Locke had tea with her, when they talked of African art and the African spirit, and of the young black writers Locke had helped to speak with a bold new voice in America. Again impressed, Mrs. Mason invited him to sit on her most precious piece of furniture, a throne-like formal chair first owned by her great-great-grandfather. She also asked to meet some of these thrilling young black men and women.

At Carnegie Hall, where Locke introduced him to Mrs. Mason, Hughes looked down at an old white-haired woman, beautifully dressed and expensively groomed, who, seated, looked up at him with such grim intensity that her eyes glittered from the effort. After a pleasant exchange, he returned to Van Vechten's party.

A bad late winter cold muffled Langston for a while, but he returned to New York on the long Easter weekend. On Friday, April 15, at a party for the black singer Nora Holt given by Rita Romilly, he met the brilliant young composer Aaron Copland, and the Welsh-born Lawrence Langner and his wife Armina Marshall, both prominent in the theater. The next day, in a meeting arranged by Locke, Hughes visited Charlotte Mason at her Park Avenue home. By this time, she had committed herself to nothing less than a personally directed and financed project to elevate African culture to its rightful place of honor against its historic adversary, which she unhesitatingly identified as the white race. With Alain Locke as her guide, Mrs. Mason's great project was to be the establishment of a museum of African art in New York. That spring, her name appeared on the stationery of the short-lived "Harlem Museum of African Art" project as vice-chairman (real-estate entrepreneur John E. Nail, the father-in-law of James Weldon Johnson, was chairman, and Locke was secretary), when its proponents began a fund-raising drive. Five controversial years later, in May, 1932, she looked back on her ambitions and her state of mind as she set out to counter the destructive influence of whites on the primitive peoples. "I believed I could find leaders enough among themselves," she wrote of blacks, "to react and publish far and wide the truth about their art impulse that daily lived in all their common acts." As Mrs. Mason dreamed of her museum of African art, she envisioned "little Negro children running in and out learning to respect themselves through the realization of those treasures. And . . . as the fire burned in me, I had the mystical vision of a great bridge reaching from Harlem to the heart of Africa, across which the Negro world, that our white United States had done everything to annihilate, should see the flaming pathway . . . and recover the treasure their people had had in the beginning of African life on the earth.''

Into this consuming vision Hughes strolled almost innocently on the afternoon of April 16. Mrs. Mason's great wealth must have been uppermost in his mind as he sat and talked of books and plays and people with her. Unlike the other rich persons he knew, Charlotte Mason was formal and intense. Her apartment exuded superb taste; her servants were English or Scandinavian, not Irish or Negro. At hand were gifted women. Cornelia Chapin, a sculptor, helped her in all her enterprises; Miss Chapin's sister Katherine Garrison Biddle, a poet, waited in attendance, with Mrs. Biddle's two little sons, Randolph and Garrison, the domestic lights of Mrs. Mason's life. With money that went back so many generations in her family that Mrs. Mason sneered at certain pillars of society as *arrivistes,* she had known some of the most powerful men and women of her time. And there was nothing faddish about her interest in primitivism. Having spent time among the Indians of the Great Plains, she later financed research that resulted in 1907 in Natalie Curtis's famous *The Indians' Book.* In her sudden enthusiasm for Africa, Charlotte Mason was an ancient volcano erupting again. She believed in cosmic energy and the intuitive powers of primitive life, from which came her drive to nourish those races still keenly in tune with spectral harmonies lost to the ears of overcivilized whites. Fearless almost to the point of fantasy, Mrs. Mason could be a terror to those who crossed her. And whenever the force of her personality or her logic began to fail, her cash usually got her what she wanted.

Hughes's visit on April 16 went very well. From her shining windows the view of Manhattan was stunning, as was the tempered beauty of her furnishings and *objets d'art.* Then, the radical force of her views on the black masses, socialism, and culture, attitudes extraordinary in one so old, so rich, so white, amazed him. In turn, Mrs. Mason found Hughes more fascinating than she had expected. What probably moved her most was something she hardly expected but saw at once—the American Indian in Langston, inherited from his mother's parents, and visible in his magnificent profile, almost straight black hair, copper-colored skin, and pacific manner. The Indian was Mrs. Mason's first love. Blacks were often far too civilized for her taste (on this score the *raffiné* Alain Locke tried her patience from the start). In Langston's winsome brown boyishness, however, she saw a noble young savage; he was the Indian child she had perhaps once wished to nurture. As the meeting drew to a close she knew that she wanted him for her very own. Just before she shut the door she gazed into his eyes and pressed a note into his hand—''a gift for a young poet.'' In his palm flowered a fifty-dollar bill.

Returning to his dingy room in Cresson Hall, Hughes thought a great deal about Mrs. Mason. She was herself ideally suited to his own more recessed needs. A woman, but old, she could be a mother, even a returned grandmother to the child never very far below his surface; rich and white, she could only be interested in him for himself. Undoubtedly, her money also attracted Langston. Although he was not venal, having veered away from materialism in making his major choices in life, Hughes also understood the value of money. Even now, his continuance at college depended almost entirely on Amy Spingarn's goodwill.

The weekend before his final examinations he returned to New York to see Mrs. Mason. Hughes's visit on Sunday, May 22 (their "first real hours together," she would remember) consolidated their friendship. By the time he left her home, perhaps chauffeured away in her limousine by her grim white driver, Mrs. Mason was his "Godmother," which was the only name she wished her close friends to use in writing to her. By this time, too, Hughes was growing accustomed to the afflated rhetoric which made each of her communications a stirring command. Langston was, she wrote on June 5, "my winged poet Child who as he flies through my mind is a noble silent Indian Chief—a shining messenger of hope for his people—and then again a precious simple little boy with his pocket full of bright colored marbles—looking up at me with his dear and blessing eyes. In all these and in the myriad other phases his 'Godmother' holds an infinite belief."

This missive reached him in the South, where he would spend the summer. At first he had toyed with the idea of going to Abyssinia with Locke, but when he received an invitation from Fisk University in Nashville, Du Bois's alma mater, to read his poems at the June commencement, Hughes decided to tour the South, encouraged by at least one letter from Zora Neale Hurston in Florida about the rich folk material everywhere. By the end of May, he had secured engagements to read in Texas. Once he was near the Gulf Coast, the Caribbean would be open, especially Cuba and Haiti, which he was eager to see. In Nashville, he was warmly received when he read on Saturday, June 4, in the Fisk University chapel. After Lincoln's all-male rowdiness, the Fisk students were impressive: "they do have an air of seriousness and refinement about them." He dined with the recently arrived white president Thomas Elsa Jones, a former U.S. ambassador to Japan, and his wife and two boys. Langston liked Jones at once, but when the president made a remark about the need to burn away black dialect, Hughes responded sharply—and reported the exchange to Godmother, whose spirit was with him from the start. "May your flight through the South be swift and clear," she had wished, "and may your Treasure house be filled with a perfect vision of your people, both in living and in aspiration."

Much as Hughes admired Fisk, he was glad to leave for Memphis, shuck his jacket and tie, and be nobody. In Memphis, he rushed straight to Beale Street and the notorious red-light district famed for the blues. Taking a room at 345 1/2 Beale Street, he explored the neighborhood. In a barrelhouse at 4th Street and Beale, then in a vaudeville show and an amateur music contest, he soon found out that every musician apparently had once played with W. C. Handy; Beale Street knew its own history a little too well. Hughes found the area positively tame compared to certain sublimely sleazy parts of Harlem and Washington. When his plan to read in Texas collapsed after the most destructive flood ever recorded in the region closed traffic to the entire Southwest, Hughes headed for the Gulf of Mexico. On June 12, attempting a grand detour around the flood, he made for Vicksburg, Mississippi. At Baton Rouge, Louisiana, he passed several days in refugee camps overcrowded with victims of the marauding Mississippi River. The racism of the authorities, especially the Red Cross, galled him; he talked to blacks laboring to save levees, forced at gun point to toil like slaves, fighting off rising waters, whites and snakes, hunger and disease. Misery was everywhere—and so were the blues. Moving quietly through the camps, Hughes tried to record the more inventive songs.

In New Orleans, which he liked from the start because it reminded him of a Mediterranean city, he went straight for the main black thoroughfare, Rampart Street, and checked into a rooming house that was part restaurant, part speakeasy, part honkytonk hotel. Rooms, including his own, were sublet by the hour in the daytime by a landlady who sold bootleg liquor and fried fish. In his pocket were several names of residents given him by friends in New York, but he called on none: "I'm having too grand a time walking around with no coat, to want to put it on to go see anybody." After a day or two he moved to the French Quarter, near the quaint Vieux Carré in front of St. Louis Cathedral, amidst filigreed iron balustrades and carved, weathered wooden facades. The old, gloomy house in which he stayed had stone floors and walls and ornate wooden shutters and no electricity or running water to mar its authenticity. Nearby on Poydras Street he browsed in the voodoo shops selling herbs, roots, conjuring powders, potions, and cures. "I bought some Wishing Powder," he would remember, "and I think it brought me luck, because the next day, quite unexpectedly, I found myself on the way to Havana." On July 9, Hughes shipped out on the *Munloyal,* a small freighter with a Filipino steward and Chinese messmen. But the trip was disappointing. In Havana's Chinatown, he visited a gloomy bordello where Cuban women catered to Asiatics. The men gathered at each door, looked at the women hauntingly, then moved on "to stand in a ring before another girl's doorway, circling thus around and around . . . in a kind of mass silence. It was the most depressing brothel I have ever seen." San Isidro Street was famous enough but, like Beale Street, clearly past its prime; he was happy to be gone.

Back in New Orleans on July 20 he tried to find Albert Edwards, a young black man Langston had first met in 1921 in Manhattan. Edwards was out of town, but Hughes met his wife, who was Walter White's sister. After three days as a celebrity, he fled New Orleans. He passed through Biloxi, Mississippi, to Mobile, Alabama. No sooner had Langston stepped off the train than he saw a familiar figure—Zora Neale Hurston, en route back to Manhattan and Barnard College driving a little car she called "Sassy Susie." Hurston had been all over Florida gathering material for the anthropologist Franz Boas at Columbia and also for Carter Woodson, Langston's old employer. Hughes and Hurston decided to join forces and, although he could not drive, share "Sassy Susie" northwards back to school: "I knew it would be fun travelling with her. It was."

If Hughes wanted to know the South, Hurston was perhaps the ideal companion. For drawing out rural folks wary of educated or "dicty" types, she was probably without equal; probably only Zora could have gotten away, as she had done, with measuring the heads of strangers on Harlem streets. Born in an all-black Florida town in 1891—but blithely passing herself off as being ten years younger—she had lived there until, heeding her mother's advice to "jump at de sun," she had quit home in her early teens to join a Gilbert and Sullivan touring troupe as a wardrobe girl. Later, Hurston entered Morgan Academy in Baltimore, where she graduated in 1919, then studied at Howard University between 1919 and 1924. A story in *Opportunity,* in which Hurston drew on the culture of her native town and emphasized its carefree character but also its rich folk language, attracted the attention of the younger literati. In January, 1925, she moved to New York, won a prize in the *Opportunity* contest that year, and secured a scholarship to Barnard, where she soon attached herself as a student to her "Papa Franz," Franz Boas. Her study of anthropology and folklore developed alongside her writing, but she was determined to complete her academic training. In May of the current year, 1927, she had married her fiancé of some years, a medical student, then left him to get on with her research.

Flamboyant in dress and speech, Hurston's personality outshone all competition. Of all the "Niggerati," Hughes would write, Hurston was "certainly the most amusing. . . . She was full of side-splitting anecdotes, humorous tales, and tragicomic stories, remembered out of her life in the South. . . . She could make you laugh one minute and cry the next." Rattling north in her car, they compared their notes. In Memphis, for example, Hughes had heard a brand new "celestial railroad" song sung by the faithful, with angels as redcaps to greet the faithful at "de heavenly depot." In New Orleans he had recorded some fine verses from a stevedore named Big Mac:

> Throw yo' arms around me
> Like de circle round de sun
> Throw yo' arms around me
> Like the circle round de sun
> And tell me, pretty mama,
> How you want yo' lovin' done.

From his notes on conjure lore (which Zora knew inside out) picked up on Poydras Street in New Orleans, Hughes recited the sonorous names of remedies, potions, and elixirs (Follow Me Powder, War and Confusion Dust, Black Cats Blood); rituals and recipes for life and death, sickness and health, love and revenge; even conjures to beat conjures. He had also been laboriously copying down all the sweet expressions he had heard. A "brick lick" was being hit by a brick; "give 'im a road map" meant make the fool run; a "hip slinger" was a sexy woman; "now you done stepped on my zillerator" meant now you done provoked me. The amazingly creative language kept Langston scribbling; Zora tossed her head and laughed and slipped him some more. Sharing their money and food, their knowledge and their ideas, they also discussed their proposed folk opera, about which both Hughes and Hurston were more enthusiastic than ever.

At Tuskegee Institute, east of Montgomery, Alabama, summer school was in session, with a Wednesday series of invited speakers in which Jessie Fauset, to Langston's delight, had just come to speak. She, Hurston, and Hughes made a solemn visit to the grave of the founder, Booker T. Washington, a cham-

pion of vocational training and a figure revered by the masses of blacks in spite of Du Bois's radical challenge to his accommodationist approaches to education and civil rights at the start of the century. Hughes also visited the most famous figure then at Tuskegee, the agricultural chemist George Washington Carver, born a slave, who had discovered hundreds of uses for the peanut and the soy bean. Almost a recluse, Carver talked quietly to Langston in his oddly high-pitched, girlish voice, and shyly showed him the paintings he created when he was not toiling in his laboratory. When Hughes sent him his two books of poetry, Carver signed his note of thanks "with so much love and admiration."

So far, Hughes had seen mainly the urban South. Eager to visit the back country, he secured permission to travel with the Moveable School Force, a school-on-wheels program started by Washington himself to take education into the countryside. Now the Force moved on a big white truck, with a Delco generator for the many places without electricity. Rolling past Montgomery north to Decatur and Huntsville, the truck turned off the main road down a dozen miles or more of dusty, red clay tracks lined by pine and oak, mulberry and chinaberry, holly and red gum. The summer fields were heavy with cotton, corn, and sweet potatoes, soybeans and sorghum, sugarcane and sunflowers. Deep in the hills lived the country folk awaiting the blend of practical and moral instruction for which Tuskegee was famous. Women were taught to care for babies and typhoid patients, to bake bread, and to can fruits and vegetables. While younger boys cleared ground, the local men constructed, under careful supervision, a toilet. Many of the people had never seen a motion picture. One evening, to disbelieving cries of "Look a yonder!," the Force showed health and education films to over a hundred people, then ended the meeting with a lecture on toilets ("But what could be needed more in the community?" Langston scribbled in his notebook).

Staying with a couple and their ten children, Hughes rose at dawn, ate hearty country breakfasts of chicken, rice, hot biscuits, and gravy, then followed the older boys on their rounds. He went swimming in the cool river, taught the younger children how to blow bubbles with spools, and held the youngest boy in his arms and rocked him to sleep. One evening he gave a speech on Great Men, "with greater emphasis on Great Negroes." Most of the children knew of no great blacks at all. One thought Abraham Lincoln was a Negro.

By August 10, he was back at Tuskegee; he gave a reading there on that day. A reporter commented on "his pleasing voice, his assured, unaffected manner"; Hughes "may not have sold his verse; but he sold himself." Hughes also received a letter, and a check, from Mrs. Mason. Telepathically, the old woman was monitoring his spiritual progress as he roamed the South: "As I watched the stars come out many fresh fields of growth of your inner being transfigured themselves before me." He must say nothing of his trip to his friends when he reached Manhattan, so that "later when you are ready to use it the flame of it can burn away the *débris* that is rampant there."

After promising to write a poem about Tuskegee to be set to music as an anthem (he would send "Alabama Earth"), Hughes headed north with Zora Neale Hurston. At a backwoods church near Fort Valley, Georgia, they heard a singer end his performance with an inspired version of the Lord's Prayer: "Our Father who art in Heaven, / *Hollywood* be thy name!" Visiting the old Toomer plantation, to which Jean Toomer had returned for the inspiration that resulted in *Cane,* they met his relatives. Learning of an excellent herb doctor and conjurer, Langston and Zora agreed it would be a pity not to conjure one of their friends in New York. In Macon, they rushed to hear Bessie Smith in concert, then discovered the next day that she was staying at their hotel. With Hurston taking the lead, Smith was warmer this time to the Bard of the Blues. "The trouble with white folks singing blues," she told him, "is that they can't get low down enough." Through a friend, he and Zora visited a railroad yard, where Zora got to sit up high in the driver's seat of a massive locomotive.

What amazed Hughes most about the South was the way black people enjoyed themselves; so much of what he had imagined was erroneous—from reading! "It seemed rather shameless to be colored and poor and happy down there at the same time," he confessed a few weeks later. "But most of the Negroes seemed to be having a grand time and one couldn't help but like them." There was surely a lesson for him in their happiness: "I'm sick of being unhappy anyway." Driving east to Savannah, Georgia, and the water, they coaxed new songs out of the guitar pickers and bluesmen on the docks. Everywhere Hughes kept his eyes and ears open for the spirit of the South—a man driving a goat cart in Columbus, gourds for bee-honey hanging high on poles, jet-black, naked children by the side of dusty roads, families toiling together under the blazing sun. A woman wanted a pistol so she could "sting" her errant husband; a blind man stroked a guitar in Decatur: "Mary had a baby, Oh Mary baby boy, What a darlin' baby! Don't you love Jesus?" One woman was so evil, "she sleeps with her fists doubled up"; they met a man in Savannah, in a chain gang at fifteen, in hiding now. They heard a sonorous conclusion to a litany of hope: "That's what Justice been tryin' to git Righteousness to do all these years!"

Near the last day of August, Hughes and Hurston crossed the Mason-Dixon line. Stopping briefly at Lincoln, they lunched with Professor Labaree and his wife, Mary. But Hughes was eager to reach Manhattan and Godmother. "Can you guess, dear Langston," she had written from her summer home in Connecticut, "what a warm welcome waits the fresh young soul who left me early in June?" They reunited in a suite at the luxurious Barclay Hotel at Lexington Avenue and 48th Street, where Mrs. Mason regularly stayed while her Park Avenue home was readied after a long absence. While she listened hungrily, Langston poured out a full report of his travels and experiences; he put in a strong word for Zora Hurston—Godmother must meet her, she was wonderful! Godmother was far more interested in Langston's plans. He must convert the exciting summer months among the black folk into art; he must write a novel!

Hughes was not interested in writing a novel; composing long literary pieces did not appeal to him. Still, he tried to summon enthusiasm for the project. Godmother urged him to leave New York and its distractions as soon as possible (she despised Van Vechten in particular). Langston must allow the inspiring power of Africa to possess him. Instead, Hughes dallied in the city. Like an explorer back from a forbidden country, he was interviewed on a radio station about the South. Du Bose and Dorothy Heyward's stage version of *Porgy* was being prepared, with apparently every black actor and would-be actor in New York, including Bruce Nugent and Wally Thurman, in the cast. At least twice, Langston visited Van Vechten: at a dinner at Eddie Wasserman's he joined Carl, Donald Angus, Zora, her husband Herbert Sheen, and Ethel Waters. Also there were Elmer Imes and his wife, Nella Larsen, the talented, half-Scandinavian black writer who would publish two novels, *Passing* (1928) and *Quicksand* (1929), with Knopf. With Hughes's enthusiasm for their blues revue rekindled by his trip, he appealed to Caroline Reagan for a retainer that would allow him to take a leave from Lincoln during the fall; but she couldn't help him. Instead, Langston took another loan from Amy Spingarn, and collected his royalties (what was left of $244.07) from *Fine Clothes to the Jew*.

Returning to Lincoln in time for registration on September 19, he was soon hard at work on the novel—or so Godmother believed. "How splendidly," she enthused a few days later, "you are conducting the pregnancy of these children of your spirit into existence." Perhaps Mrs. Mason was relying for her information on telepathy. Bent on vicarious autobiography, Langston faced the major task of blending his own early, highly atypical life with the more mature fictional interests encouraged by his journey through the South, which had furnished him with the raw material to construct a hero more representative of the race than Hughes himself had ever been. His task was to find a way to begin this conversion—to invent a plausible hero and his setting, an adhesive plot.

In early October, twin shocks jolted him—the death of the poet Clarissa Scott, the bride of Hubert Delany, and of the great musical star, discovered in *Shuffle Along,* Florence Mills. The sudden death of young Mills at the peak of her international popularity devastated Harlem; Hughes went to New York for the funerals. Later, on October 20, when Lincoln transformed itself for the inauguration of its fifth president, William Hallock Johnson, Langston slipped away from the pomp for another visit to New York. At the long-suffering credit union which had injudiciously loaned Thurman money for *Fire!!,* he paid $25 on the account. Visiting a Hall Johnson rehearsal, he discussed further settings of poems with him. Hughes had tea with Caroline Reagan and talked about *O Blues!* And Godmother had finally received Zora Neale Hurston. ("I think we got on famously," Hurston wrote Hughes. "God, I hope so!") He went to Van Vechten's for cocktails, and visited backstage with Thurman and Nugent at rehearsals of *Porgy*.

Everything, however, was secondary to his time with Godmother on Saturday—"seven hours that went like one," he marvelled. "We don't bore one another. She is so entirely wonderful." Dining on venison, they talked about the plan for a black folk opera, which Zora had developed with Godmother. About Langston's lack of progress on the novel Mrs. Mason was understanding; how could he attain "flight," her key word for creative power, while he worried about money and was distracted in so many ways at Lincoln? And yet he could not give up Lincoln. Freely Hughes confessed to her what life there meant to him in helping to heal the wounds of loneliness and alienation opened in his childhood; he held back very little from Godmother.

That weekend, before he returned to Lincoln, Godmother presented him with a plan to free Langston for the wondrous "flight" she was certain he would make. Having shelved her project for a museum of African art, she would undertake a more personal plan, beginning with him. At first for one year, but renewable, Mrs. Mason would pay him a monthly stipend of $150 or so—they would decide on the details later, with advice from Locke. While Langston would report his expenses scrupulously, the money would be his own, to encourage his art. In fact, the arrangement must be secret. He must never use her name beyond the circle she sanctioned.

The generosity of her offer startled Hughes—he could certainly use the money! The day after he dined with Mrs. Mason he had to borrow $6 from a friend's mother to get back to Lincoln. And the sum was enormous—over a year, it would amount to more than four times the sum of tuition, fees, room, and board for nine months at Lincoln. He would be free of all impediments to his creativity, free of nagging debts; in turn he would surrender to Godmother's beneficence. No, not surrender; they would merge their visions. Her money was only a token of the generosity of her spirit and her confidence in him as an artist. Here was the sacrificing, nurturing, always attending mother he had craved in his boyhood, the old, wise woman who spoke her love for him not laconically, as Mary Langston had done in Lawrence, but in rapturous polysyllables.

He returned to Lincoln to make up his mind.

7

GODMOTHER AND LANGSTON
1927 to 1930

Let me become dead eyed
Like a fish,—
I'm sure then I'd be wise
For all the wise men I've seen
Have had dead eyes . . .

"Wise Men," 1927

To celebrate Thanksgiving late in November, 1927, Hughes sent Mrs. Mason a gift of some sheaves of corn. "The corn ears are ripe now," he wrote, "the rising sun in each grain; and the promise of corn ears for tomorrow in each grain; and a secret from the earth in each grain; and a secret from God there, too." The ears of corn were "sun-colored symbols of our Thanksgiving, and symbols of the wonder and the vast power of the things we have to be thankful for." Touched by the gift, Godmother placed the yellow corn in a green dish to the right of her great chair. "Of course, Langston," she wrote, "there is no where an emblem of yourself that is so perfect to send to your 'Godmother'." She loved "the diapason tone sounding in your note—the preservation of life for the morrow—the power of the Sun-God singing a perfect song of Death & Life with the prophecy of Spring." She also sent a check—"the second foundation stone with which [to] build one of your great adventures."

A few weeks previously, on November 5, Hughes and Mrs. Mason had entered into a formal agreement. Each month he would receive the sum of $150, an amount that did not preclude other gifts for special needs. His writing during this time would be his property, not Mrs. Mason's, but Godmother expected to be consulted regularly on every important aspect of his creative flight. His single obligation was to send an itemized account of his expenses each month to Godmother. Money not actually spent was to be banked against his future needs; Hughes could use it to go on vacation after a stint of hard work, or in any other way likely to help his art by adding "to your storehouse of delightful memories."

A month later, when Zora Neale Hurston entered into her own agreement with Mrs. Mason, there were important differences. She received $200 a month, as well as an automobile and a movie camera to help in an ambitious new venture collecting folklore in the South. Unlike Charlotte Mason's understanding with Hughes (perhaps because of her suspicion of the motives of scholars such as Franz Boas), she held contractual power over Hurston's material. And although the relationship between the women was warm, either from the start or some time later Alain Locke emerged as an intermediary between them. No intermediary stood between Hughes and Godmother; she supervised him directly.

With the signing of his agreement with Godmother, Langston passed into a life as close to luxury as he would ever come. To escort Mrs. Mason properly to plays, concerts, and lectures, his aging dark suit was retired in favor of fine evening wear from shops on Fifth Avenue; for tickets to shows, including those he saw without her, Hughes had only to call or write her secretary—the seats were always the best to be had. Boxes of fine bond paper arrived at Lincoln, as well as packages of food specially chosen by the nutrition-conscious Godmother to supplement his college diet. (A list of meals from the Lincoln dining room so astonished Mrs. Mason—"the horrible menu took my breath away"—that she rushed it to Cornell University for analysis.) And ferrying him to and from her home on Park Avenue was her grim-faced white chauffeur, never more oppressed than when Hughes's friends among the black porters—some of whom were Lincoln men—crowded admiringly about him as he alighted from her glossy limousine at Pennsylvania Station.

Without the slightest apparent difficulty, Langston eased himself into the routine of privilege and a close familiarity with Godmother. Soon he was passing hours at her feet; Mrs. Mason liked to have her brown godchildren perched on a stool before her throne-like ancestral chair. Attended by her servants, they dined elegantly on meticulously prepared meals; Hughes quickly acquired a taste for fine food. When they went out together, he escorted Mrs. Mason with a mixture of bashfulness and consummate ease, as if to the manner born. He was completely confidential about their relationship, as she wished him to be. Over a year of their contract passed before Countee Cullen's best friend Harold Jackman, squiring Dorothy Peterson to hear the young contralto Marian Anderson sing, espied one of the more fantastical couples he had ever laid eyes on. "On the way," Jackman breathed to Cullen in Paris, "I saw Langston escorting a dowager (white) of ninety-eight. She must have been that or at least an octogenarian—no kidding, she was really very old. Langston was all properly 'tuxed' and the old lady handed him the carriage check and the last I saw of him he was getting into the automobile."

Christmas, 1927, was the most comfortable, and perhaps the happiest of his life; Godmother gave Hughes some present so grand that she warned him not to expect another like it in the future. Vigilant about his happiness, she sought to remove any drag on his creativity. One major source of worry was his "brother" Gwyn Clark, who was at least two years behind in school, spoiled, and rebellious. Evidently Langston placed much of the blame for Gwyn's lack of character on his mother, who thought Gwyn "one good hearted kid" in spite of abundant evidence to the contrary. In mid-January, Godmother instructed Alain Locke to find the boy a home and a school, at her expense, in New England. Soon Gwyn was living with a respectable black woman, Mrs. Eva Stokein, at 29 Lillian Street in Springfield, Massachusetts. Although this arrangement did not last long (sabotaged in some respects by Carrie, who seemed to encourage Gwyn's truculence), it only deepened Hughes's affection for Godmother, as well as his indebtedness to her.

Subtly Langston's attitude to his work and his friends changed. With his generous stipend, he soon informed Claude McKay (with whom he corresponded fairly regularly, although they had never met) that "I no longer read my poetry to ladies clubs, Y.W.C.A., and the leading literary societies in places like Columbus,—as I did for two winters." He had begun to hate his own poetry and being exhibited to schoolchildren as a model or "like a prize dog," but he did not tell McKay about Godmother. Picking and choosing his engagements, on February 3 he performed without fee in Lynchburg, Virginia, drawn by the somewhat reclusive poet Anne Spencer, whose lyrical piece "Lady, Lady" had appeared in Locke's *The New Negro*. Hughes stayed with Spencer and her husband at their home at 1313 Pierce Street, walked in her beautiful garden where she spent much of the day, and breakfasted on waffles and honey and Virginia ham. From Lynchburg he went to Manhattan, blithely taking the first week of the term off. There, on February 8, he attended first a performance of the play *Porgy* with Charles S. Johnson, then a late evening affair at Jessie Fauset's home at 1945 Seventh Avenue for Salvador de Madariaga, the Spanish diplomat and professor-elect of Spanish at Oxford. Although Hughes and Cullen read, and Bruce Nugent and Georgia Douglas Johnson added spice to the conversation, the party was a stiff affair—as parties sometimes were at Fauset's, where guests were often made to discuss lofty issues, preferably in French. When Hughes and Charles Johnson suggested a taste of Harlem nightlife, Madariaga leapt at the offer. The three men made for Small's Paradise.

Of Bruce Nugent and Wally Thurman, now at 267 West 136th Street, or "Niggerati Manor" (a rent-free haven for young artists provided by a philanthropic black woman who owned an employment agency), Hughes now saw less and less. Godmother, whose information came from Locke, frowned on them as a particular drain on Langston's precious energy. Only once did he stop in at the "Dark Tower," a sort of tea room-salon opened with considerable flair the past October 15 by A'Lelia Walker, with stylish rosewood furniture, a rosewood piano, rose-colored curtains, a sky-blue victrola, and—on opposite walls—verses by Countee Cullen and a section from Hughes's "The Weary Blues." Most of his time in Manhattan was devoted to Mrs. Mason, who had bought a victrola and a goodly stock of records by the sonorous Paul Robeson and various blues singers; as Hughes marvelled at her resilience, the old lady would tap her toes and clump her cane to the tune of "Soft Pedal" and "Salty Dog." On February 1, his twenty-sixth birthday, she gave Hughes a copy of one of Carl Sandburg's volumes on the life of Abraham Lincoln. "May the river of your life run deep & strong," she blessed him, "gathering clear mountain streams that carry our blessed child safely through to his fulfilment." A few days later, he hailed her lovingly on St. Valentine's Day (he also sent greetings to Amy Spingarn). Godmother's confidence in him only grew with their familiarity. "You are a golden star in the Firmament of Primitive Peoples," she assured him, "and by and by will come the astonishment of the suns that are contained in this star." Bearing down on him, she stressed his responsibility to the great black race threatened with cultural extinction by Europe. "You may feel Langston that I am pressing you forward very hard," she declared, "but it is because I believe you have enough truth to dare to follow the urge in this late hour of the salvation of your people."

Compared to Park Avenue luxury, Lincoln began to pall; Hughes lost almost all interest in his fellow students ("that yapping crowd," Godmother called them). In late February, when the school held a religious revival, all the shouting and singing seemed impossibly dull, "so I sat there unmoved and am still a sinner. Have mercy!" Never one to push himself forward, it was easy for Hughes to stand back, then efface himself almost entirely. Thomas Anderson Webster '31, one of the younger Lincoln poets and later a frequent host to Hughes as a distinguished Urban League officer in Kansas City, remembered him as a modest but still glamorous figure on campus: "There was always some small talk about rich people supporting him, but he was not ostentatious in any way. And nobody held it against him if rich white people wanted to help him. None of us thought "Lank" deserved anything less." Only one thing suggests that Godmother's patronage stirred feelings of guilt in Hughes: he sometimes complained about feeling vaguely ill, undernourished, and weak in the blood, terms which suggest the anemia that afflicted him in his clash with his father in 1919 in Mexico. However, since Langston enjoyed Godmother's maternal attention, perhaps he unconsciously tried to arouse it. Plying him with remedies, she fussed about "this 'hit and miss' condition of your health."

His career continued to advance. Or, more accurately, to give the appearance of doing so. In the previous fall had come one form of publication, the settings of his poetry and other lyrics to music, that Hughes would always crave. The famous Schirmer house brought out John Alden Carpenter's settings of his verse, *Four Negro Songs,* which he had heard in a recital at Aeolian Hall. From Vienna the socialist scholar Anna Nussbaum reported on her translations and publications of his poems; the following year, she would include thirty-seven poems by Hughes in her historic anthology *Afrika Singt: Eine Auslese Neuer Afro-Amerikanischer Lyrik.* The *Berliner Tageblatt* called him the leading poet of his race, and Mary White Ovington of the NAACP made Hughes ("the vagabond of Negro poets," so "handsome, charming of manner") a favored subject in her book of biographical essays, *Portraits in Color.* Solicited to write a special poem for a book of verse on Helen Keller, who apparently liked his poetry, he dutifully composed a tribute. In January, 1928, he won honorable mention in an *Opportunity* contest funded by Van Vechten, and Simon and Schuster published Alain Locke's *Four Negro Poets,* a collection of work by Hughes, McKay, Cullen, and Toomer.

But none of these achievements meant important new work. Although he was tremendously excited by the appearance that year of Claude McKay's first novel, *Home to Harlem* ("Undoubtedly, it is the finest thing 'we've' done yet"), work on his own fiction had shrivelled. *O Blues!*, the revue planned by Caroline Dudley Reagan ("that Miss Reagan," Godmother wrote sourly), was also in the doldrums. After signing a contract for the show, Paul Robeson had reneged on it following the smashing London success of the Broadway musical *Show Boat;* Reagan had promptly had him suspended by Equity in the United States. Soon Langston himself, no longer in need of money, withdrew from the troubled show, leaving the script to Rudolph Fisher. With his folk opera with Zora Hurston and Clarence Cameron White moribund, Hughes considered writing a play about the Haitian revolution but could not get going even though almost all of the new plays about blacks affronted him. In February, when Hughes attended a "*terrible*" new Miller and Lyles black musical in Philadelphia, all the hackneyed "Negro" effects were trotted out again—a listless chorus in blue overalls, the obligatory cotton field scenery, stale jokes about blacks stealing. He found no inspiration in his reading; leafing through scholarly books on African art and culture sent by Locke, Langston found them boring: "I discover therein that one had almost as well be civilized, since primitiveness is nearly so complex."

As a poet, most importantly, Hughes was definitely stalled. In *Fine Clothes to the Jew,* he had taken the blues brilliantly to their stripped and bare extreme; now he wasn't sure what to do next. For W. C. Handy, he wrote "The Golden Brown Blues"; he also briefly entertained an offer from a Manhattan music publishing firm for a regular job writing lyrics. He was a poet first of all, however, and no fine poetry came. One move seemed peculiarly regressive. Even before the appearance of *Fine Clothes to the Jew* he had begun to write sonnets. The *Lincoln News* published three of them as one long poem, "Barrel House: Chicago." The first is typical:

> There is a barrel house on the avenue,
> Where singing black boys dance and play each night
> Until the stars pale and the sky turns blue
> And dawn comes down the street all wanly white.
> They sell hard cider there in mug-like cups,
> And gin is sold in glasses finger-tall,
> And women of the streets stop in for sups
> Of whiskey as they go by to the ball.
> And all the time a singing black boy plays
> A song that once was sung beneath the sun,
> In lazy far-off sunny Southern days
> Before that strange hegira had begun,
> That brought black faces and gay dancing feet
> Into this barrel house on the city street.

"I've never tried a standard form before," Hughes explained to Van Vechten; "the idea of putting a barrel house in a sonnet rather amuses me." For a poet of Hughes's rhythmic orginality, however, this was a backward step in almost every way; perhaps it was a class exercise for his course in the art of poetry. Another unusual step came when he agreed to write a book review (a chore he disliked and would always try to avoid) for the May *Opportunity*. Praising the North Carolina sociologist Howard Odum's *Rainbow Round My Shoulder,* a book about black folkways, Hughes wrote of "the faith that keeps a homeless race alive, and singing, in a hard and white-faced world." After a decade of writing and publishing verse, and within a year of some of his finest pieces, Hughes suddenly seemed (like his rival Countee Cullen, whose work had declined in quality) to be drying up. Suddenly very sensitive, he wrote the *Crisis* begging Dr. Du Bois to publish no more of his old manuscripts in the magazine files. To his embarrassment the *Crisis* kept on printing the verse, finally responding to his protests by publishing one anonymously in the June number.

One published poem was significant. In the February *New Masses* Hughes published "Sunset—Coney Island":

> The sun,
> Like the red yolk of a rotten egg,
> Falls behind the roller-coaster
> And the horizon stinks
> With a putrid odor of colors.
> Down on the beach
> A little Jewish tailor from the Bronx,
> With a bad stomach,
> Throws up the hot-dog sandwiches
> He ate in the afternoon
> While life
> To him
> Is like a sick tomato
> In a garbage can.

Not only was the poem crude; his appearance in such a radical journal at this time (Hughes had published a page of poems in *New Masses* in December, 1926) was probably a clandestine challenge to Godmother, who abominated socialism. Her hostility came less from her class interests than from her deep belief in the primitive and intuitive, which made her reject Marxist and other claims to social science. This attitude posed a special problem for Hughes, who had to curb his political instincts while at the same time developing the extreme racial consciousness that Godmother loved. Godmother perhaps had only compounded a process started by another force. Entering college, Langston had given up much of his haunting sense of loneliness and isolation for the benefits of community. He was being tamed. Although she wanted exactly the opposite, Godmother's nurturing love was lulling her boy into a sweet drowse from which he only fitfully started.

If he was stalled as a poet, he was nevertheless still a force on the podium. On March 11, Hughes read his poems in Ogden Hall at Hampton Institute, his first appearance at the alma mater of Booker T. Washington. His host was a junior faculty member, Susie Elvie Bailey, the same bright young black woman who had invited him to Oberlin College in 1926 while she was a senior there. Sue Bailey introduced him to another teacher, Louise Thompson, with whom Hughes would be very close in the years to come. Pretty and light-complexioned, the Chicago-born Thompson had been badly scarred in her early life as a colored child compelled to live in a succession of predominantly white, often openly racist towns in Idaho and Oregon, until a move to Sacramento, California, had brought her at last into contact with a community of striving blacks. A 1924 graduate of the University of California, she had spent a year teaching at a hopelessly reactionary black school in Pine Bluff, Arkansas, before coming to Hampton at the start of the school year. Thompson was even more affronted by life at Hampton, where students had finally struck that year against the repressive rule of its mainly white administration. Both Louise Thompson and Sue Bailey were under a cloud. The independent-minded Bailey was suspected of having written an anonymous letter about conditions at Hampton to Du Bois at the *Crisis,* which took a vigorous interest in conditions at black colleges; in fact, Louise Thompson had done so, intending only to give information to Du Bois, who had promptly published the entire text. Now the two young women, glumly expecting to be fired, were planning to leave for New York. Hughes tried to cheer them up with brave talk about their prospects in the city.

Less salutary was a visit later that month to Lincoln's "father" institution, Princeton University. After protesting in vain to President Johnson of Lincoln about the propriety of Lincoln men entertaining at a school so notoriously inhospitable to blacks, Hughes went with the glee club and the quartet to perform at Alexander Hall. Ironically, a critic for the *Daily Princetonian* liked the Negro music, but thought several of Hughes's poems "somewhat ridiculous."

On Easter Monday, April 9, an again unwilling Hughes took part in the major Harlem social event of 1928, when Countee Cullen married Yolande Du Bois, daughter of W. E. B. Du Bois, before three thousand guests at his father's church, Salem Methodist. Du Bois, delving into tomes on nuptial etiquette, and drawing on his experience as the producer of a historical pageant, "The Star Of Ethiopia," envisioned the ceremony as a symbolic union of Poetry and Intellect (the bridegroom's poetry and, no doubt, his own intellect). Determined to stage the Harlem wedding of the century, he planned to crown the exchange of vows with the release of a thousand doves; he was restrained just in time. The bridegroom, too, adored a pageant. Insisting on the most formal, swallow-tail attire for the party, Countee detailed it ("pardon the impertinence of this reminder") to a disgruntled Hughes, who showed up for Cullen's "parade," as he snidely called it, in a shiny, threadbare, rented version of this costume to escort the bride's mother to her seat. He had some reason to be cynical. Except in name, the marriage lasted only a little longer than the wedding; soon, Du Bois was writing sympathetically to his son-in-law about his pampered daughter but also about the physical requirements of a marriage. On June 30, Cullen sailed for Europe with Harold Jackman. "GROOM SAILS WITH BEST MAN," one headline in a black newspaper jibed. Countee's adoptive father, Rev. F. A. Cullen, was also on board.

At the end of May, before Mrs. Mason left for Europe, she and Hughes met to decide on his plans for the summer. On her birthday, May 18, he had sent her evergreen blossoms, "these tips of budding life from a strong and beautiful tree, old and great, yet ever renewed with delicate young needles." Godmother found them, like his corn at Thanksgiving, "wonderfully emblematic." Then she soured their exchange with a cranky request that Langston bring "a summary statement of your entire bank account since November 1"; although he had faithfully sent his itemized expenses every month, "of course I made no record of it on paper nor in my mind." To Hughes, who had been filling notebooks with a litany of his purchases—fruit, paper, BromoSeltzer, subway and taxi fare, postage stamps, snacks, meals, Listerine, glue—this accounting had begun to remind him ominously of his father. But he looked ahead to the summer and the start of his novel. To stimulate his memory of childhood, Langston would venture out west to Lawrence, Kansas, then go on to the Grand Canyon (sacred Indian ground to Godmother) before returning to begin work. After intense weeks of "flight," he would then vacation briefly in Quebec. Giving Langston $450 in advance for three months' support, Mrs. Mason added $300 for his trips, then $92 for certain special expenses such as trans-Atlantic cables to her when telepathy proved inadequate. In mid-June she sent one of her favorite leather suitcases, a package of deluxe typing paper, and two loaves of bran bread and a box of Swedish bread to Lincoln, since Langston still complained of feeling poorly. On her orders he took a complete physical examination at the Life Extension Institute in Manhattan. In excellent health, according to the report, he stood just under five feet five inches and weighed 121 pounds, with a 30-inch chest and a waist of 26 inches. His vision, which had troubled him the previous summer, was now better than normal. Hughes claimed neither to smoke nor to drink—perhaps under Godmother's influence, or her orders.

When Lincoln let out, however, they agreed to a change of plans. Instead of heading west Hughes took a brief vacation at Amy Spingarn's country estate near Amenia, New York, then returned to a rented room in the small, mostly black, service village near the university. Suddenly Langston was eager to begin the novel. That year, Harlem was drenched in fiction—not only McKay's *Home to Harlem* but also Du Bois's *Dark Princess,* Rudolph Fisher's *Walls of Jericho,* Jessie Fauset's *Plum Bun,* and Nella Larsen's *Quicksand.* Also, Wallace Thurman was hard at work on a novel about color prejudice among Negroes against those with very dark skin, a subject about which Thurman, very dark himself, was hypersensitive. Sensing a shift in the ground away from poetry, Hughes did not wish to be left behind. But how to begin? For a while he thought about his past, of Lawrence, his mother and his absent father, his grandmother and uncle in Kansas City, and the Reeds (Auntie Reed had eagerly expected Langston to visit her on his way west). These were often still painful memories, flashing intimations of a deep, precocious loneliness on which Hughes depended nevertheless for the rush of bittersweet emotion he needed to create. Undoubtedly Langston also thought of his journey the previous summer in the South among the common black folk, of whom his alchemized, fictional self would be emblem. While he could never tell the plain, unvarnished story of

Langston Hughes, he would try to tell the tale of his childhood longings, of the attentive family he had wished for himself, and of the beloved brown boy he had tried to be. This family and this brown boy would stand collectively for the race, whose loving portrait would be at the heart of his fiction.

On June 21 Hughes finally slipped past fact into the mists of fiction. He started literally with a storm—a terrifying cyclone on April 12, 1911, still remembered in Lawrence, that had killed two persons and blown away his grandmother Mary Langston's fence on Alabama Street—but also dramatically like the imaginary storm that blew another Kansas child and her dog into the land of Oz. Altered, Langston's grandmother was not pale like Mary Langston but darkskinned; her speech was not clipped and polished but a molasses drawl; and he himself was a small boy named James, but known as Sandy.

> Aunt Hagar Williams stood in her doorway and looked out at the sun. The western sky was a sulphurous yellow and the sun a red ball dropping slowly behind the trees and house-tops. Its setting left the rest of the heavens grey with clouds.
>
> "Huh! a storm's comin," said Aunt Hagar aloud.
>
> A pullet ran across the back yard and into a square-cut hole in an unpainted piano-box which served as the roosting-house. An old hen clucked her brood together and, with the tiny chicks, went into a small box beside the large one. The air was very still. Not a leaf stirred on the green apple-tree. Not a single closed flower of the morning-glories trembled on the back fence. The air was very still and yellow. Something sultry and oppressive made a small boy in the doorway stand closer to his grandmother, clutching her apron with his brown hands.
>
> "Sho is a storm comin", said Aunt Hagar . . .

A noble old black woman, Aunt Hagar has three grown children, all daughters. Sandy's mother, Annjee, is sweet and simple. One of her sisters is a cold bourgeoise. The other loves singing and dancing and men, and seems headed for the gutter. Sandy's father, Jimboy (James was Langston's father's name, as well as his own), is long, lean, and golden-skinned; gifted with a rich low baritone voice, he plays a mellow blues guitar. Jimboy loves to rove, but he always comes back home to his wife and his sweet brown boy.

Methodically, Hughes drafted histories of each character and tacked the outlines to the wall above his work table. At first he tried to revise each chapter, but the writing seemed so impossibly poor to him that he decided to forge ahead to the end of a full draft. Ignoring his mail—except for Godmother's missives—he let nothing come between him and the book. Throughout the ordeal he was urged on by Mrs. Mason, who by now knew exactly which tender nerve to press. "I realize that the old hurts the old suffering are too near you," she cunningly sympathized, "but at the same moment my Boy the forces that are in those old hurts are for you to seize upon & use & thereby blot out forever all their suffering." Up to the day of her sailing he sent Mrs. Mason, at her insistence, every scrap of his handwritten draft ("most sacred to your godmother"). "It is marvellous the way you make me feel the congregating of the material & the people," she exulted; "the Gods be praised Langston your work is wonderful! How the Big Indian marches through the conception in the economy of words you use in making me see it. It is the same thing Child as when the Navajo ordered the Colorado river to cut the Grand Canyon. And then it is made whole by the negro warmth & tenderness . . . of your Precious Heart.'' Across the ocean each morning he projected his greeting, which she hungrily, telepathically received.

He turned away from the novel only to refute an ill-informed accusation by the black scholar Allison Davis in the July *Crisis* that Van Vechten had subverted his talent. The same charge had been made the previous year by the hidebound black scholar Benjamin Brawley, to whom Van Vechten had responded vigorously. Now, after the publication of McKay's *Home To Harlem,* the charge was revived and broadened in the *Crisis.* Reviewing *Home To Harlem,* Du Bois declared that McKay ''has used every art and emphasis to paint drunkenness, fighting, lascivious sexual promiscuity and utter absence of restraint in as bold and as bright colors as he can.'' In November, when Du Bois reviewed Rudolph Fisher's *The Walls of Jericho,* he would link McKay and Van Vechten as masters of a pernicious Harlem school of fiction. While Du Bois could not bring himself to attack its star pupil, Langston Hughes, whom he loved, he allowed Allison Davis to have his say. But Hughes demolished Davis's argument by pointing out that the poems in *The Weary Blues* had been written before he met Van Vechten; he might have repeated Van Vechten's declaration to Brawley that, if anything, Langston Hughes had influenced *him.* ''If I'm ruined,'' Hughes wrote Claude McKay, ''I ruined myself, and nobody else had a hand in it, and certainly not Mr. Van Vechten.''

On August 16, Langston finished the first draft of his novel. Exhausted, but tense with anticipation of the effort to come, he left on a vacation that did little for him. In Provincetown, at the tip of Cape Cod, he passed a few dull days among the usual artists and tourists. Visiting Boston, he thought of crossing the Charles to visit Harvard University, then thought again. By early September he was back at Lincoln. He had stopped only briefly in Manhattan, where the extraordinary news was that Wallace Thurman was now married. Needing a typist to help him work on his novel, he had hired Louise Thompson, who had come to New York in June from Hampton Institute. Within a short time they were married. Given Thurman's recklessness and instability, as well as his homosexual bent, the marriage seemed about as likely to succeed as Cullen's. This time, at least, Hughes had not been asked to be a member of the wedding.

Setting aside the draft of his novel for Godmother, on September 19 Hughes at last began sketching a ''singing play'' called *Emperor of Haiti.* At night in an abandoned sugar mill somewhere on the island of ''St. Domingue'' the leaders of a slave conspiracy gather, dominated by the fierce Dessalines. ''The leaders and their followers palaver. Preparations are set, the die is cast for the uprising.'' ''To the hills!'' is the cry ending the first scene, with the motley band surging off the stage to begin a night of long knives and blood, rapine and destruction. A ballet of witch dancers comprises the second scene; and so on. Meanwhile, sorcery of another kind was taking place on behalf of Hughes, who had conceived of *Emperor of Haiti* with the prepossessing Paul Robeson in mind, although Robeson showed no interest in returning from Britain, where he was a lion. However, Godmother was working night and day to subdue the star, who was more than a little terrified of her. Invoking telepathy, she passed more than one night "tearing off the veils" from his soul "that he might see something of his real self." The reason for Robeson's huge London success was clear to Godmother: "Nothing of course Langston has built these audiences of his except the coming into birth of your Singing Play. No wonder he is dazed as to what it all means."

On October 3, Hughes wrote the Library of Congress for copyright forms to register six song lyrics from the play. But when school caught up with him, he set *Emperor of Haiti* aside, unfinished. His courses presented no challenge, except perhaps for a year-long survey of sociology by the admirable Professor Labaree. In the fall, the course stressed anthropology, especially primitive life; in the spring, race relations. Labaree also encouraged empirical research for credit, an idea that definitely intrigued Langston. Suddenly he found himself once again responding warmly to Lincoln as a whole. "I was so tired of it last year," he admitted. Now, as a senior (and with Godmother in Europe) he threw himself into college life, accompanying the football team in its campaign through the mid-Atlantic coastal states. He also decided to use his considerable influence to publish a slender volume of poetry by the main Lincoln poets of his time: William Hill, Edward Silvera, Waring Cuney (who had left to study music in Rome), and himself.

So caught up was Hughes in the rituals of life as a senior that he even found himself, as in high school, a sweetheart. At Hartshorn Memorial College in Richmond, he met Laudee Williams; by mid-November, when Lincoln played the college at Petersburg, he was in love with her. Escorting her to a dance, they walked hand in hand in the moonlight; she gave him a little diamond ring, and he sent her one of silver set with Indian luck stones. Describing her to a startled Alain Locke as "my girl friend," he invited her to Washington for Thanksgiving. But his interest soon waned. For one thing, Laudee Williams did not share his view of life; brownskinned herself, she once wrote contemptuously to him about those "pitiful, dirty, sloven, boisterous and black people" crowding the streetcars. Such a remark hastened her fate. By December 5, Laudee was chafing at the lack of ardor in her "Prince of Fairies": "No word—not even a single word from you and I am wondering what's the matter? Busy? Of course. I will try to understand." Perhaps she was a little pushy. The same school magazine that called Langston "Most Popular" of the students also called him one of the three "Most Henpecked."

But if his "love affair" foundered and his poetry was uninspired, Hughes rode high publicly in the fall of 1928. Surviving the onslaught of the black press against *Fine Clothes to the Jew,* he had discovered the great American truth that all publicity is good publicity, an axiom he would frankly exploit the rest of his life, and one reinforced by his gift for counterattacking without turning others against him. On his twenty-seventh birthday his mother, Carrie, wrote with a marked sense of wonder about the way Langston met the world: "To be approved and admired has always been your position; your wonderful personality has made you many friends and benefactors, you have lived above Slander, and life for you is a clean page, to make or mar it as you will."

Although Carrie Clark never mentioned Godmother in her letters to Langston, she must have been aware of her and of the extent to which this rich old woman (old enough to be her own mother) had captivated her son and made her dispensable. With a lifetime of blasted hopes, Carrie herself was not growing old very gracefully. Now past fifty, she was still frequenting the dance halls. "Mama Clark is some chippie, yes Lord!" an embarrassed Harold Jackman wrote to Cullen in Paris; having danced with her once at A'Lelia Walker's "Dark Tower" tea room, he was astonished to have her upbraid him some time later for not calling on her or telephoning. "Why you would have thought she was a very young thing. . . ." More than once she exasperated Locke and Hughes by spiriting Gwyn Clark away from his guardian in Springfield without permission. Carrie herself blamed most of her troubles on money; every letter to Langston begged for cash. "Being poor is absolutely an irrevocable fact," she sighed. Although she loved her son ("I am hungry to see you—Love—Love—Kisses"), she saw the chasm between them. "Dear," Carrie once asked, "why don't you love me, why a'rnt we more loving and chummy, why don't you ever confide in me, I know I have no sense to help you in your work but I'd enjoy your confidence." At least once she faced up to her part in their estrangement: "Some time I feel that you & I were never as close, Heart & heart as we should be, but I have loved you very dearly and if I failed in some things it was lack of knowledge."

Hughes loved his mother, but a true reconciliation with her was impossible; his childhood loneliness made him suspicious of her and of all women. Except perhaps one. When Godmother returned to the United States in mid-autumn they had a tender reunion. Her telepathic frequencies had failed to reach Paul Robeson, but she was not daunted. Soon she had the manuscript of the first draft of the novel, typed for eight cents a page by Hughes's young schoolmate Paul Terry from Pleasantville, New Jersey. At least one section of the work she found "an amazing piece of writing"; but the whole badly needed revising. Then Godmother reasserted her authority with a firm tug at her golden hook in Langston's mouth. In his accounts for October he had not included his bank balance. "This leaves me a little uneasy," she wrote, "about your expenses and need of money, at this juncture." Again she wanted a full review of his finances. Hughes complied but despised the task and the implication that he needed to be watched; it revived his deep-seated fear that he was not being loved for himself.

The need to write, to perform for Mrs. Mason, began to tax Hughes; he was both contented and oppressed. If she made him happy and easeful, even luxurious, he was then not much moved to write, although he sympathized with the aged woman's urgent wish that he do his great deeds quickly. But while he held her at bay as best he could, Godmother would not be satisfied forever with selections from his fan mail, which he sent to her as proof of his effectiveness, or with bits of trees, sheaves of corn, valentines, or honeyed letters. Godmother wanted Langston to put out, but her intensity often made him weak and tense. At one point he confided his dilemma to Amy Spingarn, who counselled that "in answer to your question—when things go well with us—we feel no emotion—and you can't write poetry without emotion."

A clash was at hand. On December 23, Mrs. Mason made Hughes a Christmas present of a splendid leather bag. Langston either failed to thank Godmother promptly or in some other way left her large appetite for gratitude unsatisfied. Mrs. Mason was also on the lookout for signs of disaffection. Although she was now helping not only Hughes and Zora Hurston but also Hall Johnson and Alain Locke, and perhaps others, she found it increasingly hard to trust them. Why couldn't Alain Locke give a straightforward, honest answer to any question, Mrs. Mason once asked, not without good reason; and she caught Hurston in an egregious, expensive lie about the purchase of a motorcar. Sycophantic to her face, Hurston often ridiculed Mrs. Mason behind her back; Locke himself was only more elegant in his deception. Mrs. Mason sensed their manipulation and looked even more fiercely for signs of genuine gratitude. Of the three godchildren, all neurotically mortgaged to Mrs. Mason, only Hughes was both genuinely in love with her and scrupulous in his dealings. Like Lear, when the old woman vented her rage, she blasted the one who loved her most. On Monday, February 18, after a pleasant but uneventful lunch, Mrs. Mason exploded at him virtually without warning. The inadequately acknowledged leather bag had something to do with her anger, but she accused him generally of insincerity and ingratitude, of taking money and doing nothing in return. Her fury astonished him, but worst of all was the mask of weariness that settled over her wrinkled face at the end of her tirade. Hughes sank from her apartment to the lobby like a stone. Finding himself on Third Avenue without knowing why he was there, he rode the elevated railway in a daze, ending up in some corner of the Bronx where he knew no one and had never visited.

Returning to Lincoln, he passed the next few days trying to set down on paper his proper feelings. Editing, amending, discarding, he sought the right words to express his devastation. Finally, on February 23, he finished his declaration to "Dearest Godmother":

> All the week I have been thinking intensely of you and what you have done for me. And I have written you several letters that I have not sent because none of them were true enough. There were too many words in them, I guess. But all of them contained in some form or other these simple statements:
>
> I love you.
>
> I need you very much.
>
> I cannot bear to hurt you.
>
> Those are the only meaning in all that I say here. You have been kinder to me than any other person in the world. I could not help but love you. You have made me dream greater dreams than I have ever dreamed before. And without you it will not be possible to carry out those dreams. But I cannot stand to disappoint you either. The memory of your face when I went away on Monday is more than I am able to bear. I must have been terribly stupid to have hurt you so, terribly lacking in understanding, terribly blind to what you have wanted me to see. You must not let me hurt you again. I know well that I am dull and slow, but I do not want to remain that way. I don't know what to say except that I am sorry that I have not

> changed rapidly enough into what you would have me be. The other unsent letters contained more words than this one. They were much longer. They were much more emotionally revealing, perhaps. But I do not know how to write what I want to say any simpler than it is said here. Words only confuse, and I must not offer excuses for the things in which I have failed. Your face was so puzzled and so weary that day. I shall never forget it. You have been my friend, dear Godmother, and I did not want to disappoint you. If I can do no better than I have done, then for your own sake, you must let me go. You must be free, too, Godmother. At first we had wings. If there are no wings now for me, you must be free! We can still fly ahead always like the bright dream that is truth, and goodness. Free!
>
> Always my love to you, and my truth
> Langston
>
> You have been so good to me.

The same day he wrote another letter to her, containing "some little, less important things that I want to tell you." This second letter was even longer than the first, and although much of it was less intense, certain portions revealed more:

> It seems that a new life-period is centering about me. You know how I told you about the period of books,—how all my childhood, lonely, I spent reading up until the time I went to sea. Then I told you how, suddenly, the very sight of books made me ill, and I threw the ones I possessed into the sea. After that came what I guess one could call a period of solitary wandering, looking out of myself at the rest of the world, but touching no one, nothing. "The Weary Blues" belongs to that period. But I think my second book marks an unconscious turning toward this third state, for there many of the poems are outward, rather than inward, trying to catch the moods of individuals other than myself. In Washington I realized fully how terrible it was to live alone, unlike the folks about. And I came to Lincoln. For the first time in my life I found myself in some way belonging to a group. But it took me a very long time (and my work interfered often) to learn to live with them. Only now am I beginning to be at all at ease and without self-consciousness in meeting my own people. Lincoln has done that for me, and I value it immensely.

He ended with another cry from the heart: "Because I love you, I must try to tell you the truth. We agreed upon that, didn't we? Do not misunderstand me, dear Godmother. Of course, I need you terribly,—but you must be free to live for all the others who love you, too. If I am too much for you, you must not have me."

Deeply in love with Mrs. Mason, in fact he could not bear to tell her goodbye. All his self-abnegation was a plea to be recalled.

She forgave him.

This ordeal had taken place even as Hughes was preparing a highly controversial farewell to Lincoln. As an empirical exercise in Robert Labaree's sociology course, Hughes had decided to study Lincoln University itself. Assisted in his survey of Lincoln's 129 juniors and seniors by two juniors, Richard Lowery of Philadelphia and Frank B. Mitchell, Jr., of Chester, Pennsylvania, with additional work by another junior, Radcliffe Lucas, and a sophomore, Edward

S. Gray, Hughes nevertheless left no doubt about who was in charge: ''All deductions, conclusions, opinions, and comments in this paper are my own.''

Lincoln had congratulated itself on being the only major black college to escape student unrest in the previous few years, and Lincoln men were smug about the quality of their school as compared to other black institutions. Hughes set out to shatter the peace. To Van Vechten he would soon write about having upset those ''staid contented Presbyterians'' who had eased along in ''delightful mediocrity'' until Lincoln had become ''a charming winter resort and country club, but not much of a college,'' a place ''dominated by white philanthropy, well-meaning but dumb.'' Looking at every aspect of the university—its history and purpose, its physical plant, the faculty, student body, and curriculum, and its extracurricular activities, especially the clubs and fraternities—Hughes spared no one. He found not only ''obvious incompetency'' in the teaching of certain instructors and junior professors, but also ''horse play, lack of attention, bad manners and rudeness'' on the part of students in Bible classes. He deplored ''a distinct lack of what might be termed cultural courses'' in English, philosophy, and history. Too many of the student groups were moribund; the school paper featured puerile humor ''and a page or more of usually very bad, childishly conceived poetry''; the fraternities hovered over student life ''like a guiding hand, or shall we say like a dark shadow?'' Buildings were generally ''inadequate and old,'' the library stocked with obsolete books. The dining hall featured ''poor service, the food is not very good, and table manners are below the civilized level.''

Most controversial was the survey of student attitudes towards an all-white professorate (a few very bright students had been kept on briefly, without hope of tenure). Almost two-thirds of the students wanted no blacks on the faculty. The reasons most often cited were that ''favoritism, fraternity influences, and unfairness'' would follow a change; Lincoln was excellent as it was; blacks would not cooperate with blacks. Three students simply did not like Negroes, two thought Negroes would not have the interest of the students at heart, and one was sure that Negroes were not ''morally capable.''

Hughes wrote a cryptic foreword to the study:

> In the primitive world, where people live closer to the earth and much nearer to the stars, every inner and outer act combines to form the single harmony, life. Not just the tribal lore then, but every movement of life becomes a part of their education. They do not, as many civilized people do, neglect the truth of the physical for the sake of the mind. Nor do they teach with speech alone, but rather with all the acts of life. There are no books, so the barrier between words and reality is not so great as with us. The earth is right under their feet. The stars are never far away. The strength of the surest dream is the strength of the primitive world.

This was really Godmother's Credo; Hughes dropped it from his report, which ran just over two dozen pages.

By April the survey was finished and the report typed by his stenographer, Paul Terry. When a small portion was tacked up on the sociology bulletin board, someone sent a copy to the *Baltimore Afro-American,* which promptly told the black world in a front-page headline that "LINCOLN VOTED 81-46 AGAINST MIXED FACULTY." Many distinguished alumni, including the veteran lawyer and biographer Archibald Grimké, Kelly Miller of Howard University, and the newspaper editor William Monroe Trotter of Boston, spoke hotly for and against the students. Tension increased at Lincoln, where the board of trustees had just broken tradition by electing—only after the intervention of William Hallock Johnson, the president—its first black member, a New York physician. The alumni were now pressing for a greater voice in university affairs.

When Locke visited Lincoln on the weekend of April 13 he found Hughes determined not to retract a line. As for the faculty, "they are scared stiff," Locke reported to Godmother. Loving intrigue, he walked three miles away from the campus with Hughes to read the report, which they had concealed under Langston's sweater. It was "a plain naive statement of damning facts"; as for Hughes, "the quiet informal way he is doing it is masterly." Langston had once again showed the courage of the innocent; to many of his classmates, only he could have gotten away with such audacity. News of the uproar reached Claude McKay in Europe. "The only way of creating self-respect among Negroes as a group," he counselled Hughes, "is by showing that they have none."

The survey dominated Langston's last spring at Lincoln but did not lessen the thrill of graduation. In spite of his harsh criticism he was delegated to write the traditional "Ivy Day Toast," attesting to his high standing in the community. When he asked President Johnson for a room for the summer in order to write, his request was granted *gratis*. Johnson also wrote a foreword to *Four Lincoln University Poets,* Hughes's anthology with Waring Cuney, William Allyn Hill, and Edward Silvera, which would appear the following year. Grateful, Hughes thanked all the people who had helped him to succeed—Carl Van Vechten ("you're person No. 1"), Amy Spingarn (who sent a little Kodak camera as a gift), and Godmother. Unusually festive because Lincoln was celebrating its Diamond Jubilee, commencement came on Tuesday, June 4, in Livingstone Hall. Forty-one men graduated, including less than half of those who had entered in 1925-1926. Hughes received the bachelor's degree, without special honors. "Oh! Langston," his mother enthused, "I was so proud so happy over you and for you."

Waring Cuney, who had led Hughes to Lincoln, then left it in 1927 to study music in Rome, posed a good question: "Now dat yo done gat all dat edification, what you gwine do, ugh?"

The first thing Hughes did was take a two-week vacation in Manhattan, which he found oddly discomforting. The world he had once known there—the world of brave New Negroes, Jessie Fauset, Augustus Granville Dill, and Civic Club lunches, *Opportunity* and *Crisis* banquets, midnight expeditions to Harlem speakeasies—was swiftly disintegrating. The cabarets had all gone mostly white; in some, blacks were actually barred unless they came as guests of whites. Van Vechten was away in Europe. Returned from several weeks in London with the cast of *Porgy,* Bruce Nugent was strutting about in a modish English suit. About to go to Europe, Locke had suffered a mild heart attack (rheumatic fever had damaged his heart as a child) and missed his boat. Wallace Thurman's fortunes had swung wildly from high to low. Earlier in the year his sensational play *Harlem,* tailored by William Jourdan Rapp, had opened on Broadway; and Thurman's novel of color fixation, *The Blacker The Berry,* had been published in March by Macaulay. But his magazine *Harlem,* on which Bruce Nugent and Aaron Douglas also worked, had quickly folded. Worst of all, Thurman's improbable marriage to Louise Thompson had collapsed after a few weeks in a welter of recrimination, leaving his bride hurt and baffled. "I *never* understood Wallace," she would later admit. "He took nothing seriously. He laughed about everything. He would often threaten to commit suicide but you knew he would never try it. And he would never admit that he was a homosexual. *Never, never,* not to me at any rate." She was heading for Reno to seek a quick divorce; Thurman, hired by Macaulay as a reader, had sent himself an emergency telegram calling him home to Salt Lake City, but intended to escape to Hollywood.

For Hughes, a pall of bad taste seemed to hang over Manhattan. "Awfully bad colored shows are being put on Broadway every week or so now," he reported to Claude McKay. "They fail,—as they deserve to. Some of the colored victrola records are unbearably vulgar, too. Not even funny or half-sad anymore. Very bad, moronish, and, I'm afraid, largely Jewish business men are exploiting Negro things for all they're worth. You can't blame the 'nice' Negroes much for yelling, in a way, even when some of us are sincere! We do get a lot of racial dirt shoved on the market." But he, too, Hughes seemed to admit, was part of the deterioration. "I've never felt so unpoetic in my life," he told McKay. "I think I shall write no more poems. I suppose I'm not quite miserable enough. I usually have to feel very bad in order to put anything down,—and terrible to make up poems."

At Lincoln, he set up his typewriter in a room on the third floor of otherwise empty Houston Hall, where the seminary students usually lived, "the oldest, largest, darkest, and most remote" dormitory at the school. One night, just as he settled in, Hughes heard a key turn in the lock on his door. He screamed; a would-be thief fled down the corridor and out into the night. Hughes was a marked man; any student who didn't have to work in the summer must have lots of money. Godmother offered him a gun. Then she changed her mind and dispatched instead a 24-page, chapter-by-chapter critique of the novel. "The whole book," Mrs. Mason declared, "needs literary welding together. There are many good separate events, but no real sense of juxtaposition. . . . Though beauty flashes through the pages, in all your writing it is apparent that you had no training in literary expression as a little child, and in college." Mrs. Mason was referring to the childlike honesty, the absence of inflated rhetoric in the text, which she took as a fault but was in fact Langston's best style, an absolutely lucid, unaffected, American prose. She warned him, too, against allowing propaganda to mar the beauty of the scenes. "What I object to," she said frankly in one place, is "that the quality of the writing at that point becomes self conscious, and has the air of the author's propaganda."

With Godmother snapping her whip, Hughes went to work. When inspiration failed, he practiced his prose technique, scribbling entries in a daybook. More than once he commented on the oddness of his position. Outside in the sunshine the passing village boys and girls laughed languidly. They had simple souls, he reasoned, and would never amount to anything. "Well," he asked, "who cares—if they're happy!" The difference between talent and genius? "Talent is pathetic, genius is tragic when most intense. Pathos belongs to talent, true tragedy to genius alone." His ultimate hope: "To create a Negro culture in America—a real, solid, sane, racial something growing out of the folk life, not copied from another, even though surrounding race." The summer days passed, bright but heavy. "No one needs to know me," he confessed sadly; "everything I have to offer worth the offering is in my work; the rest is slag and waste."

Puzzling over a title for the novel, he sent up balloons for Godmother to prick. Not surprisingly, she disliked "So Moves This Swift World" and "Roots of Dawn." Perhaps in these flaccid titles she sensed Hughes's struggle for inspiration, the incongruity of a person with his wanderlust locked away in a cavernous old dormitory while the summer sun blazed. Stepping up her inspiration (the force of "my spirit moving with you in your work"), she cannily flashed a light into a gloomy corner of his past. "My precious boy," she urged, "no flame of chance is blowing upon you. Don't feel any hurry of the world. . . . And may I say it please Langston? Don't let one breath of your restless mother pass through you at this time. Believe in the forces that are vested in you."

On August 15, the second draft was finished. A third was needed, but he was tired. When Godmother sent $250, he headed south across the Chesapeake Bay to the remote eastern shore of Maryland, where the country was as remote as deep Georgia and (so he wrote Thurman) "every colored lady has at least six little half-breeds for practice before she gets married to a Negro. And where the word Mammy is still taken seriously, and the race is so far back that it has illimitable potential—like me!" A ferry out of Baltimore took him across the great bay. With the days burning hot and humid, he moved among the black folk in camp grounds and at revival meetings, taking notes quietly as he had done in the South two summers before.

Near September 1, he passed through New York on his way north to Canada. At Montreal he boarded a boat sailing up the scenic, fjord-like Saguenay River from Tadoussac on the St. Lawrence to see the northern lights, but the white passengers, mostly Americans, were hostile, and the steward refused to serve him until they had eaten. Angry, Hughes stormed off the boat at a stop in the mountain wilderness and made his way back to Montreal. He went on alone to the old city of Quebec. The air was already chilly as the wind swept in from the St. Lawrence and blew across the plains of Abraham and the battlements of the fort where the French had lost Canada, and Montcalm and Wolfe their lives.

Around the middle of September, Langston and Godmother were reunited in her suite at the Barclay Hotel in Manhattan. Mrs. Mason had not been well but was strong enough to go out with Langston to a concert organized by Hall Johnson, another grateful godchild (she was "first and foremost" among his supporters, he would write her). Their main business was to plan Hughes's life after Lincoln. Harlem was too distracting. Before leaving Lincoln, Langston had settled on a room at 514 Downer Street in Westfield, New Jersey, hardly an hour from Manhattan, in the home of an elderly, respectable black couple, Mr. and Mrs. J. V. Peeples. The house was large and clean, with peach trees and shrubs and flowers in a small but well-tended garden. Across an open lot was the African Methodist Episcopal Church where Paul Robeson's father, a former slave, then a Lincoln-trained Presbyterian minister, had pastored between 1907 and 1910. From a Manhattan office equipment store came a walnut table, a bronze lamp, a desk tray, one hundred manila folders, and miscellaneous other supplies at a total cost of $109.50. Three boxes and a trunk were shipped from Lincoln University. Then, either through Langston's initiative, or Locke's, Louise Thompson was summoned to the Barclay on October 2 and hired at $150 a month to work as a stenographer for him. Thompson had returned from Reno without her divorce because of the news that her beloved mother, also named Louise and known to their friends as "Mother" Thompson, was ill with cancer.

The crash of Wall Street on Black Thursday, October 24, 1929, barely rattled the exquisite china at 399 Park Avenue, and touched Mrs. Mason's godchildren not at all. Louise Thompson was surprised to find that Mrs. Mason considered her a "godchild" and expected notes of thanks for what Thompson saw as salary, but she gave in to this whim; with her mother ill, the money and the flexible hours were truly a godsend. Their large apartment at 435 Convent Avenue, near the City College of New York, became a second home for Hughes (Sue Bailey from Hampton also lived there). Lively, pretty, and intelligent, Louise Thompson had her share of suitors—of which Hughes emphatically was not one, although they enjoyed each other's company so much that a snooping reporter for the black gossip sheet, the *Inter-State Tattler,* once caught them holding hands on Seventh Avenue. Sometimes they stepped out with Aaron and Alta Douglas; in November they saw "his" Lincoln trounce "her" Hampton 13 to 7 in football at the Polo Grounds. At the Savoy Ballroom, Hughes tried valiantly to learn the Lindy with Louise, but never got beyond a shuffling two-step. Hughes also saw a great deal of Arna Bontemps, hard at work on a novel, and his young wife, Alberta, a former student at the little Harlem Academy where Arna taught. With Amy Spingarn he attended a symphony concert conducted by Toscanini, and attended her tea for Julia Peterkin of South Carolina, the white author of *Scarlet Sister Mary* and other works on black life. In Atlantic City, where Carrie's husband, Homer Clark, had surfaced again, Hughes had Thanksgiving dinner with them. But the center of his life was Godmother, with whom he went to a variety of concerts, musicals, plays, and lectures. Bruce Nugent gazed with amusement on them at least once but was hardly primitive enough for Godmother. He later dismissed her as "a cane-clumping ancient."

The best news of the fall was the acceptance by Knopf of Hughes's novel, now called "Not Without Laughter," a title that captured the power of the blues to mirror pain but also to mock it. Again Van Vechten had been helpful; before leaving for Europe, he had made Alfred Knopf himself promise that Hughes would "be treated with every tenderness and consideration." Van Vechten, just back from Europe, and Langston lunched to celebrate their latest triumph. But, to Hughes's dismay, further changes needed to be made in the novel after a reading by Alain Locke (for which Locke had been solemnly prepared by Godmother: "Have you consulted your individual antecedental Gods of Africa to save you from a false move or premise in this hour of criticism?"). Locke wanted greater stress on the hero Sandy's "inner emotional conflict" as he reached adolescence; and Godmother herself complained that "controversial detail" had come into the last draft; its "dark beauty" was being submerged "under the propaganda utterances."

With the novel almost done, Christmas should have been happy, but in the aftermath of the crash of Wall Street, Hughes could not continue to ignore the hunger and poverty raking the masses of people on the streets far below Godmother's palatial apartment—the same people at whom he peered guiltily through polished glass as he was ferried about in Mrs. Mason's limousine to Harlem or Pennsylvania Station. "People were sleeping in subways and on newspapers in office doors, because they had no homes," he later recalled of that winter, "and in every block a beggar appeared." Now Mrs. Mason's rich dinners sat more queasily on his stomach as he faced the fact that possibly only her whim kept him from joining the denizens of Manhattan flophouses. As a black, his Lincoln degree was worth little in itself; more than one Lincoln graduate was making a career of "indoor aviation," as a schoolmate mordantly called working an elevator. If there was something unsound about Hughes's situation, given his social conscience, he also knew that artists traditionally had lived by patronage. And he loved Godmother.

A further indication of his compromise came in January, when he and Mrs. Mason attended various lectures by General Jan Smuts, the former prime minister of South Africa and an ingenious apologist for the British Empire; Mrs. Mason also sent an affronted Louise Thompson to take notes. In response to his conscience, Hughes allowed his work to be included in *An Anthology of Revolutionary Poetry,* edited by Shmuel Marcus (or Marcus Graham, as he called himself); despite Godmother's anti-Semitism, he encouraged the inclusion of some of his poems in an anthology of Yiddish translations. But these were half-hearted gestures. Apart from the novel, his voice had lost its distinctive verve. A mocking reminder of this loss came when a reader for the *Nation* returned an anonymous essay to the editor with the remark that "Langston Hughes would probably do it better." The essay was by Langston Hughes.

At long last, on Monday, February 17, he personally handed the manuscript of "Not Without Laughter" to Blanche Knopf. By now, Hughes was sick of the novel. As beautiful as it was, he would never be happy about this book, which had been wrung from him by Godmother. "Now," he noted that day, "I am free for something new and better."

Turning his attention to his "singing play," he looked around for a new musician; after a summer in Haiti, Clarence Cameron White had written his own opera with another librettist, John Matheus. The mid-winter chill, the end of work on the novel, and his disappointment with Afro-American composers (William Grant Still's setting of his "Breath of a Rose" was so complex that almost nobody could sing it) convinced Langston that he should look overseas. He decided to go to Cuba, an idea Godmother backed at once—Cuba must be very primitive. As he sipped tea with her after leaving Blanche Knopf, Mrs. Mason offered him $500 for the trip. "She gives me her gift in Trust for all the people," he noted solemnly. "I am to go to Havana for rest, new strength, and contact with the song."

But the next day, at the offices of the Ward Line on Fifth Avenue, Hughes was rebuffed. Refusing to sell him a ticket, a clerk on duty cited company orders not to sell tickets to blacks. When Hughes protested, he was shown a carbon copy of inter-office correspondence to the effect that Cuba would not allow Negroes, Chinese, or Russians to land except as seamen. At the Cuban consulate, however, nobody had heard of such a regulation. Hurrying to the nearest Ward Line office at 112 Wall Street, Hughes was again refused passage. This time he proceeded to the NAACP headquarters at 70 Fifth Avenue, where he spoke with the acting secretary, Walter White, who telephoned Arthur Spingarn for legal advice. Sending wires to the Department of State of Cuba and of the United States and to the U.S. embassy in Havana, White demanded reasons why the "distinguished American Negro poet and novelist" Langston Hughes should not be allowed into Cuba to "rest." While they waited, Hughes disconsolately scribbled a note to himself. What was a Negro? Someone darker than white? A dark South American? "Make them prove that I am a Negro. If a Negro, what is their right of exclusion?"

Escorted by the very white Walter White, Langston went to the offices of the Cunard, the Panama Pacific, and the United Fruit Company lines. They were offered tickets, but decided to wait another day for answers from the government officials before paying. But when Hughes went again to Panama Pacific to buy a ticket, he was suavely told that all the berths were gone. Stepping into a nearby drugstore, he telephoned the same office.

"How many are going?"

"Three," Hughes lied.

"Very well," came the answer. "What is your name? Can you call at our office this afternoon?"

Disgusted, Hughes understood now that only White's pale skin had brought the offer of tickets previously. Next he tried the Cunard line. There, offered the choice of the remaining expensive staterooms on the *Caronia,* he took one.

The humiliation was too much for him. By the time the snow-white *Caronia* edged magnificently out of the 14th Street pier, he was coming down with a cold and had sniffled his goodbye to Carrie, who had come to see him off. As the ship steamed to the south he kept to himself, hurt and sulking, while a kindly old British steward tempted him with steaming tea and toast and marmalade. As the weather turned warm, however, and the *Caronia* slid past Palm Beach in Florida, Hughes perked up considerably. Daring now to mingle with the other guests, he found to his relief that they were mostly Jewish, affable people from the New York area, part of the winter tourist circuit. While he came back to health, he wrote fifteen letters and seven postcards.

Near noon on Tuesday, February 25, when the *Caronia* reached Havana, Hughes breezed through immigration. Using his rusty but serviceable Spanish he asked a taxi driver to take him to the most Cuban hotel he knew. When they reached the modest Las Villas inn at 20 Avenida Bélgica, Langston liked what he saw: a three-story building constructed Spanish-style, around a courtyard, with a restaurant on the ground floor and a little café-bar open to the street. He inspected a room which had no windows, but the upper half of its divided door let in fresh air at night; the walls were hung with mirrors (an amenity wasted on Hughes). In the courtyard, women and children chattered in musical Spanish punctuated by the crowing of a cock. This was perfect; Hughes was happy to be back in the tropics under a hot sun in a culture where most of the people looked somewhat like himself. "It seems years since I've felt such warmth or seen such a sky," he thanked Mrs. Mason. "I wish you were here."

Walking back from the wireless office to the hotel, he spent $5 on some souvenirs for Godmother, including a coconut straw cap and two strings of seed beads from a black man selling in front of the Teatro Nacional. Strolling past the presidential palace, Hughes was amused to see clothes hanging out to dry on top of the building. At a market, where he spent a quarter on fresh pineapple and oranges, the black young women in their bright dresses caught his eye. Finally, tired and warm, he returned to Las Villas for a siesta. When he went out later, the sun was sinking along the sea wall; darkness swept in over the water as an ocean liner drew near the shore. Alone but contented, he dined at Cafe El Paradiso near the Parque Central, close to Chinatown, which he remembered from his visit in the summer of 1927. For $1.50 he ate a good Cuban meal, which he capped with a demitasse and a small Benedictine. Strolling after dinner in the dark he heard the special music he had come to investigate for his opera—the rattles, gourds, guitars, trumpets, and drums of a typical Cuban orchestra playing in a public dance hall. Inside, he found poor blacks at their pleasure, short, shock-headed Negroes in a room festooned with carnival streamers (Mardi Gras was only days away), Japanese lanterns, and colored globes. Exhausted but expectant, he went to bed at last in his room full of mirrors, cooled by the breezes through the open upper door.

Late the next morning, after a breakfast of oranges and an omelette, he visited the leading daily newspaper, *El Diario de la Marina,* to present a letter of introduction from Miguel Covarrubias to its editor, José Antonio Fernández de Castro, "person extraordinary of this or any other world," as Hughes would call him. The remarkable Fernández de Castro, a white man, later was sometimes compared to Carl Van Vechten in his interest in black culture. The resemblance was superficial. Although they were both intellectual and cultural rebels and bon vivants, Fernández de Castro was no decadent and aesthete, but a political progressive and a student of Mayakovsky and other Soviet writers. Along with many other Cubans of all races and classes, he was a nationalist sensitive to the wrongs in his society, including racism and "Yankee imperialism," the economic exploitation of the island by North Americans, whose absentee ownership of Cuban agriculture and industry was backed by the self-proclaimed right of military intervention. His political consciousness was reflected in his considerable literary activity (he was once offered a professorship by the University of Michigan). In 1928, after reading Countee Cullen's anthology, *Carolling Dusk,* he had become the first person to translate Langston Hughes into Spanish with "Yo, también, honro a América" ("I, too, sing America"), a poem with particular appeal to hemispheric cultures unhappy with the appropriation of the name "America" by the United States.

Fernández de Castro was dictating an editorial on the second floor of the *Diario* office when Hughes walked in. Hobbled by injuries from a recent car accident, he rose on crutches to greet his visitor, then abandoned all work for the morning. Soon, he began telephoning certain people who would want to meet Hughes. There was a gifted young mulatto poet, Nicolás Guillén, already accomplished but still searching to find his authentic voice; Gustavo Urrutia, who edited the special Negro page, "Ideales de una Raza," of the *Diario;* and the Afro-Chinese writer Regino Pedroso, also prominent in the debate about race and culture and the future of Cuban writing. Pedroso was not available, but the others, along with José Antonio's young brother, Jorge, dined with Hughes early that evening at the restaurant of Lalita Zamora, reputedly the most beautiful black woman in Havana. When Langston and his new friends walked down the Prado, with José Antonio limping valiantly, they made slow progress. José Antonio knew everyone, and everyone wanted to be introduced to the greatest Negro poet of the United States of America—indeed, the greatest Negro poet in the world.

Passing in triumph to the Club Atenas, the leading Negro social club of Havana, Hughes was struck by its gleaming marble halls, sumptuous drawing rooms, handcarved bookcases generously stocked, and abundance of members; such substance and style were simply unknown among what blacks called social clubs at home. At the Club Occidente, he finally heard a full Cuban orchestra—maracas, bongos, a piano, the claves (sticks and bones), three-string guitars, guayos or long corrugated gourds, violins and flutes. Here, with the dancers either black or decided mulattos, the music was the *son,* the brisk, drum-based sound of Afro-Cuba, sensual in rhythm and ripe with double meaning, which Hughes had heard the night before but not recognized. Like the blues, the *son* was a form based squarely on the call-and-response of traditional African culture. Although no Cuban writer had yet employed it, Hughes immediately saw its possibilities as an organic base for formal poetry.

In the next few days, Fernández de Castro, who found Langston fascinating, included him in virtually every outing (in March, the Cuban would publish an article on Hughes, "Presentación de Langston Hughes," in the influential *Revista de la Habana*). They called on Conrado Massaguer, the leading editor and caricaturist of Cuba, and director of the magazines *Social, Havana,* and *Carteles,* who led Hughes on a tour of his facilities. At a special lunch at the Nuevo Mundo café the visitor met more writers and painters, including Juan Marinello, José Zacarias Tallet, Eduardo Abela, and Luis Gómez Wanguermet, the black sculptor Ramos Blanco, and a charming Mexican poet, Graziella Garbalosa, whom Hughes invited to tea the next day. He was honored with a formal reception at the black Club Atenas. José Antonio and Jorge took him with their girlfriends, who were dancers, on long drives along the Malecon skirting the water to Marianao and beyond. With Guillén, Urrutia, and others, Hughes returned almost every night to the night clubs at Marianao; Guillén would remember that Langston was a hit with the *soneros* who made the district famous for dance and music.

The only clear disappointment came when Langston met Amadeo Roldan, a leading Cuban composer recommended as a likely partner for his opera. Although he had done much to popularize Afro-Cuban music abroad, the light-skinned, freckle-faced Roldan, obviously of African descent, nevertheless resented being called a Negro and squirmed before Langston's inquiries. A little disgusted, Hughes would have to look elsewhere for a collaborator. Skeptical about the usual Cuban claims of racial harmony, he looked hawkishly for signs of racial division in work and play; he saw signs both of unity and of oppression, and even a will to revolt. In one hall, where a black band played to a crowded hall but without a single black dancer on the floor, Hughes heard "Drums like fury; like anger; like violent death."

Privately with Nicolás Guillén and Gustavo Urrutia, he sought the truth. Both men were shocked by the racial boldness, obsession almost, of this brown young man who certainly would not be called black in Cuba. For Hughes's part, when he heard Guillén's poems he was struck by their brilliance. He went to bed thinking about them (as he confided in a brief diary of the visit) and how they might be improved. Hughes by this time was essentially past major influence as a poet; not so Guillén, who was himself thinking of Langston's work and its fearless racial aesthetic. Hughes had one crucial recommendation for Guillén—that he should make the rhythms of the Afro-Cuban *son,* the authentic music of the black masses, central to his poetry, as Hughes himself had done with blues and jazz. This idea startled Guillén.

Born, like Hughes, in 1902 just as the U.S. military occupation after the Spanish-American War ended (to be resumed again in 1906) and Cuba proclaimed itself an independent republic, Guillén had attended law school in Havana before giving up his course of study—very much as Hughes had turned his back on Columbia. Also like Hughes, Guillén came from a family committed to social change; his father had been a nationalist political party leader and newspaper editor. Compared to Hughes, Guillén's growth as a poet had been more traditional: to signal his repudiation of law school, for example, he had published three sonnets ("Al margen de mis libros de estudios") in 1922. Four years of silence preceded his publication of new poems in Havana in 1927; the following year, he began to work with Urrutia on the Negro "Ideales de una Raza" page of the *Diario* and to gain some small recognition as an intellectual opponent of the dictator Gerardo Machado. A stocky, muscular man with flowing hair and brown skin, Guillén had shown signs of a steadily increasing racial consciousness, although it was almost apologetic compared to that of Langston Hughes.

After breakfast on Sunday, March 2, Hughes gave Guillén a formal interview, which appeared a week later on the "Ideales de una Raza" page. The great black American poet, Guillén wrote cheekily, looked like nothing more than "a little Cuban mulatto" ["Parece justamente un 'mulatico' cubano"], like one of the inconsequential brown dandies who whiled away their time at the university. Yet his "only concern is to study his people, to translate this experience into poetry, to make it known and loved." "I live among my people," Hughes said. "I love them and the way they're treated hurts me deeply so I sing their blues and I translate their sorrows, I make their troubles go away. And I do this like my people do, with their same ease. . . . I don't study the black man. I 'feel' him." (Oddly, Hughes denied—according to Guillén—ever having written a sonnet: "Yo tengo la suerte de no haber escrito nunca un soneto.") Mischievously, Guillén mimicked Hughes in his persistent questions about race: "Do blacks come to this cafe? Do they let blacks play in the orchestra? Aren't there black artists here?" At a black dance hall, the little North American mulatto exclaims, "My people!" Watching a black dancer, he confesses unashamedly, "I'd like to be black. Really black, truly black! [Yo quisiera ser negro. Bien Negro. ¡Negro de verdad!]."

The next day was Mardi Gras. On Tuesday, Hughes stopped by the American embassy to inquire about the alleged ban on blacks travelling to Cuba; coolly informed by officials that there was no such ban, he was discouraged from making further inquiry. Langston viewed the American embassy and the typical white American tourists and expatriates in much the same way. He avoided them and their hangouts, where they had almost inevitably brought the American vice, racism. "The Americans," he noted, "seem to clot in a dozen or so favorite places."

Too soon it was time to leave Cuba. On March 6, he went with José Antonio to change his ticket. Compelled by racism to leave New York in a Cunard stateroom, he did not intend to return that way. He traded in his ticket, paid $43.65 for third class passage on the *Essequibo,* and with the savings treated Fernández de Castro to daiquiris and a lunch of oysters and crab at the best seafood restaurant in Havana. Dinner was at Lalita Zamora's with Urrutia and Guillén, before a night on the town with José Antonio and Jorge; the evening ended happily to the sound of black music at Marianao, where Langston, received as an honored guest, was given a fine pair of maracas and a picture as gifts. The next morning, after a quick visit to the gardens of the oldest convent in Cuba, Hughes went to the *Diario* and *Social* to be photographed and to say farewell. At the home of the brothers Fernández de Castro, he was introduced formally to their mother. In his honor, Lalita Zamora prepared a special lunch of corn soup and fresh fish cooked in the most authentic Cuban style; she also gave him a picture of herself and a box of Cuban candies. At last, escorted by the brothers Fernández de Castro, Guillén, and Urrutia, all of whom gave him presents, he made his way to the third-class quarters on the *Essequibo.* The accommodations were decidedly less luxurious than on the *Caronia;* Hughes secured one of the twenty bunks in a hall that served jointly for sleeping and dining.

His Cuban friends were sorry to see him go; he was sorry to leave. In a small way he had failed: Langston knew very little more about music for his folk opera than when he had arrived. But this visit nevertheless had been important, his first overseas journey not only as a mature and accomplished man of letters but also as, self-consciously, an ambassador from black America. This aspect was very significant to Hughes. In his luggage were copies of several poems by Cubans which he would soon translate and attempt to place in North America.

On one man certainly, Cuba's future national poet, his impact was immediate. Although Guillén had previously shown a strong sense of outrage against racism and economic imperialism, he had not yet done so in language inspired by native, Afro-Cuban speech, song, and dance; and he had been far more concerned with protesting racism than with affirming the power and beauty of Cuban blackness. Within a few days of Hughes's departure, however, Guillén created a furor in Havana ("un verdadero escándolo," he informed Hughes with delight) by publishing on the "Ideales de una Raza" page of April 20 what Gustavo Urrutia called exultantly "*eight formidable negro* poems" entitled *Motivos de Son* (*Son* Motifs). For the first time, as Hughes had urged him to do, Guillén had used the *son* dance rhythms to capture the moods and features of the black Havana poor. To Langston, Urrutia identified the verse (which was dedicated to José Antonio Fernández de Castro) as "the exact equivalent of your 'blues'." Not long after, Urrutia reported that Guillén was suddenly writing "the best kind of negro poetry we ever had; indeed we had no negro poems at all" in Cuba until the new work. And when a local critic denied a relationship between Hughes and Guillén's landmark poetry, Guillén refuted him at once in "Sones y soneros," an essay published in *El País* on June 12 that year.

With such a result, Hughes's visit to Cuba had not been in vain. But what about his own future? With Godmother to guide him, Langston looked forward to the coming year of work.

8

FLIGHT AND FALL
1930 to 1931

PLANT your toes in the cool swamp mud;
Step and leave no track.
Hurry, sweating runner!
The hounds are at your back . . .

"Flight," 1930

BACK IN Westfield, New Jersey, Hughes now had company. After a stay of over two years in the South (and a visit to the Bahamas) from which she had returned with a bonanza of notes on folklore, Zora Neale Hurston had moved into a roominghouse not far from his home at 514 Downer Street. Like Hughes, Hurston was there on their patron Mrs. Mason's orders. The two writers also would share the secretarial services of Louise Thompson, hired the previous September by Mrs. Mason to assist Hughes. As Thompson later recalled, Godmother had primed her for Hurston's coming: "She used to talk about Zora, about this wonderful child of nature who was so unspoiled, and what a marvellous person she was. And Zora did not disappoint me. She was a grand storyteller." Hughes himself, having found Westfield more than a little dull, also welcomed Hurston's vital, attractive presence. While she entertained him and Louise Thompson with tales of her escapades in the Florida wilds, imitating and parodying her folk subjects with almost uncanny theatrical gifts, he languidly awaited the arrival of proofs of his novel, *Not Without Laughter,* and copies of his *Four Lincoln Poets.*

Although Hurston had reported religiously from the South to Mrs. Mason and to Alain Locke, Hughes had emerged as her most dependable ally, a calm, strong force essential (so Hurston often protested) to her growth as an artist, especially in their proposed collaboration on a folk opera. "Langston, Langston," Zora had written from Eatonville, Florida, in March, 1928, "this is going to be big. . . . Remember I am new and we want to do this tremendous thing with all the fire that genius can bring. I need your hand." Wanting him closer, she urged Hughes to join her for a carefree ramble in the South, like their journey in the summer of 1927. She offered to share fifty-fifty with him, even forty-sixty, any profits that came from their collaboration, since he was "so

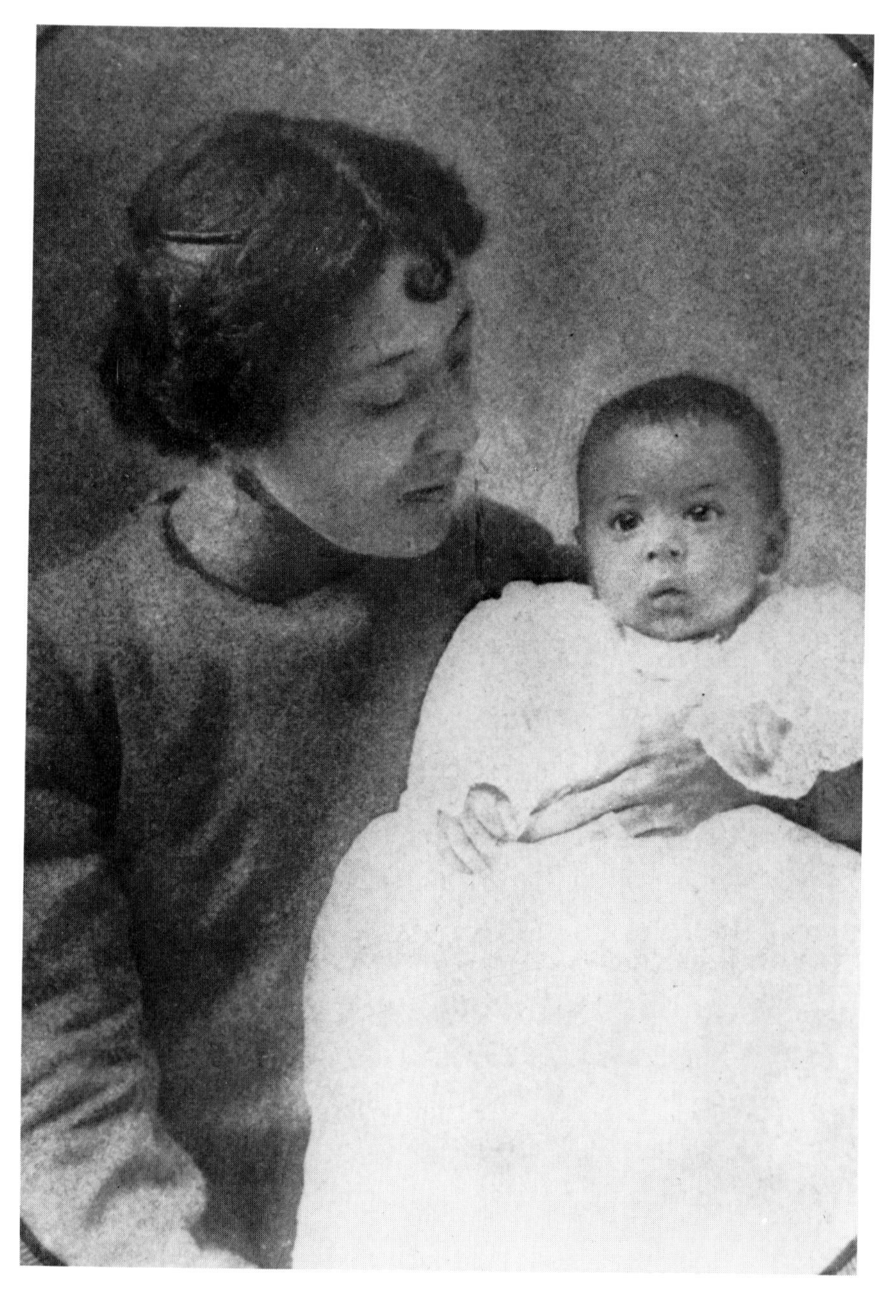

Carrie Hughes and son, 1902.

With Charles S. Johnson, E. Franklin Frazier, Rudolph Fisher, and Hubert Delany, May, 1925.

Pastel by Winold Reiss, 1925.

A busboy at the Wardman Park Hotel, Washington, D.C., 1925. *Photo by Underwood and Underwood.*

(Below) A'Lelia Walker, c. 1927. *Photo by Berenice Abbott.*

Zora Neale Hurston, Mobile, Alabama, 1927. In background, ''Sassy Susie.''

''Godmother''—Mrs. R. O. Mason (identified by Schuyler Chapin).

With Jessie Fauset and Zora Neale Hurston, at the statue of Booker T. Washington, Tuskegee, 1927.

"For Langston"—Alain Locke, 1928. *Photo by Nickolas Muray.*

Near Cresson Hall at Lincoln University, 1928.

With Carrie Clark at graduation, Lincoln University, 1929.

(*Below*) With Wallace Thurman and a friend, New York, c. 1929.

James Nathaniel Hughes, c. 1930.

much more practical than I.'' Although Hughes did not join her, Hurston kept up her adulation. ''Without flattery . . . you are the brains of this argosy,'' she insisted, ''all the ideas have come out of your head.'' Months later, she was still praising Hughes. ''You are always helpful,'' she declared, ''in fact you are the expedition.'' Excited over the prospect of an artists' colony in Eau Gallie, Florida, for Negroes (''NO niggers,'' she vowed), Hurston envisioned them living and working there in an ideal community shared by Wallace Thurman, Aaron Douglas, and other gifted members of the ''Niggerati.''

Much of this praise, in fact, was nothing more than flattery, one of the less attractive habits in Hurston's effulgent way of life. Sometimes out of necessity, perhaps more often out of her own deep-seated sense of vulnerability, Hurston tended to live by her wits—and successfully so. Once, she had moved into a Manhattan apartment ''with no furniture at all and no money'' (Hughes recalled with wonder), and ''in a few days friends had given her everything, from decorative silver birds, perched atop the linen cabinet, down to a footstool.'' One day, on her way to the Manhattan subway, Hurston fished a nickel from a blind man's cup. ''I need money worse than you today,'' she assured him. ''Lend me this! Next time, I'll give it back.'' On the other hand, Hurston matched Hughes in his deep regard for black folk culture; also, she genuinely liked him. In turn, he looked out for her interests with Godmother. When Mrs. Mason grumbled about Zora's letters, he urged Hurston to write more often and in a different tone, and tipped her off about the kind of gift sure to please Godmother—wood carvings, driftwood, orange blossoms, and melons, like his own inspired gift of corn and evergreen branch ends (or a book such as Martha Beckwith's *Black Roadways: A Study of Jamaican Folk Life,* which Hughes offered to Godmother as ''Studies from Obeah Land''). Going further, he once warned Hurston to be more discreet with Alain Locke, who sometimes betrayed her confidences to Godmother. Unlike Hughes, Locke put no one before his lucrative relationship with Mrs. Mason; finding out that Godmother disliked his friend Wally Thurman, Locke offered to arrange a sneak attack on him in a publication. ''The trouble with Locke,'' Hurston judged, ''is that he is intellectually dishonest. He is too eager to be with the winner.'' Hughes, on the other hand, could be trusted to do his best to keep her in Godmother's good graces.

In Westfield together, Hughes and Hurston built on the goodwill developed in their correspondence; they were also frequently in New York at Louise Thompson's apartment at 435 Convent Avenue. Then, almost certainly from Locke, Godmother heard that there was too much playing and too little work going on in Westfield. Her displeasure was electric, and everyone felt the shock. Instead of upbraiding Hurston, Mrs. Mason snubbed her. When Hughes tried to soothe Godmother with a telephone call, he was brushed off by someone at her Park Avenue home obviously acting under instructions. Mainly on Hurston's behalf, he quicky drafted a calming letter. When Godmother relented, he sent a longer letter of apology—for what, he wasn't sure. He had been ''terribly worried because Zora and I both felt that you had been displeased, or hurt in some way about her work.'' Zora was ''miserable,'' but he had been wrong to telephone and disturb Godmother. ''I do care for you,'' he wrote, "and when I hurt you through stupidity or error, I cannot bear it. . . . Whatever happened last week to make you unhappy must have been my fault—not Zora's. . . . But maybe I'm all wrong—all tangled up. You will forgive me if I do not understand. Perhaps it is all 'nerves' on my part. 'Emotional instability' or whatever they call it."

Although Godmother forgave him, Hughes understood that he was being watched. Mrs. Mason was angry that he had not plunged into work immediately after his return from Cuba, but the idea that she would tell him when and where to work chilled him. The novel was one thing; his writing in general was another. A short time later, he would write to Mrs. Mason about his instinctive reaction to her despotism. "So far in this world," he declared, "only my writing has been my own, to do when I wanted to do it, to finish only when I felt it was finished, to put it aside or discard it completely if I chose. . . . I have washed thousands of hotel dishes, cooked, scrubbed decks, worked 12 to 15 hours a day on a farm, swallowed my pride for the help of philanthropy and charity—but nobody ever said to me 'you must write now. You must finish that poem tomorrow. You must begin to create on the first of the month'." Godmother had gone too far.

For the moment, however, he kept his resentment to himself. Sometime in April, about a month after his return from Cuba, he began to work on a play with Hurston. Shelving their plans for a high-toned folk opera, in response to a casual complaint from Theresa Helburn of the Dramatists Guild that most scripts about black life were too grimly earnest, they decided to write a folk comedy. The basic plot came from a story, "The Bone of Contention," collected by Hurston, about two hunters who quarrel over the question of which one has shot a turkey. After one knocks out the other with the hockbone of a mule, he is tried on charges of assault and battery. Before an excitable crowd of black townsfolk, the example of Samson and the jawbone of an ass is invoked grandiloquently against the wielder of the mule bone, who is convicted and banished from the town. Uncertain how Godmother would react to such a project, Hughes and Hurston worked for a while without her permission. Hughes was responsible for "the plot, construction, and guiding the dialog toward the necessary situations and climaxes," he would recall; Hurston provided "the little story," gave the dialogue its Southern flavor and many of the 'wisecracks'." She had contributed a very brief play to *Fire!!* but did not feel very comfortable writing for the theater; with only the unfinished singing play about Haiti to his credit, Hughes nevertheless was more confident about his ability as a dramatist. Nevertheless, Hurston's contribution was almost certainly the greater to a play set in an all-black town in the backwoods South (she drew here on her childhood memories), with an abundance of tall tales, wicked quips, and farcical styles of which she was absolute master and Langston not much more than a sometime student. With Hughes and Louise Thompson laughing helplessly, she acted out male and female roles in a variety of voices. Whatever dramatic distinction the play would have, Hurston certainly brought to it. But "Mule Bone," as it was called, was at least as much a collaboration as, for example, that between Wallace Thurman and William Jourdan Rapp, who had tailored Thurman's *Harlem* into a Broadway property.

With Thompson typing almost as quickly as they invented, Hughes and Hurston made excellent progress. Then Zora seemed to grow tired of the project, becoming "restless and moody, working in a nervous manner." But, Langston reasoned, she had always been restless and moody. Moreover, while he was relatively unencumbered, Zora had to return to her massive folklore project and endure constant scrutiny from Locke and Mrs. Mason. Still, Hughes was very surprised when, with "Mule Bone" near completion but still in need of serious work, Hurston suddenly left New Jersey and moved to Manhattan. She was en route, she said, back to the South, but promised to work further on one act of "Mule Bone."

Hughes's disappointment at Hurston's departure from Westfield was completely overshadowed near the end of May, however, by the collapse of his friendship with Godmother. While the precise cause of the rupture would remain unclear even to Hughes, he took it to be a result of Godmother's displeasure over his unwillingness to return to work following his return from Cuba. Certainly the new quarrel involved a trip late in May to Washington, D.C.; when Godmother argued that he should stay home and write, Hughes rebelled against her authority and committed himself to go. This time, as if primed for destruction, the enraged Mrs. Mason went much further than she had gone before, much further than he ever imagined she would go in accusing him of ingratitude and disloyalty. "The way she talked to Langston," Louise Thompson would remember from Hughes's account later that day, "is the way a woman talks when she's keeping a pimp. 'I bought those clothes you are wearing! I took care of you! I gave you this! I gave you that!' " Godmother counted out precisely to Hughes his cost to her—$225 a month, of which $150 comprised his stipend and $75 half of Louise Thompson's time; in return, he was giving her—nothing!

Nine years later, in a reticent account of their relationship, in which Hughes never identified Mrs. Mason by name (she was still alive), he did not hide his extreme pain in their last meeting:

> I cannot write here about that last half-hour in the big bright drawing-room high above Park Avenue one morning, because when I think about it, even now, something happens in the pit of my stomach that makes me ill. That beautiful room, that had been so full of light and help and understanding for me, suddenly became like a trap closing in, faster and faster, the room darker and darker, until the light went out with a sudden crash in the dark, and everything became like that night in Kansas when I had failed to see Jesus and had lied about it afterwards. Or that morning in Mexico when I suddenly hated my father.

Shattered by her words, his body betraying a neurotic turmoil that made his muscles twitch involuntarily and his fingers curl into bizarre shapes, he rode the train to Washington. Although he made his rounds in the city (in one place going over a gift copy of Nicolás Guillén's newly arrived pamphlet of poems,

Motivos de Son, with a Cuban acquaintance), Godmother dominated his mind. To Alain Locke, Hughes poured out his hurt and confusion; listening carefully, Locke softly tendered advice. Whatever he counselled, when Langston finally wrote Godmother he begged her to release him, "or rather to release yourself from the burdens of my own lack of wisdom." Hughes flogged himself unsparingly. "The fault is mine. The darkness is mine. The search is mine. The Gods have given me the . . . light of your kindness and I do not know how to follow." He offered to settle all financial accounts ($200 in the bank, $37.94 elsewhere). Then they could go back to the happier time when money did not come between them: "The beauty of your gifts and the things of your spirit are written in my heart—not in an account book." "I should adore coming to see you when ever you will me to," he promised Godmother, "and I know you have loved the better part of my self with the most beautiful love . . . a being can know in this world."

To facilitate his financial break with Mrs. Mason he turned to Carl Van Vechten, who lent him $200, then noted Hughes's visit on May 26 in his diary: "His patron has failed him & he borrows some money."

Early in June he went out to Lincoln for his first visit since his graduation the previous August. Inwardly distracted but keeping a brave front, he stayed with Professor Robert Labaree and his wife Mary. On Wednesday, June 4, in splendid weather, Robert Labaree drove Hughes to Moylan in the beautiful Rose Valley, about a mile south of the town of Media, Pennsylvania. There, in a small complex centered in a converted mill dating back to 1840, Jasper Deeter ran his innovative theater company, the Hedgerow Players. A highly successful producer and director who nevertheless cared little for Broadway commercialism, Deeter was committed to staging experimental and socially provocative plays, including those on race; he had worked with Robeson in O'Neill's *The Emperor Jones* and also had staged the Pulitzer prize-winning *In Abraham's Bosom* by the white North Carolina playwright Paul Green. Welcoming Hughes, Deeter escorted him on a tour of the facilities. In the fall, Hedgerow would be attempting a virtual Negro season of drama, with six plays featuring black or integrated casts. Rose McClendon and Abbie Mitchell, perhaps the two finest black actresses of the age and both friends of Hughes, would be in residence. Although Deeter would depend mainly on O'Neill and Paul Green for plays, he also wanted a black playwright in residence. Hughes could come as early as August to observe a "white" production, and perhaps even act in small parts.

"Do you think the idea a good one—or would there be something dangerous about it?" Langston asked Van Vechten. But his main concern was Godmother's reply to his plea. Around June 7, bursting with righteous indignation, she had her say: "Dear child, what a hideous spectre you have made for yourself of the dead thing money!" She reminded him that on November 5, 1927, "you wanted very much to try the plan of allowance and accounting through which you knew many of the other Godchildren had succeeded joyfully in saving income toward a difficult time or a flaming need." Nothing was binding between her and "Alamari," as she sometimes poetically called Langston, so *of course* they could dispense with accounting! "I therefore enclose in this letter a check for 250.00. . . . Hail to Alamari! Love and good hope to you as you seek the sea. Success for whatever plan you make for revivification.''

Her ominous tone unnerved Hughes. On June 15, when he drafted a long letter to Mrs. Mason, he tried to explain everything. ''I was afraid it was happening all over again this time between us,'' he told her, ''the old words about things, about money—'I have given you this—now you must do that'.'' ''I love you,'' he pleaded, ''—and I do not wish to lose you. . . . I would still love you as deeply as ever because you are to me more beautiful than anybody else in the world. . . . I know that all you have ever given me came on the wings of the spirit. Only my hands have been clay—unworthy to receive the beauty—and I have not known how to keep the flight.''

> I love you, Godmother. I need you. You can help me more than anyone on earth. Forgive me for the things I do not know, the things I can not fight alone, the things I haven't understood. You know better than anyone else how stupid and unwise I am, how I must battle the darkness within my self. No one else would help me. No one else would care as you care. No one else would even try to understand. The door is never closed between us, Godmother. Only the ugly shadow of my self stands in the way now.

Hoping to draw her out, he told her about Jasper Deeter's offer. ''Could I come and talk to you about it before you go away?''

On June 17, however, a telegram from Mrs. Mason denied him an audience: ''Under present conditions it is useless for me to undertake any more than I have promised.'' A stay at Hedgerow ''could be valuable.'' She sent her love.

Suddenly it dawned on Hughes that Godmother was through with him. He had cried for his independence, but now the thought of it left him cold with fear. Moreover, something else was wrong; the air seemed suddenly thick with deception. From Alain Locke, on his way to another foreign summer sponsored by Mrs. Mason, came an odd note, almost mocking. ''Congratulations on your redemption from starvation,'' Locke wrote, referring no doubt to the check from Godmother. Almost gratuitously he let Hughes know what was happening at court, who was in, who was out. Locke was definitely in: he had seen Godmother twice, once alone, once with—Zora Neale Hurston. Wasn't Godmother lovely? ''She is really beautifully devoted to us all,'' he purred. ''I am trying my best to cause her the least possible disillusionment.''

Almost at the same time Hughes seemed to be losing Godmother, Locke, and Hurston. But could the three losses be so closely related? In any event, Langston's one desire was to be recalled to Godmother. Sending her the first copy of *Not Without Laughter,* he threw everything into his covering note: ''For the beauty of your eyes that first night at Carnegie Hall when I looked down on you; for the love of your hands on my own; for the strength of your voice; for the truth of soul, and the great freedom of your heart—I send you today my love to greet you.'' He did not achieve his goal. Although Godmother responded emotionally, her tone was elegiac: ''Oh Alamari, as the sun sets on the western slopes of Godmother's life her spirit holds in its eternal belief the morning star that Sandy was destined to carry into the hearts of his people.''

Touching bottom in his despair, Hughes felt his body wracked by the emotional strain; nausea haunted his stomach. "I ask you to help the gods to make me good," he cried out to Mrs. Mason. "I am nothing now—no more than a body of dust without wisdom, having no sight to see. Physically and spiritually I pass through the dark valley, a dryness in my throat, a weariness in my eyes, fingers twisted into strange numb shapes when I wake up at night, the mind troubled in the face of things it does not understand, the mouth silent because there is no one to talk to, the sweet air burning the lungs, the hot air cold to the body." "I am not wise, Godmother. . . . The inner soul is simple as a fool, sensitive as a shattered child, at odds too often with the reasons of the mind . . . a coward hiding in the dark, wrapping itself in the cloak of art when, too sick with loneliness, it must go out for air or die."

Miserable, he holed up in Westfield. Perhaps reflexively, perhaps hoping to please Godmother, he struggled to publish. To James Weldon Johnson, revising his important anthology *The Book of American Negro Poetry,* first published in 1922, Hughes confessed that he had nothing new (Johnson selected eleven old poems). Continuing to try to help other writers, Langston secured from Elmer Carter, now editor of *Opportunity,* a commitment to give special consideration to Latin American writers. In accepting from Hughes his translation of Guillén's poem "Black Woman," and an essay on the black Cuban sculptor Ramos Blanco, Carter lauded Hughes for taking "the first steps" toward intellectual rapprochement between people of color in the Americas. But when Hughes sent original work to Ridgely Torrence for *The New Republic,* it was rejected.

Three poems in the July *Opportunity* spoke of death or disillusion. One was "Dear Lovely Death" ("Come lovely and soothing death," Whitman had chanted in "When Lilacs Last In The Dooryard Bloom'd"):

Dear lovely Death
That taketh all things under wing—
Never to kill—
Only to change
Into some other thing
This suffering flesh,
To make it either more or less,
Yet not again the same—
Dear lovely Death,
Change is thy other name.

In "Aesthete in Harlem," Hughes wrote of searching, estrangement, and ironic discovery:

Strange,
That in this nigger place
I should meet life face to face;
When, for years, I had been seeking
Life in places gentler speaking—

Before I came to this vile street
And found Life—stepping on my feet!

And his best poem in a long time, "Afro-American Fragment," appearing in the July *Crisis,* was a requiem:

So long,
So far away
Is Africa.
Not even memories alive
Save those that history books create,
Save those that songs beat back into the blood—
Beat out of blood with words sad sung
In strange un-Negro tongue—
So long,
So far away
Is Africa.

Subdued and time lost are the drums—
And yet, through some vast mist of race
There comes this song
I do not understand,
This song of atavistic land,
Of bitter yearnings lost, without a place—
So long,
So far away
Is Africa's
Dark face.

He had other small successes, such as on July 29, when he took part in an inter-racial symposium at Columbia University. When Hughes approvingly reviewed Essie Robeson's book on her husband, *Paul Robeson, Negro,* Irita Van Doren splashed his essay on the front page of the New York *Herald-Tribune* book section. And another small book of poems was in the offing. Amy Spingarn, back in town after an art course at the University of California under the abstract expressionist master Hans Hofmann (she had helped to bring him out of Germany to the United States in 1930), proposed publishing an edition of perhaps a dozen of Hughes's poems on a small hand press she had just bought. Hughes's affection for the softly gracious Mrs. Spingarn had only grown with his troubles with Godmother; and Langston had congratulated Joel Spingarn the previous December when, not without public controversy over his race, Spingarn had been elected president of the NAACP ("GLAD THEY CHOSE YOU GLADDER STILL YOU HAVE ACCEPTED BRAVO"). With Mrs. Mason forbidding public tributes to herself, Hughes had dedicated *Not Without Laughter* to the Spingarns, and sent them the page proofs "with my deep appreciation for what you [Joel] have done for the Negro peoples, and my happiness in the friendship of you both." In turn, the Spingarns clearly respected him; when they had the ends of four huge mahogany beams at their country estate carved into human heads to represent the Nordic, Jewish, Indian, and African peoples, the model for the last was Hughes.

When his selection of poems arrived, Mrs. Spingarn at once saw death as the dominant theme; their book would be called *Dear Lovely Death.* In this grim time, little could raise Hughes's spirits, not even the fine reception of *Not Without Laughter.* Blacks and knowledgeable whites were struck by Hughes's power to sound deep racial notes without being polemical; more than any novel of the twenties, *Not Without Laughter* captured ordinary black life "without bitterness or apology and yet with truth and deep feeling," as Arthur Spingarn wrote Hughes. Professor Labaree found it "an artistic and most gripping human document," one that made him feel "more than ever ashamed of my white race" and yet (in a typical reaction to Hughes) left him "without any resentment to the man who showed us up." "Bring out the laurel wreath and drape the brow of Langston Hughes," enthused the once hostile *Pittsburgh Courier;* "to the ordinary Negro reader it arouses memories of youth, of yesterday and today. I know the people in this novel, every one of them." To Sterling Brown, the author in 1928 of a groundbreaking study, *The Negro in American Fiction,* and a fine poet who would build on Hughes's pioneering use of the blues, the remarkable simplicity of *Not Without Laughter* was truly "the simplicity of great art; a wide observation, a long brooding over humanity, and a feeling of beauty in the unexpected, out of the way places, must have gone into its make up."

Within its limitations as the vicarious autobiography of a boy told in traditional form, *Not Without Laughter* was perhaps the most appealing and completely realized novel in black fiction to that date. But in the chorus of praise Hughes found little pleasure; the one voice that mattered was silent. Between early July and very late September he heard nothing from Mrs. Mason. In the dog days of August he stalked the beach at Far Rockaway, a popular city resort near which his mother worked as a maid for white folks. Once he went north to Amenia to breathe the clean country air of Troutbeck, the Spingarns' estate, and watch Amy Spingarn smudge her fingers on her printing press. He also heard, tauntingly, from Locke in Paris. A few days before, Locke had passed the old building "where I used to climb *au cinquième* to wake you up." He had nominated Langston for a literary prize of $400; if Hughes didn't want the money, Locke added cruelly, he could write the sponsoring foundation. And as if one Paris ghost stirred another, a letter also arrived from his beloved of 1924, Anne Marie Coussey, to bring bittersweet memories of the Luxembourg Gardens and their Paris spring. Life had taken her to the West Indies, to 67 Alberta Street in Port-of-Spain, Trinidad. Now married to a prospering lawyer, Hugh Wooding, Anne had two adorable children, Selby and Ambah. She had seen a copy of Hughes's *Fine Clothes to the Jew,* but found some of the poems "not up to your usual standard."

The summer was not a complete failure, but Hughes could not conceal his abject disaffection even if he cloaked it in the guise of *ennui* to his hero Claude McKay, still in self-imposed exile in Europe. He "had gotten awfully bored with LITERATURE and WHITE FOLKS and NIGGERS and almost everything else."

When he reached the Hedgerow Theater in Moylan on September 15, there was only more disappointment. Because of the deepening Depression, which had also ravaged the book market, including sales of *Not Without Laughter,* funds for Jasper Deeter's season of Negro theater had not come through. With few options, Hughes decided to stay at Hedgerow. "It's a pretty spot down in a lovely little valley," he wrote McKay, "and I'm getting a sort of inside slant on the theatre, watching the rehearsals and the plays every night." Rose McClendon and three black actors were in residence; *Othello* and *The Emperor Jones* were in the repertory. Life at Hedgerow turned out to be pleasant enough. With Deeter himself rehearsing Shaw's *Devil's Disciple,* Hughes observed the director as often as he could but also worked several hours each day in the office, since everyone was assigned to a job off the stage. Nobody was paid, apart from the cook, but food and other basic supplies such as toothpaste and stamps were free. Hughes slept with other company members in a cottage on a hill a mile or so from the theater, over which mist and white stars drifted in the early fall nights.

Hughes was particularly pleased to be working with Rose McClendon, whose performance in Clement Wood's *Deep River* in 1926 had moved Alexander Woollcott to salute her as "the lost loveliness that was Duse"; she had won a *Morning Telegraph* acting award the following year (when Ethel Barrymore and Lynn Fontaine also won the award) for her role in Paul Green's *In Abraham's Bosom.* In *Porgy,* where she had become great friends with Bruce Nugent, McClendon had been a powerful, touching Serena, yet she had no hope of becoming a star. Ethel Barrymore, for example, had lauded her, then consented to appear blackface on Broadway with other besmirched whites in Julia Peterkin's *Scarlet Sister Mary.* McClendon's quiet revenge was to keep a scrapbook, which she shared with Hughes, of the many hostile reviews of Barrymore's performance. Langston, in turn, soon showed her something far more intriguing—his first draft of a work called "Cross: A Play of the Deep South," which included an excellent part for a mature black woman. Reading the play, Jasper Deeter pronounced the construction basically fine but the story too sensational—miscegenation, parricide, mob violence, murder. The play had emerged from Hughes's poem "Cross," on the theme of the tragic mulatto:

> My old man's a white old man
> And my old mother's black.
> If ever I cursed my white old man
> I take my curses back.
>
> If ever I cursed my black old mother
> And wished she were in hell,
> I'm sorry for that evil wish
> And now I wish her well.
>
> My old man died in a fine big house,
> My ma died in a shack.
> I wonder where I'm gonna die,
> Being neither white nor black.

The play (later called *Mulatto*) tells of white Colonel Norwood, his colored mistress, Cora, with whom he lives as in a marriage, and their children, especially Bert, a rebellious son who returns from school in the North determined to assert his human dignity. Defying his father and mother and the ways of the South, Bert brings ruin down on his family and himself. At the center of the drama is less the youth's mulatto sense of racial ambivalence than his rage for full acceptance by his father. At least three factors were behind Langston's play. The first was Hughes's ingrained sense of having been neglected by his mother and abandoned by his father, which allowed him to empathize with the tragic mulatto although he was not one. The second was the imagined relationship between his grandfather Charles Langston and Charles's father, Captain Ralph Quarles of Virginia, who had been devoted to his "wife" Lucy Langston and their mulatto children. In his autobiography, John Mercer Langston had evoked an ambiguous picture of his brother Charles: "He was not large nor apparently firm of body; but well endowed intellectually. His disposition and temper though ordinarily well controlled, were not naturally of the easy and even sort. In his constitution, he was impetuous and aggressive; and under discipline and opposition, he was always restive, yet, he yielded with reasonable docility and obedience to the training to which his father, interested in his education, sought to subject him." A third factor, Hughes's rejection by Godmother, was inevitably involved. *Mulatto* thus ventilates a rage Langston could not acknowledge in his desire to be reconciled with Mrs. Mason, who clearly dominated his thoughts and feelings during the weeks when the work was written.

Two weeks into his stay at Hedgerow, a brief, distant letter forwarded from Westfield brought only a check from Mrs. Mason and her vague good wishes; she asked nothing about his whereabouts, much less his plans. Almost at the same time he heard from Louise Thompson. On September 22, after a year of employment, she had been summoned to the Barclay Hotel by Mrs. Mason for their first meeting in several months. In a "short but excruciating" session (as Thompson recalled the following week), Mrs. Mason denounced her and all Negroes as a lost cause. Soon Thompson found herself in the lobby of the Barclay, confused and hurt. The attack had been unexpected; aside from Mrs. Mason's expectation of thank-you notes for her pay, there had never been any friction between them. "I had thought she was wonderful," Louise would remember of the "very gracious" Mrs. Mason (Thompson was not asked to call her Godmother); the old lady always inquired about "Mother" Thompson's health, which was not improving, and often pressed extra money for her into Louise's hand. Perhaps Zora Hurston would know what had brought on the attack, but Hurston's phone had been disconnected, and Louise could not reach her before writing to Hughes. (In fact, Hurston was staying in Asbury Park on the New Jersey coast, less than an hour from Manhattan.)

A new wave of illness overtook Hughes—a chronic toothache, tonsilitis, an upset stomach. With Godmother back in Manhattan after the summer, he moved cautiously. He sent her all the reviews of his novel from English and American papers; without enthusiasm, she acknowledged his letter, but advised him only to have his tonsils removed. Leaving Hedgerow near the end of October, he returned to Westfeld in time to speak at an inter-racial conference at the recently opened Riverside Church in Manhattan, which the theologian Reinhold Niebuhr also addressed. At last, after several attempts, Hughes succeeded in reaching Zora Hurston, but when he begged her to resume work on their play, "Mule Bone," she was evasive, hesitated before settling on a meeting, then failed to keep their appointments. Still Hughes did not connect Hurston's behavior to his falling out with Godmother; Zora was notorious for unpunctuality even in Harlem. In fact, Hurston was looking for a dramatic stroke to sunder their link and affirm her total loyalty to Godmother. She made this desire clear in a letter to Mrs. Mason on November 25 from Asbury Park: "You love me. You have proved it. It is up to *me* now to let you see what my behavior of love looks like. So watch your sun-burnt child do some scuffling. That is the thing that I have lacked—the urge to push hard and insist on a hearing." She would be in town soon to see what "a certain person has to say to me." Her "scuffling," indeed, had already begun.

Still Hughes hoped for a reconciliation with Godmother. Certain that he could win her heart again, perhaps he finally had his chance. More likely, he did not. "I am helping myself forget the discouraging things that have fallen on me from the Negroes," Mrs. Mason had written Locke early in August, "by talking about my Indian days." (At the same time, she sent Locke $50 as a birthday gift and $300 more for some transaction.) Nevertheless, in his autobiography ten years later, Hughes would recall a final, disastrous meeting with Mrs. Mason from which he stumbled into "the winter sunshine on Park Avenue." But nothing in Godmother's letters to him indicates any meeting later than spring. In *The Big Sea,* Hughes would compress the more than six months of pain during which he pined for a reconciliation into a far briefer space, but the length of time he suffered had much to do with the lasting impact of this crisis on him. In his autobiography, the break is clean and final; in truth, he wrote and received many letters after his expulsion. He cited a poem by him protesting the luxury of the newly opened Waldorf-Astoria Hotel as having contributed to their break—but the hotel did not open until the following year, 1931, when the poem was composed. Another, almost equally stinging poem, "Merry Christmas," an anti-imperialist parody of a *New Yorker* seasonal tradition, was composed that fall, but appeared in *New Masses* in December, after their break:

> Merry Christmas, China,
> From the gun-boats in the river,
> Ten-inch shells for Christmas gifts,
> And peace on earth forever.
>
> Merry Christmas, India,
> To Gandhi in his cell,
> From righteous Christian England,
> Ring out, bright Christmas bell! . . .

If Mrs. Mason granted him an autumn audience, it must have been very early in November; considering the state of their relationship, however, she could hardly have said much to surprise him. Clearly he tried to reach her through the mail. On November 30, she acknowledged "your two letters telling me of your sudden severe illness" and also "your Thanksgiving letter and card." She let Langston know that she had sorrows of her own. One of Katherine Biddle's young sons, Garrison, had died just before Thanksgiving.

Hughes was now definitely ill. "Violent anger makes me physically ill," he later confessed. "I didn't feel any of those things consciously—for I had loved very much that gentle woman who had been my patron. . . . But now I was violently and physically ill, with my stomach turning over and over each time I thought about that morning at all. And there was no rationalizing anything. I couldn't." On December 6 he wrote to tell Arthur Spingarn that his illness had been diagnosed by a creative Harlem physician as a Japanese tapeworm. A few days later a white Park Avenue specialist, examining Hughes (perhaps as a special favor to Spingarn) after hours, rudely dismissed the diagnosis as absurd. Langston shared the joke with Mrs. Mason, but both laughed falsely. As with his ailment in Mexico more than a decade previously, he could not bring himself to tell the doctors, or even a friend such as Arthur Spingarn, why he was ill—that a psychological disturbance taxed his body. Instead, he endured his humiliation in a charade of his own staging.

At last accepting as a fact that Godmother would not see him, Hughes planned to leave Westfield and the New York area—and without telling Mrs. Mason. After attending the usual parties in Manhattan, including one at Knopf's, he welcomed 1931 in Cleveland, to which his mother had just moved. Langston planned to stay at her home at 4800 Carnegie Avenue for only three weeks or so, then head south to Florida, with his trip financed from a check expected from Godmother early in January. But on January 10 Mrs. Mason informed him that no check would be sent—as "by mere accident it came to me that you had gone and where you were. Dear child, what am I to believe, what am I to think under these circumstances? Will the sap rise in the trees that the blossoms may come in the spring?"

Without money to visit Florida, Hughes settled in with his mother for a stay that would last three months. Cleveland was a changed city. Most astonishing of all was the spread of blacks in the decade since his graduation; from 22nd to 110th Street and beyond, mile after mile of once-white blocks now teemed with blacks. Central High School, where he had been one of a handful of colored students, was now more than fifty percent black. Returning in glory, he read his poems to the students and teachers; he also spoke to school officials in the wealthy district of Shaker Heights. Lecturing on modern writing, Hughes argued the superiority of new writers such as Hemingway, Sandburg, Millay, and Masters over virtually all the established names in American literature, excepting Whitman and Dickinson. He also visited his old friends Rowena and Russell Jelliffe, whose drama troupe, the Gilpin Players, had become probably the outstanding amateur black theatrical organization in the nation. Mrs. Jelliffe shared some good news with him. Always on the look-out for suitable new plays for blacks, she had just been offered a rich folk comedy, written by a young woman in New York, which the Gilpins would open on February 15. Within seconds, Langston recognized "Mule Bone," his collaboration with Zora Hurston. But his name was nowhere on the script sent by the reputable Samuel French play agency to the Gilpin Players.

On November 14 of the previous year, Hughes soon discovered, Hurston had delivered the play to Van Vechten for a reading. Admiring it, but without informing her, Van Vechten had sent "Mule Bone" to the Theatre Guild for consideration. An official of the Guild, Barrett Clark, certain that the Guild would not use "Mule Bone," had passed it on with high praise to the Gilpins in his capacity as an employee of the Samuel French agency. But why was Hughes's name not on the script? Only later would he discover that the previous October, while he was at Hedgerow, Zora Hurston had copyrighted "Mule Bone" exclusively in her name; she had then sent it to Van Vechten without mentioning Hughes's name. This was the start of the "scuffling" she had promised Godmother, and the main reason she had evaded Hughes in the fall.

Astonished, Hughes wrote Hurston at once for an explanation. By Friday, January 16, when she had not replied, he wrote Van Vechten for advice: "Would you do anything, or nothing, in the present situation, if you were me?" He was "amazed" by Zora; "is there something about the very word theater that turns people into thieves?" Friendly with both writers, Van Vechten advised only that Hughes had to do *something*. Hughes began gathering material and evidence; he also wrote Hurston threatening litigation.

On Saturday, January 17, Louise Thompson arrived in Cleveland. She was not there to see Langston or the Jelliffes but travelling as part of her job with the American Interracial Seminar, a non-partisan, non-propagandist group founded the previous fall with the goal of improving race relations. To the Jelliffes, Thompson backed Hughes's assertion of joint authorship. Then, later that day, Langston finally talked on the telephone to Hurston. Genuinely surprised to learn that the play was in Cleveland, she was not enthusiastic about an amateur production; on this point she had the support of Van Vechten, Fania Marinoff, Lawrence Langner, and other knowledgeable theater persons. As for the matter of authorship, it was too complicated to discuss on the telephone. She would write a letter to Hughes.

The next day, when Hurston went to see Van Vechten, she "cried and carried on no end," he wrote Hughes; years later he would remember her throwing herself to his library floor in an absolute "tantrum." But she moved more coolly on another front. Sending Hughes's complaining letters to Mrs. Mason, Hurston insisted that in "Mule Bone," "I am not using one single solitary bit in dialogue, plot nor situation from him and yet he tries to muscle in." On January 20, when Langston received her promised letter, Hurston denied his claim of co-authorship. But she also claimed grounds for grievance. Langston, she accused, had taken Louise Thompson's side against her in Westfield, bettering Louise's working hours and allowing her to be late for work; then, after Thompson had tried to cut in (so Hurston said) on future royalties of the play, he had proposed her as its business manager, when the play was produced. "By this time," Zora asserted, "I had come to feel that I hadn't much of a chance in that combination.'' She capped these charges with protestations of affection for him, but made it clear that her interest was not romantic. He could have anything he wanted. ''Tears unceasing have poured down inside me,'' she wailed. But the play was her own: ''I told Godmother that I had done my play all by myself, and so I did, and for the reasons stated above.''

This was not Hurston's first expropriation of another person's material; in October, 1927, for example, she had published an essay that was, according to her biographer, ''25 percent original research and the rest shameless plagiarism.'' But the theft of the play was probably secondary. Either mistaking or deliberately misrepresenting the relationship between Hughes and Thompson, she had characterized it as an affair to Godmother, who carefully noted Hurston's information that Langston and Louise often ''went off to his rooms'' while Zora toiled, then would return and say, ''Let's see how much you got done then—that's fine *we're* getting along splendidly.'' Unquestionably, Hurston had wanted Thompson fired. Instead, Zora caused something she probably had not expected—that Mrs. Mason would summarily banish Langston. And, in banishing him, frighten Hurston into not only shunning his company but also denying that they had ever collaborated on ''Mule Bone.'' Thus it became essential to Hurston to discredit both Hughes and Louise Thompson, although her main target was Thompson. ''It is just as we know,'' she wrote Godmother, ''Langston is weak. Weak as water. When he has a vile wretch to push him he gets vile. When he is under noble influences like yours, you know how fine he can be.'' Perhaps he should have another chance. ''He is ashamed of his attitude about the play and apologized to me for it.''

Although Hurston had acted like a lover spurned, there is no evidence either that an intimate relationship ever existed between herself and Hughes, or that Hurston even wanted one. And, according to Louise Thompson later, if Hurston spoke of an affair between Louise and Langston, ''well, she's just lying. She's just lying. Because there was never any relationship between Langston and me other than as a brother. . . . If Langston had approached me in another way, I might have been receptive, but he never did. I accepted Langston on that plane, that we were the best of friends and comrades.'' (After knowing him most of his life, she would judge Hughes to have been ''asexual''; as for suggestions that he was a homosexual, ''*that* I felt was not true. I never had in any sense any intimation that he was that way.'') Hurston's suspicion of Thompson seems to have been based on little more than a general sense of insecurity with a woman younger, prettier, more poised, and, although in a more orthodox way, as intelligent as Hurston herself.

Most interesting of all, in what was in effect an unusual emotional quadrangle, was Godmother's violent reaction to Zora's insinuations. Mrs. Mason behaved either like someone impossibly moral—or like a wronged, revengeful lover.

On January 20, the Samuel French agency wired the Jelliffes that Hurston had denied permission for their production. Langston then drew up an eleven-page letter to his lawyer, Arthur Spingarn, offering a compromise. Two-thirds of the royalties would go to Hurston, one-third to Hughes. But Zora had suddenly changed her position. Wiring her approval to the Jelliffes, she followed with two more affirmative telegrams and promised to come to Cleveland at the end of the month. Weary but relieved, Hughes wrote Van Vechten that "the split is over."

Langston had other troubles. He spent the night of January 21 in jail. With two black friends, he had left a car on busy Cedar Avenue to make a phone call. When they returned, a taxi driver accused them of denting the back of his cab. After two squad cars arrived, the police threatened the three with beatings unless they confessed. A paddy wagon took them to the station, where the owner of the car was roughed up and struck with clubs. But none "confessed"; they hadn't touched the cab. In jail overnight in a bitterly cold cell, with only a board to sleep on and no blanket, Hughes understood at last what Bessie Smith really meant in moaning of "Thirty days in jail / With ma back turned to de wall." In the morning they were released after signing "suspicious person" slips.

An even greater disappointment came later in the day. Hoping to publish his third book of verse with Knopf, Hughes had sent Van Vechten a collection of his poems, "The Singing Dark." But Van Vechten not only disparaged the title (people would call it, or Hughes, "The Singing Darkey")—the volume would add nothing to Langston's reputation, since it contained no single poem as fine as "The Weary Blues" or "Mulatto." "I shouldn't wonder," Van Vechten said pointedly, "if you are pretty nearly through with poetry." Having sent the poems apologetically, knowing well that he had written nothing really significant since the appearance of *Fine Clothes to the Jew,* Hughes was hurt but not surprised by Van Vechten's judgment. Still, he revised the manuscript and sent it as "A House in the World" to Knopf, where it was quickly rejected. An editor, Bernard Smith, was "rather disappointed." The nonracial lyrics were "neither distinguished nor important"; all the best poems were in the section devoted to racial protest, but these were not enough. After the high quality of *Not Without Laughter,* the book would be "an act of retrogression."

The solitary bright note of the winter was the news that he had indeed been given the award for which Locke had nominated him—a prize of $400 and a gold medal for distinguished achievement among blacks in literature by the Harmon Foundation, established in 1922 to promote black participation in the fine arts; Cullen had won it in 1927, McKay in 1929. The largest single sum of money Hughes had ever possessed outright, the prize smelled now of freedom. Also encouraging, if also retrospective, was a request from Effie Lee Power, a white Cleveland librarian and a national adviser on children's books, for a selection of his poems suitable for younger people, who almost always responded keenly to Hughes's work. He sent her a selection that would be published by Knopf the following year (with an introduction by Power) as *The Dream Keeper*.

With "A House in the World" refused, Hughes needed the Gilpin production of "Mule Bone" to boost his morale. Thus he was glad to know that Zora Hurston planned to arrive in Cleveland on Feburary 1, his twenty-ninth birthday. By this time, however, unknown to Hughes, she had a powerful ally against him—Alain Locke, who had just quietly and maliciously killed Louise Thompson's chances for a job with the economist Dr. Abram Harris at Howard University by describing her as indigent (as he let Godmother know). Assuring Mrs. Mason that he would back Hurston to the hilt against Hughes, Locke sadly reported the news of the Harmon award—"the tragedy is the credit will go to swell the false egotism that at present denies its own best insight." This "false egotism" was Hughes's attempts to alter his relationship with Godmother, which Locke saw as endangering all her godchildren, especially himself.

Late on February 1, Hurston and Hughes met and apparently resolved their differences. Hurston, however, had nonchalantly missed a meeting she had promised to attend earlier that day with the Gilpin Players; fed up with the controversy, the group voted to cancel the play. Hurston was deeply offended. Russell and Rowena Jelliffe then decided to make a last attempt to salvage the production. Because Hughes was bedridden with influenza and tonsilitis, a meeting with Hurston was set for the afternoon of February 3 at his mother's home. Discovering, however, that Louise Thompson had been in Cleveland, Zora arrived at Hughes's home in an evil mood. "She made such a scene as you can not possibly imagine," he wrote Van Vechten the next day; "she pushed her hat back, bucked her eyes, ground her teeth, and shook manuscripts in my face particularly the third act which she claims she wrote alone by herself while Miss Thompson and I were off doing Spanish together. (And the way she said *Spanish* meant something else)." When Mrs. Jelliffe intervened, Hurston denounced her as a trickster, then ridiculed her sixteen years of community work in Cleveland. Throughout this tirade, the stricken Hughes did little more than gasp weakly; but when Hurston made to storm from the house, Carrie Clark suddenly leaped into the fray with some blistering words of her own. "Carrie was absolutely magnificent," Mrs. Jelliffe would recall; "for once she stood up for Langston and helped him. She let Zora have it in no uncertain terms. We were all very proud of her." But Hurston had achieved her goal. At 6:30 that evening she sent a triumphant telegram to Park Avenue: "DARLING GODMOTHER ARRIVED SAFELY HAVE PUT THE PERSON ON THE RUN PLAY STOPPED LOUISE THOMPSON HAD BEEN SENT FOR TO BOLSTER CASE I SMASHED THEM ALL BE HOME BY WEEKEND ALL MY LOVE ZORA."

"Do you think she is crazy, Carl?" Hughes inquired the next day of Van Vechten.

Knowing now what he was up against, Hughes made a deathbed appeal to Godmother. His appeal was denied. "What a sorrowful misguided way to have come!" Mrs. Mason mourned on February 12. "Has not the year spent tarnishing your wings proved that 'keeping accounts' had nothing to do with your failure to do creative writing? . . . Child, why build a labyrinth about yourself that causes you to wander in a miasma of untruth?"

Knocked into a relapse by her accusation of dishonesty, Hughes felt his whole body poisoned. He blamed his tonsils, but an astute doctor recommended the sun. After a week or so, Langston wrote a very brief note to "Dear Godmother": "When the sun shines again I will write you a long letter. Now, my thanks, and my love to you, as always."

Although he would never write this "long letter," Hughes could not say goodbye. When the Harmon medals came (there were actually two) he sent them like a child to her for her admiration. But she was like stone. "Langston," Mrs. Mason admonished him, alluding to the old woman of his novel, "do hurry to recognize the travail of Aunt Hagar to bring you face to face with yourself."

At 399 Park Avenue, Locke and Hurston hammered his coffin shut. When Hughes wired his favorite Harlem photographer, James L. Allen, for a set of portraits, Locke reported to Mrs. Mason that he evidently was preparing "a signed author's portrait racket." Hughes wrote a friend in Florida about a possible visit; Locke saw "a mad careening before a big fall." As for Langston's hopes to go to the Caribbean, "isn't that shameful—or rather shameless," Locke asked. "Well—he'll have a long rope—but eventually it will pull taut." To the Labarees at Lincoln, who heard of Hughes's illness, Locke intimated that he was having a nervous breakdown. To someone Langston challenged about the unauthorized use of his work, Locke apologized for Hughes's "real mean-ness." Against Van Vechten, who still befriended both Hurston and Hughes, Locke cried out to Mrs. Mason: "Why can't he die! Nothing seems to kill him." Not to be outdone, Hurston elbowed Locke out of the way to kneel behind Godmother. "Dear Godmother, the guard-mother who sits in the Twelfth Heaven and shapes the destiny of the primitives," one March letter began. And later: "Knowing you is like Sir Percival's glimpse of the holy grail. Next to Mahatma Gandhi, you are the most spiritual person on earth." Hurston stuck by her charges of Langston's corruption. "I can't conceive of such lying and falsehood," she sighed; "but then there are many things in earth and sky that I don't know about." A year later, more amusingly: "Honest Godmother it requires all my self-restraint to keep from tearing the gin-hound to pieces. If I followed my emotions I'd take a weapon and go around the ham-bone looking for meat." And to Arthur Spingarn, Langston's lawyer, Zora suggested that he should write a play for his client, if Hughes wanted one so badly. "Or perhaps a nice box of apples and a well chosen corner. But never no play of mine."

By this time, Hurston perhaps had internalized the conviction that Hughes had had no part in the play; how she was able to do so is a mystery, perhaps a part of her imaginative genius itself. Whether she ever regretted her deception is unclear. Eight years later, Arna Bontemps would report to Hughes that she was "a changed woman, still her old humorous self, but more level and poised. She told me that the cross of her life is the fact that there has been a gulf between you and her. She said she wakes up at night crying about it even yet." Certainly Hurston had matured as a writer; in the intervening years she would publish two novels, including her masterpiece, *Their Eyes Were Watching God,* about a black woman's experience first of a near-perfect love, then of the more profound satisfaction of a noble, feminist self-reliance. Hurston's desire for a reconciliation with Hughes was probably wrecked by his droll, teasing treatment in *The Big Sea* of her quixotic behavior during the controversy; two years later, when Hurston published her own autobiography, *Dust Tracks on a Road,* she left out Hughes altogether. After "Mule Bone," the two writers apparently met only once. But, incredibly, when Hurston needed help during the most humiliating episode of her life, in which she faced prosecution on a sordid (but unfounded) morals charge, she would turn for a testament to her good character to—Langston Hughes.

Her self-deception, in any case, was in its way not much more bizarre than Hughes's pathetic enslavement by Mrs. Mason, which Louise Thompson, for one, could never understand. "He got sick," she would remember. "When I got over it, I got mad!" Mrs. Mason was "an old witch" whose behavior made Thompson "hate the power of money," the idea "that someone because they have money can do to you as they wish and talk to you as they want to. How *dare* they!" The reason for the difference between Langston's reaction to Mrs. Mason and her own, she guessed, was her mother's unceasing love of her and Langston's mother's neglect of him as a child. "Now I am ready to die," "Mother" Thompson had said when Louise graduated from college. "I can't imagine Carrie saying anything like that about Langston," Thompson judged flatly. The episode was as decisive as any other in urging her into an association for life with communism, just as it would drive Hughes almost as far to the left for many years, and in other ways affect the entire course of the rest of his life.

Nor was Hurston's self-deception less excusable than Mrs. Mason's abuse of her wealth, race, age, and intelligence in so torturing the lives of her blacks. In her case, the mitigating factors were that she exhibited a mental power so volatile, and cherished notions about Africa so novel for her class and race, that her inability to control herself was perhaps finally excusable. Only Locke's behavior was almost entirely reprehensible. For all his great learning he was a slippery character, too fond of intrigue and of the pleasures that Mrs. Mason's money assured. The tawdry "Mule Bone" affair rang down the curtain on one long era of Hughes's life. And in pitting black brother against sister, artist against artist, it marked just as conveniently the end of a cultural age. The highly sensitive Locke unconsciously caught a sense of this death of golden innocence in one of his reports to Godmother. "I hear almost no news now from New York," he confessed to her near the end of March; "a younger crowd of 'Newer Negroes' are dancing in the candle flame. The older ones are nursing their singed wings."

For Langston, the time had come for true flight. Winding up his affairs in Cleveland, he completed a revison of his play "Cross," renaming it "Mulatto." "Well," he wrote Van Vechten, "I guess my play is done. (*No woman* worked with me on this one.)" Seduced and abandoned by white philanthropy, he would return now to blackness and the sun. Hughes decided to visit the Caribbean—Cuba, Haiti, Puerto Rico, Guadeloupe, Martinique, Barbados, Trinidad. His name was honored among young black intellectuals in the English-speaking West Indies. Gordon O. Bell of Barbados, who would soon publish an article on race in the British writer Nancy Cunard's landmark anthology, *Negro,* had written to Hughes that his visit would be a sensation for the local intelligentsia awakening to nationalism and racial pride after centuries of British colonialism; from Trinidad, Olga Comma (also a *Negro* essayist) sent a similar welcome. Through the Jelliffes, Hughes found a travelling companion, Zell Ingram, a big, handsome, young black man, about twenty-one years old, who lived with his mother over a popular Cleveland hot-dog shop. A graduate of Central High, where he knew many of Hughes's old teachers and had played football, Ingram was impressive. "Zell was built like a stevedore, very muscular, very commanding," a friend, Blyden Jackson, would remember; "but he was also gentle and likable and very intelligent. I would call him one of the most intelligent people I've ever known." Ingram was also a promising artist who worked in prints and woodblocks and on theater scenery. When his mother agreed to let him have her Ford and some money if he wanted to travel, Ingram and Hughes decided to drive to Florida, then take a boat into the blue Caribbean.

At the last minute, mainly to be sure that he would not need the operation while in the Caribbean, Hughes underwent a tonsillectomy. Then, near the end of March, he and Ingram drove from Cleveland to 514 Downer Street in Westfield, where Langston made arrangements with Mr. and Mrs. Peeples to store his effects until his return. He was not sure when he would be back. On April 1, 1931, at 7:15 in the evening, Ingram fired the engine of his mother's Ford and drove away from Downer Street. A chill spring rain was falling over the fields of eastern New Jersey as the Ford rolled in the darkness away from Westfield and its painful associations for Langston.

Six days later, he and Ingram were riding a train out of Miami through groves of oranges and grapefruit to Key West. The "Nazimova" (they had christened Zell's mother's car after the Russian-born actress Alla Nazimova) had brought them safely south. Ingram proved to be a likable traveling companion, with just the right sense of history to suit Langston in his current mood; when they crossed the Mason-Dixon line at 11:20 on the morning of April 2, Zell pulled "Nazimova" over and poured the "appropriate libation" on Dixie soil. They slept in private homes and in colleges, and at least once by the sea in "Nazimova" herself, on a sandy rise along Ocean Boulevard in Jacksonville, Florida, snuggling under blankets as the moon rode through banks of clouds and the surf rustled on the beach.

At Key West they bought second-class tickets on a boat to Havana. The only other passengers in their section were a Cuban hurrying home to see his ailing mother and a fat sweating American clutching a movie camera and panting for the sex-shows in Havana. But both Hughes and Ingram were excited as they awaited, in probably Ingram's words, "Sunrise in a new land—a day that will be full of brownskin surprises, strange dark beauties, and hitherto unknown contacts in a world of color."

An immigration officer at Havana personally escorted Hughes and Ingram ashore, where at least half a dozen persons, including Nicolás Guillén, Regino Pedroso, Gustavo Urrutia, an emissary from José Antonio Fernández de Castro (who was unable to come himself), and a news photographer, welcomed Hughes on his third visit to Cuba and his second as a celebrity. From the docks the party proceeded to Fernández de Castro's office at *Orbe,* Cuba's weekly picture magazine, where he was now working. Then they went on to the same hotel,

Las Villas, where Langston had stayed the previous year; for one dollar a day, he and Ingram rented a double room.

The next morning, April 8, they found their picture on the front page of *El Diario de la Marina*. Inside was a substantial story on their arrival, "El poeta Langston Hughes nos visita de nuevo" ("Langston Hughes the poet visits us again"), by Santos Alberto, stressing the crucial difference between Hughes's racial verse and the emerging culture of the black Caribbean: "Su arte propio, nuevo, Negro requiere un mundo nuevo y propio. El negro antilliano es cera virgin que van ellos a modelar. [His own new black art requires a special new world. The black man of the Antilles is virgin wax which he will shape]." The next number of *Orbe* carried a handsome picture of the two visitors, smartly dressed and admiring a copy of the magazine. By this time, the name Langston Hughes was not unknown in Havana. Immediately following his departure the previous year, the *Revista de la Habana* had carried Fernández de Castro's "Presentación de Langston Hughes," a biographical sketch that noted his warm reception by young Cuban intellectuals in general but by black Cubans in particular. His portrait and several of his poems, translated, had appeared in "Ideales de una Raza," the Negro page (since discontinued) of the *Diario*. *Revista de la Habana* published four translations of Hughes poems by Fernández de Castro, and months later his translation of an excerpt from *Not Without Laughter*. Langston's attempts to publicize Cuban poetry and sculpture in North America had been keenly appreciated by Cubans tired of their culture being represented in the United States mainly by rum and the rumba. And Hughes's name was also linked to Guillén in the continuing controversy over *Motivos de Son,* which Guillén would fuel later in the year with his *Sóngoro Cosongo: Poemas Mulatos*. In a letter to Guillén, Hughes had praised *Motivos de Son:* "¡Hombre! ¡Que formidable tu *Motivos de Son!* Son poemas muy cubanos y muy buenos. Me alegro que tu los has escrito y que han tenido tanto exito. [Man! Your *Motivos de Son* is really great! The poems are very Cuban and very good. I'm glad you wrote them and that they have been so successful]."

But now Hughes was a reluctant lion. A year before he had disembarked from the *Caronia* with the aplomb of an ambassador, preaching the gospel of racial pride, only dimly aware of the peril in his anomalous position as the protege, or the lapdog, of Mrs. R. Osgood Mason. Now he had limped back to Cuba; bravely holding together the various parts of his public personality, Hughes was nevertheless on the verge of a collapse. His great need now was for anonymity, a quiet, sunny place where he could rest and examine his life. What he did not know, and needed to discover, was what form the future would take; to heal himself, Hughes needed nothing less than a temporary erasure of identity. His only certainty was that he would survive; his main hope, that he would prevail. Prevail not against Godmother, Hurston, or Locke, who were mere personalities, but against what he undoubtedly saw as his betrayal of the greatest truth of his life—that, in his chronic loneliness, true satisfaction came only from the love and regard of the black race, which he earned by placing his finest gift, his skill with language, in its service.

In the end, Hughes stayed about two weeks in Cuba. Sleeping late as often as he could, or dozing in the sunshine while Ingram sketched Cuban scenes and types, he began the process of self-healing. But he was too popular to be left alone. Ramos Blanco, the subject of Langston's little essay in *Opportunity* the previous November, invited them to his studio in the suburbs. Returning to the impressive black Club Atenas, he also visited various other studios and ateliers, newspapers and magazines. Once, under the auspices of a cultural group, the Cercle Français, Hughes gave a formal reading of his verse. With Guillén and Urrutia, he worked on his translations of Cuban poets, still determined to secure an audience for them in the United States. That spring, *Poetry Quarterly* of New York published together three poems by Hughes, one by Guillén (''Wash Woman''), and another by Regino Pedroso (''Alarm Clock''). Poems by Guillén (''Madrigal'') and Urrutia (''Students of Yesterday'') also appeared in the March *Opportunity* and the April *Crisis*.

Twice he had unpleasant brushes with racism. At a white and light-mulatto dance hall, he and Ingram were refused admission; their Cuban driver was furious and the high-spirited Zell willing to fight, but Langston turned away in quiet disgust. Worse was an incident at a beach leased to Americans, with pavilions and bathing houses from which colored Cubans were routinely excluded. At the gate, the cashier demanded an entrance fee of $10 each from Hughes and Ingram, although the posted rate was much lower. When a white American bouncer and the white American manager menaced them, they refused to budge. Soon policemen arrived ''with drawn sabres . . . as if they intended to slice us into mince meat,'' Hughes would recall. More police came to cart Hughes and Ingram off to a police station where the officer in charge, ''an enormous, very dark colored man,'' sympathetically refused to jail them: ''This is outrageous, but it is what happens to colored people in Cuba where white Americans are in control!'' But they were charged with trespassing and disturbing the peace and forced to post bond of $50. At their trial the next day, witnesses for the Americans lied that Hughes and Ingram had entered the beach café behaving badly and using profane language—although neither had been allowed past the gates. The judge, an old mulatto, listened carefully to the case before dismissing the charges and rebuking the plaintiffs.

Although Hughes might have preferred to drift and doze in the Cuban sun, he could not ignore either his recent troubles with Mrs. Mason or events in Cuba. The nation was lurching towards an attempted revolution that year against the dictator Machado, who would fall two years later. Hughes penned an attack on American economic imperialism so blunt that it startled his friends, who warned him that such words were dynamite in Cuba. The poem was in tribute to the little seaside fort of San Lázaro in Havana that had helped Cubans withstand pirates of earlier centuries—

> But now,
> Against a pirate called
> THE NATIONAL CITY BANK
> What can you do alone?
> Would it not be

Just as well you tumbled down,
Stone by helpless stone?

During this visit, Hughes was certain that he was followed by a government spy, perhaps after visiting Ramos Blanco's studio, a place notorious with the police as a gathering place of radicals. Later, Hughes turned the surveillance into a deliberately Hemingwayesque tale ("my most verbatim story") called "Little Old Spy," published in 1934 in *Esquire* magazine. The piece touched on Cuban censorship, conspiracy, and the will to revolt—"But the [American] tourists seemed blissfully unaware of all this. They still flocked to the casino, wore their fine clothes through quaint streets of misery." And yet Hughes in April, 1931, was in no mood for open opposition to imperialism. If one poem accurately caught his current demoralized state, his despair at the impotence of the imagination, it was most likely "Havana Dreams":

The dream is a cocktail at Sloppy Joe's—
(Maybe—nobody knows.)

The dream is the road to Batabano.
(But nobody knows if that is so.)

Perhaps the dream is only her face—
Perhaps it's a fan of silver lace—
Or maybe the dream's a Vedado rose—
(*Quien sabe?* Who really knows?)

After a ride across the island in an ancient bus loaded with peasants and squawking chickens, Hughes and Ingram boarded a boat at Santiago bound for Port-au-Prince, Haiti. The accommodations were crude, but after the years of comfort at Lincoln and luxury with Godmother, Hughes needed to rip the starched collar from his neck and descend once again to the level of "my people," just as he had done after quitting Columbia. For three days they rode on the open deck with black Haitian cane workers going home from the fields of Cuba. When a storm came up on the last night, the two young Americans were admitted to the dank hold, where they slept among the black crew.

At Port-au-Prince, before he could land, a white U.S. marine thumbed his passport on board the ship; this was his introduction to the most historic black culture in the New World, the republic of Toussaint L'Ouverture, Jean-Jacques Dessalines, and Henri Christophe, leaders who had defied Europe to establish the only independent black republic to emerge from the centuries of slavery in the Americas. The fact that his mother's uncle John Mercer Langston had twice represented the United States there had made Haiti an enchanted land for Hughes in his childhood. But Port-au-Prince was a disappointment. Instead of a vital metropolis, he found a squalid collection of tin-roofed wooden huts. The Presidential Palace gleamed white under the sun but only heightened the drabness of Haitian poverty. Begging children, many apparently diseased, dogged the tourists on the streets; the shops were pitifully stocked and very expensive. Haiti seemed out of step with history—in the Champs de Mars "the palace band plays immortally outworn music while genteel people stroll round and round the brilliance of the lighted bandstand.'' But the vast majority of people lived desperately.

Especially distasteful were the ''groups of marines in the little cafes, talking in 'Cracker' accents, and drinking in the usual boisterous American manner,'' or swaggering at night through the badly lit streets, enforcers of the U.S. occupation now almost a generation old but under a treaty never ratified by the U.S. Senate. In that time, ''about all for which one can give the Marines credit are a few decent hospitals and a rural health service.'' The roads were impassable, schooling was inadequate, and the need for economic reform urgent while foreigners plundered the economy and ''the officials of the National City Bank, New York, keep their heavy-jawed portraits in the offices of the Banque d'Haiti.'' Black Haiti ''has its hair caught in the white fingers of unsympathetic foreigners, and the Haitian people live today under a sort of military dictatorship backed by American guns. They are not free.'' Replacing the heroic armies of L'Ouverture, Dessalines, and Christophe were the hordes of black people without shoes. The average wage in Haiti was thirty cents a day, but ''all the work that keeps Haiti alive, pays for the American Occupation, and enriches foreign traders—that vast and basic work—is done there by Negroes without shoes.'' Scorning the colored Haitian ''leadership,'' Langston described it as surviving mainly on sinecures while its members composed silly poetry in the French academic manner, whined about the American presence, and wasted their energy on petty party strife and intrigue. ''The result,'' he wrote the *New Masses,* is ''a country poor, ignorant, and hungry at the bottom, corrupt and greedy at the top—a wide open way for the equally greedy Yankees of the North to step in, with a corruption more powerful than Paris-educated mulattoes had the gall to muster.''

> The people without shoes cannot read or write. Most of them have never seen a movie, have never seen a train. They live in thatched huts or rickety houses; rise with the sun; sleep with the dark. They wash their clothes in running streams with lathery weeds—too poor to buy soap. They move slowly, appear lazy because of generations of undernourishment and constant lack of incentive to ambition. On Saturdays they dance to the Congo drums; and on Sundays go to mass,—for they believe in the Saints and the Voodoo gods, all mixed. They grow old and die, and are buried the following day after an all-night wake where their friends drink, sing, and play games, like a party. The rulers of the land never miss them. More black infants are born to grow up and work. Foreign ships continue to come into Haitian harbors, dump goods, and sail away with the products of black labor. . . . The mulatto upper classes continue to send their children to Europe for an education. The American Occupation lives in the best houses. . . . And because black hands have touched the earth, gathered in the fruits, and loaded ships, somebody— across the class and color lines—many somebodies become richer and wiser, educate their children to read and write, to travel, to be ambitious, to be superior, to create armies, and to build banks. Somebody wears coats and shoes.

Although he had several letters of introduction, Hughes refused to use them. After a few days, he and Ingram boarded a bus festooned with chicken coops and packing cases and set out for Cap-Haïtien on the north coast. His real destination was the great Citadel La Ferrière on a 3,000-foot peak near the Cap, an authentic wonder of the New World begun by Dessalines and completed by Henri Christophe. While the trip to the Cap normally took twelve hours, three weeks passed before Hughes and Ingram saw the Citadel. At the town of St. Marc, halfway up the coast, torrential rain had washed away the only road. Determined to see the Citadel even if it meant the end of his Caribbean voyaging, Hughes rested at St. Marc for three weeks while the road was repaired. When the road finally opened, the first bus to Cap-Haïtien ran out of gas and left Hughes, Ingram, and the other passengers to spend a night exposed to the cold mountain air. A torrential downpour soaked them to the skin, since the vehicle was open at the sides. At last, descending from the hills overlooking the sea, they reached the Cap. The town was not much of an improvement over Port-au-Prince, but it was smaller and less crowded. At $25 a month for room and board, they lived at the International Hotel, a humble establishment catering mainly to visiting seamen and fugitives from neighboring Santo Domingo. Their room looked out on the sea.

As soon as he could, Hughes made the long, steep ascent to the ruins of the Citadel. Designed to hold 10,000 soldiers, the fortress boasted walls 140 feet high and 16 feet thick at their base, defended by two-ton cannon hauled by hand up a steep trail still in use. Of the 200,000 blacks who had worked on its construction, more than 20,000 had died; many had been slaughtered as an example to other workers. Near the base of the mountain were the ruins of Sans Souci, the magnificent palace built by Henri Christophe, with mountain streams diverted under marble floors to cool the rooms on summer days and nights. Here, in 1820, facing defeat in a civil war, Christophe had shot himself with a silver bullet. Borne to the Citadel, his body was buried in a vat of quicklime. His epitaph was perhaps an augury for Hughes and the black race: "Here lies Henri Christophe, King of Haiti. I am reborn from my ashes." Even in ruin the edifices were beautiful, graceful of line and in surviving details among windy battlements, towers, and terraces, broken staircases, chambers, and cellars. Hughes was so moved by this monument to lost black grandeur that he decided to go no further in the Caribbean; he would rest in the shadow of the Citadel to contemplate its meaning.

"Stronger, vaster, more beautiful than you could imagine," he wrote in *New Masses* of the Citadel, "it stands in futile ruin now, the iron cannon rusting, the bronze one turning green, the great passages and deep stairways alive with bats, while the planes of the United States Marines hum daily overhead." Three times, aided by donkeys in the arduous climb, Hughes made a pilgrimage to the "splendid, lovely monument to the genius of a black king." Although he snapped photographs like any other tourist (with the Kodak given to him as a graduation present by Amy Spingarn), these visits were not for ordinary sightseeing. Hughes had fled unerringly from the greatest hurt of his adult life, the collapse of his relationship with Mrs. Mason, towards the mightiest symbol of black power in the Americas, the Citadel. Clearly he saw himself as a prodigal son who, having wasted his birthright in a foreign clime, had now limped home for forgiveness. Clearly, too, he thought of himself as having just betrayed some basic allegiance, of having been in some way deeply dishonorable—as when he had lied to the black congregation and to Auntie Reed as a boy in Kansas. Now Godmother had banished him, just as his father and mother had done in inflicting the great, unhealable wound of his life. Only his race, black and itself abused, never banished or abandoned.

While his dead skin peeled away, the days at Cap-Haïtien passed sleepily. In the bar of the International Hotel the few patrons sipped iced drinks, quietly played cards and dominoes, and yawned away the hours. For a while the only other guest was a very fat "revolutionary" from Santo Domingo who translated Langston's Spanish into the local Creole. Ingram sketched and painted, and sculpted heads in Haitian wood. Two brown bar butterflies alighted on the young Americans. Coloma, the more beautiful, chose Ingram; her little friend Anna took up with Hughes. The two American visitors, friendly with the fishermen and market women of the town, scandalized the mulattoes by sitting on the ground to talk with vendors, teaching games to the naked little black children, and playing dominoes with barefoot dockworkers on the sea wall. Normally friendly, the manager of the hotel objected "violently to our carrying groceries *in our own hands* down the street." They sailed out in the fishing boats with the black matelots, who took them on Saturday nights to the "Congo" dance ("a sex dance undisguised," Langston saw, although the partners never touch) held outdoors under a canopy of banana fronds, with the only instruments the cowskin drums of various sizes. At the first dance an intoxicated Zell Ingram entered the lists and outdanced the locals—or so Langston thought. Hughes bought a Mama drum and a Baby drum to take back to the United States; the Papa drum was hard to come by, since both the priests and the Marines discouraged the music and disapproved of the "Congo" dances. "But the black Haitians of the soil seem to remember Africa in their souls and far-off ancestral tribes where each man and each woman danced alone."

When weeks had dragged by, Ingram became restless but Hughes made no move to leave. The hotel meals were monotonous—rice, beans, and soup served sometimes three times a day; they bought canned goods and prepared their own food. Every bar in town became familiar, including one frequented by U.S. Marines, where no colored person was served until after the military curfew each day. Hughes developed a special loathing for one of the soldiers; they glared at each other when they met in the streets. Langston's most abandoned moment came one night when he drank too much absinthe, made teetering passes at several women, stumbled toward the exit but sat down heavily on the floor, then tried to climb into his hotel room like Douglas Fairbanks in the movies, mounting from outside. Insisting on a nighttime swim in the ocean, he ran giggling in the dark to the sea over a narrow, treacherous path difficult even in broad daylight. Somehow he made it safely to the beach, where he passed out. In his current mood, Hughes cared nothing about respectability. Haiti, poor and black, was quietly recharging his political and racial zeal. "It was in Haiti that I first realized how class lines may cut across color lines within a race, and how dark people of the same nationality may scorn those below them. Certainly the upper-class Haitians I observed at a distance seemed a delightful and cultured group. No doubt, many of the French slave owners were delighted and cultured, too—but the slaves could not enjoy their culture.''

Ingram's sketching and sculpting dwindled to fitful efforts. He was also bickering with Coloma, who had at least one other lover. Hughes worked not at all, and wrote virtually no letters. ''I haven't done any work,'' he finally admitted to Van Vechten. ''Been trying to wear my troubles off my mind.''

He answered one letter, of almost diabolical timeliness, that he had carried around since the middle of January, when his father had written for the first time to Langston since the spring of 1922. James N. Hughes had been moved to write his son after reading *Not Without Laughter,* which he found ''very amusing''—although Langston should avoid using dialect in the future. His father had not mellowed: Langston should marry soon—but not allow ''some nigger wench'' to squander his money; instead he should find someone ''brought up in the washtub.'' Although he still read the *Crisis,* he did so only to ''keep up with the doings of you African monkeys.'' And he wanted to know if Langston actually made a living as a black writer.

How Langston had fared as a writer was indeed the question of the hour. Replying, he detailed some of his publishing successes to his father, then ended with a sentence that at best was half-true: ''In one way or another, directly or indirectly, enough money has come to me from my books to live.''

Near the end of June, his Harmon prize money began to run out. Giving away many of his nicer items of clothing to poor Haitians, Hughes packed up his drums and other souvenirs and personal effects. But the first ship to arrive was full. They shipped some boxes to New York but could not travel themselves. If Hughes was indifferent to the news, Ingram was not. ''This is a hell of a country,'' he scribbled. ''I wonder why we ever came here and when we are going to get out.'' On July 9 they finally left on a freighter, the *Amassia,* with Coloma and Anna weeping on the shore. ''Farewell to Haiti,'' Ingram jeered, ''and all its children of eternal blackness.''

At Port-au-Prince, Hughes finally used one of his letters of introduction to call on Haiti's best-known poet, Jacques Roumain, who hurried home from his executive office at the Department of the Interior after being summoned by his wife, Nicole, to meet Langston Hughes. Tall, copper-brown like Hughes, and strikingly good-looking, Roumain had, according to Langston, ''the deep fiery eyes of a picture-book poet.'' Five years younger than Hughes, the son of a wealthy landowner and grandson of a former president of the republic, Roumain had been educated in Switzerland and Spain before he was twenty, and was fluent in at least three languages. As with Guillén in Cuba the previous year, Hughes was meeting a future national writer at a crucial moment of transition; a polished versifier, Roumain had just begun to write poetry about the black masses of Haiti, inspired in part by the writings of Hughes and other black Americans (although he would never experiment in native Haitian musical forms, as Hughes and Guillén had done). His opposition to the American occupation would speed his entry into the Communist Party and result in a life, ending prematurely, of imprisonment and exile. Anxious to devote himself to writing but feeling himself fatally hindered by his class and education, Roumain saw Hughes as the personification of the freedom he craved. In turn, Langston was very impressed with Roumain after their conversation in rusty French and halting English; from Roumain's home, "I descended the hill thinking that if the delightful Roumains were typical representatives of the Haitian elite, then I regretted not having met more of them."

Back later on his ship, the *Amassia,* preparing to leave as a steerage passenger, Hughes had stripped himself of his shirt and undershirt and put on his shabbiest slacks when "I saw approaching a long line of elegantly dressed gentlemen, some in tail coats and gloves, followed by a number of dark porters and barefoot boys bearing parcels and baskets." To Hughes's profound embarrassment, Roumain had brought a company of important citizens to say goodbye: "I was caught greasy-handed, half-naked—and soxless," Langston would recall, "by an official delegation of leading Haitians." But Roumain made a gallant little speech saluting "the greatest Negro poet who had ever come to honor Haitian soil." The gentlemen each bowed solemnly and presented Hughes and Ingram with gifts of books, fruit, handicrafts, and bottles of Barbancourt rum. Zell Ingram, not impressed by much in Haiti, thought Roumain "one of the finest people" they had met on the trip. Soon Roumain would send his new friend "Langston Hughes," a poem "inspiré par votre existence magnifique et aventureuse (au sens le meilleur de ce mot) [inspired by your magnificent and reckless—in the best sense of this word—life]."

At Santiago, Cuba, after travelling on deck, they made the mistake of declaring themselves to be "artists"; the immigration officials promptly quarantined them at the Cayo Duan station. Ingram fell sick—not from some tropical disease, Hughes would claim impishly in his autobiography *I Wonder As I Wander,* but from excessive bouts of lovemaking at Cap-Haïtien with the demanding Coloma while they had waited and waited for the ship. A day later, Hughes and Ingram abandoned the vessel, the *Amassia,* and proceeded to Havana by train. By this time only ten cents of the Harmon prize remained. At Las Villas Hotel the manager lent Langston five dollars, and José Antonio Fernández de Castro arranged for a doctor to treat Ingram. As soon as he was able to travel, Zell and Langston left for the United States.

After almost four months away, Hughes consciously saw his return as marking a fresh start in his life. And, in an oddly prophetic way after their brief conversation, Jacques Roumain captured the spirit of Hughes's past and even foreshadowed his beginning future. Roumain did so in his poem "Langston Hughes":

> At Lagos you knew sad faced girls.
> Silver circled their ankles.
> They offered themselves to you naked as the night
> Gold-circled by the moon.
>
> You saw France without uttering a worn, shop-made phrase;
> *Here we are, Lafayette!*
> The Seine seemed less lovely than the Congo.

Venice. You sought the shade of Desdemona.
Her name was Paola.
You said: *Sweet, sweet Love!*

And sometimes
Babe! Baby!
Then she wept and asked for twenty *lire*.

Like a Baedeker your nomad heart wandered
From Harlem to Dakar.
The Sea sounded on in your songs—sweet, rhythmic, wild . . . .
And its bitter tears
Of white foam blossom-born.

Now here in this cabaret as the dawn draws near you murmur . . . .
The blues again play for me!
O! for me again play the blues!

Are you dreaming tonight, perhaps, of the palm trees, of Black Men down there, who paddled you down the dusks?

9

STARTING OVER
1931 to 1932

Listen!
All you beauty-makers,
Give up beauty for a moment.
Look at harshness, look at pain,
Look at life again . . .
"Call to Creation," 1931

On July 22, tired and broke, Hughes and Zell Ingram reached Miami to find their motor car, "Nazimova," with all its tires flat and one ruined. A telegram to Ingram's mother in Cleveland brought only an irate order to her son to *"get in my car and drive on back home."* Chastised, they pawned everything of value they had, except for the two drums bought at Cap-Haïtien in Haiti, which Langston flatly refused to yield. For his Kodak camera, he accepted three dollars. Dogged along the way by tire punctures, the travellers reached the city of Daytona Beach up the Atlantic coast. There, at the college she had founded in 1904 as the Daytona Normal and Industrial School for Negro Girls (later Bethune-Cookman College), "America's leading Negro woman," as Hughes would call Mary McLeod Bethune, "received us cordially, sat us down to dinner, and cashed a thirty-dollar check without question." Soon the fifty-six-year-old Mrs. Bethune made a startling suggestion to Hughes and Ingram: "I was intending to go North myself in a few days by train, so I might as well ride with you and save that fare."

"What luck for us!" Hughes would recall of an encounter so timely as to be almost providential. "We shared Mrs. Bethune's wit and wisdom, too, the wisdom of a jet-black woman who had risen from a barefooted field hand in a cotton patch to be head of one of the leading junior colleges in America, and a leader of her people." At the crucial moment of his return to the United States, he had found the perfect figure to counteract in his mind the still unvanquished image of Godmother. Big-boned, black-skinned, and dynamic, at once both commanding and maternal, Mrs. Bethune seemed to personify for Langston, in this pivotal moment of his life, what their race might be—and what he himself might be in his relationship to it. Born in South Carolina as the fifteenth child of former slaves, Mrs. Bethune had gone far in the world against the odds. At twenty-nine, with one dollar and fifty cents in declared capital, she had founded her school for black women; soon the institution had over 250 students, including men, and had become the focus of community life for blacks in Daytona Beach. A brilliant organizer, Mrs. Bethune quickly gained recognition from prominent black leaders such as Booker T. Washington and Mary Church Terrell of the National Association of Colored Women, which Bethune twice headed with great effectiveness in the nineteen twenties. Appointed first by Calvin Coolidge, then by Herbert Hoover, to the National Child Welfare Commission, she gave advice to both presidents on race and education; eventually she would serve five presidents in this way.

If Mrs. Bethune's motto (as Hughes often repeated it) was that the reward of service is more service, she was not servile; on the Daytona Beach campus, small but picturesque with its palms, poinsettias, and hanging moss, she tolerated no segregation even as she solicited support from local and visiting whites. At least once, she organized black voters in defiance of the Ku Klux Klan. Nor was she pompous, as some race leaders were. "She was a wonderful sport," Hughes would recall, "riding all day without complaint in our cramped, hot little car, jolly and talkative, never grumbling." With friends and admirers all along the route, she ensured that Langston and Ingram were housed and fed well. Chickens fled at Mrs. Bethune's coming, people said, because "they knew some necks would surely be wrung in her honor to make a heaping platter of Southern fried chicken."

As he headed north, Hughes was fairly certain that the next step in his career would be to write a sequel to his novel, *Not Without Laughter*. Before leaving for Haiti he had raised the subject of a fellowship with Walter White of the NAACP and Charles S. Johnson, now a professor at Fisk. Both had sent encouraging news. Nathan W. Levin of the Julius Rosenwald Fund (founded in 1913 by the president of Sears Roebuck, and a major source of private support for black education) was "almost certain" that Langston could get a Rosenwald fellowship; in mid-May, Henry Allen Moe of the Guggenheim Foundation had "practically assured" Johnson that Hughes would win a Guggenheim if he applied for one. And Amy Spingarn had let Langston know that "yes, I'll be very glad to help you next year."

Mrs. Bethune led him to consider a very different plan. At a reading in the summer school of Coulter Academy in Cheraw, South Carolina, arranged by her, she "made a warmhearted little talk, then introduced me as a poet whom she wanted the South to know better." His reading went so well that she pressed him further: "You must go all over the South with your poems. People need poetry." At first, Langston balked at the idea. His "discoverer" Vachel Lindsay had made a career of readings, and the fine black tenor Roland Hayes once had toured the country singing in black high schools and churches. But could Langston live by reading poetry—*his* poetry—to black audiences in the chronically poor South in the midst of the Depression? Mrs. Bethune was certain that he could, just as, ironically, another woman had urged Hughes once "to make a loafing tour of the South like the blind Homer, singing your songs. . . . You are the poet of the people and your subjects are crazy about you. Why not?'' This was Zora Neale Hurston, in their days of friendship.

Around August 1, ''Nazimova'' reached New York. When he checked into the YMCA at 181 West 135th Street, Hughes had come full circle. He was back where he had started on September 5, 1921, ten years before, when he had come from Mexico to attend Columbia University.

Also living on the fifth floor, where Hughes and Ingram shared a room for about three months, was young Blyden Jackson, just arrived from Louisville, Kentucky to begin graduate study in English at Columbia. Later an able scholar and teacher, Jackson remembered Ingram as ''very affable and extrovert in a good way. He had very strong, very positive feelings about the people.'' On the other hand, Hughes seemed reticent: ''Langston came as close as a very successful person can come to true modesty. He claimed nothing because of his fame. And you couldn't get him to say a bad word about anybody.'' To a New York *Amsterdam News* reporter, however, the fit and tanned Hughes was more assertive; he blamed United States imperialism for the abysmal conditions in Haiti (''there is not in the entire country a single bookstore'') and in Cuba, where the American sugar tariff had created a national crisis. About his own plans, Hughes was less forthcoming. Overshadowing his return was one very sad event, the death of the heiress A'Lelia Walker, felled by a stroke in the middle of her life. At her funeral service, Mrs. Bethune spoke in her deep voice, a nightclub quartet sang Noel Coward's ''I'll See You Again,'' and someone recited a poem by Hughes, ''To A'Lelia.'' Her death seemed to him to mark the passing of an age, ''the end of the gay times of the New Negro era in Harlem.''

Langston made no attempt to see or write Mrs. Mason, Zora Neale Hurston, or Alain Locke; that part of his life was over. Then Wallace Thurman announced the jolting news that he had been hired by an agency to revise the play ''Mule Bone,'' and that Hughes's name was nowhere on the script. Hustling down to the agency, Langston was rebuffed—or so Hurston reported. ''All he got for his trouble,'' she assured Mrs. Mason, ''was to be called a vicious liar to his face, a sneak and a weakling.'' Hughes countered with a long letter to the Dramatists Guild, and he permanently chilled his friendship with Thurman, who chose to give the impression that he didn't know exactly whom to believe in the ''Mule Bone'' dispute. Since Thurman still claimed to love his wife, Louise Thompson (from whom he was never divorced, despite living apart), perhaps his attitude was to be expected. On the other hand, the following year Hughes would praise Thurman's novel of Harlem life, *Infants of the Spring* (''YOU HAVE WRITTEN A SWELL BOOK PROVOKING BRAVE AND VERY TRUE YOUR POTENTIAL SOARS LIKE A KITE BREAKING PATTERNS FOR NEGRO WRITERS'') and laugh off its portrait of him, at a ludicrous Harlem literary salon, as an inscrutable little fellow named ''Tony Crews.'' ''Smiling and self-effacing, a mischievous boy,'' Crews had published two books of verse ''prematurely.'' He winked and smiled a great deal, but said absolutely nothing; he was either utterly shallow or "too deep for plumbing by ordinary mortals."

Hughes concentrated now on expressing his radical opinions about the economic crisis gripping the country. A recent Urban League study showed national black unemployment at disastrous levels. With a population only a quarter Negro, half the unemployed in Houston were black; in Little Rock, Arkansas, blacks formed twenty percent of the city but more than half of the unemployed; in Memphis, almost twice as many blacks as whites were without a job. The collapse of cotton prices had brought four million black farmers near bankruptcy. As for Harlem—in the decade since Hughes's arrival, the once beautiful Manhattan district had deteriorated sadly. Behind the elegant, Stanford White-designed townhouses on 139th Street, the alleys were "dilapidated and garbage-strewn," as Claude McKay, back from exile, would ruefully observe; on the so-called "Block Beautiful," at 130th Street between Lenox and Fifth Avenue, "the neat fences are broken, the gates unhinged and leaning awry, the sidewalks unkempt." The Depression had only accelerated the community's downward slide. The previous year, 1930, James Weldon Johnson wrote of a Harlem "still in the process of making." But Claude McKay would regret only that calling "all of Harlem a vast slum" would offend the "upper class" of blacks, "who must needs make their lives worth living there. It is their tragedy that externally their individuality is almost effaced in the rough scramble of the mass. Slum dwellers do not always see themselves as others see them. Little foxes leap and fleas jump. But both must live in holes and nests."

The national crisis only stiffened Hughes's resolve to try something dramatic: he decided to accept Mrs. Bethune's challenge. "I'd like to make a reading tour of the whole South," he wrote Walter White formally on August 5, "in order to build up a sustaining Negro audience for my work and that of other Negro writers and artists. I want to create an interest in racial expression through books, and do what I can to encourage young literary talent among our people." To free him from the railroad lines, Hughes would need "a small Ford." "I feel that a consecutive tour of the South would do a great deal to link the younger writers with the black public they must have, if their work is to be racially sound." His goals beyond the tour were to work on a second novel and on his plays. And Hughes did not repeat to Walter White the half-lie he had told his father. "So far," he admitted flatly, "I have never been able to make enough from my books to live."

That day he read to a convention of YMCA officials in New York—an important event, since the support of the national organization would make his venture much easier. Apparently his performance was flawless; charming and unthreatening, yet powerful, Hughes seemed to one white official "a rare personality. His sheer simplicity captivates one." Even white Southerners in the audience seemed to like him.

But most of his friends were appalled by his plan. "Lots of New Yorkers," he would soon recall, "tried to tell me that I would have no audiences in the South, or understanding." Borrowing a *Negro Year Book,* however, he drew up a list of school officials in the South and began writing to them. Then, on September 17, the president of the Rosenwald Fund, Edwin Embree, informed Hughes that he had been awarded $1,000 for the project. When a check for $600 came, Langston put down $225 on a new Model A Ford sedan, arranging for twelve monthly payments of $38 each to complete the purchase (including the cost of a learner's permit wasted on the mechanically inept Hughes, who would never learn to drive). Other money went for posters, circulars, portraits, typing, postage, and the purchase of books for a small traveling exhibition of black literature. Before the start of the tour he had spent $608.40, according to his careful accounting. Seeking a chauffeur (a "business manager"), who would receive forty percent of the profits, Hughes found Zell Ingram either uninterested or unable to take the job. He approached W. Radcliffe Lucas, a schoolmate at Lincoln, where he had played the tuba and occasionally published Hughesian verse ("I am Africa, / Black— / Jet Black"); he had also worked on Langston's senior survey. A part-time redcap at Pennsylvania Station, Lucas accepted Hughes's terms at once: "It beats being a Red Cap in these tipless times. Let's go."

In the three months between Langston's return to New York and his departure for the South he also began a dramatic move to the far left—so far, in fact, that for the rest of his life he would hear the accusation, which he would always deny, that he had joined the Communist Party of the United States. His main link to communists was his association with the John Reed Club of New York, which the party controlled, and with *New Masses* magazine, which operated in these months as virtually the club organ (the *New Masses* office was on the second floor at 63 West 15th Street, with the John Reed Club on the first and third floors). Formed in the fall of 1929 as a center for radical writers and artists, the club was further shaped by the six principles that emerged from the international conference of writers at Kharkov in the Ukraine in November, 1930, including the need to "Fight against white chauvinism (against all forms of Negro discrimination or persecution)." By 1931, the club numbered about a dozen branches across the United States. Whether Hughes had attended meetings of the club before he left for Haiti is unclear. His two autobiographies would deliberately hide his ties to the American left; perhaps Hughes also expunged some evidence of leftist involvement from his private papers. But before his departure for Haiti, *New Masses* (in which he had published previously) carried in its December, 1930 number his anti-imperialist poem "Merry Christmas," and followed in February with four equally caustic pieces of verse.

In the second half of 1931, however, *New Masses* became Hughes's major outlet. In July, it published "A Letter from Haiti"; in August, his poem "Justice"; in September, another poem, "Union"; in October, his passionate report on Haiti, "People Without Shoes"; in November, "Scottsboro, Limited," a Marxist one-act play; and in December, the long, bitterly anti-capitalist verse parody "Advertisement for the Waldorf-Astoria." But his even deeper involvement with the far left was signalled in the September *New Masses* with the news that he would be one of four directors to lead the newly formed New York Suitcase Theater, "a group of proficient actors who will travel with a minimum [of] equipment and a repertory of working class plays to be given before labor organizations." The other three directors were to be Jacob Burck, a staff artist on the communist *Daily Worker;* Paul Peters, a Kentucky-born party member and playwright, whose *Stevedore* (formerly *Wharf Nigger*) would be a major stage success in 1934; and Whittaker Chambers, recently appointed editor of *New Masses* after the success of four short stories hailed in the Soviet Union as setting a new standard of Bolshevik art in America. Chambers would shortly go underground for the party, break with communism near the end of the decade, then feature years later in the celebrated espionage controversy involving Alger Hiss. Hughes, Peters, and Chambers apparently had high hopes for the Suitcase Theater, although it would come to nothing. Years later, Peters reminded him nostalgically about "the days when you and Whittaker Chambers and I thought of starting something called the Suitcase Theater."

In the fall of 1931, the driving public force in Hughes's move to the left was certainly the Scottsboro controversy. Even among blacks, he was hardly alone in favoring the Communist Party's response to the case, which vastly increased the organization's prestige in Harlem. On March 25, just before Hughes and Zell Ingram headed for Florida, two white officers and an armed white mob had dragged nine black youths off a train at the tiny town of Paint Rock, Alabama, and taken them to the nearest jail big enough to hold them—in Scottsboro, the seat of Jackson County, Alabama. The blacks were accused of raping two young white women on the train after a fist fight between blacks and whites, all riding the train illegally, had ended with the routing of most of the white men. Following a tense trial in Scottsboro, with National Guardsmen keeping watch outside on a potential mob of almost ten thousand, one of the black youths was sentenced to life imprisonment, the other eight to the electric chair. To impartial observers, the verdict was a travesty of justice, but while the NAACP hesitated to act, the International Labor Defense, the legal defense arm of the Communist Party, threw its energies into appealing the case and mobilizing public support for the defendants. Its tactics of confrontation gained communism a hearing in places where it had been unwelcome.

Attempts at cooperation between the International Labor Defense and the NAACP collapsed, with Association leaders hurrying to protect their traditional support among black ministers and church members. Hughes's friend Walter White led the NAACP counterattack on the communists. "With jesuitical zeal and cleverness," he would write in December, just before the NAACP angrily pulled out of the case, "the American Communist agitator . . . resorted to every possible means to impress upon the Negro that he had no stake in his own land, that a philosophy of complete despair was the only sane and intelligent attitude for him to take. All this was centered about the Scottsboro cases as the basis for a highly emotional appeal." Certainly Hughes's emotions were touched. The black poet Melvin B. Tolson, just arrived in New York from Wiley, Texas, to study for a master's degree at Columbia, vividly remembered Langston's anguish as he sped away one day in New York through driving rain to attend a public meeting about Scottsboro: "There is a tenseness, an agony in the Poet's face. It seems that his life depends on getting to that meeting in time. . . . The Poet talks passionately of the Scottsboro boys. They are innocent. They must go free.''

For the Communist Party, often frustrated in its attempts to recruit Negroes, Scottsboro was a great opportunity. After a decade of recruiting, the black party membership in Harlem numbered fewer than two dozen. In 1925, the party had responded to this indifference by forming the American Negro Labor Congress, designed to bring the race problem into sharper relief. Three years later, at the Sixth World Congress of the Cominterm, Joseph Stalin had identified black participation in the party as a major issue, and controversially helped to set the goal of self-determination for Afro-Americans as a commitment of the communist effort in the United States. Soon, five blacks were appointed to the central committee of the U.S. party. In November, 1930, at a convention in St. Louis, Missouri, the now moribund American Negro Labor Congress was reconstituted as the League of Struggle for Negro Rights, with the crucial call for black self-determination prominent among its articles.

Improbably, Hughes at some point became president of the league. Exactly when he accepted the honorary post is not clear, but it was probably not before 1934. Although he later would be identified as its president from 1930, his name was not mentioned that year in the dozen or so dispatches, including several by the veteran black communist Cyril Briggs, sent from St. Louis to the *Daily Worker*. Langston did not attend the convention, which took place after his break with Godmother, while he was ill. Nevertheless, he shared certain attitudes of the league from the start, and in particular its hostility to most current black leaders. Convention speakers who had denounced Du Bois, Marcus Garvey, A. Philip Randolph, and others as vacillating helmsmen of the race would relish at least one of Hughes's four poems in the February, 1931 *New Masses,* ''To Certain Negro Leaders'':

> Voices crying in the wilderness
> At so much per word
> From the white folks:
> ''Be meek and humble,
> All you niggers,
> And do not cry
> Too loud.''

In ''Tired,'' Hughes called for revolution:

> I am so tired of waiting,
> Aren't you,
> For the world to become good
> And beautiful and kind?
> Let us take a knife
> And cut the world in two—
> And see what worms are eating
> At the rind.

"Call to Creation" exhorted fellow artists to "Give up beauty for a moment. / Look at harshness, look at pain, / Look at life again." And "A Christian Country" found Langston unafraid to challenge religion and, presumably, the black ministers who dominated the general leadership of the race:

> God slumbers in a back alley
> With a gin bottle in His hand.
> Come on, God, get up and fight
> Like a man.

Though grateful for Walter White's help, Hughes detested him for opposing the communists. If Langston had been speeding, in Melvin Tolson's story, to a meeting at St. Mark's Methodist Episcopal near mid-October, then the meeting was never held: White had telephoned the pastor, Lorenzo H. King, and pressured him into barring the International Labor Defense. Also firmly in the NAACP camp against the communists was Rev. F. A. Cullen, the poet's father, whose Salem Methodist Episcopal Church had hosted at least one mass meeting, late in June, in direct challenge to a rally staged elsewhere by the party. In 1931, Hughes did not attack White openly. About a year later, however, he would write two articles highly critical, even nasty in tone, about White and the Cullens (ironically, Countee would soon be converted to the Communist Party). As for Du Bois—the special if unspoken relationship between Langston and the great man endured; in April of the following year, Hughes would respond to a call to help the ailing *Crisis* with a check and a message: "The CRISIS must live for the sake of the Negroes of the world."

A poem early in the fall, "October the Sixteenth" ("Perhaps today / You will remember John Brown"), about the anniversary of the raid on Harpers Ferry, underscored the extent to which Hughes, who had slept under Lewis Leary's bullet-pierced shawl as a child, saw himself involved in a drama of radical renewal. His two major poems that autumn were unequivocal. One, published in the December *Opportunity*, was "Scottsboro":

> 8 BLACK BOYS IN A SOUTHERN JAIL.
> WORLD TURN PALE!
>
> *8 black boys and one white lie.*
> *Is it much to die?*
>
> *Is it much to die when immortal feet*
> *March with you down Time's street,*
> *When beyond steel bars sound the deathless drums*
> *Like a mighty heart-beat as They come?*
>
> *Who comes?*
>
> *Christ,*
> *Who fought alone.*
>
> *John Brown . . .*

The other was provoked by a plush advertisement occupying several pages of the October *Vanity Fair* for the opening of the twenty-eight-million-dollar Waldorf-Astoria in Manhattan. Outraged by its boasts of luxury, Hughes picked up a pencil and some loose sheets of paper and soon produced a sardonic parody of the advertisement. He dispatched the work to Whittaker Chambers at *New Masses,* for which Langston was about to become a contributing editor; his parody ranged over two pages in the December number. The first section was **"Listen Hungry Ones!"**

> Look! See what **Vanity Fair** says about the new
> Waldorf Astoria:
> **"All the luxuries of private home . . ."**
> Now, won't that be charming when the last flophouse
> has turned you down this winter? Furthermore:
> "It is far beyond anything hitherto attempted in the
> hotel world. . ." It cost twenty-eight million
> dollars. The famous Oscar Tschirky is in charge
> of banqueting. Alexandre Gastaud is chef. It
> will be a distinguished background for society.
> So when you've got no place else to go, homeless and
> hungry ones,
> choose the Waldorf as a background for your rags—
> (Or do you still consider the subway after midnight
> good enough?)

The last section was a timely "Christmas Card":

> Hail Mary, Mother of God!
> The new Christ child of the Revolution's about to
> be born.
> (Kick hard, red baby, in the bitter womb of the mob.)
> Somebody, put an ad in **Vanity Fair** quick!
> Call Oscar of the Waldorf—for Christ's sake!
> It's almost Christmas, and that little girl—turned
> whore because her belly was too hungry to stand
> it any more—wants a nice clean bed for the
> Immaculate Conception.
> Listen, Mary, Mother of God, wrap your new born
> babe in the red flag of Revolution:
> The Waldorf-Astoria's the best manger we've got.
> For reservations: Telephone
> **ELdorado 5-3000.**

Hughes's third major piece of writing that fall was "Scottsboro, Limited: A One Act Play." Previously he had asked Wallace Thurman to collaborate with him on such a play; following the friction over "Mule Bone," Thurman withdrew (but only after attesting in writing, doubtless on Langston's insistence, that the idea had been Hughes's). Requiring only one chair on a platform,

"Scottsboro, Limited" called for "Eight Black Boys, A White Man, Two White Women, Eight White Workers, Voices in the audience." In jingling rhymes and driving rhythms, the white man and the boys confront each other with accusations and denials. Then, when the electric chair is imminent, "Red voices" exhort the boys to fight, and offer the promise of unity between blacks and whites. The blacks smash the electric chair and join the white workers, while the aroused audience chants: "Fight! Fight! Fight! Fight! (*Here the* Internationale *may be sung and the red flag raised above the heads of the black and white workers together*.)"

Hughes's surging radicalism evidently did not kill his interest in Broadway; he agreed to work with Kaj Gynt, a gentle but very determined playwright who had grown up in Stockholm as a friend of Greta Garbo, and had enjoyed some success with her work on the 1927 black musical comedy *Rang Tang*. For the past three years, Gynt had been laboring on "Cock o' the Walk" (later, "Cock o' the World"), a musical vehicle intended for Paul Robeson. Hughes agreed to revise her outline about the life and times of the son of "Mammy Bless 'em." Cocko "went out in rags to conquer the world—and came back in rags; found love, found beauty, adventure, gold—and lost them all each time . . . but from the broken heart of all his lost dreams, he makes a song—MY SONG—the only thing he brings home to Mammy." Carefully, Hughes registered his revised outline with the Dramatists Guild.

Pushing ahead with his plans for the tour, he convinced Knopf to run off a cheaper, one-dollar edition of *The Weary Blues* for sale on the road. Few schools leaped at his first request for a fee of $50; in many places, the Depression had crippled enrollment. Hughes's fee dwindled sometimes to $10, but he was prepared to speak for nothing but hospitality and a share of the gate. Smoothly he informed his poorest prospects "that we were planning to stop in their town overnight, and so would—contingent, of course, upon being permitted to offer my books for sale—give a *free* program for culture's sake. When all other offers failed, that one almost always worked."

Hughes also entered into a significant new venture, founding a small publishing house in partnership with Carl Van Vechten and Prentiss Taylor, a promising white artist five years younger than Langston. Although he would write of Taylor as a Southerner, he was in fact a third-generation native of Washington, D.C., with no family ties to the rebel South. Taylor's sensitivity to racism owed much to a black woman he called "Cookie Bell." She had been not only "a very valiant and forthright servant in my family but also—she was very much the matriarch—a great help to a large number of other people. Through her, I became aware even as a boy of all sorts of racial obtuseness and racial disadvantages in Washington." Trained as an art student at McKinley Manual Training School in Washington, by 1927 he was living in New York trying to become established as a theater designer. For a while he designed books in styles "charming and very English in their feeling." But Taylor also began to think of himself as a radical at heart, even if he disliked the party and its propaganda. When Hughes, back from Haiti, turned to Van Vechten for help in producing a book of poetry, or of "dramatic recitations," Van Vechten suggested the services of Taylor and lent them $200 to found the Golden Stair Press at Taylor's home, 23 Bank Street in Greenwich Village.

For some time, Langston wrote Taylor, he had believed that "the modern Negro Art Movement" (in which he had been a star) was "largely over the heads, and out of reach, of the masses of the Negro people." If the masses didn't care for jazz poetry or novels of Harlem follies, "one can't blame them much—since they usually have such things all too well in life." He had noticed in recent black poetry "a distinct lack of rhymed poems dramatizing current racial interests in simple, understandable verse, pleasing to the ear, and suitable for reading aloud, or for recitation in school, churches, lodges, etc." He had written a group of poems "in this unpretentious fashion." A two-dollar book was out of the question; the Golden Stair Press would have to do it cheaply. As for sales, he himself would take the book to the people.

Hughes recognized that the black masses, especially in the conservative South, would have little to do with radical socialist verse, attacks on religion, or Marxist appeals to them as a class. Marxism helped him understand the black masses, Langston believed, but did not help the black masses understand themselves. He would speak to them in language they understood, which was neither that of the blues nor of revolution (both anathema in most black homes), but of the genteel tradition, adapted to the special features of black culture. "What the black masses need," the anti-communist black writer George Schuyler wrote sensitively to him, "is hope and pride, an understanding of their potentialities, a reinforcement of dignity. I believe you can be of great service to this end because you have the proletarian point of view."

As noble as his aims were, Hughes was thus also placing an enormous strain on his integrity as an artist. The defiant spirit of "The Negro Artist and the Racial Mountain" was essentially no more. Like many writers responding to the Depression, Langston was altering his aesthetic to accommodate social reality. Unlike most white artists, however, he faced a paradox: to reach the black masses, his writing had to be not radical but genteel, not aggressive but uplifting and sentimental. In effect, he was becoming at least three different writers—radical, as in the *New Masses;* commercial, as in the Kaj Gynt musical; and genteel, if also racial, as in the poems proposed to Prentiss Taylor. Additionally, Langston had come back from Haiti determined to live by his writing; he refused to teach school and write on the side, as Cullen was trying to do. But in throwing himself at the mercy of the marketplace, which he would later sometimes seem to confuse with the people, Hughes further endangered his independence; from this moment, he would have to count very carefully the number of jobs at hand, the contracts signed, the sums advanced against royalties. Perhaps this is why Hughes eventually would portray himself, in explaining his decision in 1931 to be a professional writer, as a victim of his own ineptitude, who had chosen writing as a career because at the age of twenty-nine "I did not know how to do anything else."

On September 11, with Van Vechten as witness, Hughes signed a contract for the booklet of poems with illustrations by Prentiss Taylor. Its title was taken from one of its poems, "The Negro Mother," whose fifty or so lines extolled the black woman as the hope of the race:

> Children, I come back today
> To tell you a story of the long dark way
> That I had to climb, that I had to know
> In order that the race might live and grow.
> Look at my face—dark as the night—
> Yet shining like the sun with love's true light.
> I am the child they stole from the sand
> Three hundred years ago in Africa's land.
> I am the black girl who crossed the dark sea
> Carrying in my body the seed of the Free.
> I am the woman who worked in the field . . .

With this poem, inspired by Mary McLeod Bethune (and women like Prentiss Taylor's "Cookie Bell"), Langston sought to banish Godmother and restore the black mother to her rightful place. The poem also suggests a crucial modulation of identity on his part. Where once he had faced the race as a child, seeking its love, now Langston also felt ready to address it, in his new, painfully earned maturity, as a nurturing black mother.

"He never gave away much or talked much," Taylor would remember of him; "but he was discreet in a commendable way, not secretive. He knew many people but he never dropped names the way other people did. And he certainly was very honest and efficient." In October, the Golden Stair Press announced the publication of *The Negro Mother and Other Dramatic Recitations,* with illustrations by Prentiss Taylor and printed by William J. Clark of 601 Sixth Avenue. When Langston took 100 copies home to the YMCA, they went swiftly: one elevator boy sold eighteen between floors. (*The Negro Mother* would have seven printings and sell about 1700 copies; seventeen copies were hand-colored by Taylor and signed by him and Hughes for future collectors.) Also intended for the tour, a quantity of posters and single-poem broadsides were produced by the Golden Stair Press.

The Negro Mother, in spite of its mawkishness, attracted some sophisticated admirers. At the end of the year Van Vechten would write Hughes that Aaron Copland had been so moved by the title poem that he wanted to set it to music. Apparently nothing came of the idea, but by this time Hughes had another volume of verse published. On October 31, he received the first bound copy from Amy Spingarn's Troutbeck Press of *Dear Lovely Death,* printed on handmade paper in a signed, numbered edition of one hundred copies. The cover was designed by Zell Ingram; the frontispiece of Langston was by Amy Spingarn herself. In some respects, this humble effort would remain his favorite book.

Limbering up for the coming tour, he shared a program with a prize-winning black classical pianist, Sonoma Talley. At the popular Barron's restaurant on Seventh Avenue in Harlem, Blyden Jackson, Zell Ingram, and another friend from the YMCA gave a farewell dinner for Hughes and Radcliffe Lucas. Then, at five o'clock on the morning of Monday, November 2, Hughes rose in the dark at the YMCA. Disaster almost struck. Lucas had just come up to Langston's room when someone sounded an alarm: their Ford was on fire. Racing downstairs, they saw wisps of smoke curling from the dashboard. Lucas quickly disconnected a radio which had just been installed. Loading the car with luggage, stacks of *The Negro Mother* and *The Weary Blues,* posters, broadsides, and books for their travelling exhibition of black writers, they set out for the South.

The tour opened the next morning a few miles west of Philadelphia, at Downingtown Industrial and Agricultural School for Boys. Then, on successive days, Langston read at Morgan College in Baltimore, Virginia Union in Richmond, and Virginia State College in Ettrick (where Hughes gazed on a portrait of John Mercer Langston, its first president), before a Saturday, November 7, appearance at Hampton Institute.

Tragically, Hughes's visit to "the most beautiful of Negro campuses" was also "the beginning of a realization that I was in the South, the troubled Jim Crow South of ever-present danger for Negroes." The football coach of Alabama A. & M. Institute at Normal, a recent Hampton star, had just been beaten to death in Birmingham after mistakenly parking his car in a "white" lot while attending a football game; and Juliette Derricotte, dean of women at Fisk but known nationally as a YWCA executive, died after being refused admission to a white hospital following a car accident. Distraught, some of the students asked Hughes to help with a protest. "I was deeply touched," he remembered, "and we began to lay plans for the organization of a Sunday evening protest meeting, from which we would send wires to the press and formulate a memorial to these most recent victims of race hate." But the dean of men, a black, forbade any gesture. "That is not Hampton's way," Hughes heard. "We educate, not protest." Langston was introduced to what he would soon identify as the state of systematic repression and cowardice at black colleges. But to several hundred students jammed into Ogden Hall for his reading, he seemed the epitome of freedom. "This young artist, swaying the emotions of his large audience, now into hilarious guffaws, then into a death-like silence," the Hampton *Script* reported, dominated his audience. That night, a Liberian student wrote to Hughes that his words on conditions in Africa had "burned" him.

By this time the general outline of his program was set. First came "Life Makes Poems," in which Hughes related certain lighter poems to his personal experiences, outlining a sort of autobiography. Then, wherever possible, came an interlude of music. The local talent almost inevitably played classical pieces or spirituals; most likely, no one ever ventured to sing the blues, as Langston would have preferred. Then came "Negro Dreams," poems with weighty, uplifting, sometimes tragic themes; any restlessness was quickly quelled with a poem on miscegenation such as "Mulatto" or "Cross." He always ended with his poem that began "I, too, sing America." Once the audience got over its surprise that the great New Negro poet was short, slight, and not particularly African, and that his voice was light and sing-song, they usually surrendered to his strengths—Langston's childlike, ingratiating charm and humility, the fair degree of emotional resonance in his verse, and his outstanding courage.

From Virginia, Hughes and Lucas continued south to Chapel Hill, North Carolina, and to the only white school where he had a firm booking, the University of North Carolina. Previously, he had written to the sociologist Howard Odum (Langston had reviewed his recent book, *Rainbow Round My Shoulder*) and to the Pulitzer Prize-winning playwright Paul Green. Odum had passed the letter on to his colleague Guy B. Johnson, then teaching a course on black culture. Although UNC was probably the most progressive white university in the South, for a black speaker to be featured there was extraordinary. Johnson also freely admitted that "most of us white folks are either too hypocritical or crowded, or both" to put Hughes up for the night. But two students (perhaps the total membership of the local John Reed Club) contacted Langston independently. Anthony J. Buttitta and Milton Abernethy, editors of the progressive magazine *Contempo: A Review of Books and Personalities,* with an editorial board including Paul Green, Ezra Pound, Louis Adamic, and Norman Thomas, saw a chance to rouse the campus. (Buttitta, from nearby Raleigh, was not enrolled in school that year; Abernethy, from Hickory, N.C., had just matriculated as a first year law student.) As editors, they had read with envy Hughes's "Scottsboro, Limited" in that month's *New Masses.* On September 15, they had published a small poem by Hughes on the front page of *Contempo,* next to an essay by Ezra Pound. Would Langston now send something on Scottsboro?

He responded with "Christ in Alabama":

> Christ is a Nigger,
> Beaten and black—
> *O, bare your back.*
>
> Mary is His Mother—
> *Mammy of the South,*
> *Silence your mouth.*
>
> God's His Father—
> *White Master above,*
> *Grant us your love.*
>
> Most holy bastard
> Of the bleeding mouth:
> *Nigger Christ*
> *On the cross of the South.*

He also sent them a taunting essay about the Scottsboro case, "Southern Gentlemen, White Prostitutes, Mill-Owners, and Negroes." Let the Alabama mill-owners pay white women decent wages "so they won't need to be prostitutes," he urged. "And let the sensible citizens of Alabama (if there are any) supply schools for the black populace of their state, (and for the half-black, too—the mulatto children of the Southern gentlemen. [I reckon they're gentlemen.]) so the Negroes won't be so dumb again." As for the jailed men—if blacks didn't howl in protest "(and I don't mean a polite howl, either) then let Dixie justice (blind and syphilitic as it may be) take its course."

What probably saved Langston from a mob was the neat timing of the editors in releasing the new *Contempo* just as he arrived for his brief visit; he slipped in and out of Chapel Hill before the yell went up. But Buttitta and Abernethy made the most of his stay. With Hughes's essay and poem (illustrated by a splendid sketch by Zell Ingram) blazoned on its front page, they printed five thousand extra copies of *Contempo*. Guy Johnson was outraged. The executive secretary of the university denounced the editors as "half-baked, uneducated, and wholly reprehensible adolescents." Frank Porter Graham, the new president, having boldly asserted the need for freedom of speech at Chapel Hill, felt betrayed; but he made no move to bar Hughes, who read on the night of November 19 to an excited crowd in the Greek Revival Little Theater, while a contingent of police stood guard outside. The next day, flashing his most disarming grin and behaving exasperatingly like a gentleman, he visited some classes in sociology. Hughes slept at the home of a black businessman, but also ate in a restaurant as the guest of the *Contempo* editors and other daring students. Shortly after, Buttitta recalled the visit for Nancy Cunard's anthology, *Negro:* "We had Langston Hughes for dinner at the snappiest cafeteria in town . . . everyone looked and looked for surprise, and what was the idea of it all . . . the Negro waiters and waitresses were particularly pleased in seeing the event. Much comment arose." At a soda fountain, "the cheap, southern soda jerker took Hughes for a mexican or something and let it go at that, but since he found out that Hughes was a 'nigger' and he had given him service the way he would have a white man, he got angry and attempted to catch us in a place or two and sock us in the jaw. . . . That is not all. . . . A certain wife of a certain nephew of good old Woodrow Wilson—she can't have kids, by the way—thought Hughes was darling, cute, and wanted to have him out to dinner in her Dreamshop." Larry Flynn, a student reputedly the nephew of millionaire Andrew Mellon, organized a party for him at the landmark Gimghoul Castle—with "a million bottles of home brew," Hughes reported with pleasure.

The powerful *Southern Textile Bulletin,* denouncing "the insulting and blasphemous articles of the negro, Langston Hughes," also deplored the fact that the *Daily Tar Heel,* the college newspaper, had referred to him as "Mr." Hughes. Three hundred persons signed a protest to the state governor against "the angels of darkness," Hughes and Bertrand Russell (another critic of Southern justice), whose "lives and writings violate the very decencies of life and the most sacred things of our heart." A prominent citizen wrote about the *Contempo* poem that "nothing but a corrupt, distorted brain could produce such sordid literature." "It's bad enough to call Christ a bastard," a local politician fumed about "Christ in Alabama"; "but to call Him a nigger—that's too much!" When the Associated Negro Press took the story even further afield, Hughes's northern friends grew nervous. "I have been a little fearful about your safety," Elmer Carter wrote Langston from his office at *Opportunity;* "the furor . . . might cause some of the more hot-headed of the cracker type to attempt to do you bodily harm." Worried about his plans to go to Scottsboro, his mother begged him "Please Please *don't go*—cancel the engagement." She had prayers said for him in a Cleveland church.

Langston, however, moved like a man possessed; he had had "a swell time" at Chapel Hill, he assured Walter White. He was also gambling that the publicity would seal the success of his tour, if he handled it right—and if he wasn't killed first. One school official indeed mentioned the "furore in North Carolina and in fact in all Dixie" in predicting success for his coming visit. But Hughes also had higher motives, as he explained to a writer about his *Contempo* pieces: "I believe that anything which makes people think of existing evil conditions is worth while. Sometimes in order to attract attention somebody must embody these ideas in sensational forms. I meant my poem to be a protest against the domination of all stronger peoples over weaker ones."

With his booklets, pamphlets, and broadsides selling briskly, the tour was beginning to pay for itself. Langston's entire store of *The Negro Mother* was gone just one week after the start; Prentiss Taylor rushed 450 copies to Durham in time for a major reading, as Hughes worked the school circuit in North Carolina until early December. On December 1, Taylor acknowledged receipt of an "enormous check"; soon after, Hughes sent Van Vechten $48 on his loan, and the printer, W. J. Clark, was paid off before Christmas. Langston's local sponsors were meeting their obligations. In fact, after one department reneged on its promise, only the white University of North Carolina of all his sponsors failed to pay his fee. Hughes was also meeting some interesting people. Between engagements in South Carolina, he visited the hospitable, eccentric old white folklorist Dr. E. C. L. Adams, author of *Congaree Sketches,* who received him just as he would have welcomed a white caller (Adams was "exactly my idea of what a *true* Southern gentleman should be"). On the other hand, when Langston tried to visit Julia Peterkin's home at her plantation "Lang Syne" in Fort Motte, South Carolina, he was rudely turned away by a white man. Sipping tea at Amy Spingarn's home in New York, Peterkin had invited Hughes to come visit her any time he was down South; but that had been in the North.

The roads were dusty and demanding, the roadside services often insultingly Jim Crow. Once, when he entered a waiting room to buy a copy of the *New York Times,* a policeman obtusely insisted that Hughes leave via the railroad tracks, since blacks were forbidden to use the front door, and there was no other. Yet Langston gave no thought to quitting the tour. In the black faces turned upwards to him he read trust and admiration and even love; the wounds of the previous year began to close. Now and then a critic quibbled about his poems, but in general the tour was a victory. "What has been said of Hughes many times is true," one newspaper report concluded. "He has broken away from the Negro poetry which has filled a large part of the gap from Dunbar to the Younger Group. He does not speak for the Negro. He speaks as a Negro. And one accepts his poetry not as propaganda but as a finished artist's contribution to America's writings."

Nevertheless, Langston was relieved on December 18 when he ended a week of readings at Spelman College, Morehouse College, and Clark University in Atlanta, and began his Christmas vacation by driving to Oakwood Junior College near Huntsville, Alabama. Here, at the home of Arna and Alberta Bontemps and their two children, Joan and Paul, he rested. Mailing boxes of pralines and cotton as presents to the Spingarns and other friends, he also caught up on his correspondence. But the visit would prove to be more than a respite from the road. This reunion with his look-alike, Arna Bontemps, would be another major step toward his future. Although the young men had known each other since the end of 1924, their meeting in December, 1931, would mark the start of a deeper relationship, virtually a marriage of minds, that would last without the slightest friction until Hughes's death thirty-six years later. By that time, although they had collaborated on only a few projects, the two men had exchanged almost 2500 letters.

Like Hughes, Bontemps in 1931 was enduring perhaps the greatest upheaval of his life, when the wisdom of his decision to leave his father's home in Los Angeles in 1924 and go to Harlem seemed most in doubt. Restless in spite of his outward calm (at one point he had been on the brink of going to sea, like Langston), Bontemps had endured seven years as a teacher at the little Harlem Academy run by the Seventh Day Adventist Church. Estranged from his religion, he had found deep satisfaction in writing poetry and in starting a family following his marriage in August, 1926, to Alberta Johnson, an eighteen-year-old student at the school. (In fathering several children within a marriage, he succeeded where almost every other member of the Harlem Renaissance failed—including Hughes, Cullen, Hurston, Thurman, Locke, Jessie Fauset, Dorothy West, Harold Jackman, Bruce Nugent, Rudolph Fisher, Nella Larsen, and Aaron Douglas.) Later, Hughes would teasingly recall Bontemps in the nineteen twenties as "quiet and scholarly, looking like a young edition of Dr. DuBois," and as "the mysterious member of the Harlem literati" because no one ever saw his wife in public. Alberta Bontemps was "a shy and charming girl," who greeted Hughes from time to time, on his infrequent visits to their home, with "a new golden baby, each prettier than the last—so that was why the literati never saw Mrs. Bontemps."

Genial and well liked, Bontemps was nevertheless also a contemplative and even somewhat reclusive man, especially as compared to bohemian personalities such as Thurman, Hurston, and Bruce Nugent. Unlike Hughes, who perhaps had already lost the habit of deep reading, Bontemps tried conscientiously to keep up with the latest in literature, history, and current affairs. In spite of Harlem, his poetry had little to do with cabarets or the blues, or brassy protests against injustice. Where others thundered, Bontemps had written sensitively of pacifism and nonviolence, although he was very much a race man. Nevetheless, his talent was appreciated; twice he had received the Pushkin Prize of *Opportunity* magazine, and he had also won the first *Crisis* poetry prize with the powerful "Nocturne at Bethesda," about the modern loss of spiritual values:

> I thought I saw an angel flying low,
> I thought I saw the flicker of a wing
> Above the mulberry trees; but not again.
> Bethesda sleeps. This ancient pool that healed
> A host of bearded Jews does not awake.

. . . .
There was a day, I remember now,
I beat my breast and cried, "Wash me God,
Wash me with a wave of wind upon
The barley; O quiet One, draw near, draw near!
Walk upon the hills with lovely feet
And in the waterfall stand and speak." . . .

A turn to fiction resulted in the controversial publication in 1931 of a novel, *God Sends Sunday* (Louise Thompson had been his typist). To the zealots at the Harlem Academy, who had long disapproved of his association with the "Niggerati," his use of "God" in the title was blasphemy. They had banished him to Oakwood Junior College, which he saw for the first time after a dismal train ride from New York the previous summer, while Hughes had been in Harlem planning his tour.

"We lived in a decaying plantation mansion for a time," Bontemps would recall of the house Hughes visited, "entertaining its ghosts in our awkward way." Oakwood was beautiful but sinister. Magnolia trees shaded a green lawn, and a rose arbor garlanded a fountain; but the old slave block and slave huts were still standing from the days before the Civil War. The "Old Mansion," where the Bontemps lived, had been the main plantation house; a farm near the college was run by whites along lines shamefully close to slavery. Moreover, Bontemps's undeserved reputation as a Harlem radical had preceded him; one day, the president of the school, a crude white Southerner, burst into the "Old Mansion" to demand that Arna burn his library of secular works. Oakwood was "the world's worst school," Bontemps judged, bound by a choking religious discipline that included a rigidly enforced vegetarian diet and a reactionary curriculum. He faced the grim likelihood that, after seven years of work, he would have to quit Alabama and limp with his wife and children back to his father's home in Los Angeles.

Just before leaving Harlem, Bontemps and Countee Cullen had finished a dramatization of *God Sends Sunday* that would reach the stage many years later as *St. Louis Woman*. Now Hughes and Bontemps decided to try their hand at collaborating on a little story for children. Such a book would both satisfy Langston's desire to write for black children and, he hoped, bring some quick money. Already in contact with a publisher, Macmillan, Bontemps brought Hughes in mainly because of his colorful Haitian material but also because the young men were at ease with one another. "I find him a steady and consistent writer with a style that is at once simple and emotionally effective," Langston wrote from Oakwood to the Rosenwald Fund, as he angled for a fellowship for Arna; before leaving the "Old Mansion," he wrote Macmillan giving Arna complete authority to handle all financial matters concerning the book, *Popo and Fifina*. (In March, they would each receive $150 as an advance; the book, which appeared the following year, 1932, would stay for over two decades on the Macmillan list.) The Christmas visit was a great success—except for the diet that insulted Hughes's craving for meat; he stared at mashed potatoes shaped into meat patties and vegetables disguised as rib roast. Now and then, he would recall, Alberta Bontemps took pity and slipped him a real pork chop.

On Saturday, January 2, with Bontemps in the chair, Hughes resumed his tour with a reading at six o'clock in the morning in the college chapel. The Bontemps family saw him off with some trepidation, since Langston had decided to visit the town of Scottsboro and, later, the prisoners themselves. Cautious, Arna Bontemps resisted the urge to attend the trials in nearby Decatur: "My wife did not favor this, and neither did I on second thought." (Alabama inspired caution: when Hughes spoke the word "Russia" at a gas station, a black youth almost panicked: "Sh-SSS-SS-S!" he begged Hughes. "You can't talk about Communism here.") Only urgent pleas, however, had prevented Langston from seeking an interview in Huntsville with Ruby Bates, one of the two white women involved in the incident. About twenty miles east of Huntsville, he and Lucas at last reached Paint Rock, where the previous year the nine black boys had been dragged from a train. Rolling slowly through the town, Hughes and Lucas picked up the Southern Highway trailing the railroad tracks to Scottsboro, about twenty-five miles away. There, Hughes stayed only long enough to send Van Vechten a note (he would have written a letter "if I dared remain that long") and a brief poem:

Scottsboro's just a little place:
No shame is writ across its face—
Its court, too weak to stand against a mob
Its people's heart, too small to hold a sob.

After engagements in the Miami, Florida, area, including one program held in a funeral parlor because local blacks had no hall, Langston reached Daytona Beach to read at a benefit performance for Mary McLeod Bethune's college. Twelve hundred people, about half of them whites from nearby hotels, heard a student choir of a hundred voices sing black music backed by a full orchestra. "Mrs. Bethune knew how to get things done," he would write in admiration. Accepting only $50 for his appearance, Hughes was happy that Mrs. Bethune made considerably more for her college. If Jessie Fauset had once been his "brown goddess," Mary Bethune had become the black deity of his new life. The greatest moment of the entire tour—perhaps his finest moment since his last, dreadful visit with Godmother—came when Hughes finished reading his poem "The Negro Mother." " 'My son, my son!' cried Mrs. Bethune, rising with tears in her eyes to embrace me on the platform." With her great black arms about him, Langston at last had come home to his people.

After further readings in Jacksonville and Tallahassee, he and Lucas drove north over the Georgia border to Albany, then to Fort Valley, where they spent a long weekend at the black college there. One sad task for him was to compose a tribute to the man who had tried to help him in 1925, Vachel Lindsay. "His work was a living encouragement of the original and the American in the creation of beauty," Hughes wrote. "Thousands feel now a sense of personal loss. We miss him as we miss a friend. A fine poet, and a great heart, has left us for the universe of stars." After achieving extraordinary popularity as an apocalyptic poet-prophet, at the age of fifty-two Lindsay had put an end to bad health and growing poverty by drinking Lysol.

Leaving Fort Valley, Hughes headed back to Alabama with Scottsboro weighing heavily on his mind. At most of the black schools, the case could not be mentioned. "A great many teachers and students knew nothing of it, or if they did the official attitude would be, 'Why bring that up?' " But he wanted, at the very least, to see for himself how Southern justice worked. On January 20, he reached Birmingham, where the trial of Willie Peterson, yet another black man trapped by Southern racism, was in session. Charged with the murder of two white women, Peterson survived an attempt to lynch him, then was shot in jail by the brother of one of the victims. The case against him was so meager that at his first trial the all-white jury failed to reach a verdict. About one o'clock in the afternoon Hughes slipped into the line of colored folk along the left side of the corridor on the sixth floor of the Jefferson County courthouse. On the other side were the white folks, who all seemed relaxed; the word "nigger" slid like stale chewing gum from their lips. The blacks were downcast and obviously poor. Where was the black middle class? An old woman in a ragged overcoat kept up a sad refrain: "It sho is pityful." "I wonder where the well-dressed Negroes of Birmingham are," Hughes scribbled. "Or are only the very poor, friends of this man going through hell in the Jefferson County Courthouse this week in Birmingham? Don't the college men care? Don't the professional men care? When you get smart and learned, do you cease to care? Why is it that none of them are . . . at the trial today?" Guarded by a phalanx of white men, Willie Peterson approached—tall, gaunt, ashen-skinned, and bug-eyed with fear. A black woman fainted. "No time for fainting now," a cracker wit drawled. A few whites guffawed. "Pityful! Pityful!" the old woman in the overcoat lamented. "It sho is pityful."

Fifty miles east of Birmingham, in the Blue Ridge Mountains, Langston found the professors tweedy and obtuse when he read at Talladega College, located curiously near the state institute for the deaf and blind. Angry at their indifference, he composed a sneering sonnet, "Ph.D." ("And all the human world is vast and strange— / And quite beyond his Ph.D.'s small range"). "Dear Negro Leaders—Whom do you lead?" he asked elsewhere, picking up the question he had posed before in *New Masses.* "Not the millions and millions of black ones who have never heard your name? Not the porter, peasant, scullion, cook, day laborer, maid-servant, cleaner after a nation's dirt?" Nor could Hughes forget himself. "Dear Negro Leaders, who would starve tomorrow if your salaries and train fares and lecture fees were not paid by white folks (maybe I would starve, too; maybe that's the tragedy and waste; maybe there's no other way . . .) but I ask you, whom is it that you lead, and where is it you are going?"

The following Sunday, he visited the Scottsboro boys in Montgomery, Alabama. Escorted by the black man who often led them in religious services, Hughes entered Kilby Prison on the ugly outskirts of the city. Steel gates opened and clashed, opened and clashed behind him until he reached a two-story building deep within the compound. On the second floor were the death cells. One room held sixteen cages, all occupied by blacks. Beyond the hallway, behind a green door, was the electric chair. A flash of terror struck Hughes. "For a moment the fear came: even for me, a Sunday morning visitor, the doors might never open again. WHITE guards held the keys. . . . And I'm only a nigger poet. Nigger. Niggers. Hundreds of niggers in Kilby Prison. Black, brown, yellow, near-white niggers. The guards, WHITE. Me—a visiting nigger." After prayers, when he read to the prisoners, his poems sounded "futile and stupid in the face of death." None of the Scottsboro boys (one was fourteen) seemed even slightly interested as he spoke of the greatness of the black race and the desire of many people to see them free. Someone asked for cigarettes. When he told Andy Wright that he had seen his mother, who often spoke for the International Defense League, Wright smiled and got up to shake his hand.

At the State Teachers College later that day no one wanted to speak about the case; it was unspeakable. Morosely, Hughes and Lucas drove northeast to Knoxville, Tennessee, then westward to Nashville. Here Langston's gloom lifted. His reading on January 29 in Fisk University's Memorial Chapel, introduced glowingly by James Weldon Johnson (since 1930 a professor at Fisk), was a tremendous success; at least fifteen times the packed hall erupted into applause. The only blight on the Nashville visit came when Thomas Mabry, a young white Southerner and an instructor in English at the arch-conservative Vanderbilt University across town, was pressured into cancelling a little party, to which he had invited Hughes and Johnson, by Mabry's well-known colleague, the poet and critic Allen Tate. The main theorist among the so-called "Fugitives" defending the Southern way of life (*I'll Take My Stand* had appeared the previous year), Tate not only refused to attend but circulated his letter of refusal (a "vacuous and insulting essay on race relations," Mabry called it) among members of the department. Although he might meet the two blacks ("both very interesting writers and as such I would like to meet them") in the North or overseas, Tate would not meet them in the South; he compared attending a party with Hughes and Johnson to meeting socially with his black cook. Southern custom was "unfortunate," but Tate was not willing "to expend any effort" to change it. Normally sweet-tempered, Johnson was crushed to read the letters. "The south as an Institution," he wrote privately, "can sink through the bottom of the pit of hell."

The chasm between the races in the South appalled Hughes. And yet in his first reading in the most backward state in the union, Mississippi, he was respectfully introduced by the patrician William Alexander Percy, acclaimed as its leading poet on the basis of his three collections between 1915 and 1924, *Sappho in Levkas, In April Once,* and *Enzio's Kingdom.* Exactly the opposite of Hughes as a poet, Percy made little use of his native South in his verse, but often sang elegiacally of earlier or mythic times and places. In addition, he was a leading planter of the region, even if no one thought of him and his family, who had opposed the Klan and demagogues such as James Vardaman, as typical of Mississippi. Nevertheless, at the black St. Matthew's A.M.E. church in Greenville, Will Percy "introduced me most graciously" to an audience of both races; one black man would remember Percy's reference to Hughes as "my fellow poet." Years later, Percy would send him a copy of his autobiography,

Lanterns on the Levee, "and over the years that followed I had several beautiful letters from him. But I met less than half a dozen such gentlemanly Southerners on my winter-long tour. Instead, I found a great social and cultural gulf between the races in the South, astonishing to one who, like myself, from the North, had never known such uncompromising prejudices." For one other reason, Hughes no doubt treasured Percy's gesture. The reading in Greenville took place on February 1, 1932, Langston's thirtieth birthday.

The next ten days were passed in Mississippi, with Hughes reading in and around Jackson, Meridian, and Vicksburg. At Alcorn College, as elsewhere, he found the attitude of black administrators outrageous. At the end of his lecture, the chairman tapped a bell to release first escorted women, then those unescorted, then young men, then teachers: "Such regimentation as practiced in this college was long ago done away with, even in many grammar schools of the North." In general, many black schools "are not trying to make men and women of their students at all—they are doing their best to produce spineless Uncle Toms, uninformed, and full of mental and moral evasions." A nationally known black professor refused to discuss communism on the campus without "a letter first from the president and the board of trustees." Hughes concluded that, even on matters such as facial cosmetics and smoking, "freedom of expression for teachers in most Negro schools . . . is more or less unknown." When he paused near Vicksburg long enough to draft an introduction to *Negro Songs of Protest* by Lawrence Gellert, he hit again at the cowardice of most educated blacks. "Fear has silenced their mouths. . . . The tossed scrap of American philanthropy has bribed their leaders. . . . The charity-luxury of Fisk, Spelman, Howard, fool and delude their students. . . . The black millions as a mass are still hungry, poor, and beaten." Two years later, in the *Crisis,* Hughes would remember this tour and conclude that "American Negroes in the future had best look to the unlettered for their leaders, and expect only cowards from the colleges."

Still, he looked eagerly for signs of change among the young. On the evening of February 11, after he read at New Orleans University before "a grand high yellow audience that seemingly never heard of Harlem," a pretty but nervous sixteen-year-old brownskinned girl in a white two-piece surplice-front knit outfit made three sallies through the reception line before timidly offering Hughes her manuscript of verse. "I was so very nervous," Margaret Walker would recall. "Langston Hughes was an absolute hero to me. Somebody had given me a copy of *Four Lincoln Poets* and I just wore that booklet out! Just wore it out! And he was so kind. He stopped what he was doing, put my poems up on the piano where my sister had sung during his program, and proceeded to go through them one by one for about an hour, talking about them, how they might be improved. He said I had talent. He encouraged me tremendously. He even urged my parents to send me to school in the North, where I would have more freedom to grow." Walker's parents, a professor of religion and philosophy and a music teacher, had helped bring him to their school. His friendship with her would continue. A decade later, when she won the Yale Younger Poets Prize for her volume *For My People,* a title that might have come from Hughes, Walker sent him a copy "in gratitude for his encouragement even when the poems were no good."

In few places were the contradictions between education and leadership clearer than at Tuskegee Institute, which Hughes had visited in 1927. The wealthiest of black schools, with an enrollment of 2600 students, Tuskegee professed to educate blacks but—to his horror—"maintains a guest house on its campus for *whites only!*" To Claude McKay, who had come years before from Jamaica to attend Tuskegee, Hughes had written earlier of "the nice Negroes living like parasites on the body of a dead dream that was alive once for Booker Washington." When he sent an anthem requested during his visit in 1927, he had praised not the school but its indomitable founder, in "Alabama Earth: At Booker Washington's Grave":

> Deep in Alabama earth
> His buried body lies—
> But higher than the singing pines
> And taller than the skies
> And out of Alabama earth
> To all the world there goes
> The truth a simple heart has held
> And the strength a strong hand knows,
> While over Alabama earth
> These words are gently spoken:
> Serve—and hate will die unborn.
> Love—and chains are broken.

By 1932, Hughes was totally disenchanted with Washington's successors. His most moving moment came away from the school, when he went with the actress and singer Abbie Mitchell, now employed at Tuskegee, to hear a massed choir of some four hundred black children sing "with such great depth and sincerity and such elemental emotion," he wrote, "that after listening to two songs one could scarcely stand any more."

As the winter drew to an end, Hughes's audiences seemed only to increase. In Birmingham, copies of *The Negro Mother* "sold like reefers on 131st Street," he joked to Van Vechten, as in one ten-day period at the end of February Hughes read in seven different states. In Memphis, he was happily upstaged by his friend W. C. Handy and his trumpet; in St. Louis with Hughes for a March 1 reading, Handy brought down the packed house with a rendition of his "St. Louis Blues." A few days later, when Langston sent an interim report to the Rosenwald office, he declared 54 paid lectures and a net income of $1337.83 shared between Lucas and himself. "I feel that for the first time, I have met the South," he announced. "I have talked with many white Southerners, and thousands of Negroes, teachers, students, and towns people. I know now attitudes and complexes I had not realized before." The cynics in New York were wrong: "I've never had a finer response anywhere, or met more beautiful people." He had come to believe that "poetry can mean something to uneducated people. Even in the backwoods, they seemed to know what I was talking about, and to appreciate it.''

The following two weeks took him mainly out of the South. At the University of Iowa, he was introduced by the accomplished historian of journalism Frank Luther Mott. Then, after reading in Des Moines, Hughes turned southwest for a major personal event, his first visit to Lawrence, Kansas, since 1915. He stayed with ''Auntie'' Reed, now Mrs. Mary Campbell, and went with her to St. Luke's A.M.E. Church just as they had often gone when he was a boy. Also in town was Langston's mother, Carrie Clark, who stirred memories in the local black newspaper of the days forty years before when Carolina Mercer Langston was ''the belle of the community.'' On March 9, sponsored by the black sorority Alpha Kappa Alpha, Hughes made a triumphant appearance at Fraser Hall at the university; he was also feted by the college's Poetry Society. ''Our needs are many,'' he insisted to a black reporter. ''One need is leadership: a Voice to speak to us, and for us; a new Frederick Douglass thundering across the land; a new Harriet Tubman . . . a new Sojourner Truth, unafraid. . . . Where is the Voice that will awaken the man in the cotton fields, the woman in the kitchen, the laborer in the cities? . . . Or are we all too educated to give a damn about the poor, the ignorant, the oppressed? . . . Who cares? Does anybody? I mean, enough to make the world know it.''

Ten days of hustling about Kansas and adjoining Missouri, then south to Langston University in Oklahoma, brought Hughes to the point of exhaustion. ''This tour m'en fiche!'' he groaned to Prentiss Taylor in New York. Moreover, Radcliffe Lucas was pining for his girlfriend in the east and wanted to get married. Then Hughes received an urgent summons from Kaj Gynt, his collaborator on the musical ''Cock o' the World.'' Paul Robeson, in the U.S. on a concert tour, was on the brink of a decision about appearing in the show; George Gershwin himself might do the music. Could Hughes return to New York at once? Although Van Vechten advised caution, Hughes decided to cancel some engagements and dash to New York.

Weary, Hughes reached the city on Thursday, March 24, just before the Easter weekend, but at the apartment building where Robeson's white manager lived, he was refused admission. The irony was almost too much for him: Paul Robeson's manager lived in a building where blacks were refused even as visitors. Then he discovered that Robeson had decamped and gone back to London, nor was the star much interested in their play; Kaj Gynt had been carried away by her hopes. An angry Hughes may have blown up at Gynt, then received a phone call on Gynt's behalf from Zora Neale Hurston—who informed Mrs. Mason that Hughes had been polite, even cordial to her, and had sent Godmother his regards. Alain Locke, who that year used a review of his Howard University colleague Sterling Brown's fine book of verse, *Southern Road,* to dismiss Hughes's historic treatment of black culture (''it was essentially a jazz version of Negro life . . . and though fascinating and true to an epoch this version was surface quality after all''), put Langston's tour in perspective for Mrs. Mason. Hughes was ''burnt out and trading on the past.''

But the visit to New York, far from being wasted, brought Langston closer to a major development—a visit to the U.S.S.R. Early in March, Louise Thompson had alerted him about a motion picture on American race relations proposed by Soviet authorities and backed by the leading black American Communist, James W. Ford, who would be a candidate for the U.S. vice-presidency in the coming elections. While in the Soviet Union, Ford had been asked to find a party of black Americans willing to travel to Moscow to help make such a film. The radicalized Louise Thompson, who had recently helped found a Harlem branch of the Friends of the Soviet Union, was in charge of the search for performers who could pay their way to Moscow, where their expenses would be refunded. "I am sure the whole plan would gain a remarkable stimulus," she had written Hughes, "if it were known that you were to be a member of the group." On March 11, he had wired his willingness both to serve on a committee and to consider going to Russia, but requested secrecy on the latter because of the prospects of "Cock o' the World." Hughes was appointed to the "Sponsoring Committee for Production of a Soviet Film on Negro Life" along with Whittaker Chambers, Malcolm Cowley, Floyd Dell, Waldo Frank, Rose McClendon, Louise Thompson, and others; he was also mentioned in a notice about the film on March 16 in the *Amsterdam News*. After learning from Thompson in New York that few blacks had signed up, he wrote at Easter to at least two persons in the South on behalf of the project.

Visiting the Knopf office, he admired the drawings by the illustrator Helen Sewell for his children's poetry book, *The Dream Keeper*, which he dedicated to his "brother," Gwyn Clark. Langston had begged Sewell to avoid "the usual kinky headed caricatures" used so often to represent blacks; "I hope that they can be beautiful people that Negro children can look at and not be ashamed to feel that they represent themselves." Her illustrations proved to be very sensitive. Hughes and Prentiss Taylor also agreed on the next venture of the Golden Stair Press—a slender book comprising the play "Scottsboro, Limited," four related poems, and four lithographs on Scottsboro by Taylor. Curiously, Hughes had wondered whether the play was "too red to be included," but Taylor insisted on it. With Van Vechten's financial backing once again, they planned a booklet to be sold at fifty cents a copy, plus a signed edition of one hundred copies at three dollars each. This special edition soon shrank to thirty copies but promised to be superb, published on *papier de rives* with the lithographs printed from the original stones. "I'm more excited about this Scottsboro booklet," Hughes would write Taylor, "than anything I've ever had published."

The most important task on the Easter visit was to find a replacement for Radcliffe Lucas. James Ronald Derry, a mighty athlete for Lincoln in more than one meeting of the Penn Relays, begged for the job: Hughes was heading for Los Angeles, the site of the summer Olympics. Hiring him, Langston raised the driver's share of the profits to fifty percent because the tour was more than half over. Then Hughes discovered that Derry could not drive; Derry, when he learned that Langston could not teach him, burst into tears. George Lee, once a mathematics major at Lincoln, now unemployed, became "tour manager," or driver; Derry, unpaid as Hughes's "personal secretary," went along for the ride.

On Monday, March 28, Langston and his entourage swept out of New York in the gallant Ford, headed for the plains of Texas. In the first three weeks of April the men covered the vast state, which Hughes had not seen since 1920, and then mainly from the window of a train going to Mexico with his father. The isolation of the black communities scattered across the immense land, the presence of Indians and Mexicans who often formed a novel underclass to the black communities, the stark desert landscapes, piercing skies, and the heat, all revived his passion for the road. Starting with a YMCA reading in Dallas on April 1, he went south to Prairie View College near Houston, where he met the very pretty, stylish, and intelligent young academics Youra Qualls and her twin sister Ina, down from Texas College in Tyler to hear him. He passed westward to Austin, returned to Houston, then set out on the long, dusty drive to remote San Antonio, "where women shoot pool and rattlesnakes walk the main streets." Then he and his party drove into southern New Mexico and Arizona through hot desert country he loved at once, and at last saw California.

To Hughes and his friends, Los Angeles seemed more a miracle than a city, a place where oranges sold for one cent a dozen, ordinary black folks lived in huge houses with "miles of yards," and prosperity seemed to reign in spite of the Depression. His host was Loren Miller, a young black lawyer ("my representative on the coast") originally from the Topeka, Kansas area, who wrote a column for the local black newspaper, the *California Eagle*. Handsome and brown, with straight, shiny black hair, vaguely Indian in his features, Miller looked not unlike Hughes. A brilliant speaker, merciless in debate, Miller was committed to radical socialism without being a party member; the resolutions of the Sixth World Congress of the Comintern, when the cause of black self-determination had been first raised by communists, had become his Bible. He was prominent in the John Reed Club of Los Angeles, which had just been prevented by the police from staging Langston's "Scottsboro, Limited."

His spirits high, Hughes moved into Miller's home at 837 East 24th Street—"the swellest apartment in a street of palms and flowers for only $25 a month"; to be perfect, "all it needs now is a cocktail shaker." On Thursday, April 21, when he read at a church under the auspices of the integrated Los Angeles Civic League, Hughes seemed "like a big kid—just smiling," with a "big kiddish grin" and a manner "chock full of boyish amusement that warmed his audience." He had "the non-plussed attitude of a child." A few days later, in a room vibrant with revolutionary posters, he lectured on "The Social Implications of Current Negro Literature" to the John Reed Club. Loren Miller's political intensity (he openly dismissed almost all the writing of the Harlem Renaissance school, including much of Hughes's work, as worthless) and the futuristic atmosphere of Los Angeles sparked Langston's pen to write a "mass chant" about the most famous prisoner in California, Tom Mooney, to be read from a megaphone. (Later, Hughes would visit Mooney at San Quentin, where he was serving time for allegedly taking part in a bombing in 1916 in San Francisco.) In Los Angeles, Langston also composed one of his most aggressive poems, "A New Song," in which the refrain "That day is past" ends with a clarion call to black and white workers to "Revolt! Arise!":

The Black
And White World
Shall be one!
The Worker's World!

The past is done!

A new dream flames
Against the
Sun!

In "An Open Letter to the South," accepted by *New Masses,* a black worker expresses to white workers of the South the hope

That the land might be ours,
And the mines and the factories and the office towers
At Harlan, Richmond, Gastonia, Atlanta, New Orleans;
That the plants and the roads and the tools of power
Be ours.

Let us forget what Booker T. said:
"Separate as the fingers."
He knew he lied.

Let us become instead, you and I,
One single hand
That can united rise
To smash the old dead dogmas of the past—
To kill the lies of color
That keep the rich enthroned. . . .
. . . .
White worker,
Here is my hand.

Today,
We're Man to Man.

Catching up on his mail, Hughes finally answered the poet Ezra Pound, who had written him from Rapallo, Italy, about the chance of black colleges supporting an English translation of the pro-African archeologist and anthropologist Leo Frobenius, the author of *The Voice of Africa* and founder of the Institute for Cultural Morphology in Frankfurt; Pound worried that black colleges were ignoring African culture in favor of a dying Europe. Although Hughes had written to Fisk and Howard about the matter, he had little hope that black schools would be interested in Africa. "Many of your poems insist on remaining in my head," he ended vaguely to Pound, to whom he also sent three of his books, "not the words, but the mood and the meaning, which, after all, is the heart of a poem."

Joining Miller and the white, Oregon-born poet Norman Macleod, then twenty-six years old and a student at the University of Southern California, Hughes went east to the construction site of the massive Boulder Dam (later named for Herbert Hoover) to investigate charges of racial discrimination on the federal project, which employed no blacks. Appropriately, they arrived on Labor Day. The next evening, after talks with officials, they left for the Grand Canyon by way of Zion National Park just over the Utah border (a visit Hughes had once expected to make at Godmother's expense). They returned to Los Angeles in time for his star appearance on Sunday, May 8, at a mass meeting on Scottsboro, with a production of "Scottsboro, Limited" and the Tom Mooney mass chant, in which the big crowd joined enthusiastically. The following day, he read his poems at the Philharmonic Auditorium downtown, sponsored by the Liberal Forum, with the black Hollywood actor Clarence Muse singing some blues and spirituals.

Enjoying his visit, Hughes might have stayed longer in Los Angeles, but on May 9, a telegram arrived from Louise Thompson. Only ten persons were committed to go to Russia, but the Soviet film company would settle for having others come later. Langston and Loren Miller wired their commitment to join the group. They would drive together to New York, from where the party would sail in about one month.

To close out their tour, Hughes and George Lee headed north four hundred miles towards San Francisco Bay. On Saturday, May 14, "America's Foremost Negro Poet" appeared at the Berkeley High School auditorium at the corner of Alston Way and Grove Street under the auspices of the Acorn Club, an Oakland-based group of socially conscious young black men. Introducing Hughes was an old friend of Louise Thompson's, Matt Crawford, a cheerful, light-complexioned chiropractor and insurance clerk who liked poetry in general but revered "The Negro Speaks of Rivers." ("That poem had a tremendous impact on me and all of us," he later recalled. "For the first time, it showed us a perspective far beyond the United States.") Crawford's wife, the former Evelyn "Nebby" Graves, had been Louise Thompson's best friend ever since their schooldays in Oakland and Berkeley; she had lived for some time with the Thompsons and had met Langston in New York while visiting them. At the Crawfords' comfortable home at 1920 Acton Street in Berkeley, Hughes told them about the trip to Russia.

The following afternoon, Sunday, May 15, he moved to far more luxurious accommodations at 2323 Hyde Street, San Francisco, the old Robert Louis Stevenson residence high on Russian Hill, now the home of one of the city's most prominent bachelors, forty-three-year-old Noël Sullivan. From an instructor in Pine Bluff, Arkansas, and his wife, Arthur and Juanita Williams, Hughes had heard of him as an extraordinary man ("the only Noël Sullivan in the world," Langston would salute him shortly). Sullivan was wealthy; his grandfather had founded the Hibernia Bank of San Francisco, and his uncle, James D. Phelan, had been a mayor of the city, a state senator, and a major patron of the arts. A lover of music, Sullivan had passed several years studying voice and the organ in Paris before returning home in the mid-twenties to assume the leadership of his family. He was an odd mixture of indulgence and asceticism; a tall, sometimes gaunt man with an increasingly priest-like manner, he was a devout Roman Catholic with a political consciousness sharpened by his knowledge of Irish history. ''Deep down in my heart,'' he once confessed, ''though grateful for the many unsought blessings that have come to me, I have always been ashamed of privilege.'' In response, he generously supported a wide range of liberal causes, from the protection of animals to the abolition of the death penalty. He employed blacks in his home—Juanita Williams's sister, Mrs. Eulah Pharr, was his housekeeper; in addition, to the astonishment of many people, not least of all blacks, Sullivan also treated some Negroes as social equals. And although a few were famous—Roland Hayes, Marian Anderson, Dorothy Maynor, Bill ''Bojangles'' Robinson, and Duke Ellington either stayed or were entertained repeatedly at his home—he was also concerned with the welfare of much more ordinary folk. Hughes, however, arrived as a celebrity. Early in March, Sullivan had written to invite Langston to stay with him; previously, he mentioned, they had missed meeting in New York at the home of a mutual acquaintance. A dedicated basso, he had also just sung in concert a Hughes poem set by John Alden Carpenter. He ''would feel very much honored'' to have Hughes as a guest; ''I have long been a great admirer of your poetry.''

By mid-May, Langston had washed away the grime of the road and was happily installed at 2323 Hyde Street high above the bay. The gardens on Russian Hill were frescoed in blossoms; the mansion was sunny and splendid. He unpacked in the ''Negro Room,'' which was decorated with appropriate paintings and an exquisite little porcelain jazz band. Sullivan's housekeeper, Eulah Pharr, and her husband, Eddie, were attentive but discreet, with just the right mixture of racial familiarity and respect; the butler and the chauffeur were also colored. Ed Best, Sullivan's secretary, offered his services as stenographer. The mansion swarmed with pets—canaries, a parrot, and many dogs, including a handsome young German Shepherd named Greta (after Garbo), a bushy Chinese Chow called Turandot, and several waddling dachshunds who leaped into Hughes's bed to nibble at his breakfast. Also a house guest was the film star Ramon Novarro—a ''very amusing boy,'' Hughes informed Van Vechten. They made a party to attend the local opening of Marc Connelly's acclaimed musical play *Green Pastures* (Hall Johnson had been its musical director), with Richard Harrison in his famous role as ''De Lawd.'' Langston lectured to the colored Dumas Club at the YWCA on Sutter Street, in ''an evening of spirit, a temperateness and a sense of humor that were very appealing,'' according to a reporter.

To Eulah Pharr, who would know him for the rest of his life, Hughes was a rare guest: ''He seemed amazed about everything. For example, he loved food, and the meals were always good and the service formal and fine. These things were actually touching to Langston. He always seemed to appreciate what you did for him. His eyes would actually sparkle with pleasure and gratitude.'' Years later, one could still see ''the youth in Langston. He always seemed like a young person starting out in life. He had such a naive attitude; it was very beautiful to watch.'' Hughes's immediate impact on Noël Sullivan was more dramatic. ''Without any possible disloyalty to the great people of your race that it has been my privilege to know,'' Sullivan wrote on his way briefly out of town, ''I want to assure you that never has anyone of them inspired in me the unqualified regard and admiration that I feel for you. Indeed, had I been told ten days ago that your integrity (and by that word I think I mean more than is implied in its customary usage) existed anywhere—I should have been skeptical. . . . The contact with you has extended immeasurably for me a sense of tenderness toward everyone.''

On May 23, Hughes and George Lee left for Portland and Seattle. On the way they were refused accommodation in an Oregon town, until the police placed them in a seedy hotel haunted by prostitutes. In Seattle, Hughes was the guest of a community leader, Madge Cayton, and met her very intelligent brother, a young policeman named Horace Cayton who would become an important sociologist (co-author with St. Clair Drake of *Black Metropolis*) and a settlement house leader in Chicago. The Caytons were old friends of Louise Thompson and her mother. Langston returned to the San Francisco area to give one of his major readings of the tour, at the University of California in Berkeley. He also discovered that Matt Crawford was going with him to Russia. Since arriving from Anniston, Alabama, in 1911 as a boy of eight with his family, Crawford had never left California. ''One evening at dinner I was telling my wife, Evelyn, how exciting the trip sounded, how great it would be to travel with Langston and Louise and their friends,'' Crawford remembered of a decision that would change his life (within two years he would be on the national council of the communist League of Struggle for Negro Rights). ''She said, Matt, if you really want to go, call Louise. Maybe she can help you. I got up from the table and telephoned her in New York and that was that!'' Pressed for travellers, whether or not they could sing, dance, or act, Thompson signed up Crawford via long distance.

With Noël Sullivan, Hughes went down to the Monterey Peninsula to read in the village of Carmel-by-the-Sea, known as a haven for writers and artists; not far away was the Carmelite Convent that Sullivan had built (one of his sisters was a Carmelite nun, Mother Agnes) in memory of his father. Singing spirituals in his sometimes wayward *basso profundo,* Sullivan shared the stage with Hughes in a program at the Community Theater sponsored by the local John Reed Club, with the lively leftist Ella Winter, the Australian-born divorced wife of the old muckraker Lincoln Steffens (with whom she still lived) in the chair. The next day, Hughes visited Steffens, the sage of Carmel. He found a frail but still jaunty fellow with twinkling eyes, who stroked his goatee and advised Hughes to expect *not* to like anything about the Soviet Union; then perhaps he might see the future at work. ''Be sure to take soap and toilet paper for yourself—don't part with it!—and lipsticks and silk stockings to give the girls,'' Steffens wisely counselled, ''and maybe you won't be too unhappy.'' At Tor House, the stone retreat built by the poet Robinson Jeffers, who had attended his reading with his wife, Una, Hughes sipped tea and talked about poetry. The master photographer Edward Weston, another friend of Sullivan's, photographed Langston.

On June 3, after a very satisfying stay, Hughes headed south for Los Angeles. At Bakersfield, he officially ended the tour with a program for $15 at the Cain A.M.E. Church there. The next day, Langston, Loren Miller, Matt Crawford, and George Lee left for New York. Ron Derry loyally followed them eastward in another car with a party of young women, until at Monrovia the Russian-bound travellers said goodbye to their friends.

Apart from nasty encounters with Jim Crow, as when they tried to eat in a restaurant in Yuma, Arizona, the first night out, and a mechanical breakdown in McAlester, Oklahoma, which cost $37 for bearings and piston work, the drive across the continent was swift and uneventful. On the way, Langston finally opened a package sent by Prentiss Taylor. Inside he found a copy of the special *Scottsboro Limited: Four Poems And A Play In Verse*. ''DE LUXE EDITION MOST BEAUTIFUL BOOK I HAVE EVER SEEN,'' Hughes wired from New Mexico.

Passing through Cleveland, Hughes left his mother all the money he could spare. Then he hurried on.

With only hours to spare, they reached New York on the night of June 13. Hughes bade farewell to a skeptical Van Vechten, who predicted that the intrepid band of blacks, swollen at the last moment to twenty-two, would walk back to Harlem from the Soviet Union. Prentiss Taylor was astonished to be left with a huge pile of unsold *Scottsboro Limited,* the cheap edition having reached California too late for Hughes to sell many.

Near midnight on June 14, accompanied to the wharf by his ''Aunt'' Toy Harper and staggering under the weight of his bags, his typewriter, a victrola, and a box of blues and jazz records, Hughes was the last passenger to climb the gangplank of the *Europa* before it slipped out to sea. But he was elated to be sailing to Russia, the motherland of radical socialism. ''YOU HOLD THAT BOAT,'' he had implored Louise Thompson in a telegram from darkest Yuma, Arizona, ''CAUSE ITS AN ARK TO ME.''

10

GOOD MORNING, REVOLUTION
1932 to 1933

> Good-morning, Revolution:
> You're the very best friend
> I ever had.
> We gonna pal around together from now on . . .
>
> "Good Morning Revolution," 1932

THE BAND of twenty-two travellers crossed the Atlantic in calm water, but with the weather somewhat cool and the sun shining through only on one day. Exhausted by his long ride from California, Hughes sank into almost two days of sleep. Then, refreshed and relaxed, dashing in light gray flannel slacks and striped sailor's jersey (always his favorite garment), and sporting a rose in his jacket lapel, he posed with the others for a group picture on deck. Looking on, also smiling, was a dapper figure from his past—Alain Locke, daintily down from first class, on his way to another summer in Europe at Mrs. Mason's expense, faithfully reporting to her every encounter with Hughes and Louise Thompson. Langston had returned Locke's cheerful greeting coldly. He declined even to offer his hand, and answered "in a mock jaunty tone," or so Locke complained to Godmother. Locke was ignored again when he brought his brilliant but aloof young colleague at Howard University, Ralph Bunche (grown "stuffy" and "a frightful bore," according to Louise Thompson) to meet Hughes, ostensibly at Bunche's request. Mrs. Mason was assured that the movie enterprise was headed for disaster.

Certainly the group was an odd mixture. Two were *Amsterdam News* journalists: sober little Henry Lee Moon, who had helped publicize the venture and had known Hughes from their teenage years in Cleveland, and the fun-loving Ted Poston. One, besides Hughes, was a writer: Dorothy West, a charming naïf from New England who showed promise in poetry and fiction. Curiously, only two were experienced actors. Plump, powerful Sylvia Garner had been a member of the supporting cast of the blackface *Scarlet Sister Mary,* and Wayland Rudd had performed in more than one Eugene O'Neill play and in *Porgy* (they were "the only really mature people in our group," Hughes would recall, "everyone else being well under thirty and some hardly out of their teens").

However, Juanita Lewis had sung with Hall Johnson's choir. Thurston McNairy Lewis, once a member of the American Communist Party, called himself an actor but could name few roles. The only member of the Communist Party of the United States was Alan McKenzie, a somewhat erratic salesman from Long Island who astonished everyone by bringing along a white woman, who was not his wife; McKenzie would ally himself with Hughes, Loren Miller, Louise Thompson, and Matt Crawford to comprise the radical core of the group. Homer Smith, whose degree in journalism from the University of Minnesota had earned him a clerk's job in the postal service, hoped to settle in Russia; also interested in politics was Laurence Alberga, an agricultural worker from Jamaica in the West Indies. But many in the group were neither artistic nor political. Constance White of Massachusetts and Leonard Hill from Howard University were social workers; Katherine Jenkins had been drafted from Louise Thompson's Convent Avenue home, where she roomed, and had brought along her fiancé, George Sample. Also in the group were Frank Montero of Howard University, Mollie Lewis from Teachers' College of Columbia University, and the beautiful Mildred Jones, lately an art student at Hampton Institute, from which Lloyd Patterson, a lanky youth, had come directly following his graduation there. Some members had approached the trip very seriously, "but most of the twenty-two simply thought they had found an exciting way to spend the summer," Hughes wrote later. "An exciting summer it turned out to be, too."

Differences surfaced quickly between the radicals and those who, bound for the Soviet Union, nevertheless resisted any identification with the left. The increasingly confident group leader, Louise Thompson ("a magnificent person," the playwright Paul Peters had called her in a recent letter), found her authority tested from the start. A proposal to send a cable of support to Mrs. Ada Wright, the Scottsboro mother then touring Europe, was defeated in a vote, and while all had solemnly agreed to represent the race creditably, some had already reneged. "We have some giddy people along," Thompson wrote home, referring no doubt to Thurston Lewis and Ted Poston, both of whom seemed unhinged by their unusual freedom and, in particular, the many white women on board. On the other hand, Henry Moon was strangely quiet: "Henry seems about the most subdued person on the boat, for what reason I don't know." Still, the enterprise had started reasonably well. In the afternoons, Thompson organized discussions of Soviet history and institutions and lessons in Russian; Hughes also practiced German, a smattering of which he had picked up in Mexico. During the rest of the time, the men and women ate, played cards, lounged in the well-stocked bars, or lolled on deck, anticipating their arrival in Russia and a quick start to the film.

On Wednesday, June 22, when the *Europa* dropped anchor in its home port of Bremerhaven, Germany, a representative of the Meschrabpom film company welcomed them, then escorted the group to a special train that sped them eastward to Berlin, which they reached that night. They were introduced to high-ranking members of the Berlin-based Workers' International Relief, parent organization of Meschrabpom Film, but a mix-up over their Soviet visas delayed their departure for the Baltic. Hughes found the slum-ridden, prostitute-infested city, wracked by years of economic depression, distressing in the scale of its misery; "the pathos and poverty of Berlin's low-priced market in bodies depressed me. As a seaman I had been in many ports . . . but I had not seen anywhere people so desperate as these walkers of the night streets in Berlin." Anxious to leave, he was almost left behind. Hughes was at a museum when the visas cleared and the group departed for the train station. His speeding taxi nearly killed a delivery boy, but Langston boarded the train just before it pulled out for Stettin on the Baltic. There, the group boarded the spotlessly clean and white Swedish vessel *Ariadne* for the journey to Helsinki. Sailing in superb weather, in the season of the "white nights," when at midnight Hughes could read by natural light and the sun rose only two hours later, he remembered an almost "fairy-tale journey on a boat filled with amiable people." On Friday, June 24, the *Ariadne* reached Finland. Helsinki, "a plainly pleasant town with music and dancing in the parks in the long white twilights," was crisply cool in spite of the bright sun as Hughes and his friends toured the city in charming *droskis,* or low, open carriages. That night, they boarded the train for the Soviet Union.

When the sun came up, Hughes was wide awake to greet it. Neither his first visit to Manhattan in 1921 nor to Africa in 1923 had surpassed in excitement this approach to the Soviet Union. Near the border, when the train slowed almost to a halt, Hughes stared through the window at young soldiers of the Republics, some white, some Asian, their caps surmounted by the celebrated red star. High above the railroad tracks a banner proclaimed: WORKERS OF THE WORLD UNITE. Memories of Central High in 1917, when many of his schoolmates, the children of immigrants, had thrilled to the news of the Russian Revolution, flooded Hughes; in fifteen years the fledgling Soviet Union had grown to be the greatest antagonist of capitalism and racism. In those same years Hughes himself had oscillated between radical fervor and compliant sentimentality, between worship of the masses and an attachment to luxurious patronage, between radical socialist discipline and an almost aimless anarchy of consciousness. Now he was in the motherland of world revolution, where political indecisiveness had no place. Just before the train braked to a stop, Langston leapt from the carriage. Stooping to the earth, he walked back to his friends bearing a handful of the sacred soil of Russia.

In Leningrad, a brass band blared out the "Internationale" as delegations from Meschrabpom Film and the city itself greeted the gallant band of artists come at great sacrifice from across the ocean to represent the downtrodden Negro workers of America. Although most of the twenty-two blacks could not answer to this description, they warmed to the role. At the October Hotel they feasted on caviar, roast chicken, fresh vegetables, coffee, and ice cream—a meal Hughes was frankly astonished to find in the notoriously famished Soviet Union. There was just enough time for a tour of Leningrad, the storied St. Petersburg of Pushkin and Turgenev, Dostoyevsky and Tolstoy. On Prospect Nevski they looked in at the censorious Museum of Religion and Atheism, hurried past Monferrand's mighty St. Isaac's Cathedral (celebrated subversively by Claude McKay, who had addressed the Comintern as a radical, in a stirring sonnet: "Bow down my soul in worship very low / And in the holy silences be lost"), crossed the many bridges and canals that gave the city its expansive Venetian air, admired the grey and pink granite embankments of the Neva River, and walked reverentially in the square fronting the Winter Palace, on which the cruiser *Aurora* had fired to begin the October revolution and Lenin's rise to power.

A day later, at ten o'clock on the morning of Sunday, June 26, Hughes and the group reached Nikolayevski Station in Moscow. Once again they were warmly greeted; a flock of incongruous Buicks and Lincolns sped the visitors to their hotel for a gala breakfast. Normally reserved for visiting upper-rank provincial officials, the Grand Hotel stood only a block from the Kremlin and the fabulous domes of St. Basil's on Red Square. Soon, Meschrabpom sent film contracts there to be signed. The terms were certainly generous; each player would receive for four months' work (or until October 26) four hundred roubles a month. As a writer, Hughes was offered even better pay, "in terms of Russian buying power . . . about a hundred times a week as much as I had ever made anywhere else." But he antagonized his hosts by refusing to sign any contract not in English. Hollywood contracts were notorious, he declared.

> "Don't mention Hollywood in the same breath with the film industry of the Workers' Socialist Soviet Republics," shouted the Meschrabpom executive with whom I was dealing. "That citadel of capitalist escapism—Hollywood! Bah!"
>
> "Don't yell at me," I said. "I'll go right back home to New York and never sign your contract."

The translation arrived, and the flow of roubles began in time for Hughes to celebrate the Fourth of July, and his recovery from an attack of influenza, under the magnificent pink marble columns in the dining room of one of Moscow's more opulent hotels, the Metropole. There, dining alone, he dispatched an entire roasted chicken in a meal that cost over thirty roubles.

At the Grand Hotel, Hughes, Loren Miller, and Matt Crawford shared a suite handsomely furnished in pre-Czarist style, with ancient beds, heavy drapes, and deep rugs. Not all the arrangements were perfect: toilet paper (as Lincoln Steffens had warned Hughes) was in short supply, laundry facilities uncertain, and only one meal a day was guaranteed by the film company. Langston took charge of the little household. Noël Sullivan had packed a hamper of marmalades, jams, and teas; with normally elusive items such as butter and eggs sent freely by Meschrabpom, in addition to the readily available sausages, canned fish, and brown bread, Hughes was able to offer hearty breakfasts. Other meals were more troublesome. Tomatoes were two roubles each, or four times the cost of a taxi to any point in Moscow; fresh fruit cost five roubles a kilo. At Torgsin, the store for foreigners trading in *valuta,* Hughes stocked up on canned peaches and bought an electric pot for cooking in his room.

These inconveniences seemed unimportant. "Of all the big cities in the world where I've been," Hughes would write, "the Muscovites seemed to me to be the politest of peoples to strangers." As visiting blacks, in fact, the twenty-two Americans were instant celebrities. Lines retreated impulsively before them, seats emptied as they approached. "Let the *Negrochanski tovarish* go forward!" was the cry. "Please! Visitor to the front!" Muscovites loved most foreigners, but the Scottsboro case gave American blacks an instant prestige; twice in the first few days the black comrades, or the leftists among them, took part in demonstrations in Moscow's Park of Culture and Rest (later Gorky Park). In a country gripped by economic austerity, the perquisites of stardom were freely available. Meschrabpom found a house for ten of the group to live together, with free cooking, cleaning, and transportation provided. The Americans were living "like royalty used to be entertained in Tsarist Russia, I imagine," Louise Thompson wrote home to her mother. "Everywhere we go we are treated as honored guests . . . and offered the best. It will really be difficult to scramble back into obscurity when we return to the old U.S.A. I suspect."

Unquestionably, Hughes was their star. When the English-language Moscow *Daily News* ran a welcoming story, his portrait appeared separately, as befitting the "famous Negro novelist and poet who will aid" in making the picture, called "Black and White" after a famous Mayakovsky poem, which Langston would soon translate. No one was more appreciative than Hughes of Soviet hospitality. Inconveniences bothered him not at all; he kept in mind how life in America was daily blotched by racism. "Quite truthfully, there was no toilet paper," he admitted of the Grand Hotel. "And no Jim Crow." In the central square of Moscow stood a great statue of the national poet, Alexander Pushkin, a black man. As the center of global revolution, Moscow in 1932 was without doubt "the greatest city in the world today." Stalin's first Five-Year Plan imposed burdens, to be sure, but the nation was on the move. Certain foods were expensive and in short supply, but there was no sign of malnutrition and hunger; dress was often shabby, but nobody seemed to envy the Americans their New York finery. Workers thronged the concert halls, theaters, and museums; also unlike their sisters in the United States, Soviet women worked alongside men in transportation, construction, and medicine. Everything in Moscow seemed to contrast with the disastrous Depression at home.

On the film "Black and White," however, little progress was visible. The director, Karl Junghans, an energetic, ruddy, young German, was inexperienced; his only picture had been *Strange Birds of Africa* (later called *Fleeting Shadows*), an anti-imperialist documentary set in Africa. Junghans spoke neither English nor Russian well, and needed interpreters to talk to Hughes and the actors. The scenario was slow in arriving; exasperatingly, it arrived finally in Russian. While Hughes waited for a translation he filled his days constructively; many members of the group, however, idled away the hours in the bar of the Metropole, where a little band played jazz wretchedly, and mysterious women (spies, not prostitutes, it seemed clear) invited attention. Thurston Lewis and Ted Poston, who fancied himself a dazzling ballroom dancer, pursued the women crudely; Lewis especially (so Thompson wrote home) "has disappointed everybody and acts like a perfect fool." Certain women in the group had their own problems. A lesbian affair would lead to heartache; someone sipped potassium formaldehyde in a theatrical attempt at suicide that embarrassed the group after her screams echoed through the Grand Hotel. And one question was asked of the group too frequently. Were they all really Negro? Some could pass for white. "We have had to argue at great lengths," Louise wrote home, "to tell them we are all Negroes, and to try to explain just what being a Negro means in the United States."

Hughes, for one, tried to keep busy. With his novel, *Not Without Laughter,* nearing publication in Russian by Goslitizdat of Moscow, he met its translator, Vera Stanevic, and also Julian Annisimov, who was working on some of his poems. At *Izvestia,* he talked about writing for the newspaper with the distinguished editor Karl Radek, a debonair, goateed person who reminded him of Lincoln Steffens. In addition, among the radical American colony in Moscow (so large that baseball was played regularly in the Park of Culture and Rest) Hughes became friends with Walt Carmon, formerly of *New Masses,* now in charge of the English language edition of *International Literature*. He also met the journalist Anna Louise Strong and the talented dancer John Bovingdon, a sinuous fellow who affected Greek tunics instead of trousers and improbably had been picked by the director of "Black and White" to act as a tough labor organizer (" 'Vot ist matter?' Junghans quizzed Hughes, Bovingdon "nich look like American worker?' "). And passing regularly through Moscow were important visitors, including the black educator John Hope of Atlanta University, Olin Downes, the music critic of the *New York Times,* and the drama critic of the Cleveland *Plain Dealer,* who reported meeting an obviously much admired Langston Hughes at a Moscow party.

Finally the translation of the scenario arrived from Meschrabpom. But Hughes did not read very far before he set it down in disbelief. "At first I was astonished at what I read. Then I laughed until I cried. And I wasn't crying really because the script was in places so mistaken and so funny. I was crying because the writer meant well, but knew so little about his subject and the result was a pathetic hodgepodge of good intentions and faulty facts." The scenario depicted the struggle of black steelworkers in Alabama against racism and class antagonism. Threatened by white Southerners, rich Birmingham blacks use their private radio station to summon white workers from the North. The whites rush to the city to defend their black "brothers." Both in broad outline and detail (a rich young white Southerner, for example, dances with a black maid in front of his friends at a party), the text betrayed the fact that its author knew little about the United States. Lovett Forte-Whiteman, a black American consultant who had met Hughes and the others in Leningrad, had either not been consulted or was an egregious Uncle Tom. "It would have looked wonderful on the screen, so well do the Russians handle crowds in films," Hughes would declare. "But it just couldn't be true. It was not even plausible fantasy—being both ahead of and far behind the times."

Meschrabpom met Hughes's objections with dour defensiveness. Since the script was ideologically correct, were these tiny details so important? Perhaps he would rewrite the script. Hughes refused: he knew nothing about steel mills. Karl Junghans, who desperately wanted to direct a full-length film, undertook to rewrite it himself. To hold the actors at bay, he set up screen and singing tests. On the screen, however, most of the Negroes looked obstinately refined;

worse, Junghans "nearly became a nervous wreck after that first singing rehearsal," Hughes recalled, "—because almost none of the Negroes in our group could sing." With only a hazy notion of spirituals, they also manifested a near-complete ignorance of black work-songs.

The future of the film became a subject of speculation. In the Metropole bar, Eugene Lyons, the correspondent of the United Press, and Ralph Barnes of the New York *Herald-Tribune,* took a keen interest in the group's troubles. One day in the bar, Colonel Hugh Cooper, the American chief engineer of the Dniepostroi Dam, on which Stalin was depending for rapid Soviet industrialization, reportedly told Alan McKenzie that he strongly disapproved of the film. With the United States apparently on the brink of finally recognizing the U.S.S.R. (a move deeply desired by the Soviet Union), Cooper's disapproval was ominous. On July 25, perhaps to keep the group occupied, Hughes and Louise Thompson organized a concert before a packed house at the Library of Foreign Literature. Loren Miller spoke and Langston read poems, but the laurels unquestionably went to the singers Wayland Rudd and Sylvia Garner, who also sang trios with Juanita Lewis and the fearless Louise Thompson. Telling of the steady evolution of his poetry toward the revolutionary viewpoint, as the *Daily News* put it, Hughes recited his "Waldorf-Astoria" parody and Tom Mooney mass-chant.

But dissension was splitting the ranks of the twenty-two. Because of her continued defense of Meschrabpom's intentions, Thompson would earn the nicknames "Madame Moscow" and "Glupie," which her opponents told her was Russian for "stupid." Louise, Langston, Loren, and Alan McKenzie were linked as the " 'ell-raisers"; they were seen as too trusting of the Russians, too radical for their own good. Leading the opposition against them were Henry Lee Moon, so unusually quiet on the *Europa,* Ted Poston, Thurston Lewis, and the sometimes intemperate Jamaican, Laurence Alberga. In addition, McKenzie repeatedly proved to be a poor, confused ally in spite of his membership in the party. The situation deteriorated further when Junghans's revised script was curtly rejected by Meschrabpom. Then Hughes learned that no version of the script had ever been accepted by the company; no scenery, sets, costumes, or other preparation yet had been attempted, although the brief Moscow summer was running its course. Decades later, summing up his reaction to the group's journeying to Moscow without contracts, and the film company's faith in accepting them, Hughes would be "amazed at the naïveté shown on both sides. But I must say there was never any temporizing regarding work or money," even though "our expedition ended in an international scandal and front-page headlines around the world."

Meschrabpom rescheduled the start of filming for August 15, the last date possible because of the Russian weather, then announced that the group was going to the Black Sea at the invitation of the Theatrical Trade Unions. Certain crucial outdoor summer scenes would be shot there; Odessa had provided the setting for parts of several successful movies, including Sergei Eisenstein's famous *Potemkin.* After irritating delays, which were now freely attributed to Meschrabpom's inefficiency, all but two members (one of whom was Henry

Lee Moon) left Moscow by train on August 3 for the south. Strangely, neither the director Junghans nor any high company official went with them. And although the days passed pleasantly enough, no work was done on "Black and White." Instead, the group sunbathed on the beaches, with Poston and Thurston Lewis willfully defying local custom by stripping off their swim suits and playing "naked as birds and as frolicksome as Virginia hounds" on the sand. On August 9, Hughes and the others left on the cruise ship *Abhazia* for a tour of selected Black Sea ports—the Crimean towns of Sebastopol and Yalta, and on the Georgian coast Gagry and Sukhumi, reaching almost as far as the Turkish border to the south. By this time Thurston Lewis was uncontrollable, raining choice Afro-American profanities on their Soviet political leader and on the interpreter, Lydia Myrtseva. Louise Thompson wearily wrote home to her mother about Lewis's "disgraceful conduct" during the entire Soviet stay. His ally, Ted Poston, had also been a "malicious trouble maker"; he was "thoroughly irresponsible and Thurston really incorrigible." Especially concerning white women, they were "like two puppies, chained for a while and then let loose." Henry Moon, who had stayed behind in Moscow, had "not acted in the disgraceful manner" of the two, but had repeatedly backed them.

Returning to Odessa, the group was surprised to find Moon waiting for them. "Comrades!" he cried. "We've been screwed!" He waved at them a copy of the Paris *Herald-Tribune,* dated August 12. One headline was telling: "SOVIET CALLS OFF FILM ON U.S. NEGROES; FEAR OF AMERICAN REACTION CAUSE." The paper reported a "sensation" among Americans in Moscow caused by "the sudden collapse" of the project. The future of the band of twenty-two was uncertain, including the fate of Langston Hughes, who was scheduled "to play the principal part in the film." A United Press News Letter, dated August 9, cited as the cause of the cancellation "the fears that it [the film] might be considered an intrusion into the internal affairs of the United States." Behind the abandonment was the strong desire of the Soviets for diplomatic recognition by the United States.

"No Negroes went bathing on the Odessa beaches that day," Hughes later wrote. "Instead, hell broke loose. Hysterics took place." The group quieted long enough to hear Boris Babitsky, a high Meschrabpom official, assure them that, no, the film was not cancelled, only postponed, and only because of the quality of the script. Obviously this particular group would not take part in the revived project. They would be offered a tour of the country, then could either return to the United States or stay as immigrants in the Soviet Union. The rest of their contracts would be honored. Babitsky's news was too much for some in the group. A few sobbed despondently; others, seeing their movie careers vanish in the Odessa haze, raged at the injustice of it all. Thurston Lewis, Poston, Moon, and Laurence Alberga denounced Meschrabpom, the Soviet leadership, and even Stalin himself for selling out the black race. On the defensive, Louise Thompson argued that Meschrabpom was not a Soviet organization but an agency of an international body. All agreed on one point: Odessa had lost its charm. Hurrying back to Moscow, they reached the capital on August 20 to find that their rooms at the Grand Hotel had been reassigned. The film company finally placed them at the Mininskya, adequate but inferior, in spite of its location across the street from the main gates of the Kremlin.

Three days of bitter arguments followed, in sessions lasting up to five or six hours. By the third day, Hughes, Thompson, and their allies had swung the majority around to their position. They would criticize Meschrabpom for inefficiency, pointing out the likely unfortunate effect of the postponement on the party in the United States, especially on the electoral chances of James W. Ford, the black vice-presidential candidate and the main U.S. communist behind the film. At a meeting with company officials on August 22, Loren Miller presented these arguments. He had barely finished before Ted Poston jumped up to read a prepared statement. "Rejecting as unsound, insufficient, and insulting to our intelligence the reasons offered" by the film company, Poston, Lewis, Moon, and Alberga charged that the visiting blacks had been sacrificed by Meschrabpom in an act of "Right Opportunism" that hindered revolution not only among Afro-Americans "but all the darker exploited people of the world." Poston and his friends further charged the company, "and any other organization which may support it in this stand," with "base betrayal of the Negro workers of America and the International proletariat" and "with sabotage against the Revolution."

These extraordinary accusations were repeated before the prestigious Comintern, which acceded to a request for a meeting. In a gloomy room fronted by a long table behind which sat several grim Bolsheviks, Hughes joined Miller in presenting the case for the continuation of the project. Meschrabpom had shown "gross administrative inefficiency in the discharge of its duties" almost to the point of "direct sabotage" of "Black and White." But the abandonment would badly damage communist prestige in the United States and elsewhere. The scenario was now acceptable; why not proceed? In their turn, Poston and Moon gave the minority position, attacking Meschrabpom and hinting at collusion with the Kremlin itself in anti-revolutionary activity (to Hughes and Loren Miller, their performance was "a shameful, foolish experience" that showed "unbelievable ignorance" of communism). In the end, the Cominterm ruled that Meschrabpom indeed had been inefficient and that the picture should be made, but after a year's postponement. Moon, Poston, Alberga, and Lewis, joined by George Sample and Katherine Jenkins, next sent a letter to Stalin himself, which resulted only in a statement by Tass, the Soviet news agency, reaffirming the stated reasons for the failure to produce the film. At their last group meeting on the matter, on August 23, bitterness boiled over. Hughes, Thompson, Miller, Crawford, and Alan McKenzie were denounced as stooges of communism and traitors to the group. Turning on Hughes, whose success in Russia evidently had rankled him, Ted Poston sneered that Langston had come to Russia because he was all washed up at home: Langston Hughes was an opportunistic "son-of-a-bitch" and a communist Uncle Tom. Leaping to his feet, Hughes denounced Poston as yet another "son-of-a-bitch." On this inspired note the colloquy on "Black and White" ended.

Meanwhile, the American press, white and black, reported colorful versions of the group's problems. The twenty-two were reportedly stranded, starving, and unpaid. "NEGROES ADRIFT IN 'UNCLE TOM'S' RUSSIAN CABIN," the New York *Herald-Tribune* jeered. In the Metropole bar, the American correspondents laughed when Hughes, amazed at this duplicity, confronted them. "It was the first time I realized that a big-name correspondent would deliberately lie to conform to an editorial policy." Moon and Poston had also sent hostile reports to the United States, which they would confirm to the New York *Amsterdam News* following their return home: "The two assailed Joseph Stalin's international policy, and said that the Soviet dictator had subordinated revolutionary activity to court favor with the capitalist countries." When Thompson cabled a long statement, the *Crisis,* for one, took its time to print it; on September 3, the *Pittsburgh Courier* published what Loren Miller hotly attacked as "a strange concoction of lies, half-truths and abysmal ignorance."

With "Black and White" dead, Hughes turned his attention to going on the tour that had been offered as compensation. No pleasure trip, the journey would afford a study of Soviet treatment of its racial minorities in Central Asia, which was divided into the autonomous republics of Turkmenistan, Uzbekistan, and Tadzhikistan. At first, few in the group expressed interest; but the mention of a special car and magical names such as Tashkent and Samarkand filled the carriage at the last moment. Tension abated; the group began to joke about its grand "son-of-a-bitch" meetings. Once the five dissidents, including Poston, Moon, and Thurston Lewis (the "Black White Guard," Matt Crawford dubbed them) had left for the United States, hostility died away. But the parting was ugly. As it had promised, Meschrabpom refunded the cost of passage to the Soviet Union. Lewis, whose ticket to Russia ironically had been subsidized by a number of people, including Hughes and Sylvia Garner, almost slipped away with their refund. A heated argument broke out; it ended only when Garner whipped out a knife on the stunned Soviet paymaster, who at once saw the wisdom of her position. The remaining members celebrated the departure with a victrola party in Louise Thompson's room far into the night. It was sad, Louise wrote her mother, that "a few of us had to go act a nigger and try to give a black eye" to the only country that treats Negroes as humans.

"O, Movies. Temperaments," Hughes punned wearily. To him and other radicals in the group, "Black and White" had shown less the incompetence of certain film officials than the perversity of human nature as displayed by the dissidents—although the radicals themselves must have suspected that the Soviet officials had been swayed by their desire for U.S. recognition. Given the needs of the young nation, a Soviet change of heart for this reason would have been defensible. "The fact of the case," Matt Crawford wrote home to his wife, "is that these Negroes have never lived as well in all their lives. Just imagine being here for almost four months without doing a bit of actual work; being paid regularly—some of them drawing more than their salaries. Living in Moscow's best hotels and traveling all over the Soviet Union with all expenses paid. If that isn't a freebee I want to see one." Crawford had been no radical when the trip started, but the Russian experience had already changed him. "I realize now how much being in Russia has affected me," he confided to his wife.

"Unconsciously I have lost that depressing subconsciousness of being a Negro. The ever-present thought that my dark skin must circumscribe my activities at all times. I was a bit surprised how absolutely normal my moving about the Russian people has become. All of the antagonism which I have always felt among ofeys [whites] at home has left me. I can understand why the masses of Russian people are willing to endure any sort of hardship during this transitional period from capitalism and slavery to socialism and freedom."

Hughes felt the freedom of Russia at least as intensely; for the first time in his life he could live by his writing, and handsomely so. Deciding to stay awhile in the Soviet Union, he signalled his excitement to the world with a poem that bravely parodied Carl Sandburg's "Good Morning, America."

> Good-morning, Revolution:
>   You're the very best friend
>   I ever had.
> We gonna pal around together from now on.
>
> Say, listen, Revolution:
> You know, the boss where I used to work,
> The guy that gimme the air to cut down expenses,
> He wrote a long letter to the papers about you:
> Said you was a trouble maker, a alien-enemy,
> In other words a son-of-a-bitch.
> He called up the police
> And told 'em to watch out for a guy
> Named Revolution. . . .

Inexplicably, he sent this poem home to the bourgeois *Saturday Evening Post,* which promptly rejected it (*New Masses* took it). Hughes then testily lashed at the journal in a poem that would haunt him for the rest of his life. Having said good morning to revolution, he next bade goodbye to Jesus Christ. Pent up in him since that evening as a child when Jesus had not come to him, and augmented by his great betrayal by Charlotte Mason, Hughes's resentment overflowed now in "Goodbye Christ":

> Listen, Christ,
> You did alright in your day, I reckon—
> But that day's gone now.
> They ghosted you up a swell story, too,
> Called it Bible—
> But it's dead now.
> The popes and the preachers've
> Made too much money from it.
> They've sold you to too many
>
> Kings, generals, robbers, and killers—
> Even to the Tzar and the Cossacks,
> Even to Rockefeller's Church,

Even to THE SATURDAY EVENING POST.
You ain't no good no more.
They've pawned you
Till you've done wore out.

Goodbye,
Christ Jesus Lord God Jehova,
Beat it on away from here now.
Make way for a new guy with no religion at all—
A real guy named
Marx Communist Lenin Peasant Stalin Worker ME—

I said, ME!

Go ahead on now,
You're getting in the way of things, Lord.
And please take Saint Gandhi with you when you go,
And Saint Pope Pius,
And Saint Aimee McPherson,
And big black Saint Becton
Of the Consecrated Dime.
And step on the gas, Christ!
Move!
Don't be so slow about movin'!
The world is mine from now on—
And nobody's gonna sell ME
To a king, or a general,
Or a millionaire.

Otto Huiswood of Dutch Guiana, a leading black communist well known in the U.S. but then in Moscow, where he had befriended the group, secured a copy of the poem. He was a contributing editor to *The Negro Worker,* which the radical George Padmore of Trinidad edited for the Hamburg-based International Trade Union Committee of Negro Workers. In July, the journal had reprinted Hughes's revolutionary "Open Letter to the South" from *New Masses.* Apparently without Hughes's knowledge, Huiswood sent "Goodbye Christ" to the magazine, along with "The Same" ("It is the same everywhere for me . . . / Black: / Exploited, beaten, and robbed"), which appeared in the September-October issue.

. . . Better that my blood makes one with the blood
Of all the struggling workers in the world—
Till every land is free of
    Dollar robbers
    Pound robbers
    Franc robbers
    Peseta robbers
    Lire robbers
    Life robbers—

Until the Red Armies of the International Proletariat,
Their faces black, white, olive, yellow, brown,
Unite to raise the blood Red Flag that
Never will come down!

In the next issue, November-December, "Goodbye Christ" appeared. Away in Central Asia by then, Hughes probably did not see either number of the magazine until the winter.

While he waited with his friends for the tour to commence, he also looked forward now to leaving them. Karl Radek at *Izvestia* had retained him to do a series of six to ten articles on his impressions of Soviet Asian life, for which he would be paid two thousand roubles. His plan was to leave the group at Ashkabad, the capital of Turkmenistan. Radek's offer confirmed Hughes's increased prestige in Moscow. His *Scottsboro Limited* and the Tom Mooney mass-chant, recently revised, were being translated. He was further preparing two volumes of poetry for translation. The well-known critic Lydia Filatova was planning a long essay on the evolution of his radical sensibility as a writer for *International Literature*. And on the eve of his departure, the first copies of *Smech Skvoz Slezy* (his novel, *Not Without Laughter*) reached Hughes at the Mininskya. "He is certainly doing very well here," Louise Thompson wrote her mother, "and in fact everything he wants seems to be available to him."

He was also doing well at home, where *The Dream Keeper,* his selection of poems for younger readers, had appeared from Knopf to generally good reviews, followed a few weeks later by the unanimously praised *Popo and Fifina,* the children's book he wrote with Arna Bontemps, with handsome sketches by the black artist E. Simms Campbell. The *New York Times Book Review* called it "a model of its kind" that tempted one to think that all children's books should be written by poets.

On the afternoon of September 22 a special railroad car left Moscow heading south on the five-day journey to Tashkent. Only eleven of the original twenty-two were on board: Hughes, Thompson, Miller, Crawford, and Alan McKenzie; Katherine Jenkins and her fiancé George Sample; and Connie White, Mollie Lewis, Juanita Lewis, and Mildred Jones. With them was a small supporting group, including their interpreter Lydia Myrtseva, a political guide, and Otto Huiswood of *The Negro Worker*. "To an American Negro living in the United States," Hughes reported to his Russian readers, "the word *South* has an unpleasant sound, an overtone of horror and of fear." Travel by train in the United States almost invariably meant humiliation, especially in the Jim Crow cars of the South. In Central Asia, racism and segregation had once prevailed; now, under Soviet rule, Jim Crow was dead. The other passengers certainly made the Americans feel welcome. By the third day, "mandolins and guitars, balalaikas and accordions had set everyone to dancing in the aisles, and friendships had become more intimate across political and national lines." Hughes could play nothing, but he had his victrola and his records of Louis Armstrong and Ethel Waters, which were a hit as the train headed toward Central Asia.

A Leningrad librarian was returning to Tashkent to continue work on building collections in the now Latinized alphabet of various regional languages. A Red Army officer told him about the cosmopolitan military school at Tashkent. A Russian merchant spoke about industrial growth in the once backward Asian countries. Two Komsomol, or Communist Youth League, poets hoped to stimulate creative writing among young Asiatics. A brown young man dressed in tan trousers and a casual gray coat (an Asiatic factory worker, Hughes guessed) revealed that he was Sodok Kurabanov, twenty-eight years old: "He is the mayor of Bokhara, the Chairman of the City Soviet! I make a note in the back of my mind: 'In the Soviet Union dark men are also the mayors of cities'." Kurabanov's life exemplified the prospects of Soviet Asians. The son of peasants, he had known no school until he was fifteen. At eighteen, he went to Tashkent to study languages in the new state facility there; after joining the party, he had risen steadily in an Uzbekistan transformed by the revolution. Kurabanov was so taken with pointing out Soviet progress to the Americans that he decided to travel with them in Tashkent and Samarkand, and on to Bokhara.

Near sundown on their second day out, they crossed the Volga. When the train passed into the great desert steppes and the Kazakstan Republic, the ancient and modern world, East and West, began to mix. Camels loped by, overtaken by motorcars. Sometimes modern tractors, sometimes camels, ploughed the land. The colorful desert tents of the Kazaks were pitched next to red-brick cooperative structures designed to wean the nomads away from their ancestral roving. On the fourth day, Hughes glimpsed for the first time the austere, silver gleam of the Aral Sea in the sunshine across the Kirghiz steppes. Later, at sunset, on the platform of the little railway station in Kazalinsk, he took part in a ceremony unimaginable in the United States. That day, September 25, the Soviet Union celebrated the fortieth anniversary of the start of Maxim Gorky's career. At sixty-four, Gorky was revered as the father of Soviet literature, the champion of the working classes, a friend to Tolstoy and Chekhov, an early and constant supporter of the revolution. First, one of the Komsomol poets, a Russian Jew, spoke; then a Polish porter on behalf of the train crew; then someone from the station itself; and last of all, Langston offered a few tributary words, his speech translated for the attentive crowd into Russian and Kazak. "We sent a telegram to Comrade Gorky from the passengers of the train, and another from our Negro group. And as the whistle blew, we climbed back into our coaches, and the engine steamed on through the desert pulling the long train deeper into Asia. It was sunset, and there was a great vastness of sky over sand before the first stars came."

The next day they entered an oasis where cotton flowered and green trees drooped with fruit. This was the land of "White Gold," the counterpart to the "Black Belt" of the American South where cotton was king and where, Hughes told the Russians, "the colour line is hard and fast, Jim Crow rules, and I am treated like a dog." Just after seven in the evening of September 26, they reached Tashkent, center of the Asian Republics, where the group was met by a delegation led by John Golden, a prominent member of a colony of American Negroes from Tuskegee and other schools helping to modernize the Soviet cotton industry. Also present, splendidly incongruous in a tuxedo, was a young Howard University engineer, Bernard Powers.

With the arrival in Tashkent began long days of banquets and speeches that tired the more fickle members of the group but left the radicals elated. "If our [tour] continues to be as interesting and as pleasant as it has been here in Tashkent," Matt Crawford wrote to his wife, "there shall certainly be plenty to tell you when I get back." Central Asia was crucial to believing in Russia: "There is an exact parallel between the condition that these people were under during the Czarist regime and the position of Negroes in the States now." The first full day included a visit to an agricultural machinery factory, interspersed with long speeches rendered consecutively in Russian, Uzbek, and English, and salted freely with statistics. The evening was spent at the Uzbek Opera, another dividend of the revolution, since theater was forbidden by the old religion. The following day, the Women's Club of Tashkent welcomed them, with more speeches in triplicate. Hughes the journalist peered into nurseries and classes in literacy that trained women only recently released from virtual slavery; in the mud huts of the Old City he could still see the squalor that the revolution hoped to erase. The warm afternoon found them at the House of the Trade Unions, where the group posed for a picture outdoors amidst lush foliage under a banner that proclaimed in Russian: WELCOME TO OUR NEGRO COMRADES. The president of the half-million-member All-Uzbekistan Trade Unions spoke of the changes since 1917. Where once there had been an almost total absence of schools and hospitals, Uzbekistan now boasted compulsory education, open universities, and a program of industrialization. Hughes took notes for his *Izvestia* articles.

The next morning, at the Central Council of Trade Unions of Central Asia, more long speeches spelled out the progress of the workers, and the connective role of trade unions between government, the party, and the masses. In the afternoon, Langston and the others were formally greeted by the President of the Central Committee of the Soviets of Uzbekistan, Akhun Babaef. He rehearsed the history of Turkestan before 1917, dwelling on the role of Russian, British, and other imperialisms in subjugating the people. He was happy, he said, to be welcoming Americans on an equal footing. Then Jahan Obidova, vice-president of the committee, rose to speak. Only thirty-two, she had been sold at eleven to a kulak in urgent need of a fourth wife. With the revolution, Obidova ran away from her husband. Protected by Russians, she entered school for the first time. In 1923, she made herself a target of vengeance by becoming only the second woman in all Middle Asia to unveil herself. Here Hughes took his turn in replying on behalf of the visitors. That night, their last in Tashkent, they attended a gala dinner that included recitations by Uzbek poets. Aided by translators, Langston read his own work to the gathering.

Before leaving Tashkent, Hughes agreed to have a book of his poems published in Uzbek by the state press, for which he would receive a substantial advance. Then, at six p.m. on September 30, they left Tashkent by train.

The following morning, banners and a brass band welcomed them to the historic city of Samarkand. They spent the morning touring its monuments, incomparable in all of Central Asia for its medieval Oriental architecture—the noble azure mosaics surmounting the tomb of Tamerlane, the turquoise cupola of the Bibi Khanym Mosque, with its majolica tiles and gold tracery, the ruins of the hill-top observatory built by Tamerlane's scholarly grandson, Ulugh-Beg, and Registan Square, the center of medieval Samarkand and the crossroads of ancient trading routes in what was once called the Eden of the Orient. The major meeting here was at the Silk Factory, built four years before to encourage the liberation of women, who comprised over three-quarters of the twelve hundred workers. The visit included a concert of Uzbek, Jewish, Tadjik, and other music. Then the Americans visited the Uzbekistan Pedagogical College, with its 3,000 students from over a dozen ethnic groups. Before 1917, there was no college in Middle Asia; now, thirty-four flourished in Uzbekistan alone. Of great interest to Hughes was the liberation of the Jews, who previously could not own land, attend school, live within the city itself, or enter public restaurants.

On October 3, accompanied still by Sodok Kurabanov, mayor of Bokhara, the party reached his city. Matt Crawford thought Bokhara "the most typically Asiatic of any of the cities we have visited" and "the most interesting in many respects." The capital of the emirate until the last emir was routed in 1920, Bokhara had been the most reactionary of Middle Asian cities, and the greatest center of Moslem faith. In the narrow streets the people still wore the dress of the past—the women under *boerks* (tall, cylindrical, multicolored hats adorned with silk, coins, and amulets), their faces obscured by many veils, their arms flashing with jewels and bracelets. Donkeys and camels ambled down streets so narrow that two cars could not pass each other. At the Museum of the Revolution, where Hughes saw whips and chains and other tools of state repression, the old man in charge showed him his own scars from torture. In days gone by, concubinage and illiteracy kept women enslaved; minorities could not enter the army; pogroms were organized against Jews, who could ride donkeys but not horses, and wear only strings around their waist, not belts. Near the palace of the Emir Alum-Khan, built only eighteen years before, they saw the tower of the mosque from which his enemies were sometimes flung to their death. Now the palace housed students; the emir's summer palace and estate outside the city were to provide a rest home for workers.

One afternoon, Langston and the others visited a twelve-hundred-hectare *sovkhoz,* or soviet farm, then moved on to the center of a great collective of one hundred and sixty farms, an excellent example of the new thrust in agriculture. In the warm, dry weather, the sirocco of speeches began to make even the radicals drowse. But when they sat down on rugs to enjoy an Uzbek feast, with bowls of thick cream, fresh bread and farm butter, melons and other fruit, they were glad they had come. An evening concert at the Bokhara Public Hall was followed the next morning by lectures on history and archeology at the Communist Party headquarters, then a tour of a silk factory, a small Jewish *kolkhoz* growing cotton and grapes, and a caracul factory, where sheepskin and fur garments were prepared.

After four days days in Bokhara, as they prepared to leave Uzbekistan, the Americans drew up a message, which Hughes signed first (but almost certainly did not write), "to the Workers and Peasants" of the Republic:

> We have been able to see the practical application of the Leninist national policy successfully converting Middle Asia from a czarist colony of oppressed peoples and an undeveloped country to an industrialized country under working class rule. The emancipation of women, the complete elimination of national antagonisms, the stimulation of national proletarian culture, the proletarianization and collectivization of workers and poor peasants we have found realized in Uzbekistan since the Revolution. . . . We shall carry the warm proletarian greetings of the various workers and peasants of Uzbekistan to the black and white workers and peasants of the United States. . . .
>
> Long live the workers and peasants of Uzbekistan!
> Long live the international solidarity of workers of all nationalities!
> Long live the Soviet Union!

Unquestionably, Hughes and his fellow Americans had been astonished by the Russian treatment of darker-skinned minorities in Central Asia. "The past ten days," Louise Thompson told her mother, "have been the greatest of my whole life."

Nevertheless, because they were exhausted and running behind schedule, the group voted ten to one to cancel the tour of Ashkabad, where Hughes was to leave them. Langston was furious; he would be dropped off and left on his own in the most desolate of the Soviet Central Asian republics. "I seldom get angry. When I do, I usually don't say a word." He smelled a hidden motive among the Soviets in the party, who also wanted to bypass Ashkabad—perhaps they disapproved of his mission. When the train stopped in the middle of a lonely Turkmenistan plain, beside a small, almost deserted station, "I jumped off the train without handshakes or farewells. A couple of fellows were decent enough to toss my bags down. . . . Dark heads from open windows cried, 'Goodbye! So long!' But I did not say a word. . . . As the last coach rattled by, to myself I said, 'To hell with all of you! I hope I never see any of you again'."

Fortunately, the lone station attendant spoke Russian. He secured Hughes a room at the *dom sovietov,* or official guest house, in Ashkabad. Langston's rage evaporated. As much as he loved company, solitude was his oldest friend. But he was not alone for long. A bright young Red Army captain, Chinese-Negro in appearance, with dark skin and Oriental eyes, took to him. Shaarieh Kikiloff, the gentle, little, smiling president of the Turkoman Writers Federation, came by to pay his respects. Although what Kikiloff actually wrote was never clear, he spoke Turkoman and Russian, and thus could help Hughes. A Ukrainian, Kolya Shagurin, who had been a vagabond sailor, and sported a salaciously mobile nude on his biceps, was at the *dom,* as well as a Ukrainian film director and a Red Army colonel from Chinese East Turkestan, who at fifteen had seen his parents beheaded by the *Basmachi,* or marauding counter-revolutionaries.

Hampered by his lack of languages, Langston wrote very little. In fact, boredom had almost set in when one day there was a knock on his door. Outside stood a young white man in European garb, who later described hearing through the closed door the voice of Sophie Tucker singing "My Yiddishe Momma" on a victrola. (He was dead wrong, Hughes insisted.) The picture of a young European intellectual, intense, with sharp features and oily dark hair, Arthur Koestler delighted Hughes by introducing himself in very good English. Langston Hughes in Ashkabad! "It was difficult not to say 'Dr. Livingstone, I presume'," Koestler would remember; he had read and liked some of Hughes's poems. Only twenty-eight, brilliant and restless, he was already well known as a reporter, having journeyed to the North Pole with the Graf Zeppelin. Koestler, born in Budapest of a Jewish father, had traveled to Palestine as a youth before emigrating to Germany. There he had become a radical; a year before meeting Hughes, he had joined the Communist Party. Now he was in Turkestan partly because he had long heard that the ancestral origin of the Hungarian people was in the old Turanian Basin, now West Turkestan. In turn, Hughes explained his own reasons for being in Ashkabad. Koestler at once found Hughes attractive; he would remember Langston as moving "with the graceful ease of his race." When Koestler discovered that the poet had made little progress on his *Izvestia* articles, he offered to help. Together they would dig up facts and stories and share their findings. "A writer must write," the tireless Koestler reminded Hughes.

With Kikiloff ("a timid little mouse of a man with a wizened Tartar face," Koestler later recalled) and Kolya Shagurin, Hughes and Koestler formed an "International Proletarian Writers Brigade." They attended conferences, visited trade union sessions, and toured factories and schools. Inspired by Koestler's determined note-taking, Langston bought half a dozen notebooks of cheap yellow paper, covered in oilcloth. Kikiloff translated into Russian, which Koestler rendered into English for Hughes's benefit. On his own, however, Hughes stumbled onto an event that would prove very important to Koestler. A trial was taking place at the City Soviet. A former chairman of the town council, Atta Kurdov, once a member of the Central Executive Committee of the party, had been charged with various crimes involving his administration. What Koestler saw and heard at the trial disturbed him; he "was so fascinated by this sleepy-eyed trial," Hughes later wrote, "in which everyone looked half hypnotized, that he stayed until court closed." In an atmosphere of listless resignation, party self-righteousness appeared to meet cultural customs ingrained over the centuries in Central Asia. To Koestler, the human was being ground down in the name of ideology and power; the trial was a deadly charade in which the accused, doubtless guilty of certain crimes, were scapegoats to cover up the failure of collectivization programs, as well as to destroy individualism and ethnic chauvinism.

Koestler was so upset that he could not continue to attend the proceedings. Seeking a similar resonance in Hughes, he found nothing but a teasing acquiescence to the aims of power. "Atta Kurdov looks guilty to me," Langston would candidly remember giggling to Koestler, "of what I don't know, but he just *looks* like a rogue." Such an insipid reaction deepened Koestler's gloom. Here, at these trials, Hughes would mark the start of Koestler's famous turn away from communism towards his repudiation of the party in *Darkness at Noon.* Here, too, Langston might have marked for himself the beginning of a certain fatalism, a brittleness of spirit, that would replace the shining optimism of his past life. The fact that the Soviet Union had outlawed racism certainly made him eager to forgive its mistakes, but his refusal to look where Koestler pointed, even to refute him, probably had deeper roots. "I knew mine was not proper reasoning either and had nothing to do with due process of law," Hughes would concede. "But when I saw that it upset him, I repeated that night just for fun, 'Well, anyhow Atta Kurdov does look like a rascal'."

Hughes's increasing fatalism was almost inevitable following his break with Mrs. Mason, and yet many observers still saw him as sometimes magically childlike; just before Langston's reading tour of the South, for example, the writer Melvin Tolson had watched as Hughes listened raptly to a Harlem matron tell him how to dress for his audiences: "The Poet listens like a child whose mother is telling him where to find the cookies." If Koestler did not see Langston as a child, he later judged him to have been "a poet with a purely humanitarian approach to politics—in fact, an innocent abroad." Hughes's pro-Soviet zeal, however, had not derived simply from idealism. Behind "Good Morning Revolution" and "Goodbye Christ" was also the pain of his headlong fall from grace with Mrs. Mason, an embittering experience that consolidated in Langston his tendency towards withdrawal from emotional involvement. He had moved, moreover, towards a secession of the heart that would also affect the creative play of his intelligence. Emotion could not be diminished in Hughes without a corresponding loss of poetically effective intelligence; in its desire to prevail, his will measured emotion and intelligence precisely. Now he would venture to express less; one might say that in becoming somewhat more cynical, he reflected one of the less attractive but authentic features of the blues culture he so profoundly admired. He himself would write of the always questioning Koestler as one of the "emotional hypochondriacs" he had known and liked but refused to emulate: "I often feel very sad inside myself, too, though not inclined to show it. Koestler wore his sadness on his sleeve." Of Langston, Koestler would remember that "behind the warm smile of his dark eyes there was a grave dignity, and a polite reserve which communicated itself at once. He was very likeable and easy to get on with, but at the same time one felt an impenetrable, elusive remoteness which warded off all undue familiarity."

On October 19, after almost two weeks in Ashkabad, Hughes moved on with his fellow writers and a film unit to the oasis of Merv, where the Murgab River from the mountains of Afghanistan branches and attenuates before disappearing into the Kara Kum desert. Once, Merv was thought to have been the cradle of mankind, the site of earthly Paradise; now it was the major cotton region of Turkmenistan. The old city was off-limits to the writers. The new Merv, which they reached on a bitterly cold night, seemed to Hughes "as drab and ugly a city as I have ever seen." Three in a bed, the four men shared one room so filthy that Langston quietly cleaned it the next day to spare the sensibilities of Koestler, who was obsessive about hygiene. Visiting the Aitakov Cotton Collective nearby, they watched the women at work picking cotton; the men were also supposed to be working, but Koestler caught them smoking in the tea house. Some things no revolution could change. Hughes tried to draw out the women, but they were painfully shy. The men, in any case, were more interesting. To a formal meeting in the visitors' honor, for men only, came big, striking fellows in padded robes and baggy trousers, and "paler lads who looked like Persian figures on old vases or drawings on parchment scrolls." The shepherd boys seemed to have stepped straight from the Book of Moses; their black caracul hats "alone saved them being Sunday-school characters of my Protestant youth, perhaps bearing the Lamb of God." Two bards arrived to sing and play what sounded to Langston like a blend of Chinese tonality and the flamenco. Dinner was a sheep slaughtered in their honor, the meat swimming in fat-laden gravy, with one spoon for all to use and one tea cup passed around from mouth to mouth. (" 'Slobbering in each other's bowls,' Koestler muttered, 'a bloody disgusting filthy habit!'." Langston sipped nonchalantly.)

Lingering near Merv for several days, he went with Koestler to Permetyab, some fifty miles to the south, across almost impossible terrain, where they found a village of refugee Afghani and Baluchi tribesmen in dire straits; the children were filthy and diseased, the women like broken animals. Syphilis was the house disease; ninety percent had caught it. "Koestler almost keeled over," Hughes would recall. "I was a bit upset myself." Near the end of October, the men reached Bokhara. On this second visit Hughes began to fall in love with the old city, the fabled setting of *A Thousand and One Nights*. Leaving research to Koestler ("always, if he had any notes I wanted, Koestler would share them with me"), he drank like a native in the wine houses, strolled in the bazaars, and watched auctioneers sell their quilted robes, brass, jewelry, and water pipes. Charmed by the tinkle of camel bells and the rough voices of the drivers in the dark, he walked at night to the edges of the city. He and Koestler visited a synagogue a hundred years old, where the halls were hung with rich tapestries and rugs. Old bearded men, dressed in patriarchal robes, showed them around the Hebrew school. Many synagogues had fared less well. "I saw other former synagogues, which had been turned into storage houses, offices, or trade-union centers, and where, over the Star of David had been superimposed the Hammer and Sickle."

Koestler left for Samarkand, but Hughes decided to stay where he was. Although no one spoke English, and almost no one spoke Russian, he seemed to get along fine even at movies and plays. More than once he dined with Mayor Kurabanov, whom he had met on the train from Moscow. At the Tea House of the Red Partisans, he interviewed one of the emir's former wives, finding a still beautiful woman with ivory skin and soft gray eyes. The emir had accepted her when she was twelve, passing her on to his prime minister when she was fifteen. Now she was married to a workman, had shed her veil, and earned her own living at the Tea House. When the last emir was routed by the communists, Hughes noted with amusement, he abandoned most of his women but fled with his harem of boys.

Near the end of November, Hughes caught up with Koestler in Tashkent. By this time Koestler was completely depressed; Tashkent was as dirty and ugly as the rest of Central Asia. Famine was scourging the land, but no Soviet official would admit it; the jails were full of people like the unfortunate Atta Kurdov. Langston listened to his complaints, then went off to find a room at the *dom sovietov*. His money was running low; certain promised funds had not come. (According to Koestler, he himself secretly complained in a long telegram to the Comintern that the distinguished Negro writer Langston Hughes "had been lured under false pretences to Central Asia and simply forgotten there.") Suddenly Hughes's contract for a book of poems in Uzbek with the Uzbekistan State Publishing House was ready, along with a check for six thousand roubles, the largest sum he had ever received for writing. Then he fell seriously ill. His stomach was like a rock, his head ached. A bottle of bright green medicine, prescribed by a doctor, led to nausea and vomiting. After some visiting Tashkent writers sounded the alarm, the personal physician to the president of the republic hurried to examine Hughes. The bright green medicine turned out to be arsenic, sometimes used for heart trouble, never for the stomach. Langston went on a diet of soups, fowl, and custards.

The writers moved the invalid upstairs to share a large room with a muscular, jolly young man, as brown as himself, named Yusef Nichanov. Now the city director of physical culture, Nichanov had once been an Oriental vagabond, one of the *besprizorni,* or homeless youths, who still roamed the Soviet Union in the wake of the revolution. (The *besprizorni* were an insouciant lot; Hughes once offered his hand to a youth, who insisted on a rouble before he would shake it.) Cooking for Hughes was an unhappy old lady who had lived in the *dom* since its days as a grand hotel. Now the residents threw their garbage in the toilets. Openly detesting the revolution, she refused to answer to *tovarish,* or comrade, begrudgingly accepting instead *grasdani,* or citizen. Hughes felt sorry for her. But, as with Koestler's complaints, he could not prefer the past. "Something hard and young in me," he remembered, "could not help thinking, now had come the hour of those from the desert. . . . Now it is the turn of those who in former days had to beg of the Cossacks, 'Please, master! No more lashes, please! White master, no more! Please!' " Still, the past fed his imagination: "As dusk approached, that vast hotel room where I lay would fill with the shadows of Cossacks and soldiers, adventurers and barons, tax collectors and colonial officers from St. Petersburg . . . the men who had once wined, dined, danced with, and courted this little old lady who now brought my dinner and remained to talk with me as I ate, mourning the passing of the Romanoffs, the old culture and the days of no tin cans in the toilets."

A procession of Red Army officers, writers, sportsmen, and assorted folk passed before his invalid bed, entertained mainly by his jazz records. A charming Georgian sculptor, Nina Zaratelli, used Nichanov as her model for a gigantic statue of "Uzbek Youth" for the new Tashkent stadium. Hughes also worked to gather and translate some pieces of Uzbek poetry. When he was sufficiently recovered, he resumed his journalism. At Chirchikstroy, about an hour away, he took notes on the construction of a massive electrochemical plant. Langston also interviewed the renowned Uzbek dancer Tamara Khanum, the first Uzbek woman to perform in public, who explained something of the Uzbek dance technique and aesthetic, with its delicate traceries of wrist and hand, fingers, eyebrows, eyes, and mouth. Then, the day before Christmas, Langston went with Bernard Powers (the Howard engineer who had greeted the film group in a tuxedo) about forty miles from Tashkent to spend Christmas with a small band of black Americans and their families living on a cotton collective. Made to leap from a train that slowed but would not stop (they landed ankle-deep in slush), Hughes arrived in a sour mood: "I am afraid I did not act civilly. . . . I just sat glumly by the fire, nursed my cognac, and wished I was home in the U.S.A." But Christmas Day was "wonderful"—a stocking full of halvah, cashew nuts, and pistachios hanging over his bed, and pumpkin pie for dessert. Outside the house "I saw padding around the stables in the snow some tall brown Uzbeks who looked like the pictures of my Sunday-school cards in Kansas when I was a child. In their robes these Uzbeks looked just like Bible characters, and I imagined in their stable a Manger and a Child."

Leaving Tashkent, Hughes paid a brief but memorable visit to Samarkand, where the ruins of Ulugh-Beg's ancient observatory and the tomb of Tamerlane represented its old splendor, and a Soviet university, a medical school, a hydroelectric station, and a silk factory exemplified the new. At the office of the City Soviet, Hughes gathered the statistics and other facts of Soviet progress, but spent his last evening in the city of Genghis Khan and Alexander the Great lingering outside Tamerlane's tomb:

> The gates were locked, but I looked into the charming courtyard that I had often visited before. Birds were nesting in the trees and a little grey lizard scurried across the ground. The setting sun gleamed warmly on the lovely old tiles, and on the great patches of sun-dried brick where the tiles had fallen away. Near the walls, beneath the mulberry trees, a little stream of water gurgled. Across the stream a child ran after a bantam chicken to put it to roost. In the courtyards round about, people were lighting fires and cooking.
>
> As I walked back toward the center of the town, darkness began to fall. . . . In a dark grove of trees, I heard music—the high string lutes of the East. I went into this grove of darkness, where the glow of charcoal fires made the only light. I mounted one of the outdoor platforms, sat cross-legged on a rug and called, *'Chai,'* the Russian word for tea. But before the attendant reached me, several bowls of tea were offered me in the half-darkness . . .

In the depth of the Russian winter, near the end of January 1933, and just after his all-important travel permit expired, he returned to Moscow. After an excruciating search in the cold for a hotel room, Hughes found a vacancy at the Savoy Hotel. But the clerk in charge, pointing out that his permit was invalid, absolutely refused to give him a room. A shouting match gained Langston only the right to stay overnight in the lobby. Angry but exhausted after his long journey, he fell asleep in a chair, barricaded by his luggage. The next morning, when he visited Walt Carmon, editor of *International Literature,* who promptly interviewed him for the Moscow *Daily News,* he was still seething; Hughes specifically denounced the staff of the Savoy Hotel. But he was full of praise for Soviet Central Asia. "I know what imperialist exploitation of the dark races means," he told Carmon. "I have seen the dark races exploited unmercifully the world over. And now . . . I have at last seen a country where the dark peoples are given every opportunity." Uzbek literature was "vital and growing," and the Tashkent theater a brilliant example of a culture " 'national in form and socialist in content'."

Carmon invited Hughes home to spend a day or two; Langston was there a month, frustrated in his search for his own place by the Soviet bureaucracy. In a situation normal in Moscow, where housing was critically scarce, Walt and Rose Carmon had one of four rooms in an apartment they shared with another American couple and the couple's daughter and her husband. The son-in-law was an Afro-Chinese named Jack Chen, who had a brother and two sisters in Moscow. The previous fall, his sisters, Yolande and Sylvia Chen, had called on Hughes at the Grand Hotel to talk about Harlem. "Langston was charming, listening very seriously to our naive questions," Sylvia Chen would recall; then another fellow strolled into the room, took one look at the stunning Chen sisters, "and gasped, 'Oh my, just *look* at what blew in on you, Lang—I must go tell the news to the other guys.' This made us all laugh and forget our Harlem investigation." But Hughes did not forget Sylvia Chen. Jack, Percy, Sylvia, and Yolande were the children of Eugene Chen, who was born in Trinidad of Chinese immigrant parents, but served at one time as English secretary to Sun Yat-Sen, founder of the Chinese republic. After Sun's death, Chen had become internationally known as foreign minister of China and an articulate mainstay of the left wing of the Kuomintang. Sylvia Chen had grown up with her mother in Trinidad and England, where she developed an interest in dance and pantomime. In 1927, after visiting her father in China, Sylvia (or Si-lan, as she would call herself) became a radical in politics and dance; she developed a repertoire that included both Chinese dances and the modern Western choreography of Isadora Duncan. When the leftists were routed by Chiang Kai-Shek, she and others in her family fled to Moscow, where she studied choreography with a pupil of Fokine. By 1933, she was known in Moscow dance circles as an interpreter of Asian and other exotic forms; perhaps Hughes attended her successful recital advertised in the newspapers for February 14 at a Moscow theater.

"My admiration for his integrity and art increased each time we met," Sylvia Chen would write of Hughes. "He was such a jolly person and so natural. Langston had been a sailor and he walked like one; I remember him sloshing around in white corduroy pants in the middle of a Russian winter. I said to him, 'Langston, how can you wear white pants in this slush?' 'Oh,' he replied. 'They're so easy to wash'." To Hughes, in turn, Sylvia Chen was a stunning vision among the drab Muscovites. "A delicate, flowerlike girl, beautiful in a reedy, golden-skinned sort of way," she favored long, tight, high-necked Chinese dresses with a little slit at the side "showing a very pretty leg." Well-mannered but also high spirited, she spoke English with a precision like that of Anne Marie Coussey in Paris nine years before. Also a proper young woman with marriage and motherhood on her mind, Sylvia Chen found the handsome, brown, young revolutionary poet more than acceptable; he became a regular visitor to her room at the Metropole Hotel. Once he sent her a jaunty but warm little poem:

> I am so sad
> Over half a kiss
> That with half a pencil
> I write this.

Sylvia Chen was "the girl I was in love with that winter." "Langston was the first man I was ever intimate with," she would acknowledge. (Around the spring of 1931, she had married and quickly divorced, but apparently never lived with, a manic-depressive Russian director named Mikhail Mikhailovich, who later hanged himself.) Hughes wrote home to tell Van Vechten about his new girlfriend.

Sylvia Chen's presence probably helped Langston decide to stay in the Soviet Union; he found a room at the New Moscow Hotel, near Red Square, and settled in. At home, the Depression was deepening. "People are just committing suicide for fun," his mother wrote; *"Can I come to Russia?"* Some of the original "Black and White" party were still in Moscow—Dorothy West, Mildred Jones, Homer Smith (hired as an expert by the postal service at 500 roubles a month), Lloyd Patterson (an interior decorator at Meschrabpom), and Wayland Rudd (a student at the Meyerhold Theater). On the other hand, Louise Thompson, now in New York, wanted him back: "There is so much to do here, Lang, and we need you." Finally, however, he stayed mostly because he was making more money as a writer than ever before. In 1932 in America, his main source of income had been the backbreaking tour; Knopf had sold that year only 1420 copies of his four books. In addition, a bill had reached Hughes in Moscow demanding the payment of $162 owed by him and Prentiss Taylor to their printer in connection with the Golden Stair Press. Still, Langston thought wistfully at times of America. "Hey! Hey!" he jived to Taylor, "I would like to see Harlem once more before everybody starves to death."

In his *Izvestia* essays, published in Moscow as a small book, *A Negro Looks at Soviet Central Asia,* Hughes contrasted Soviet justice with the disgraceful treatment of blacks in the United States. Except for gently questioning the official disinterest in the old folk arts, the articles were uncritical and superficial. He made no mention of bureaucratic obtuseness or incompetence, rumors of famine, political prisoners in jail, or the fact that the Soviet Union was not exempt from the rule that (as he later observed) "even in places where there is almost nothing, the rich, the beautiful, the talented, or the very clever can always get something; in fact, the *best* of whatever there is. . . . Some can always get cream while most drink milk, some have wine while others hardly have water. The system under which the successful live—left or right, capitalist or communist—did not seem to make much difference to that group of people." But Hughes concentrated on praising the considerable Soviet achievement. Having witnessed the rising dark-skinned workers of Tashkent, Bokhara, Ashkabad, and Merv, "now I know why the near-by Indian Empire trembles and Africa stirs in a wretched sleep."

Writing in *International Literature,* the critic Lydia Filatova stamped him as, at last, a revolutionary writer. After the "bourgeois aestheticism" of *The Weary Blues,* the ignoring of class interest in *Fine Clothes to the Jew,* and his failure to recognize the revolutionary potential of the white masses in *Not Without Laughter,* his art had come to "a decided turning point" in 1931. He had entered "the revolutionary period. His new poetical *credo* is the total negation of his former creative position. . . . The writer has faith in the strength of the working class, in the ultimate victory of the revolution."

"Never must mysticism or beauty," Hughes solemnly wrote to Prentiss Taylor, "be gotten into any religious motive when used as a proletarian weapon." He saw his current work as such a weapon. Early in March, when he sent Blanche Knopf and Van Vechten a manuscript of poems entitled "Good Morning Revolution," containing the best of his recent radical verse, he assured them that the collection had been expertly criticized in Moscow. But he had badly misjudged the strength of socialism at home. Van Vechten replied with a sharp slap: "The revolutionary poems seem very weak to me: I mean very weak on the lyric side. I think in ten years, whatever the social outcome, you will be ashamed of these." Moreover, he wrote Hughes, it was ironic to ask a capitalist publisher to bring out a book "so very revolutionary and so little poetic in tone." When Blanche Knopf later refused the book, Hughes frankly blamed Van Vechten: "Blanche bases her note to me on your reactions." It was "swell of you to write so frankly about my poems," he declared, mixing a little bohemian diffidence with much more radical defiance. "I agree with you, of course, that many of the poems are not as lyrical as they might be—but even at that I like some of them as well as anything I ever did." He was willing to revise the work, "but I would not like to change the general ensemble of the book. I think it would be amusing to publish a volume of such poems just now, risking the shame of the future (as you predict) for the impulse of the moment. And if Knopf's do not care to do it, they have a perfect right to refuse."

With Lydia Filatova's help he completed his translations of Vladimir Mayakovsky's poems "Black and White" and "Syphilis"; he also met Boris Pasternak and translated some of his poems. Next he worked on modern Chinese poems, aided perhaps by the Chens, and on work by Louis Aragon of France, who visited Moscow and met Hughes. But Hughes's most powerful work was original, and appeared in *International Literature* (published six times a year, in English, French, German, and Russian). In the first number of 1933, which also carried a photograph of Hughes, Koestler, Kikiloff, and Shagurin together, came "Our Spring" ("Bring us with our hands bound, / Our teeth knocked out / Our heads broken"); and a highly sensational attack on the United States, "Columbia":

> Columbia,
> My dear girl,
> You really haven't been a virgin so long
> It's ludicrous to keep up the pretext.
> You're terribly involved in world assignations

And everybody knows it.
You've slept with all the big powers
In military uniforms,
And you've taken the sweet life
Of all the little brown fellows
In loin cloths and cotton trousers.
When they've resisted,
You've yelled, 'Rape,'

. . . .

Being one of the world's big vampires,
Why don't you come on out and say so
Like Japan, and England, and France,
And all the other nymphomaniacs of power? . . .

The third number carried "Moscow and Me," a personal essay on his experiences in the capital. In the fourth was a similar essay, "Negroes in Moscow," as well as a translated poetical fragment by Louis Aragon. Perhaps his strongest poem of the year, "Letter to the Academy," also appeared in *International Literature*.

The gentlemen who have got to be classics and are now old with beards (or dead and in their graves) will kindly come forward and speak upon the subject

Of the Revolution. I mean the gentlemen who wrote lovely books about the defeat of the flesh and the triumph of the spirit that sold in the hundreds of thousands and are studied in the high schools and read by the best people will kindly come forward and

Speak about the Revolution—where the flesh triumphs (as well as the spirit) and the hungry belly eats, and there are no best people, and the poor are mighty and no longer poor, and the young by the hundreds of thousands are free from hunger to grow and study and love and propagate, bodies and souls unchained without My Lord saying a commoner shall never marry my daughter or the Rabbi crying cursed be the mating of Jews and Gentiles or Kipling writing never the twain shall meet—

For the twain have met. But please—all you gentlemen with beards who are so wise and old and who write better than we do and whose souls have triumphed (in spite of hungers and

> wars and the evils about you) and whose books
> have soared in calmness and beauty aloof from
> the struggle to the library shelves and the
> desks of students and who are now classics—
> come forward and speak upon
>
> The subject of the Revolution.
>
> We want to know what in the hell you'd say?

In addition to these Moscow pieces, his work also appeared at home in the *Crisis,* to which he sent a revised "A New Song," first written in California:

> . . . Black world
> Against the wall,
> Open your eyes—
> The long white snake will strike to kill!
> Be wary—
> And be wise! . . .

The *Crisis* also took the brief "Black Workers":

> The bees work.
> Their work is taken from them.
> We are like the bees—
> But it won't last
> Forever.

These *Crisis* pieces are decidedly more racial than the poems published in Russia, where they would have been criticized as chauvinist; to some extent, Hughes was speaking with a forked tongue. To Amy Spingarn, however, his recent works seemed "to have a new ring to them, as tho' you were tremendously alive."

Indeed he was tremendously alive. With critical and financial success that had come not from godmotherly patronage but from radical work, Hughes enjoyed a surge of confidence not felt since the year following the publishing of *The Weary Blues;* it gave him the drive to pursue Sylvia Chen ardently enough for them to think of marriage. But he later recalled another involvement that spring, almost certainly after Sylvia Chen left Moscow on tour, with a married Russian actress. "Natasha" had "a buxom body, a round smiling face, Slavic—not beautiful, not ugly—and was very healthy-looking." Her "one-track mind" led straight to bed. Hughes was also involved in some way, though probably not sexual, with his youthful companion in the film group, Dorothy West. In an impulsive letter on May 26, West proposed marriage to him. She loved him, she wrote, but above all wanted a child; then she would leave Langston free to roam the world like the eternal boy she knew he was meant to be.

In spite of this warm season, however, Hughes knew he could not marry. Certain only of the solace of poetry and motion, he would not encumber himself. Which is perhaps the same as saying that he could not give himself, and thus could not finally possess any of the women who were drawn by his successes, his handsome face, and his winning ways. Later, although Hughes would pursue her again, a hurt and neglected Sylvia Chen ("I'm writing you as to a ghost") would claim never to have been fooled; charming but also superficial and insincere in his courting, Hughes was always "flitting around being amiable."

Insincere in love or not, Hughes was clearly ripe for inspiration when, at the suggestion of Marie Seton, an English friend of Paul and Essie Robeson, he started reading D. H. Lawrence's collection of stories, *The Lovely Lady* (apparently he had never read Lawrence before). The effect was overwhelming. Both "The Rocking Horse Winner" and the title story "made my hair stand on end." The resemblance between the grasping old woman, Pauline Attenborough, of "The Lovely Lady" and his own Godmother, Charlotte Mason, riveted him: "I could not put the book down, although it brought cold sweat and goose-pimples to my body." In Lawrence's stories Hughes saw not only something of the face of his tormentor, Charlotte Mason, but also glimpses of his own neuroses. Setting aside his current work, he turned to write his first short stories since 1927, which was also the last time he had touched on strongly sexual, possibly autobiographical themes in his writing. Now he stressed the volatile mixture of race, class, and sexuality behind not only his troubles with Mrs. Mason, but also the rituals of liberal race relations in the United States.

"They were people who went in for Negroes," he opened his first story, "Slave on the Block," about a patronizing, fraudulent white couple in Greenwich Village who collide with reality in the form of their resentful black cook and a physically impressive but insolent black youth, whom the wife (an artist) insists on posing without his shirt. Another story, "Cora Unashamed," is a dramatic picture of black isolation in the white Midwestern world; at forty, black Cora has had only one lover, and he was not Negro. Hughes contrasted this black woman's emotional power, albeit frustrated and incoherent, with the suppressions of the bourgeois white heart. In "Poor Little Black Fellow," adopted black Arnie grows up unhappily in a wealthy, white New England household that is cold, racist, and self-righteous. On a family trip to Paris, he discovers his own people and the possibilities of freedom; for the first time, "someone had offered him something without charity, without condescension, without prayer, without distance, and without being nice." Repudiating his family, Arnie breaks with the old, false life. Quite apart from their social criticism, all three stories reflect Hughes's personal sense of isolation and betrayed trust; additionally, in Cora's sexual starvation and, obversely, the black boy's sexual confidence in "Slave on the Block," as well as in Arnie's firm decision to break with his "family" and possess the white woman who wants him, one sees a thinly veiled fictive admission by Hughes of the psycho-sexual cost of his own isolation.

Almost certainly he did not try to find out what Lydia Filatova and other Muscovite critics thought of these pieces. Just before mid-March, he sent the stories to Blanche Knopf, asking her to offer them to H. L. Mencken's *American Mercury*. She sold two at once to the magazine for $100 each; not long after, she turned over the business of placing his manuscripts to a literary agent named Maxim Lieber, of whom Hughes had heard only fine reports from American friends in Moscow. With this start, Langston began to plan his return to the United States. Intent on returning via war-torn China, he began a trek, almost as trying, through the sludge of Soviet bureaucracy. Red tape was never more odious than at Intourist, the official travel agency, which Hughes despairingly would call "the world's worst travel bureau" because of the rudeness and ignorance of its workers. On March 11, he went to Riga, Latvia, to renew his American passport. The vice-consul, who was from Lawrence, Kansas, showed him "a swell time." (Sylvia Chen later remembered her amusement when he left for Riga with "a suitcase crammed with tin goods so that he would have plenty to eat. But he didn't bring back any of it.")

With an abundance of roubles and the power of his membership in the International Union of Revolutionary Writers, Hughes enjoyed the Moscow spring. Many of his evenings were passed in the crowded theaters; unlike the cinemas, which were usually unheated and smelled "like very pungent kennels," they were warm and clean. Hughes's Russian was now good enough for him to follow the formal plays; only the music hall and vaudeville skits were beyond him. His favorite theater was Oklopkov's Krasni Presnia, where experimentation, especially for audience involvement, had almost no limits. But he also liked Stanislavsky's traditional Moscow Art Theater, where plays by Chekhov, Gorky, and Strindberg drew capacity audiences. At the esoteric Kamery he saw a staging of *All God's Chillun Got Wings,* in which the set literally throbbed as in O'Neill's stage directions. At the Vakhtangov, Hamlet was a man of action, entering on a white charger. Hughes even had the pleasure of seeing his friend Wayland Rudd act a small role at the Meyerhold. Langston himself performed on April 2 in a program shared with the respected American-based opera singer Sergei Radansky and the Moscow Philharmonic Orchestra.

He was still in Moscow in May, when tulips bloomed in the Park of Culture and Rest and the trees began to bud around the Kremlin walls; the Moscow River cracked and melted, the Lenin Hills rose through a blue spring haze above the river banks, and lines lengthened outside Lenin's tomb as warm winds swept Red Square. On May Day, he viewed the great parade from a privileged position not more than a hundred yards from the reviewing stand and Stalin, Molotov, Kalinin, and Voroshilov. The stones rumbled and thundered under boots and steel-shod hooves and wheels; to the pacifist Hughes, however, more important than the military might was the international parade of workers. "This has even the hundreds of tanks and soldiers and airplanes beat for impressiveness," he wrote home.

On May 19, Hughes and Wayland Rudd gave a recital at the Foreign Library. Rudd sang the ever-popular spirituals; Hughes read poems and talked about his dilemma as a writer, torn between radicals who never found him radical enough, and bourgeois critics who called his radical work a deterioration. He left no doubt which opinion mattered more. Just about this time he finished at least three articles on black American writers seen from the Marxist viewpoint. For Claude McKay, "the outstanding living Negro writer," he had great praise—although McKay's current writing was hardly radical (but then, Max Eastman and Floyd Dell, the old *Liberator* tigers, were about "as revolutionary as kittens now"). On Countee Cullen ("OUR KEATS AND SHELLEY") and Walter White ("A WHITE NEGRO") Hughes heaped scorn. Cullen was "an inconspicuous little brown fellow, the color of coffee in a Moscow restaurant. He is inclined to stoutness, and has a tiny bay window of a stomach." Wanting to be a teacher, he had no job in sight and lived at home. His father's church was "as much like a circus as a church"; money collected there went to support many things, but "not to Jesus." Unexpectedly, Cullen had come out for the communist ticket in the last elections, but whether this change of heart would improve his poetry remained to be seen. (The previous year, as Hughes probably knew, Cullen had ridiculed in his novel *One Way to Heaven* a New Negro teacher whose pupils "could recite like small bronze Ciceros" Langston Hughes's poem "I Too Sing America," but were "unaware of the contributions of Longfellow, Whittier, and Holmes to American literature." Negro poets—those like Hughes—sound one theme only: " 'Niggers have a hell of a time in this God-damned country'.") On Walter White, openly anti-communist because of Scottsboro, Hughes was even harder. White was "a red-baiter"—but then, there was money in baiting communists. "What the class-conscious workers think of him, I will not repeat here. It would be too obscene. I would not like to have my articles censored for vulgarity."

(Hughes wrote White about the essays, then sent copies to him via Louise Thompson. An affronted White had his secretary return them—after making copies, one of which he sent to Joel and Amy Spingarn.)

To the end, Moscow remained exciting. Langston found kindred spirits, many of them American, when over 200 artists from all over the world assembled for the first annual Workers' Theaters International Olympiad. Some of the Americans reported having seen his mother on stage in a small part in Hall Johnson's *Run Little Chillun* ("Yes your mother is an actress at last," Carrie Clark had written, "the dream I dreamed as a little child is very near realized"). At the conference that followed the Olympiad he made friends with several revolutionary writers, including Seki Sano and Yoshi Hijikata, who promised to alert their colleagues at home about Hughes's coming trip.

Early in June, when his visas and letters of permission finally came through, the war in China had forced a change of plans. Japan had bombed Shanghai and stepped up its military operation in Manchuria. From Vladivostok, Hughes would have to proceed to Japan before going to China. A popular American couple, Fred and Ethel Ellis, threw a farewell party for him. Sylvia Chen was not there, having left the Soviet Union on a dance tour; she and Hughes had parted without any firm commitment on either side, which meant mostly that she had given in to his vagueness and indecision. A crowd of over twenty saw him off on the 5500-mile journey on the Trans-Siberian Express across the Volga and over the Ural Mountains to Lake Baikal, and on through the raw wilderness of the Soviet Far East. Few prospects were more appealing to Langston than that of a long train ride alone, but suddenly, according to him, his buxom actress lover, "Natasha," appeared; she had left her husband, taken her savings, and bought a ticket to Vladivostok to be with him to the end. Hughes was dismayed: "All the peace and happiness and freedom of the first half hour of that train ride gone!" At last, after a tearful scene, he convinced "Natasha" to leave the train. "I felt unkind, ungallant, embarrassed and unhappy," he recalled. "I felt very sad, very bad—yet very glad she didn't go any further."

On June 17, he reached Vladivostok. The setting for his farewell to Russia was forbidding, "a gray fortress of a town edging the water at the foot of a series of scraggly hills . . . dismal, damp, depressing and dirty." Much of the town was off-limits, the prices high, the services poor. On Lenin Street, he found newspapers in Japanese and Chinese. Elsewhere, he stumbled on the dead body of an Asiatic man, but no one else seemed even slightly disturbed by the corpse. On this macabre note Hughes said goodbye to the U.S.S.R., after almost exactly one year there.

The weather was warm, the journey to Korea across the Sea of Japan blissful; through green fir trees on hilly little islands, red temple roofs gleamed. But Hughes was a marked man. At Chongjin on the Korean peninsula he had been followed everywhere by men he took to be police agents. At Honshu, the Japanese police questioned Hughes and other passengers who had associated with him. Then he took the train to Kyoto, where he reunited with the ship's captain, whose hospitality apparently included taking Hughes first to a geisha house, then to a brothel. The next day Langston toured Kyoto, regarded as the heart of the Japanese national culture. Impressed by its many tombs, castles, palaces, and manicured parks and gardens, he nevertheless found the city "as impersonal as a Technicolor movie." He also found the degree of militarism frightening. The country was covered with restricted zones; foreigners were watched closely. "The Japanese militarists are quite open about all this," Hughes wrote. "They make no secret that they are shadowing you, and that they are suspicious of everyone."

In Tokyo, with a favorable rate of exchange for his few American dollars, he stayed at the Imperial Hotel, designed by Frank Lloyd Wright. At the Tsukuji-za Theater he was almost mobbed; it was as if O'Neill or Shaw had entered, he remembered. Seki Sano, its former director, had written glowingly from Moscow about his coming visit. In the next few days, the theater was his host. Like every other progressive organization in Japan, however, it was in grave danger. The Proletarian Revolutionary Theatre, the Proletarian Writers Federation, and the Proletarian Artists Federation were all closely watched by the authorities. All scripts had to be approved; if a performance deviated, a censor in the house could stop a play at once. To these groups, Hughes was glamorous—a revolutionary fresh from Moscow and a Negro American (they had recently staged *Porgy*). On the night of June 29, he was made an honorary member of the Tsukuji-za. Hughes enjoyed both the attention and the sense of danger. One night he heard what he thought were tractors passing his hotel, "but they were tanks, more than a dozen of them. Where they were going down a big city street in the middle of the night, I do not know." He was having "the grandest time imaginable," he wrote home from Japan. "You may be killed, but you'll never be bored."

On that day, he was a guest in the Imperial Hotel at a luncheon of the prestigious Pan-Pacific Club, where he sat next to the wife of E. L. Neville of the American embassy. One feature of the luncheon was an explanation of the Japanese tea ceremony. Afterwards, Mrs. Neville, apparently a native of a certain Southern city, made a little speech about the ceremony that Hughes thought condescending and possibly hypocritical; when he was last there, no downtown restaurant would serve tea or anything else to colored people. When his turn came, he spoke of race relations in America and of black American sympathy for the Japanese nation, "the only large group of dark people in the world who are free and independent." Blacks needed the psychological assurance that some dark people "are not down and oppressed. So the American Negro is glad that Japan is able to enjoy her ceremonial tea without the unwelcome intrusion of the imperialist powers of the west." When he sat down, Mrs. Neville apparently accused him in a whisper of doing his country a disservice. She left without saying goodbye to him. Later, someone reported to the Federal Bureau of Investigation that in his speech at the luncheon Langston Hughes had "predicted that there would one day come a war in which all the colored races, black, yellow, and red, would join in the subjugation of the whites. HUGHES stated that there was a natural bond between these colored races and that their opposition to the white race should be expressed in combat." The report became a damaging part of his official file.

The next day he sailed for Shanghai, the most populous city in Asia, and one sectioned since the nineteenth century into British, American, French, and other zones that boasted separate courts, police, and even armies. "Arms bristle everywhere, on everybody, on all nationalities," he wrote, "—except the Chinese whose land the foreigners have taken." Up and down the Yangtze and the Whangpoo rivers passed the major portion of Chinese trade, including the supremely lucrative traffic in opium, heroin, and other narcotics. To gain access to this market, the Japanese had invaded the city. In "that vast international powderkeg of a city," Hughes would report, "the Japanese marines patrol the streets in fives, marching slowly and gravely, armed, swinging little sticks on constant patrol." He stayed at a small Chinese hotel; the leading hotels excluded all colored guests, including Chinese, and the YMCA was also segregated. With Helen Chung, an American-born YWCA worker, he visited factories and sweatshops, many exploiting small children. Against all advice, he roamed past the barbed wire of the International Settlement to visit pagodas, teahouses, and bazaars. Requesting an interview with the most famous resident in the city, Madame Sun Yat-Sen, the widow of the founder of the Chinese republic, and the sister-in-law of Chiang Kai-Shek, he received instead an invitation to dinner. As Soong Ching-ling, she had been educated at Wesleyan College in Macon, Georgia, and thus was familiar with American racism. Hughes found Madame Sun Yat-Sen, who knew Sylvia Chen and her family well, "as lovely to look at as her pictures, with jet-black hair, soft, luminous eyes and a complexion of delicate amber. She was a relaxing and delighful woman with whom to talk.''

He visited Nanking and Sun Yat-Sen's tomb in the Purple Mountains overlooking the city, but a trip to Peking was forbidden by the Japanese army. Back in Shanghai, a group of journalists, writers, and progressives (including a man translating his *Not Without Laughter* into Chinese) gave a farewell party for him. Langston sailed from Shanghai just after mid-July on the Japanese ship *Taiyo Maru,* bound for San Francisco via Kobe and Yokohama. At Kobe and at Yokohama, he was questioned by the police about his visit to Madame Sun Yat-Sen. Since the ship would be in port several days, he took the train to Tokyo, arriving there on Sunday, July 23, and registered once again at the Imperial Hotel.

He was watching a showing of the movie *Mata Hari,* starring Greta Garbo and Ramon Novarro, when a telephone call brought an invitation to lunch the next day with several writers. Then some visitors were announced; Hughes descended to the lobby to find three Japanese men, one of whom presented himself as a famous pacifist author who had been recently jailed for his views. Hughes had not heard of his release. Soon it became clear that the two other men were his guards, and he their prisoner. The attempt to trick Hughes having failed, the police agents withdrew. The next day, however, a delegation from the Tokyo police escorted him to its headquarters, where he was questioned about all his activities since Vladivostok. The interview dragged on into the afternoon, with the head of the Euro-American Affairs section accusing Hughes of being a courier for the socialists, and of recently securing an important list of names. He was left alone for about an hour, although he was assured that he was not under arrest. Finally Hughes was told that he was *persona non grata* in Japan, and should leave without communicating with any Japanese.

On his return with the police to his hotel, his luggage was searched. A fan given to him by the theater group, inscribed with names, was confiscated; this was the subversive ''list.'' After the police left, a pack of newsmen, including Stanley Wood of the *New York Times,* descended on Hughes. He informed the Japanese that he was forbidden to speak with them, but that they could listen to him talk to Wood, to whom he denounced his detention by the police. Then Wood informed him that the eleven Japanese waiting for him at the luncheon had been arrested. Later, at dinner, he was watched by two policemen. He ate alone until a young white American introduced himself—Alex Buchman, a graduate of Central High in Cleveland, come to pay his respects. Hughes warned Buchman about the police, but Buchman insisted on keeping him company.

Hughes remained under police scrutiny until July 26, when he reboarded the *Taiyo Maru* to leave Japan. In his absence from the ship, his luggage had been thoroughly, though very neatly, searched by the authorities. The voyage across the Pacific was uneventful, but at Honolulu, back at last on American soil, he was met by an FBI agent and several newsmen. To them, he indignantly categorized Japan as a fascist country. Hughes then learned that Alex Buchman had just been deported from Japan, presumably for speaking to him.

A telephone call to the leading Afro-American in the islands, Nolle Smith, a member of the Hawaii legislature, led to a tour of Waikiki Beach, the Moanalua Gardens, the heights of Pali, and also, more painfully, the courtroom where the infamous Massey trials had taken place, in which local beach boys had been accused of raping a white woman, the wife of a Navy officer.

Then, garlanded with leis, Langston waved goodbye to Smith and his family from the deck of the *Taiyo Maru*. He began the last leg of his fourteen-month journey around the world.

11

WAITING ON ROOSEVELT
1933 to 1935

The pot was empty,
The cupboard was bare.
I said, Papa,
What's the matter here?
  I'm waitin' on Roosevelt, son,
  Roosevelt, Roosevelt,
  Waitin' on Roosevelt, son . . .
    ''Ballad of Roosevelt,'' 1934

NEAR ten o'clock on the morning of August 9, 1933, the *Taiyo Maru* steamed through the magnificent Golden Gate to the piers of San Francisco. At the dock to meet Hughes was Noël Sullivan's employee, Eddie Pharr, who drove him in Sullivan's cream-colored limousine up Russian Hill to the mansion at 2323 Hyde Street, overlooking Fisherman's Wharf and the grand vista of San Francisco Bay. Soon Hughes was warmly welcomed by Sullivan, his housekeeper Eulah Pharr, and other members of the staff in the home Langston remembered with pleasure from his visit in the late spring of the previous year, after his gruelling tour of the South. Relaxing on the terrace in cool, fog-shrouded San Francisco summer weather, he sipped California wine, lunched on abalone and artichokes, strawberries and cream, and savored his first day in almost fourteen months on his native soil. As Sullivan listened to details of Hughes's adventures in the Soviet Union, Korea, China, and Japan, ''saying little, but radiating kindness, good will and sympathy, I thought of a phrase in one of Walt Whitman's poems as applying to him, 'I and mine . . . convince by our presence'.''

Broke and without serious prospects, Hughes was relieved to discover that Sullivan's invitation the previous year for him to stay a while in California, at Sullivan's expense, still stood. Langston had been burned by one patron, Mrs. Mason; but he understood that if he were to remain a professional writer, some form of patronage would be necessary. While in the Soviet Union, he had not neglected his wealthy California admirer. ''Our acquaintanceship was for such an hurried moment,'' Hughes had regretted to Sullivan from Moscow. ''I like you very much. I think, Noel, that life has given us the same loneliness.'' Now the two men quickly worked out an agreement. In a few days, Hughes would leave for Carmel-by-the-Sea to live in Sullivan's cottage, ''Ennesfree.'' With him would go Sullivan's beautiful three-year-old German Shepherd, Greta (the reclusive Garbo often hid out in Carmel, allowing only her opponent to watch her play tennis). Along with the cottage came free groceries and utilities and the services of a German cook (replaced later by a Filipino couple), who lived elsewhere. Hughes would have a year in which to consolidate his career as a professional writer.

Before leaving the San Francisco area he reunited happily with Matt Crawford, back almost a year from the Soviet film venture, and his wife, Evelyn, at a reception given in Hughes's honor by some young blacks. In these first few days back home, Hughes also showed that he did not intend to go into retirement. He fired off a sharp letter of protest to the U.S. Department of State about his treatment in Japan. (Washington replied that since the Japanese police had acted in compliance with their laws, the United States could not approach their government about the matter.) Next, welcomed as a radical writer fresh from the Soviet Union and the Far East, he spoke at a public meeting under the auspices of the Northern California branch of the National Committee for the Defense of Political Prisoners, an organization of leading American artists and intellectuals to which both he and Noël Sullivan belonged. He attended two meetings of the group, including one at the home of the prominent San Franciscan writer and social activist Dorothy Erskine. Hughes was soon aware of the grave extent to which the social, economic, and political crisis facing the nation as a whole in the depths of the Depression threatened the state of California.

The previous winter had been the most disastrous in American economic history. The mood of the United States had sunk to its most pessimistic since the beginning months of the Civil War in 1861, with industry and agriculture still showing no signs of recovery from the crash of Wall Street in October, 1929. Since then, the national income had dropped by more than half. Salaries had fallen by forty per cent, dividends by almost sixty per cent, and unemployment, bankruptcy, and foreclosures had become epidemic. ''The country faces the gravest crisis in its peace-time history,'' the *Nation* warned; the new president, Franklin D. Roosevelt, must be prepared to try ''new leaders and new ideas, and to venture boldly into untrodden paths.'' Although Roosevelt, who had carried California overwhelmingly, seemed ready to do whatever was necessary to end the crisis, his talk of a New Deal, through the National Industrial Recovery Act and the National Recovery Administration, had amounted to little or nothing by the fall of 1933. California, long the land of promise to the rest of the nation, was absorbing a deluge of migrants from the drought-stricken Midwest, the chronically poor South, and Mexico. The state banking system, as everywhere, was in disarray; in March, a six-day bank holiday had been ordered to give the government time to devise protective legislation. The previous year, 1932, a series of strikes had seriously disrupted industry and agriculture; California had become a major battleground between industrialists, property owners, ranchers, and farmers, on one hand, and the progressive or even more radical leadership of a diverse labor force, on the other, including fishermen and cannery workers in Monterey, cherry pickers in Santa Clara, cotton workers near Visalia, and longshoremen on the docks of San Francisco. Just before Langston left for Carmel, labor took a decisive step with the formation of the Cannery and Agricultural Workers' Industrial Union, which promptly targeted the important cotton industry in the San Joaquin Valley for a major test of strength in the fall.

Such concerns, however, may have seemed some distance away when Hughes took up residence near the first day of September at "Ennesfree" on the corner of 13th Street and Carmelo in the village of Carmel-by-the-Sea. Recently acquired by Sullivan, "Ennesfree" was a comfortable cottage of two bedrooms, with another sleeping area in the basement; the dining room, large and sunny, included a fireplace and a generous bay window that offered an unobstructed view of the blue Pacific. On September 6, the Carmel *Village Daily* noted his arrival—or sounded an alarm: "Colored Poet Here."

Certainly Carmel was an odd home for a black poet. Fringed by an arc of snowy beach and rustling Pacific surf, the village seemed remote not only from racial struggle but from all struggle itself. To the north rose the headlands of the Monterey peninsula with its celebrated cypresses and lush golf courses, notably Pebble Beach, that served mainly the peninsula rich, of whom there were many. To find a decrepit neighborhood, and a few black faces, Hughes would have to hunt in certain remote parts of nearby Pacific Grove and the city of Monterey, the ancient capital of Alta California until the Unites States captured it from Mexico in 1846. An air of make-believe, not to say deceit, clung to Carmel. Although the musty, eighteenth-century mission of San Carlos Borromeo preserved the remains of the California pioneer Father Junipero Serra, the village itself was very much a creature of the twentieth century. Founded as an "artists' colony" on a neglected tract of hilly land, Carmel had long outgrown what had been, for the most part, a real-estate promotional ruse. Some painters called the town home, but few made a living by selling their shiny "marines" to the throngs of tourists strolling up and down Ocean Avenue. On the other hand, Carmel's literary associations were stronger. Robert Louis Stevenson and Richard Henry Dana had admired the region and drawn on its history and character. Jack London (one of the first residents, lured with a free lot by the developers), William Rose Benét, Van Wyck Brooks, Witter Bynner, Sinclair Lewis, and the ubiquitous Mabel Dodge Luhan all had Carmel associations. The local library boasted 200 books written by residents or regular visitors; forty Carmelites were said to be in *Who's Who In America*. When Hughes arrived, the resident genius was clearly the poet Robinson Jeffers, a Carmelite since 1914; his Celtic tower of stone, Tor House, which he had built with his own hands, rose strangely by the ocean. Less an artist but even better known was Lincoln Steffens, the old muckraker and political visionary, recently in ill health but still a presence that attracted a steady stream of visitors to his home at the corner of San Antonio Street and the main thoroughfare, Ocean Avenue.

With Radcliffe Lucas on tour, 1931.

On tour, 1931.

(Above) Prentiss Taylor, with entertainer Gladys Bentley (left) and Nora Holt. *Courtesy of Prentiss Taylor.*

Scottsboro Ltd., 1932. *Courtesy of Prentiss Taylor.*

With Louise Thompson on the Europa, June 1932. *Courtesy of Louise Thompson Patterson.*

(Below) En route to Moscow, June 1932.

In Central Asia, 1932.

In a Beluchi turban at a kolkhoz near Merv, Turkmenistan, 1932.

With Kolya Shagurin, Shaarieh Kikilov, and Arthur Koestler, Central Asia, 1932.

Sylvia Chen in Uzbek dance costume, n.d.
Courtesy of Si-lan Chen Leyda (Sylvia Chen).

With Russian translation of *Not Without Laughter*, Moscow, 1933.

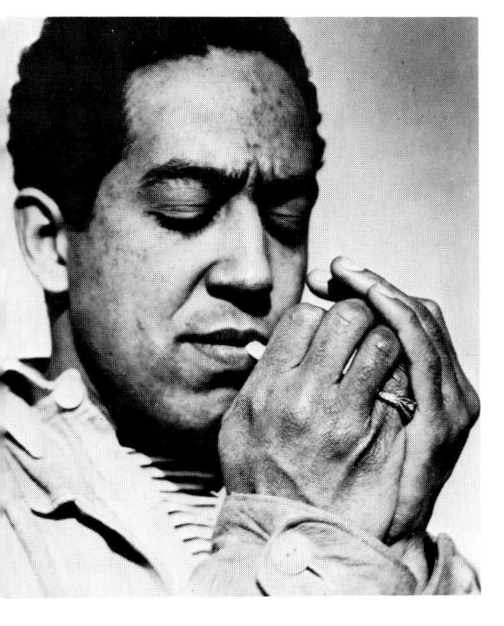

Carmel, 1933. *By Consuela Kanaga.*

(Right) With Noël Sullivan, 1932.

On the beach at Carmel, 1933.

In Honolulu, August 1933.

Thirty-second birthday. With Una Jeffers, February 1, 1934, Big Sur, California.

Carmel, c. 1935. *By Consuela Kanaga. Courtesy of John D. Short, Jr.*

Gwyn "Kit" Clark, c. 1937.

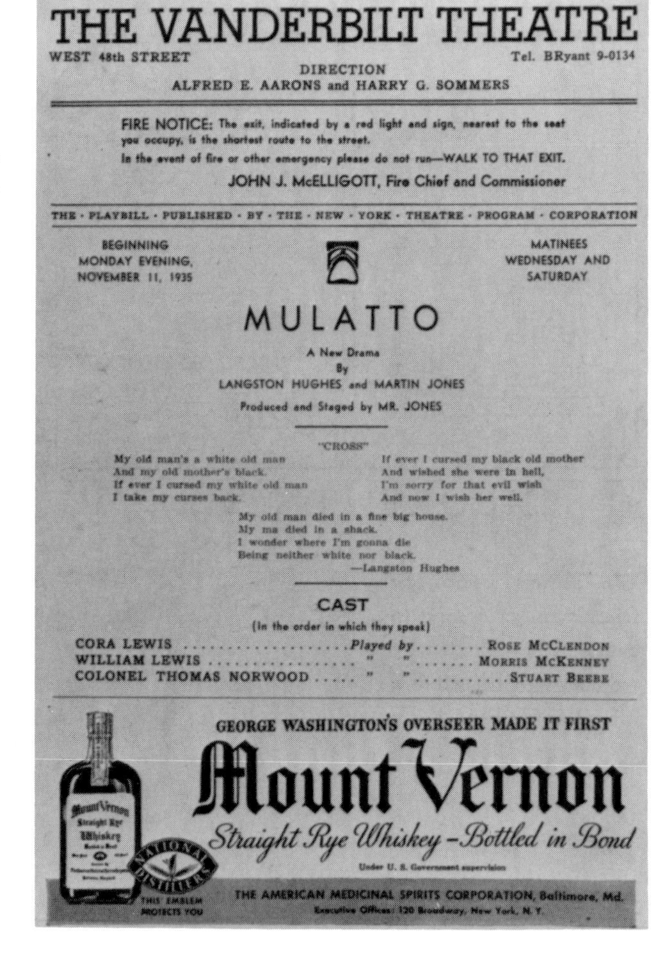

THE VANDERBILT THEATRE

WEST 48th STREET — Tel. BRyant 9-0134

DIRECTION
ALFRED E. AARONS and HARRY G. SOMMERS

FIRE NOTICE: The exit, indicated by a red light and sign, nearest to the seat you occupy, is the shortest route to the street.
In the event of fire or other emergency please do not run—WALK TO THAT EXIT.
JOHN J. McELLIGOTT, Fire Chief and Commissioner

THE · PLAYBILL · PUBLISHED · BY · THE · NEW · YORK · THEATRE · PROGRAM · CORPORATION

BEGINNING
MONDAY EVENING,
NOVEMBER 11, 1935

MATINEES
WEDNESDAY AND
SATURDAY

MULATTO

A New Drama
By
LANGSTON HUGHES and MARTIN JONES

Produced and Staged by MR. JONES

"CROSS"

My old man's a white old man
And my old mother's black.
If ever I cursed my white old man
I take my curses back.

If ever I cursed my black old mother
And wished she were in hell,
I'm sorry for that evil wish
And now I wish her well.

My old man died in a fine big house.
My ma died in a shack.
I wonder where I'm gonna die
Being neither white nor black.
—Langston Hughes

CAST

(In the order in which they speak)

CORA LEWIS . . . . *Played by* . . . . ROSE McCLENDON
WILLIAM LEWIS . . . . " " . . . . MORRIS McKENNEY
COLONEL THOMAS NORWOOD . . . . . " " . . . . STUART BEEBE

Playbill for *Mulatto,* November, 1935.

With Thaddeus Battle and Bernard "Bunny" Rucker of the Abraham Lincoln Brigade, Spain, 1937.

With Mikhail Koltzov, Ernest Hemingway, and Nicolás Guillén, Madrid, 1937.

To Langston, "Con Amore!" Elsie Roxborough, 1937.

(Below) From Nancy Cunard, "with love and admiration." Paris, 1938.

With Matt and Evelyn ''Nebby Lou'' Crawford, Berkeley, California, 1939.

With Alberta Bontemps and children, Paul, Camille, Connie, Joan, and Poppy, July, 1939. *Photo by Arna Bontemps.*

With Etta Moten and Arna Bontemps, Chicago, 1940.

Determined to be quaint, Carmel resolutely refused to incorporate as a town. As a village, it forbade paved streets, sidewalks (thus sacrificing the home delivery of mail), or gas stations within its limits; modern architecture was shunned in favor of gingerbread cottages with twisted chimneys and towers. Studiously casual hedges were set off by a controlled profusion of flowers and plants; geraniums flared tastefully under fragrant eucalyptus trees, and petunias, roses, and fuchsia flourished in the impeccable climate. In spite of its efforts, however, Carmel was not exempt from the pressures affecting the rest of the country. The Depression had hit hard at the village's main sources of income—tourism and real estate. Now quaintness sat poorly with most of the members of the commercial community, who had long argued in favor of expansion, modernization, and progress. By the time of Hughes's arrival in 1933, a deep fissure had opened between these leaders, who were almost all political conservatives, and the local liberals and radicals. The latter were very much in the minority. In November, 1932, in spite of his dismal record and the state's overwhelming endorsement of Roosevelt, Carmel had voted strongly for Herbert Hoover. The next year, for the first time in Carmel history, no artist was elected to the village council. The "artists' colony" had become, in the words of a San Francisco newspaper, "just one more California mecca for prosperous retired people, who are always conservative."

Outnumbered, the left nevertheless was well represented by the local chapter of the John Reed Club, which had sponsored Hughes's reading the previous year. Although its actual membership was no more than twenty, no one questioned the radical credentials of a chapter over which one of John Reed's mentors, Lincoln Steffens, though not a member, presided in spirit. His youthful former wife, the contentious, Australian-born Ella Winter, with whom he still lived, was a founder and a crucial force in its affairs. Winter, a graduate of the London School of Economics, had been to the Soviet Union; her *Red Virtue: Human Relationships in the New Russia,* a breathless tribute to the Soviet system, had just appeared when Hughes arrived. The playwright Orrick Johns was secretary-treasurer, and other members or sympathizers included another popular playwright, Martin Flavin, the fine photographer Edward Weston, the acclaimed sculptor Jo Davidson, and Albert Rhys Williams, who, in witnessing the historic attack on the Winter Palace in 1917, had earned a measure of immortality for himself in John Reed's *Ten Days That Shook The World.* The sheer glamour of such members frightened the village conservatives. One newspaper had greeted the formation of the chapter with a call for "a good old-fashioned tar-and-feather out-of-town-on-a-rail party." The Carmel *Pine Cone* advised the categorical rejection of its offer of books to the village library, since "gifts of snakes and poisoned candy shall always be refused."

Perhaps unaware of the depth of hostility to the club and the left, the "colored poet" settled into a routine. In the morning and early afternoon he wrote, then made time for exercise on the beach, long walks with the energetic German Shepherd, Greta, and cocktail parties and teas at which he got to know better his fellow writers, artists, and intellectuals in the village. Hughes was not isolated in Carmel. Every other weekend or so, Noël Sullivan drove down from San Francisco, and his friends passed through fairly regularly. Langston had barely arrived when the film star Ramon Novarro wrote to ask him to receive two friends for a brief stay. On his daily trips to the post office Hughes often passed the twinkle-eyed, goateed Lincoln Steffens, frail and recovering that fall from a serious operation, taking the sun on his porch but eager to talk (you tell the blacks to ask for *everything,* he once piped to Hughes; then maybe they would get *something*). Ella Winter was herself extremely friendly. Although Robinson Jeffers's stern narrative poems, with their contempt for human beings in comparison to the noble grandeur of nature, stood a world apart in vision from his own work, Hughes also felt welcome at Tor House. He soon recognized that the only truly severe aspect of the shy, deep-voiced Jeffers was his poetry; Una Jeffers, as outgoing as her husband was reclusive, also became a friend. With Steffens, Winter, Jeffers, and Sullivan encouraging him, Langston soon believed that he had found a home in Carmel. California was a "grand place to live and write in," he wrote contentedly to Harold Jackman in October; "the blue curve of the bay with its rocks and pines here at Carmel is something to go crazy about." "I am living so much like white folks these days," Hughes joked to Van Vechten, "that I'm washing my hair with Golden Glow [a popular shampoo]."

The local blacks were few, and Hughes soon knew them all. His favorite was a local cook, the large, gregarious Mrs. Willa Black White, the mother of two young sons, Nat and Cliff; she threw frequent parties, for which she prepared great quantities of excellent food. Hughes also befriended Carl Harris, who ran a machine shop and the gas station that served the village, and his family, including his brother-in-law, Al Byrd; through these Carmelites, he quickly learned where to find other blacks scattered through Pacific Grove and Monterey. Carmel in 1933 seemed free of racial prejudice; open bigotry would come later, when the war brought an influx of Southerners to the military base nearby. Yet, even as Hughes tried to relax, the isolation from typical American social problems vexed him a little. "The remoteness of this part of the world," he wrote Countee Cullen in November, "is amazing." To Van Vechten (busy in New York trying to get the first black on an American postage stamp) Hughes remarked on the number of "rich young kids who were bored with life" and seemed to be without a sense of purpose in Carmel.

To counter this inertia, he reported, the local John Reed Club was doing "noble work!" Its Sunday night meetings, held that fall in a revolutionary-postered barn at the corner of Fifth and Junipero (until the hostile fire chief padlocked the doors to prevent the showing of a Pudovkin film and Soviet newsreels), were almost always stimulating. Joseph Freeman, editor of *New Masses,* spoke early in the autumn on "American Intellectuals and the Crisis." Later, Loren Miller, confirmed as a radical socialist by his months in the Soviet Union in connection with the "Black and White" film project, came up from Los Angeles to address the chapter. On October 22, Hughes himself, introduced by Lincoln Steffens, lectured on Soviet Central Asia. But the members did more than talk. Two days after his lecture, he and others celebrated a union victory of fishermen and cannery workers by going out on a sardine run at night, in heavy seas and fog, on a fishing boat, the *Amazon;* they wanted to see how the underpaid workers, mainly Italian immigrants, earned their living. Hughes, Sullivan, Winter, the poet Marie de Lisle Welch, and other liberal and radical sympathizers travelled once in a group to Visalia in rural Tulare County, a major arena of labor unrest. "We could not stay in Carmel and merely read about it," Winter would write. "Our little bunch gathered one early dawn and traveled to the valley in three automobiles, led by Noël Sullivan in his Cadillac." There, the labor leader Caroline Decker, only twenty-one years old, who had recently addressed the John Reed Club in Carmel, had organized 8,000 Mexicans, blacks, and whites against the cotton growers ("a very thrilling struggle," Hughes told his friends in the East). And on Sunday, December 3, Langston took a prominent part in an anti-war rally sponsored by the Peninsula-American League Against War and Fascism. The day before, in the main publicity parade, he had sat up high in an open car under a ghastly skull and thumped a big drum ("Bear in mind / That death is a drum," he had once written, "Beating for ever / Till the last worms come / To answer its call . . .)."

Hughes, however, had come to Carmel not to agitate but to write. Setting himself the task of completing one short story each week, he wrote steadily. He had a taskmaster now, Maxim Lieber of 545 Fifth Avenue in New York, the literary agent to whom Blanche Knopf had turned over his stories sent from Moscow. Born in the Warsaw ghetto, uncolleged but well read and shrewd, Lieber had started his agency in 1930 after losing his job as an editor with the publishing department of Brentano's when the department collapsed. Openly a socialist, Lieber looked for fair payment for his clients from the bourgeois publishers but also reminded the writers of their political obligations. "I was never personally interested in making money," he would say of his philosophy as a literary agent. "I liked books. If Erskine Caldwell's *Tobacco Road* hadn't become such a successful play, I probably would never have made any money at all. I would send lots of stories to little magazines that paid little or nothing. And I was very much engaged politically in those days. I represented practically the entire left among the writers in America." (In fact, Lieber would later be identified as an underground operative of the Communist Party, and as a spy for the Soviet Union, by Whittaker Chambers, Hughes's friend through *New Masses* and the John Reed Club of New York in the fall of 1930. For some time, Chambers worked in Lieber's agency.) His authors included or would include Nathanael West, Erskine Caldwell, John Cheever, Richard Wright (before, but not including *Native Son*), Thomas Wolfe, Carson McCullers, Tess Slesinger, and Josephine Herbst. He took on Hughes eagerly. "By all means send me all of your material," he had urged. "I shall be very glad to represent you."

Lieber worked boldly to place Hughes's work. To Langston's surprise, he sent "Slave on the Block" (one of the tales written in the Soviet Union) to the venerable *Scribner's* magazine; the journal published it in September. "A Good Job Gone," about an aging white bon-vivant who falls to pieces after an affair with a Harlem temptress, was pressed on the very nervous founder of a new magazine in Chicago. "I have tried to assure him that it will be perfectly safe," Lieber informed Hughes about Arnold Gingrich of *Esquire*. In a publicity ploy, the struggling *Esquire* teased its readers with a synopsis, then asked coyly whether this sordid tale of miscegenation—"the kind of story no commercial magazine would touch with a ten foot pole"—should be published. Letters for and against publication showered on the magazine ("the violence of both the Yeas and the Nays is amazing"). In a long telegram Noël Sullivan ("BEING ENTHUSIASTIC ADMIRER OF LANGSTON HUGHES AND CONSIDERING HIM IN THE SHORT STORY THE EQUAL OF FLAUBERT AND WILLA CATHER") pledged to take out ten yearly subscriptions if the story appeared. It was published in the April number, but other stories were harder to place. In "Home," Hughes wrote with power about the lynching of a sensitive, gifted, and deeply ill young black classical musician who had returned to the South after many years away. "The Blues I'm Playing" concerns a white patron, clearly modelled on Mrs. Mason, and a rebellious, earthy young black woman (this heroine, in rebelling against her patron, loving the blues, and marrying a medical student, curiously combined elements of the lives of both Hughes and Zora Neale Hurston). Reading these two stories, Lieber warned Hughes that he couldn't be "too sanguine as to their sales possibilities. I enjoyed both of them, but I can imagine how shocked every bourgeois editor in town would be with either of these pieces." With the arrival of "Red-Headed Baby," in which a white man confronts his mulatto child and its mother, and "Passing," where a man passing for white writes to his mother after he has refused to speak to her publicly, Lieber threw up his hands: "Nothing I can do will stop you. But you may as well know that editors are frightfully squeamish and will not welcome such pieces."

His warning was well founded. "Home" visited *Scribner's, Forum, Harper's, American Mercury,* and *Atlantic Monthly* before *Esquire* took it the following year. "Why is it that authors think it is their function to lay the flesh bare and rub salt in the wound?" an *Atlantic Monthly* editor fumed; the tale was "both powerful and delicate," but "most people read for pleasure, and certainly there is no pleasure here." Apparently there was even less pleasure in reading about miscegenation. "Red-Headed Baby" was refused by *American Mercury, Scribner's, Story, Harper's,* and *Abbott's,* as well as *Esquire* and *New Masses.* As if affronted by these rejections, Hughes went a step further. In "Mother and Child," he cheerfully told of a married white woman, her black lover, and their defiantly healthy baby boy (in American literature, mulatto babies were almost always sickly and doomed). Worse, the local blacks are prepared to fight if the whites give them any trouble over the couple. This willingness to fight was probably the only reason that *New Masses* took the story, which Hughes liked so much that he turned it into a little play.

The whirlwind of fiction that started in Moscow with his reading of D. H. Lawrence carried Hughes through long sessions at his typewriter during the fall in Carmel. At last, on December 7, he sent Van Vechten and Blanche Knopf twelve stories for inspection. He called the manuscript "The Ways of White Folks" (no doubt after Du Bois's epochal *The Souls of Black Folk*). Their response was all that Hughes could have hoped for. Van Vechten read it at one sitting "and was THRILLED. I think it is superb from beginning to end (including the magnificent title). In fact I think it's the best thing you have done and I am PROUD of you." Accepting it for publication, Blanche Knopf praised the stories as "absolutely top notch and superb."

Having succeeded so dramatically, Hughes covered his typewriter and prepared to enjoy the holidays in San Francisco and Carmel. He needed to rest; a doctor treating him for a stomach pain diagnosed a severe ulcer. The highlight of the vacation was probably a surprise birthday party for Noël Sullivan given a day early, on Christmas Eve, by several of his black friends; Etta Moten came up from Hollywood to sing for him. On Christmas, Langston dined first with Lincoln Steffens and Ella Winter, then again with Sullivan, drank eggnog with the village auto mechanic Carl Harris and his family, and ended the day with champagne at "Ennesfree." His Christmas gift to Sullivan was a manuscript copy of "The Ways of White Folks." "These stories are for you," Hughes inscribed. "You helped me with them, have listened to many of them before they were ever written, have read them all, have given me the music, and the shelter of your roof, and the truth of your friendship, and the time to work. You're a swell fellow. And having cast your bread upon the waters, it comes back to you (this time)—manuscript."

Pleased with what he had accomplished, and eager to keep on writing, Hughes nevertheless gave up some of his time to a different project. After many months in the death house at Kilby and in a Birmingham jail, the Scottsboro Boys were scheduled to face new trials. In December, Langston spoke at a rally in their support, then decided to use his prestige to raise money for their legal fees. Under the letterhead of the local chapter of the Committee for the Defense of Political Prisoners, he appealed for donations to a large number of writers "who have based their inspiration (or a portion of it) on the life of the Negro peoples." To Eugene O'Neill, for example, he acknowledged "the beauty you have given them in your writing. I feel you would not want these nine Negro boys to die." He sent a similar letter to a large number of blacks and whites, including Van Vechten, Cullen, Paul Green, Edna Ferber, and Julia Peterkin.

Encouraged by their response, Hughes next decided to try an auction of manuscripts, books, pictures, and other gifts from writers and artists, to be held in San Francisco in February. Once again, his personal appeal attracted wide support. Noël Sullivan, who gave a tea for a visiting Scottsboro mother, Mrs. Janie Patterson, offered the use of his Hyde Street mansion; the movie star James Cagney, a liberal admirer of Lincoln Steffens who had dined at least once in Hughes's company at "Ennesfree," agreed to be the auctioneer. (Within months, Cagney would be denying potentially disastrous newspaper reports that he was a communist sympathizer.) John Dos Passos sent several manuscript pages from his latest book, *1919,* W. C. Handy offered an autographed score, and donations also came from Ezra Pound, Julian Huxley, Shaw, Bertrand Russell, McKay, Cullen, Anita Loos (a page from, of all things, *Gentlemen Prefer Blondes*), and a host of other artists answering the call of indisputably the leading black writer in America. A preview at Noël Sullivan's home drew a large crowd to Russian Hill to inspect the items and the mansion. By the time the tireless James Cagney jumped down from the podium at the Women's City Club on Post Street, where the auction itself was held, he had sold over $1400 in articles. Hughes, who had hoped to take in $1000, began to plan another auction, to be held in Los Angeles.

The acceptance of his book of stories and the success of the Scottsboro appeals boosted Hughes's confidence enormously; by almost any measure, his career as a writer was proving a solid success. Over Christmas, he let it be known that he might hire a private secretary. On New Year's Day, Roy Blackburn, handsome, brownskinned, and twenty-two years old, arrived to seek the job. "I didn't have an invitation, or even an address, but I had his box number," Roy Blackburn would recall. "Those were hard times. I took the bus down from Oakland and went to the post office to find out where he lived." Blackburn, who wanted to become a chartered accountant and had graduated from a business college, knew no shorthand, but claimed speed and reliability as a typist. In turn, Hughes offered not a salary but room, board, and twenty per cent of the net income on the sales of all the stories, articles, and poems written in the following six months, exclusive of books published or plays sold. Blackburn accepted the terms at once. On January 8, he and Hughes signed a formal agreement for the following six months.

With a resident secretary, Hughes set out to make 1934 an especially productive year. He was determined to revise and publish his once-rejected manuscript of radical verse, "Good Morning Revolution." With Blanche Knopf's blessing, he was also working on an autobiographical account of his year in the Soviet Union, called "From Harlem to Samarkand." A third project was entirely fresh. Hughes and Ella Winter decided to collaborate on a play based on the San Joaquin Valley cotton strike and the heroism of the young labor leader Caroline Decker, to be called "Blood on the Fields." (With the memory of Zora Neale Hurston and his "Mule Bone" disaster still vivid, Hughes decided to take no chances in this collaboration. "At least," he assured Van Vechten, "I shall hire two lawyers and buy a gun at once.") These projects and his short stories kept Hughes and Roy Blackburn busy; gradually their working hours fell deeper into the night, after the teas, cocktail parties, and dinners that Langston hated to miss. "We often worked until two or three in the morning," Blackburn later recalled. "Mr. Sullivan's dog would be asleep in front of the fireplace in the diningroom and we could hear the ocean nearby. Usually Langston dictated from his own typescript or from notes, and I typed. The sound of the machine didn't bother him at all; he had excellent concentration. Each draft was revised and then retyped, sometimes four or five times, until he judged it ready for the mail. Langston was very scrupulous about his work. And he was also a very thoughtful, very kind person to work for."

His efforts continued to pay off. "Rejuvenation Through Joy," a burlesque inspired by a famous Westchester County religious scam of the nineteen twenties involving a messianic charlatan, the "Great Om," was taken by the *Brook-*

lyn Daily Eagle and included in *The Ways of White Folks.* Then Lillian Mae Ehrman, a Carmel acquaintance whose brother-in-law, Ivan Kahn, worked as a Paramount executive, suggested that Hughes send him the story for consideration as a film. Very eager to break into Hollywood, Hughes mailed it off at once. (Eugene Lesche, the confidence man in "Rejuvenation Through Joy," was almost certainly based to some extent on Jean Toomer, or Eugene Toomer—Toomer's correct name—who had brought the word of Gurdjieff from Europe to Harlem, before taking it to more lucrative places among whites. At the end of the tale, a newspaper has reported that Lesche is a Negro, passing for white. Toomer, who had lived briefly in Carmel in 1932 with his first wife, Margery Latimer, a white woman, now adamantly denied that he was a Negro. He had also declined to contribute to Langston's first Scottsboro appeal, claiming that he was broke.) The appearance in the March 17 *New Yorker* of "Why, You Reckon," Hughes's deft short story about a Harlem hold-up of a young, rich, thrill-seeking white man by two hungry blacks, brought eager inquiries from other major magazines to Maxim Lieber about material by Hughes. When Langston sent him five prose pieces on the Soviet Union, presumably for leftist journals, Lieber shrewdly decided to try them first on the capitalist editors. Sex won out brilliantly over fear of Marxist ideology when *Woman's Home Companion* bought "From an Emir's Harem," a slightly titillating article based on Hughes's interview with a former wife of the emir of Bokhara. Although his short stories typically sold for under $150 ("Why, You Reckon" fetched the good price of $135), the harem piece brought $400, by far the largest check to date for a single work by Hughes. "Whoopee!!" Lieber exulted.

In breaking so consistently into the major magazines, Hughes was achieving what no black writer of fiction had done since Charles Chesnutt's more limited vogue late in the previous century. Apparently, he had arrived as a professional writer. Suddenly there even was interest in "Mulatto," the play written in 1930 at Jasper Deeter's Hedgerow Theater. The Harlem Experimental Theater, a new project in which Harold Jackman was involved, wanted to stage it, and a Workers' Theater group in Paris had a similar plan (alas, they had no black actors); a San Francisco amateur group, the Wayfarers, also sought permission for a performance. Hughes, moreover, apparently had arrived as a professional writer without sacrificing his radical principles, as his John Reed Club activity and his Scottsboro efforts showed. Nor had he lost the nerve he had shown on his tour of the South. He could still risk physical danger for the cause, as when he went on January 21 to San Jose with Ella Winter and another John Reed Club member, Marie Short, into the teeth of hostile police and counter-demonstrators to speak out at an anti-lynching rally; Hughes displayed such consummate fearlessness that a quaking Marie Short could scarcely believe it.

Although he was now publishing very little verse, most of it was fiercely radical—"Wait" ("I, silently, / And without a single word, / Shall begin the slaughter"); "Revolution" ("Great mob that knows no fear— / Come here!"); and "Ballads of Lenin" ("Alive in a marble tomb, / Move over, Comrade Lenin, / And make me room"). Most sensational of all his verse was a song composed for a Scottsboro rally, but used also to mark the eighth convention of the Communist Party of the United States. Hughes called it "One More 'S' in the U.S.A.":

> Put one more *S* in the U.S.A.
> To make it Soviet.
> One more *S* in the U.S.A.
> Oh, we'll live to see it yet.
> When the land belongs to the farmers
> And the factories to the working men—
> The U.S.A. when we take control
> Will be the U.S.*S*.A. then . . .

With such writing and his related actions, Langston confirmed himself as the most glamorous black leader sympathetic to the leftist cause. Not surprisingly, when the Communist Party looked for a nominal leader of its major black front organization, the League of Struggle for Negro Rights, he quickly emerged as the leading candidate (proposed no doubt by his friends Loren Miller and Matt Crawford, who were both members of its national council). Knowing very little about the League, and without once setting foot that year in its headquarters on Seventh Avenue in Manhattan, Hughes willingly became a figurehead for the general secretary, Harry Haywood, a communist. In a draft program published while Hughes was president, the League reaffirmed its old, controversial position calling for "the confiscation without compensation of the land of the big landlords and capitalists in the South," with the land to be given to black and white small farmers and sharecroppers, and for "the complete right of self-determination for the Negro people in the Black Belt with full rights for the toiling white minority."

Hughes's acceptance of the presidency of the League of Struggle for Negro Rights was further evidence of how confident he had become about his ability to sustain his literary career. Another sign was more intimate. Apparently, he began to think a little, or as seriously as he could ever do so, about marriage. On January 1, after a silence of about seven months, he sent a short note with an invitation, albeit casual, to his old Moscow sweetheart, the beautiful Afro-Chinese dancer Sylvia Chen. "Dear Sylvia," he began, "—Think about you all the time and wish you were here. Come on over this way, will you?" After news and requests for news, he closed affectionately: "Write to me. Love to you (lots!) Langston."

From the Hotel Metropole in Moscow, Sylvia Chen replied to "Langston—Langston—": "However did you happen to remember me after all this time—I am surprised! I suppose one of your New Year duties is to write to all your girl 'friends' who are scattered over the world, and tell them you remember them. I see I happen to fall into this category—how flattering. . . . Sorry you were silent for such a long time. I was hurt—I liked you." Hughes received this letter, but did not reply at once. Then, three months after breaking his silence with the card, he sent her a telegram: "COME OVER HERE THIS IS SERIOUS I LOVE YOU." In Moscow, a startled Chen believed him for just a moment before seeing the joke. The telegram had been sent on April 1. From Norway, where she was on tour, she chided him a little, but also confessed that she was very happy to have received the telegram. Indeed, she had been thinking of coming to the U.S.A. Perhaps they might meet.

Hughes then let their correspondence almost die. On June 26, after receiving only a copy of one of his books, Sylvia Chen swore affectionately never "to write you anymore, honest. I've written lots of times & received in return—a Xmas card, a telegram, and a book—now I ask you is that the way to treat a girl!" But early in July, when Hughes felt bold enough to write her again, he displayed far more feeling than he had shown before:

> Darling kid,
>
> You know what would happen if you came over here? I would take you and keep you forever, that's what would happen. And even if you didn't come over here and I ever found you anywhere else in the world—I'd keep you, too. So you see, I love you!
>
> Swell to have your letter. But listen—were you any more serious than I was? And how did I know I was so much going to *miss* the hell out of you after you went South to dance or I came half a world away from you? But I do miss you—lots more than you miss me, I guess,—and I want you, Sylvia baby, more than anyone else in the world, believe it or not. I love you.
>
> But to change the subject. . . .
>
> Please, dear kid, believe what I say about how much I like you. If you want me to say it over and *over* and more and *more,* just act like you don't believe it in the next letter you write me. . . .
>
> Two tons of love to you,
> Write soon,
> Langston
>
> Wish I could kiss you! Do you?

Hughes was serious enough about Chen to ask Marie Short's husband, Douglas, a San Francisco lawyer, to look into the legal aspects of her possible emigration to the United States. When Short reported that she would have a much better chance coming to the United States as a black than as a Chinese, because of the exclusionary laws, the men discussed the effect that marriage to Hughes would have on her chances of immigration. But Hughes never communicated this degree of seriousness to Chen, or made a decisive move to claim "the sweetest little girl I have ever ever seen!" Even with his new confidence, and in the congenial atmosphere of Carmel, he could not fool himself for long into thinking that he was ready for the challenge of love and marriage. When Chen wrote affectionately about her desire for a brown baby, Hughes joined in the fun. "What nationality would our baby be anyhow?" he joked. "Just so he or she is anti-Fascist!" "I love you," he wrote in October. "And what else is important?" "Be a good girl and a great dancer, and *Sylvia,*" he ended his letter, after Chen had wondered about her mixed race and her possible career in America, "don't be nobody else—cause I don't want nobody else—but you. Love and—kisses. x x x x x x These are the kisses. x x x x x x. " But, in the end, Sylvia Chen's patience brought nothing. Nor did her sharper words, as when she probed him about his superficiality. Was he really like that, "or is it habit or maybe protection from sentiment or—well it could be a hundred different things. . . . I find I can't be natural with you, because you treat me and most people . . . so artificially."

A fumbling, half-hearted lover, by the late autumn Hughes, in fact, had lost Sylvia Chen. Just before hearing from him the previous January, she had met on a Moscow street a bashful but smart young white American from Dayton, Ohio, Jay Leyda, who had come to the Soviet Union to study the cinema. While Hughes had been neglecting to write, Jay Leyda had fallen in love. Nevertheless, Chen had kept the door open to Hughes. That summer, however, just after Leyda returned to Moscow from a theater festival in Georgia, sunburnt and more exuberant than Sylvia Chen had ever known him, they were married.

Hurt by his treatment of her, Chen did not reveal her marriage at once to Hughes. A year later, when she received from him another book, but no letter, she wrote to Langston "as to a ghost, it's such a long time since I've heard or seen you, yet still I like to keep in touch with you to remember that you're there somewhere, flitting around being amiable—but you know Langston although you tried to do the same with me don't you think you had me fooled, no sir, I know you—and, I may add, being frank as usual, like you in spite of knowing you!" In a letter one year later, she finally broke the news about Jay Leyda to Hughes: "Yes I got tired of waiting on you to propose so got myself a consort. Maybe you'd like to get on my waiting list for future consideration." "Langston wrote me a very angry letter in response," Si-lan Chen Leyda later remembered. "He was furious with me for not waiting on him. I destroyed the letter at once." In any case, Hughes eventually recovered from his wounds; he was most polite to the Leydas when they finally arrived in the United States, on St. Valentine's Day in 1937.

Long before his inept "courtship" had played itself out in 1934, however, his old restlessness, the old contradiction between his love of laughter, talk, and company and his sense of aloneness had overtaken Hughes in Carmel. "Seemingly there are too many people in my life," he complained early that spring. "Carmel is not at all a remote village on a rocky and wind-blown coast. It is 59th Street and Fifth Avenue. And every hour somebody from London, Paris, Chicago, or New York arrives and a cocktail party is given for them." Cocktail parties were a poor, mocking version of the intimacy Hughes evidently wanted but could not realize. One cause of his growing discomfort must have been the sheer whiteness of the Carmel community. Given his love of his race, he could hardly have failed to notice that, both in his radicalism and his privileged living, he had strayed a fair distance from the black masses. The local blacks liked him but distrusted his politics. Once, after he had spoken in praise of the Soviet Union at the black First Baptist Church of Monterey, Hughes found himself badly embarrassed by a white man, California State Senator E. H. Tickle, who painted such a gloomy picture of religion in Russia that, at a reception afterwards, Hughes felt the cold disapproval of the congregation. On certain occasions he felt almost at home in Carmel. On his thirty-second birthday, for example, he was happy when Robinson and Una Jeffers and Noël Sullivan drove with him down to majestic Big Sur, the only wild coast left in California, for a picnic of wine and salami sandwiches among the lilacs, sagebrush, and live oaks high above the ocean. And later that day, his black friends honored him with a dinner of ham and greens. But none of these friends, black or white, woman or man, became a true intimate. Gregarious in the eyes of the world, Hughes remained a man apart.

What he did for sex was, as usual, a mystery. Inevitably his name was joined in gossip with the aggressive Ella Winter, who was reported once in a newspaper as having slept with twenty-nine black men (''Why not thirty?'' Lincoln Steffens yawned). And he was also unquestionably fond of the talkative Marie Short, a former San Francisco debutante turned Carmel intellectual and socialist, whose marriage to Douglas Short was collapsing, and who sometimes seemed confused and forlorn. Accepting her friendship, and quick to join her in gossip about their neighbors, Hughes shared many hours with Marie Short on the Carmel beach and in her home. Still he apparently remained unconnected. Roy Blackburn would remember that at ''Ennesfree'' the topic of romance never arose with Hughes. ''Langston never brought it up,'' he recalled. ''We never talked about sex or the various ladies in the village in any way, even joking. He was an extremely private person, but he also seemed to have no interest whatsoever in the subject. Even then, it seemed to me odd, this lack of interest. We were both young men, and young men usually talk about women. There were times when I wondered whether he had any sexual feeling at all. I'm really not sure he had any.'' To John Short, one of Marie Short's sons, Hughes was ''always the center of lots of laughter. One could sense under the laughter an undercurrent of loneliness, or a deep quietude, but he was warm and loving, and liked to touch and to hug, without the slightest inhibition in any way. On the other hand, one didn't get a sense of sexuality. As affectionate as he was with all of us, especially my mother, I felt then and later that Langston was asexual. I'm sure he was not a homosexual.''

Although homosexuality was largely an unspoken word in public, Noël Sullivan's status as an inveterate bachelor, and his relationship then and over the coming years with the equally resolute Hughes, may have caused some speculation. According to Noël Sullivan's great friend Ben Lehman, a prominent professor at the University of California, Sullivan ''was totally uninterested in women physically,'' often confessing ''frankly and clinically as a doctor might'' that his interests were in the opposite direction. Lehman did not speculate on the extent to which Sullivan fulfilled his interests; certainly Sullivan was restrained by his devout Catholicism. But if Hughes and Sullivan were thought to be involved, or to have been involved, the idea was not to be taken seriously. One irrepressibly homosexual intimate of the large Sullivan household (in which Hughes lived again later) would find astonishing the idea that Langston might have been homosexual. Amiable, fun-loving, Hughes was yet a sexual blank; his libido, under stimulation or pressure, seemed to vanish into a void.

What he could depend on for satisfaction was his work. In May came the latest dividend of his efforts, the ten author's copies of *The Ways of White Folks*. Beneath his dedication to Noël Sullivan was an epigraph taken from the story

"Berry": "The ways of white folks, I mean some white folks. . . . " The reviews, by and large, were very favorable. In the *North American Review,* Herschel Brickell called the tales "some of the best stories that have appeared in this country in years." Elsewhere, Horace Gregory praised Langston for a "spiritual prose style and an accurate understanding of human character" that suggested genius. In the *Saturday Review,* Vernon Loggins, the author of the fine study *The Negro Author: His Development in America to 1900* (1931), proclaimed the collection Hughes's strongest work. High praise also came from Alain Locke, though not without a murmur that "greater artistry, deeper sympathy and less resentment, would have made it a book for all times." Hughes's harsh treatment of whites did not escape notice. "My hat is off to you in relation to your own race," Sherwood Anderson saluted him in the *Nation,* but not in relation to the whites, who were caricatures. And Martha Gruening, a prominent white liberal, deplored the fact that he saw whites as "either sordid and cruel, or silly and sentimental."

With this collection, Hughes repeated in the short story the feat he had already accomplished both in poetry and in the novel; he set a new standard of excellence for black writers. In emphasizing the folly of liberal whites, or whites who involve themselves with blacks mainly to exploit them, he was exorcizing the fiercest demon in his own past; the entire venture had started with his recognition of Mrs. Mason's terrifying white face in a story by D. H. Lawrence, and with Hughes's immediate recognition that he could lash back at her behind the disguise of fiction. Unlike the sweet little boy at the center of *Not Without Laughter,* the presiding consciousness in this collection is adult and unromantic, disillusioned and frankly bitter. Instead of staying in the black quarter, as in his novel, the narrator risks the barbed wire and the minefields where the two hostile races clash. Compared to *Not Without Laughter, The Ways of White Folks* is far more adult and neurotic, more militant and defensive, and thus more modern and accurate as a description of the Afro-American temper as it was emerging. For all its original beauty, the novel *Not Without Laughter* is sometimes a genre piece; for all its inconsistencies of quality, tone, and spirit, as well as its unevenness as art, *The Ways of White Folks* is a striking original, daring to say what had never been said so definitively before.

As spring came to Carmel, he pushed his latest projects forward. Using a twelve-pound package of his notes mailed from Moscow, Langston worked hard on his Soviet book, "From Harlem to Samarkand." Not surprisingly, given the almost daily influence of Steffens and Winter, as well as his own pro-Soviet disposition, he was determined not to criticize the U.S.S.R. in any way. Instead, Hughes stressed its great progress and its illustrious future. He soon mailed the finished manuscript to Blanche Knopf. On April 3, he began "Blood on the Fields," his collaboration with Ella Winter. They met virtually every day for about a month to hammer out its basic structure from a mountain of information she had assembled, including a detailed chronology of events from the announcement of a strike in the San Joaquin Valley by the Cannery and Agricultural Workers' Industrial Union on September 18, 1933, through the killing of two strikers by anti-labor vigilantes. Drawing on drama techniques he had seen in Moscow, Hughes proposed a sort of cavalcade of three acts and about a dozen scenes, with a huge cast (eventually about 125 characters) that would include poor whites, Mexican migrant laborers, blacks, policemen, scabs, owners, vigilantes, and union organizers. At the center would stand a character based on the blond young heroine of Tulare, Caroline Decker. "The play follows the strike almost exactly," Hughes informed Lieber. "The result," he hoped, "ought to be a sound and provocative drama of the actual struggles of agricultural and migratory workers in California (and in a sense, all America) today."

With late April and May, however, came distractions. The weather was spectacular even for Carmel, golden days that encouraged not writing but picnics and parties and hours on the beach. May also brought lively visitors to "Ennesfree," including the quixotic black cabaret star Nora Holt, a fleshy favorite in New York and Los Angeles, who swept into Carmel for a week and left the white village in shambles. Having arrived in a humble Ford, Hughes noted gleefully, Holt had departed in a Cadillac—most likely driven by Noël Sullivan, who had hustled down from San Francisco to meet her; Edward Weston photographed her about forty times, Langston counted. One other visitor stood out for Hughes. On May 5, Wallace Thurman arrived at "Ennesfree" with a friend from Hollywood, where Thurman was working as a screenwriter. Sick with tuberculosis, drinking heavily, depressed, Thurman was a ghost of the bright black boy Hughes had met in New York ten years before. His talk now ran hollow and bitter; the self-hatred was naked. Thurman was also broke. Hughes lent him money and was relieved when he left Carmel on Sunday evening. Soon Langston was pondering the meaning of a telegram sent by Thurman announcing that he had just married someone named "Fay"—a curious piece of news, since he was still married to Louise Thompson.

But the greatest distraction from work on "Blood on the Fields" came, ironically, with the worsening of the labor situation in California. On May 9, under the leadership of the Australian-born radical Harry Bridges, almost twelve thousand members of the once toothless International Longshoremen's Association began a bitterly contested strike that would close down ports from Seattle to San Diego for ninety days. With cargo piling up on the docks, and harvests rotting in the fields, virtually every California community was seriously affected; newspaper editors and other leaders charged that communism was behind not only the dock strikes but all labor unrest in California. The *Carmel Sun* and the *Pine Cone,* frantic about the loss of tourist revenue and the decline of real-estate values in the area, were quickly among the papers reporting rumors of communist plots to destroy the harvests. For both papers, the time had come for a showdown between local patriots and the radicals in the John Reed Club. The radicals, for their part, seemed ready to fight. They had celebrated May Day with a stirring evening of "proletarian" poetry at the Steffens cottage, during which Langston read his translations of two Mayakovsky poems. Totally in support of the strike, the club collected clothes, food, and money for the stevedores; the members also sent telegrams to the governor, the mayor of San Francisco, the secretary of labor, and President Roosevelt himself, demanding that the strikers be protected from vigilante action and overzealous police.

As the most obvious symbol of the left in Carmel, Hughes felt the sudden rise in tension almost at once. Pressured by village merchants and other employers, virtually all the other blacks in the area banded together to affirm their allegiance to the U.S. Constitution and to oppose the strikers and the John Reed Club. Disgusted, Hughes went the opposite way. Joining in the public welcome accorded the Soviet consul to San Francisco on a visit to Carmel, he also visited the city for the much-advertised opening of a new Soviet consulate there. But he was probably also glad that he was scheduled to leave Carmel soon. His subsidized year was coming to an end; moreover, Noël Sullivan, short on cash because most of his fortune was tied up in real estate currently depressed in value, had decided to renovate "Ennesfree" to make it his main residence. On June 11, some of the younger blacks in the Carmel area gave Langston a farewell stag party and presented him with a purse. A few days later, before he left "Ennesfree," the carpenters arrived to begin work.

But he was still in Carmel on July 2 when, after an electric meeting of the John Reed Club addressed by an organizer for the striking longshoremen, the radio stations broadcast the news that the police intended to question the "radical capitalist" Noël Sullivan, Lincoln Steffens, and others in the club about the telegrams sent in support of the strikers. One persistent report had Ella Winter as the conduit by which money flowed to the strikers from the Soviet Union. The next day, fighting broke out on the San Francisco piers. An attempt by the pro-business Industrial Association to open the docks with strike-breakers led to a four-hour running battle with the strikers that involved guns, gas bombs, bricks, and rocks. The Fourth of July brought a lull in the fighting, but the day after, when the Industrial Association, backed by the police, again attempted to move cargo, the waterfront erupted into violence unequalled in San Francisco since the earthquake of 1906. By nightfall, two strikers were dead, a hundred persons were injured, and no end to the disturbance was in sight.

In Carmel, when the John Reed Club prepared leaflets supporting the call by the longshoremen for a general strike, members of the arch-patriotic American Legion destroyed all the copies they could find. On July 8, at a stormy public meeting attended by both Hughes and Sullivan, the majority of residents voiced bitter opposition against the club and its supporters. Soon after, seventy-six residents gathered to found a Carmel post of the American Legion. July 17 was a day of terror for socialists in San Francisco, where vigilante groups attacked workers' halls and the police arrested nearly 400 persons affiliated with the left. At a meeting called by the John Reed Club to discuss the situation, American Legionnaires and villagers now openly hostile ("one's laundry man and one's druggist," Hughes noted, "real estate men and hotel keepers, and the young man who sold papers in front of the Post Office") vastly outnumbered leftists in the jammed hall, and an American Legion stenographer conspicuously took down all the remarks. To his special dismay, Hughes noticed something else: "I was the only Negro there."

A few days later, the Carmel village council ordered the purchase of tear gas in order to be ready for civil insurrection. A citizens' force, backed by the Legion and armed with riot guns, began to drill on the local polo grounds. "There is trouble in store for those who persist in communistic activity," the *Sun* warned. "And that is as it should be."

"I, as a Negro member of the Club, seemed to be singled out as especially worthy of attack," Hughes would soon note. "Rumors of malicious intent filled the town: that I was frequently seen on the beach and in cars in company of white women, that I called them by their first names, that I was a bad influence on the Negroes of the town." When a club member was told that vigilante action was being planned, "Carmel bubbled with a kind of hysteria." A leading vigilante leader passed the news to Hughes through another local black "that I *was* in physical danger, and that there was no way of assuring anyone safety." On July 20, Marie Short telephoned Hughes to warn him that an attack was imminent. The next day, he sent Roy Blackburn home to Oakland. On the evening of July 24, after shipping a large parcel of papers to Maxim Lieber, taking with him as much material as he could carry and hiding the rest, Langston slipped out of town. Passing over the heavily forested Santa Cruz Mountains, he reached San Francisco by way of the Skyline Drive. He left Carmel, Hughes would write Sylvia Chen, "not wishing to be tarred and feathered."

In San Francisco, he continued his work on "Blood on the Fields." On August 8, he finished the play and sent it at once to Lieber. At the last minute, to Hughes's surprise, Ella Winter asked him to leave her name off the title page. Hughes could flee at any time, but she had to live in Carmel with Steffens and their ten-year-old son, Pete. Permissively reared, Pete was once attacked in the *Pine Cone* for urinating in public; Winter was generally accused of being a failed wife, a bad mother, and a pernicious influence on the community. Her dairyman had expressed an interest in taking care of the members of the John Reed Club with a machine-gun. Nevertheless, on August 21, Winter and Hughes sent a collaboration contract (fifty-fifty) to the Dramatists Guild in New York.

By this time, "Blood on the Fields" had become a crucial venture for Hughes, whose apparently well-established career had suddenly begun to fall apart. In spite of the good reviews, *The Ways of White Folks* was selling poorly except in San Francisco; instead of sending an advance on Hughes's Soviet book, as he had hoped she would, Blanche Knopf had rejected "From Harlem to Samarkand" outright. The manuscript was innocuous, "charming and pleasant, but . . . not fresh and not new." She had wanted an important book—about his love affairs with Russian women, "or a very definitely Communistic book, or many other things," but not "a pleasant book about Russia." His career had reached "a very important moment." "I want to go on with you, without question," she assured Hughes (who probably understood that his head was on the block). Earlier in the year, she again had turned down "Good Morning Revolution," in a decision backed by Van Vechten in spite of Hughes's personal appeal to him on behalf of the radical poems. "I know you *don't* like them," he had pleaded, "but I *do* like them." Van Vechten had insisted that politics had nothing to do with it; the poems were "lacking in any of the elementary requisites of a work of art." Not long after Blanche Knopf's rejection of the Soviet book, Hughes lunched uncomfortably with her at the Mark Hopkins Hotel on Nob Hill in San Francisco. As polite as ever, she was also just as unsentimental.

With the breaking of the strike in San Francisco, the vigilante threat blew over. But Hughes was still subdued when, on August 13, he returned to Carmel with Roy Blackburn and Eulah Pharr, Noël Sullivan's housekeeper, to find the cottage beautifully transformed by the painters and carpenters, who had converted part of the attic into a cozy sleeping area reached by a retractable staircase. A cold, or a disinclination to go out, or both, kept Langston in bed for a few days, where he read a book about Nijinsky (which he later mailed to Sylvia Chen); he recovered in time to attend a picnic on the beach with a crowd of local blacks. At the height of the summer season Carmel seemed at last back to normal. Residents and tourists alike were filling the beaches and golf courses, crowding Ocean Avenue and the Seventeen Mile Drive through the heart of the peninsula and along the coast, and preparing for a five-day celebration of the sesquicentennial of the death of Father Junipero Serra, to be capped by a pageant in which Noël Sullivan would take part.

Hughes had just begun to catch something of the holiday spirit when, on August 23, he opened the *Sun* to find himself singled out for scorn in an essay by its editor, E. F. Bunch:

> . . . Langston Hughes has been a very "distinguished" guest in Carmel—not that the town is proud of it. He has been the guest of "honor" at parties. Whites were invited—a "select" few—to bask in his wisdom. White girls have ridden down the street with him, have walked with him, smiling [into] his face. And the Woman's Home Companion and other magazines which should be standing for America, have spread his Communistic doctrines to the thousands and thousands of American homes. Russia would be a good place for Hughes. . . .

Stunned, Hughes had not recovered from this attack when, a week later, the *Sun* insulted him again with a letter to the editor that Bunch himself introduced. Hughes was "a tireless and violent advocator for Soviet Rule, for Communism, in this country." The article on the emir's harem was a "subtle defamation of American womanhood." As for the mingling of races in Carmel, that was "a very difficult subject. So difficult for Carmel as to have been in the past . . . odorous, to say the least."

Hughes knew now that he had stayed too long in Carmel. Behind its smiling mask had lurked all along the old, ugly face of American racism. These humiliating public attacks, the stinging publishing defeats, and the fact that his year of patronage was over pushed him into deep gloom. From which Noël Sullivan tried to lift him. On the first day of September, Langston, Noël, and Sullivan's very close friend, the big, brassy actress-singer Elsie Arden, drove north from Carmel past Sacramento; the next day, they reached the country home of another friend of Sullivan's, Ernestine Black, a socially prominent *San Francisco Call* columnist, on Echo Lake high in the Sierras just south of Lake Tahoe. Here, about 7,500 feet above sea level, the friends idled and enjoyed the magnificent weather. They went out rowing and swimming, and slept outdoors on pallets strewn on a large verandah under star-filled skies.

Hughes's mind, however, was on his future. Blanche Knopf was right about one thing; he was at a crucial stage in his career. Only four years after his break with Godmother he was down low again; if the element of specific personal rejection was now missing, he needed nevertheless, once again, to make a fresh start. He also knew from bitter experience what he had to do immediately. He needed to retreat; he also needed to feel, like broken bones in his flesh, the twinned pains of isolation and poverty, the forces that had shaped his life and his art. And, with that sense of suffering, to return to his greatest obligations, from which he had been led away, or had strayed, in Carmel. Perhaps it was at this point, or not long after, that Hughes sketched an outline called "Me and the White Race," intending it either as a plan for an essay or simply as a random accounting, not always clear in meaning, of his beliefs concerning race and poverty. Certainly Hughes seemed to be responding to the insults in the *Sun*. He did not hate whites, he wrote, and he was not bitter. One note said simply "white prostitutes from Cleveland to Vicksburg" (prostitutes he had patronized? or who had propositioned him?). He admired many whites. If white liberals were not brave, certainly white radicals were. On the other hand: "I would hate to *have* to marry a white person." "Personally I am not attracted by whiteness in either males or females." How silly the blacks and the whites who wear their friends of other races like jewels, always talking about them! The Jews were great internationalists (like himself?). Rich Jews and rich blacks could buy their way out of racial trouble; "therefore, to realize a new brotherhood, I count out the rich of all races. They don't suffer enough. The poor, when they get wise to themselves, will accomplish the real internationalism. Then me and the white folks will be Us."

Another note perhaps told tersely of his determination to act, or simply recorded a remark he had heard and now considered very important: "You got to be hard inside. Maybe you might be soft outside but you got to be hard inside."

In a decision probably made without much premeditation, Hughes decided to go to Reno, Nevada, a haven of legal gambling and prostitution and quick divorces, and look around for a place where he could live cheaply and anonymously. On September 4, he rose very early at Echo Lake and caught the 5:55 train to Reno. Once there, he liked what he saw—the minimalist desert landscape of sand, dirt, red rock, and blue sky, but with wooded mountains not far away; the weather hot—like Africa, he noted. Reno was hardly Africa, but Hughes, as he had done in fleeing to Haiti three years before, perhaps wanted to be reminded of blackness. The porters on the train directed him to 521 Elko Street, where a fat, black, respectable lady with a cute Pekinese dog ran a boarding house. The house itself was clean and white, and fronted by a neat lawn. For $2.50 a week, "James Hughes" rented a room; adding some money to a pool, he shared meals with the porters and cooks boarding there. With a base in Reno, he quietly explored the town. In one of the two casinos that admitted blacks he dropped a few dollars; loitering around a gymnasium where the brilliant Cuban prizefighter Kid Chocolate was training, he was allowed to keep time while the Kid sparred.

Satisfied that Reno would do, at least for a while, he returned to San Francisco on his first airplane ride, a thrilling flight over the Sierras into the Pacific sunset. The next day he went down to Carmel for a beach party with Sullivan, Marie Short, Albert Rhys Williams, Ella Winter, and other friends. All were mortified by the attacks in the *Sun,* but their talk struck no sparks. Collecting his mail, Hughes shared his latest bad news with Winter. Paul Peters and Margaret Larkin of the left-wing Theater Union in New York had found "Blood on the Fields" grandiose and lacking in cohesion. Hughes and Winter agreed to continue work on the script, but aided by Ann Hawkins, an experienced director who published from time to time in Lincoln Steffens's *Pacific Weekly.* Langston would send his changes to Ella Winter, who would pass them along to Ann Hawkins.

A few days later, Hughes was in San Francisco for a mid-afternoon reading at Paul Elder's Gallery on Post Street that lifted his spirits; so many people wanted to hear him that he had to read twice. In the second audience was twenty-six-year-old William Saroyan, who later introduced himself and gave Hughes a copy of his new book, *The Daring Young Man on the Flying Trapeze.* Afterwards, Langston, Saroyan, and young Tillie Lerner (later Tillie Olsen) went out together for a quick supper. Lerner, one of the communists jailed in the recent police round-up, excitedly showed Langston a big contract she had just received from Random House to write a book about her adventures. Hughes found Saroyan "a very Armenian looking young man, Americanly nervous at the same time something like James Cagney," and almost as quick to laugh (Saroyan told him he had once worked for an undertaker whose slogan was, "We give you lots for your money"). Insisting on seeing Langston to a streetcar, Saroyan put him on the wrong one. As it pulled away, he admitted later, he didn't know whether to stand there like a fool, or run away.

Back in Carmel, Hughes twice heard the Indian mystic Jeddu Krishnamurti speak. On the cliffs at Big Sur, above fields of fog, he passed a glorious afternoon with Douglas and Marie Short and their friend, the photographer Constance Kanaga. Later, he posed for Kanaga, who captured a well-dressed, sleek, brown man, all smooth skin and slicked hair, a little plump but still handsome, still young. But if Hughes seemed a portrait of sepia success, his bank account was now overdrawn by $5.80. A check arrived, but only for $13, from *The New Republic,* for a clever blues satire, "Ballad of Roosevelt" ("I'm waitin' on Roosevelt, son, / Roosevelt, Roosevelt, / just waitin' on Roosevelt"). Delaying his return to Reno, and enjoying Noël Sullivan's hospitality, he finished "On The Road," a short story inspired by rugged Reno and perhaps by his own increasingly desperate circumstances. A big, black, hungry man, who takes refuge on a snowy night in a hostile white church, hallucinates about dragging the building down, then converses with Christ before being thrown in jail.

Neither the anemic state of his bank account nor the recent humiliations prevented Hughes himself from hitting out. In "The Vigilantes Knock At My Door," written in Carmel for *New Masses,* he ridiculed the tactics of the right wing. He saved his last words of contempt, however, for four local blacks allied to the conservatives—"political opportunists hoping for a rake-off from somebody after elections"—who had opposed the John Reed Club in spite of its stand on Scottsboro, and opposed the longshoremen's union despite its goal of ending discrimination on the docks. For the tenth anniversary of the communist *Daily Worker* he sent a ringing endorsement: "Every Negro receiving a regular salary in this country should subscribe to the *Daily Worker,* and share it with his brothers who are unemployed." And he was resolute again when news came that Jacques Roumain, the cultivated Haitian poet he had visited in Port-au-Prince in 1931, who had become increasingly radical in the years since, had just been sentenced to three years in jail for alleged pro-communist activity. Outraged, Hughes drew up and circulated a stern demand for action: "As a fellow writer of color, I call upon all writers and artists of whatever race who believe in the freedom of words and of the human spirit, to immediately protest to the President of Haiti and to the nearest Haitian consulate the uncalled for and unmerited sentence to prison of Jacques Roumain, one of the few, and by far the most talented of the literary men of Haiti."

Evidently Hughes's opinion, especially on such matters, carried weight nationally. In an article in the New York *Daily Mirror* that would be cited for many years, the prominent historian Charles A. Beard named him (along with Franklin Roosevelt, Will Rogers, Charlie Chaplin, H. L. Mencken, Edith Wharton, Sinclair Lewis, and others) as one of the twenty-five "most interesting" persons in the nation, each a leader "to whom in times of stress we can turn and who is socially conscious." The notice countered Hughes's inclusion, along with Steffens, Winter, and many wildly improbable people, in Elizabeth Dilling's *Red Network,* a guide to American communists and their dupes.

Beard's notice was flattering, but it paid no bills. In a quiet moment, Hughes read Saroyan's book outdoors in the September sunshine and tried to look beyond Reno into his future.

He decided to apply for a fellowship from the Guggenheim Foundation. In the following days, he secured promises of endorsement from Robinson Jeffers, Amy Spingarn, Noël Sullivan, Ben Lehman, Blanche Knopf, Mary McLeod Bethune, and Elmer Carter, the editor of *Opportunity.* Heeding the advice of Blanche Knopf, he proposed to write another novel, about "urban Negro life in America, to be complete in itself as to interest and story, but at the same time to form a continuation of *Not Without Laughter,*" which he planned to "develop into a trilogy of Negro life covering the childhood, youth, and manhood of a black boy in this country."

Near mid-October, Langston returned to Reno. By this time he had made another decision: he would go to live, at least for a while, with his mother, who was now residing near distant relatives in Oberlin, Ohio, where her parents had lived in the middle of the nineteenth century. He mailed her a large package, almost certainly the bulk of his files. Hughes arrived at this decision reluctantly. Although Carrie Clark had been urging him for some time to join her, he distrusted her motives. To his annoyance, she had refused to go on welfare; instead, she hinted very broadly that Langston had been hiding his riches from her. Whatever her delusions, however, her situation was not good; he would go home to help her. Langston had barely reached Reno when an offer came from Noël Sullivan to send thirty dollars each month for a while to support Carrie—although, Sullivan insisted, "the world owes a lot more than this meagre security to the woman who bore you.'' Embarrassed, but accepting the first installment, Hughes gently scolded Sullivan that there was ''no reason'' at all ''why you should take care of 90% of the troubles of the world.'' And especially when Carrie was being difficult. To Maxim Lieber, Hughes complained about having a family that ''strangely enough insists on *not* going on relief, having a son who is a 'great' writer—although I am about to go on relief myself.'' Later, when Langston sent Carrie $81.30 in four amounts, she not only failed to acknowledge receipt of any money but reproached him, when she at last went on relief, about how ''the Negroes exploded on the fact I had a 'wealthy' son, that you were in Hollywood almost playing with Joan Crawford and Clark Gable.''

Back in Reno, where only Noël Sullivan and Blanche Knopf had his address, he began to revise his manuscript on the Soviet Union. Although Mrs. Knopf's criticism of it as innocuous had stung, he knew that somewhere on the road from Moscow he had lost his voice. ''I think my *métier* is protesting about something,'' he conceded. But the book went nowhere. ''From Harlem to Samarkand'' was to be autobiography, and Hughes on Hughes was deeply recessive. Partly a political treatise, such a book demanded a trenchant intellectualism that his emotional reticence made very difficult. Men as different as W. E. B. Du Bois and Arthur Koestler kneaded their loves and hates, vanity and vulnerability, into their intellectual positions. Hughes tried to keep his private emotions and his display of intellect far apart; moreover, his private emotions, in so far as one could guess them, seemed to be distressingly benign. The result could only be a book without force. Despairing of improving the manuscript, and badly in need of cash, Hughes begged Lieber to sell pieces to the black press, or to ''anyone for 7 or 8 bucks.''

Not surprisingly, a melancholy mood swept over him. Almost every day, after hours at the typewriter, he took long walks into the nearby hills, from which he looked down on the city and the Truckee River. The leaves were turning red and gold; everything conspired to make Langston sad, and think of death. One day, during a stroll, he came upon ''a forlorn little mountain cemetery with a log fence around it.'' ''I was startled to see on the cemetery gate—with no house or living person about—what looked to be a mailbox.'' A closer inspection revealed only a piece of warped board. ''As I came back down the hill in the sunset, I thought to myself, 'What a striking title for a story, 'Mailbox for the Dead'.'' That day, October 22, Hughes started a story called ''Mail Box for the Dead'' (soon changed to ''Postal Box: Love'') about a man who writes daily to his wife for twenty years after her death, until one day he receives an answer in the form of a pregnant young woman and her lover, whom he takes in. ''Several times during the evening as I wrote,'' Hughes later wrote of what he considered a psychic event, ''I kept thinking about my father. This was unusual, as I did not think of him often.''

Langston, in fact, had a reason to think about his father. The previous month, James Hughes had sent Langston one of his infrequent letters. In it, his father had restated to Langston his basic opinion about the American black: ''I have no faith in him; never did have, and never will. The ignorant uncultured Mexican Indian has got more sense than the American negro,—college bred or otherwise.'' Opposition to his father on this point had anchored Langston in his poetic career and formed the basis of his adult life. Now, starting with ''Mail Box for the Dead,'' he was raising that anchor. Having finished about a half-dozen stories, which he sent to Lieber with the apology that they were not very commercial, he had begun a radical change that suggested that perhaps, after all, his father had been right. Writing about blacks was a waste of time; Langston's new idea was to do a series of ''white'' stories, using a pseudonym.

Begging Noël Sullivan not to tell anyone (''They might think I was trying to pass!'') he alerted Lieber about his plans: ''I have hit upon an idea. Tell me what you think of it: to do every week one story with sentiment, love, romance, and a happy ending, *under a pen name;* the race, the colored race, conspicuous by its absence, and no problems involved except the all-eternal problem of LOVE. Maybe we could even sell them once in a while to the movies. . . . Would you be willing to handle them for me? Give me your opinion on the project? And not tell anybody about it?'' He nervously justified himself: ''Since I must make a living by my pen—typewriter, to be exact—and since the market for Race and Russian stuff is distinctly limited, I see no reason why I should not weekly turn to LOVE, and Love in the best American Caucasian 100% slick paper fashion. Do you? I shall await your reaction, and advice.''

''Mail Box for the Dead'' included white characters only, and had nothing to do with either race or poverty. In addition, its style was deliberately maudlin; for the first time in his life, Hughes worked as a hack. ''Then Philip's father died,'' he wrote in one place. ''But little more than twelve months after his mother left, the second caller from eternity came and took his father away, too. For awhile, grief shattered Philip's world of test tubes and love. But now he and Sallie felt that they must create a family of their own to bring new life back into the big house, to fill their days, and their hearts with ever more of love.''

On November 5, a letter from Mexico, sent on October 22 by one of the three Patiño sisters, his father's old friends, finally reached him in Reno. James Hughes was gravely ill, and would probably die. Wiring Mexico City for further information, Hughes asked Noël Sullivan for help: ''Of course, if my father is still living, I would like to go to him if he wishes it, and give him that evidence of friendship.'' He also wrote Blanche Knopf for a loan. The next day, a second letter from Mexico City, delivered promptly, announced that James Hughes had died on October 22, after an operation from which he had seemed to be recovering. He had died while Langston had been thinking of him and composing ''Mail Box for the Dead.'' If Langston felt a deep sense of loss at the news, he did not show it. He left a display of grief, or self-pity, to his mother. ''All of the misunderstandings are now wiped away,'' Carrie wrote him, ''and I sorrow much at his passing. . . . I feel always now that there is nothing for me ever of happiness.'' For Langston, his father had died in the summer of 1919, when the young poet, in writing ''The Negro Speaks of Rivers,'' had first discovered the artistic dividends of patricide.

Citing complicated business matters, the Patiño sisters urged Hughes (''es *necesario*") to come to Mexico City; from Indianapolis, Aunt Sallie Garvin, anticipating a share of her brother's fortune, insisted on it. Hughes, too, apparently had his hopes. When Blanche Knopf authorized a loan to him of $150, he joked to her that "it may be that I am an heir, and will never again have to worry publishers about money before it is due!" His father's brother, John Hughes, living prosperously in Los Angeles on money from Oklahoma oil wells, promised $100 to Langston for the visit, and Noël Sullivan deposited $50 in his account. Sending off four stories to Lieber, he rushed to finish an interview with Oscar H. Hammonds of the U.S. Weather Bureau, the only black weatherman in the nation, then headed for San Francisco on November 15 to secure a visa.

The document was two weeks in coming. Fortunately, by the time it arrived, the mail had also brought a check for $135 from *Esquire* magazine for "On The Road." At last, in bitterly cold weather, Hughes headed south in a car driven by a close friend of Noël Sullivan's, Mario Calderon. They reached Los Angeles on the afternoon of December 3. First, Langston visited his uncle John Hughes and Flora Coates, a cousin from the Hughes side of the family, now living in Los Angeles. Then he hurried over to call on Arna and Alberta Bontemps, who had just moved from Oakwood Junior College in Alabama, where Langston had visited them at Christmas, 1931, to the black Watts section of Los Angeles, where Arna's father owned a modest house. With them were their children Paul, Joan, and the infant Poppy. Langston and Arna agreed to work on a series of children's stories as soon as Hughes returned from Mexico, as he expected to do, early in the new year; at Macmillan, their *Popo and Fifina* was still selling excellently.

Then, accompanied by Bontemps and John Hughes, Langston was off to the train station, where he paid $116 for a round-trip ticket to Mexico City. At Nogales on the border, however, the Mexicans refused to admit him because his visa did not specify that he was colored. Several days passed before he was allowed to proceed.

Surprisingly little had changed in the Patiño household since his last visit in 1921 to Mexico City, although they now lived on the Calle de San Ildefonso. The three sisters, Fela, Cuca, and Dolores, who had known Langston first as a child of five, had aged but very gently, as it seemed to him. Still deep in mourning, black mantillas drawn against their olive skin, they all burst into tears at their first sight of him. After the years of estrangement from his father, he was glad that he had made the trip. "I was happy myself to be back with them in their big, dark, old apartment after many years—with their saints, madonnas, candles, stiff white lace curtains, bitter black coffee and scurrying Indian servants. The three old women made me as welcome as a son, and vied with each other to see that I was comfortable." He was given his father's old room.

Langston discovered quickly that his father had left a far smaller estate than expected. James Hughes had died without recovering fully from the stroke that had crippled him in 1922; most of his money had gone for medical treatments, including costly trips to Germany. The ranch in the mountains, the high place from which he had once tempted Langston with his Mexican world, had been sold. A few pieces of property were left, along with some 8000 pesos, about U.S. $2000, in the bank. On August 29, 1925, when James Hughes had drawn up his last will, he left everything to his very loyal friends, the Patiño sisters. By this time he was separated from the German housekeeper, Bertha Schultz, later Bertha Klatte, whom he had married in 1923. Frau Klatte was mentioned in his will (as she had to be, under Mexican law), but left nothing. Langston was not even mentioned.

Although the sisters wanted him to share equally in the estate, Langston resisted their offer as going against his father's clear wishes. Finally he agreed to accept a quarter of the cash in the bank. His aunt in Indianapolis, showing all of her brother James's refinement of spirit ("I am afraid of those Mexicans they are treacherous and sly"), pressed him to contest the will, but he refused to do so. Hughes visited Frau Klatte, who received him warmly but declared her intention to fight the will. He also paid a respectful visit to his father's grave in the Panteón Moderno in Mexico City.

In no hurry to have the will probated, a process that would take six to eight weeks, he put off going to Toluca. By Christmas, he was completely at home with the Patiños, "perfect darlings" who waited on him hand and foot—or their servants did, especially the cooks. At last, two days after Christmas, he went with Dolores, the main businesswoman among the three sisters, to the Hughes house in Toluca, where they located the original of the will. Browsing in James Hughes's meager collection of books, Langston noticed that they were all grimly practical—on rifles, irrigation, horse breeding, geology, swine husbandry, peach culture, and electric power plants. In the files, he found all of his letters to his father.

Hughes made no move to return to the United States. At the earliest, the little money from the Patiños would not come for several weeks, and he now owed substantial sums to Uncle John, Aunt Sallie, the Knopf firm, and Noël Sullivan. A check arrived from *The New Yorker* for one of his stories, "Oyster's Son," a vigorous Depression tale that mixed terse dialect and racial protest with pro-communist zeal. Otherwise, the news from Maxim Lieber, at first encouraging, was bad. Lieber liked the four "white" stories written in Reno so much that he wanted Hughes to offer three ("Eyes Like A Gypsy," "Hello Henry," and "A Posthumous Tale") under his own name. The fourth was poor enough to be given to "David Boatman," which Hughes had chosen as a nom de plume. But each story would repeatedly be rejected; Wolcott Gibbs of *The New Yorker* urged him to give up such frivolous writing. A long, expert "Frankie and Johnnie" ballad, "Death in Harlem," was rejected by seven major magazines before appearing, for a modest fee, in the June issue of *Literary America*. As a creative writer sailing on the big sea of life (an image he himself reflected in his choice of a pen name), Hughes was now dead in the water.

January also brought grim news about old friends. On December 22, Wallace Thurman, almost entirely neglected in a charity hospital on Welfare Island in New York, where he had languished since the previous July, had died of tuberculosis. Their friendship had never recovered from the "Mule Bone" affair, when Thurman had agreed to rewrite the play without Hughes's consent. Langston had asked Carl Van Vechten to send magazines to Thurman, but he himself seldom replied to the pathetic letters written from Thurman's sickbed. Next came news that on December 26, Rudolph Fisher, the genial and brilliant young medical intern, fiction writer, and wit Hughes had first met and admired at Georgia Douglas Johnson's salon in 1925, and who had opened a successful practice as a radiologist in Harlem, had died of cancer. The Harlem Renaissance was indeed over.

Nursing his own energy, he waited patiently for the sun to turn toward him. As he had done previously, in similar episodes of depression, Hughes took on something of the guise of a child. In spite of his thirty-two years, he obediently observed the Patiño curfew of nine o'clock at night, when the house was shut tight. Like a schoolboy, he took lessons in Spanish from a Profesora Esperanza Aguilar; she set him to reading the multi-volume *Don Quixote* of Cervantes, which he found "most amusing" and "so simply and charmingly written." Dutifully Langston filled notebooks with ladders of vocabulary. Mostly he scribbled his fiction by day and, propped up in bed by pillows and under warm blankets, read his *Don Quixote* by night. He finished three stories within a month, but without being very happy about any of them.

From this secluded spot, the radical poet surveyed the local political scene, which was far more bloody and chaotic than its American counterpart, as the recently elected president Lázaro Cárdenas sought to bring an end to two decades of bloodshed and instability in Mexico. Near New Year's Day some young Fascists, firing on a crowd at a socialist rally, killed five people; one Fascist, who showed up too late for the slaughter ("he must have been colored!" Langston quipped), was promptly lynched. In the aftermath, when the city rocked nervously with marches and rallies, Hughes kept his distance. He joined a hiking club, Los Dragones. He also attended a boxing match between the promising young black American boxer Henry Armstrong, then unknown, and a local yokel (the yokel won). And Hughes discovered at the Sunday bullfights that he was still an *aficionado*. His favorite fights involved the reckless young *novilleros,* paid the trivial sum of 50 pesos a week to risk death in the unfashionable plaza of Vista Allegre, while they dreamed of promotion to the grand Torea.

At last, a few days after the Patiños marked his thirty-third birthday with a day of feasting, Hughes broke cover with a phone call to his Cuban friend and first Spanish translator, José Antonio Fernández de Castro, now an officer in the Cuban embassay and married to a Mexican. Miguel Covarrubias, the Mexican-born illustrator of *The Weary Blues,* was also alerted. Soon the invitations began to arrive. On February 23, the editors of the influential review *Crisol,* which in March, 1931, had published a laudatory article, "Langston Hughes, El Poeta Afro-Estadounidense" by Rafael Lozano, held a dinner in his honor. Soon after, in the important daily *El Nacional,* Fernández de Castro and the Guatemalan writer Luis Cardoza y Aragon proclaimed Hughes the quintessential "poeta militante negro" and "el poeta de los negros." (Later would also come articles on him by Xavier Villaurrutia, José Mancisidor, and Rafael Heliodoro Valle.) Trying to alert the Mexicans to its significance, Fernández de

Castro put forward Hughes's work as a poetical apotheosis of blackness that combined raw African strength with the godlike power of laughter to transcend circumstance. His revolutionary quality had been recognized before in occasional translations published in Mexico, but the new articles had a more immediate effect. From the mainly apolitical *Contemporáneos* group to the League of Revolutionary Artists and Writers, he was welcomed by the most accomplished Mexican writers and painters. Among the latter, he met the melancholy Orozco, the mountainous, dark-skinned Diego Rivera (a Negro grandmother, Rivera claimed proudly), Siquieros, Izquierdo, Tamayo, and Montenegro, and was taken up by the flamboyant Lupe Marin, Rivera's estranged wife and his favorite model.

In turn, Hughes was impressed by the vitality of Mexican art and culture, which had been transformed from the pallid uncertainties of the previous generation mainly by the inspired blending of European modernist technique and a nationalism based in large part on the unprecedented acknowledgment of the Indian roots of the national culture. With his typical selflessness and strong internationalist sense (and also because he was now doing no great work himself), Langston decided to try to place translations of as many of the fiction writers as he could, especially the radicals. He virtually stopped his own writing to perform a service from which he expected to receive next to nothing, or nothing. Later, with the aid of Fernández de Castro, who suggested some Cuban pieces, and a white woman from Arkansas who cheerfully typed according to Hughes's orders, he finished a collection called "Troubled Lands." Langston would persist in his effort to place the stories even after Maxim Lieber, to whom they were sent, dismissed all as inferior to Hughes's own work.

Out in the open now, he enjoyed a way of living he had not really known before. "I never lived in Greenwich Village," he would write, "so its bohemian life—in the old days when it was bohemian—was outside my orbit. . . . The nearest I've ever come to *la vie de boheme* was my winter in Mexico when my friends were almost all writers and artists." Leaving the Patiños, Hughes roomed with a shy Frenchman, six years his junior, who had once hoped to be a painter but was now devoted to photography. Henri Cartier-Bresson, it was rumored, was the son of a very rich Frenchman who detested his son's bohemianism. All Hughes knew for sure was that his quiet young friend was unpretentious, loved jazz and folk music, had been to Africa, and knew Langston's old idol, Claude McKay. In 1931, after a year in the Ivory Coast, during which he contracted a tropical disease, Cartier-Bresson had turned to photography; he had met McKay while recuperating in Marseilles. In 1932, his first photographs were exhibited in New York in a show organized by Harry and Caresse Crosby; two years later, he had come to Mexico with an ethnographic expedition. Together with the good young Mexican poet Andres Henestrosa, who looked to Hughes like a white man but was in fact Indian, Langston and Cartier-Bresson shared a place. Hughes would call it a *vivienda,* or apartment, with a walled courtyard which Cartier-Bresson insisted on photographing for a coming exhibition. "It was no apartment," Cartier-Bresson himself later recalled. "It was a shack. We lived in a very humble place near the Lagunilla market and the little bars where the mariachi bands played. It was very cheap because we didn't have any money. We pooled what we had and worked a little and entertained our girl friends there and enjoyed life a great deal.'' When the grime was too much for Hughes, he slipped away to the cool, clean rooms and abundant pantry of the Patiños. But poverty was part of what Langston needed at this point of his life, and he accepted it as a blessing.

''Langston was a fine man, the perfect companion,'' Cartier-Bresson later remembered of a friendship that would endure. ''He was a noble human being. But he was also natural, always smiling, always good-humored, even if you could sense that he kept something in reserve.'' Respecting Langston's sensibility in spite of his ignorance of photography, Cartier-Bresson asked him to write an introduction for the brochure of his showing at the Palacio de las Artes, which he shared with a Mexican photographer, Manuel Alvarez Bravo. Although Hughes's technical knowledge was limited to the little Kodak camera Amy Spingarn had given him as a graduation present in 1929, he was sensitive to form. ''A wall can be merely a wall,'' he wrote bravely, ''but in some of Cartier-Bresson's photographs the walls are painfully human, and live and talk about themselves. There is that vulgar wall behind the man in the brass bed; that great lonesome wall of broken paint and plaster along which some child is wandering; there is a huge sun-bright wall of a prison or an apartment house with a boy who is like a shadow at its base. . . . There is the clash of sun and shadow, like modern music, in a Cartier-Bresson picture.''

Between the shack and the embassy parties, Hughes was enjoying his visit. Nevertheless, exactly as in Carmel the previous spring, he began to complain about the distracting number of dinners and parties and people in his life. Although he wrote to Marie and Douglas Short that he could ''go on staying here for years,'' so much did he love Mexico, Hughes looked forward to leaving the country at mid-year. He was freed by the very good news that the Guggenheim Foundation had awarded him $1500 for nine months' work on his novel. Immediately securing a postponement until July 1, he took a small but comfortable apartment for $11 a month on the seventh floor of the highest apartment building in the city, the *Ermite,* so that he could complete his translations and other work in comfort. He postponed taking the Guggenheim fellowship mainly in order to work for a few weeks with Arna Bontemps on children's stories in Los Angeles. Then, taking up the fellowship, he would begin research on his novel in Chicago.

Home was certainly calling. That year, both Harlem and Chicago were inflamed by riots that resulted in death and widespread destruction. Hughes could not stay away from the American fire forever, nor could he fan it from a safe distance as he had done from Mexico on January 22, as a signatory to a ''Call for an American Writers' Congress'' in *New Masses,* joining Steffens, Ella Winter, Nathanael West, Dreiser, Erskine Caldwell, Malcolm Cowley, and young Richard Wright in urging the formation of a League of American Writers affiliated with the International Union of Revolutionary Writers. Or when, at the first session of the Congress in April in New York, his extremely militant statement ''To Negro Writers'' was read in his absence. ''There are certain practical things American Negro writers can do through their work,'' Hughes had sternly insisted, including the radicalization of the black masses, the education of whites about ''those Negro qualities which go beyond the mere ability to laugh and sing and dance and make music,'' and the building of black-white unity ''on the *solid* ground of the daily working-class struggle to wipe out, now and forever, all the old inequalities of the past.'' Black writers could expose ''the lovely grinning face of Philanthropy—which gives a million dollars to a Jim Crow school, but not one job to a graduate of that school''; white union leaders who exclude blacks from jobs; ''the sick-sweet smile of organized religion—which lies about what it doesn't know, and about what it *does* know. And the half-voodoo, half-clown, face of revivalism, dulling the mind with the clap of its empty hands.'' Black writers could expose the shame of their compromised leaders, the ''Contentment Tradition of the O-lovely-Negroes school of American fiction,'' and ''the old My-Country-'Tis-of-Thee lie.''

Such a lacerating message should be delivered either in person or not at all. He would go home, as he had planned before his father's death, to his mother. For three years, Hughes had wandered among Germans and Russians, Central Asians, Chinese, and Japanese, Californians and Mexicans; it was time now to go back to whatever he could discern as his beginnings. And his mother had a special need of him now. Gwyn Clark, her stepson, had drifted off to Cleveland. Homer Clark, her husband, having surfaced again, had vanished again. ''I wish I could write you lovely, sunny letters,'' Carrie Clark had sobbed in one letter, ''but Langston my life is so drab.'' Worst of all, she had also discovered a tumor in her breast, which she was too terrified to have examined.

For all her importunings and flightiness, and his despair at having been abandoned by her, Langston loved her. Having lost his father without the absolution of a final visit, he would not let the same fate overtake him with his mother.

12

STILL WAITING ON ROOSEVELT
1935 to 1937

Say who are you that mumbles in the dark?
And who are you that draws your veil across the stars?
"Let America Be America Again," 1935

NEAR the end of the first week of June, Hughes returned so quietly to Los Angeles that for a while only his uncle John Hughes, his cousin Flora Coates, and Arna and Alberta Bontemps and their family knew that he was back after six months in Mexico. At first intending to stay only a few days, he would pass almost three months in California. The main reason was the presence of Arna Bontemps; much of Hughes's time would be spent at 10310 Weigand Avenue, the home of Bontemps's father and stepmother in the mainly black Watts district of Los Angeles. Langston's brief visit the previous December had only strengthened his interest in working with Arna, who was living in cramped quarters (his third child, the baby Poppy, slept in a bureau drawer) under the critical eye of his father, an often haughty, dominating man who had given up bricklaying for the life of a lay minister in the Seventh Day Adventist Church. In a world that seemed determined to discourage them as writers, Hughes and Bontemps turned more and more to one another for comfort. Langston remembered staying at the house on Weigand Avenue for at least a few days: "We each put what we had into the pot and we lived—but on very simple fare."

Bontemps's determination to write had helped to push him out of Oakwood Junior College in Alabama, as it had dislodged him earlier from the Harlem Academy in New York; he had seen clearly "that an able-bodied young Negro with a healthy young family could not continue to keep friends in that community if he sat around trifling with a type-writer on the shady side of his house when he should have been working or at least digging in the yard and trying to raise something for the table." Undeterred, Bontemps had kept at his writing; in Alabama, he wrote several short stories that would be published later in his collection *The Old South.* In 1934, to very good reviews, Morrow had brought out his children's book *You Can't Pet a Possum;* he had also started a second novel (to be published as *Black Thunder* by Macmillan in 1936), about the celebrated slave insurrection led by Gabriel Prosser. These achievements only deepened his increasingly remarkable sense of ease with himself. If Hughes met the world behind a brilliant smile, Bontemps offered a massive calm. "He seems to preserve a core of serenity that transcends circumstance," an observer would write four years later. "He has dignity, reserve, restraint; makes no effort to impress; yet to meet him is to be struck with the certainty, 'Here is a *person*'." The sociologist Horace Cayton, who would recall Bontemps as "extremely good looking, a beautiful brown color," also remembered that "above all he was calm, deliberate, and well tempered in his attitude toward life and race." Near the end of his life, Bontemps himself would look back placidly on more militant black writers: "I don't have that fury. I'm not a furious writer and I'm not a furious man, perhaps."

Dropping into a tight schedule, the two writers raced to turn out a children's story set in Mexico—"a fantasy with lots of humor," as Hughes proposed, modelled on their *Popo and Fifina.* This time they were after not the trickle of royalties but the lump sum of a publisher's advance. If the Mexican book sold quickly, Hughes would hurry to complete another, set in Russia, that the publisher Macmillan had asked for. Then he might write others on North Africa and Spain. Anxious not to take up the Guggenheim fellowship prematurely, Hughes asked Maxim Lieber in New York to look for some lectures for him. Meanwhile, he and Bontemps worked hard; after ten hard days at their typewriters, "The Paste Board Bandit" was completed. Mailing the manuscript and two typical Mexican paste-board toys to Macmillan in New York, the team turned next to begin a series of stories for children called "Bon-Bon Buddy," based on Bontemps's affectionate memory of his own "Uncle Buddy," on whom he had drawn in creating the irrepressible hero, Little Augie, in *God Sends Sunday.*

Near the Fourth of July, Hughes at last heeded Noël Sullivan's entreaties and travelled north for a short visit to Carmel-by-the-Sea. There he was very warmly received at "Ennesfree" by Sullivan himself, Marie Short, Robinson and Una Jeffers, Ella Winter, Lincoln Steffens, Martin Flavin, Albert Rhys Williams, and others in the old circle. The political tension of the previous summer had faded away. The John Reed Club itself was dead—killed, ironically, not by vigilantes or the American Legion but, across the nation, by decree of the Communist Party, which suddenly had found the club no longer a suitable vehicle for its radical goals. This action, and others like it, cost the Party many friends. One was Noël Sullivan, who had grown hostile to radicals ("in their way they are as selfish and ruthless as [the] bourgeois, and will take everything you have to give"); he especially resented Ella Winter and her sometimes reckless articles in the *Pacific Weekly.* Hughes listened to Sullivan's complaints, murmured soft assent, then surrendered to the sun and the company of Marie Short on the Carmel beach. He met the latest addition to the "artists' colony"—Myron Brinig, whose most recent novel, hardly a masterpiece, apparently had fetched an advance of $25,000. Reflecting on the inequities of America, Hughes swallowed his astonishment and tried to enjoy his brief vacation. To Marie Short, he seemed contented in Carmel. "You really should just settle down here," she soon urged Langston; "doesn't it now seem [like] 'home' to you?" Carmel did not, however, seem like home, even if Noël Sullivan was keeping a large suitcase for Hughes, and a cabinet of his files. A few boxes of his possessions had also been stored, since 1931, at the Peeples's house in Westfield, New Jersey. Neither place was home. Carmel, in fact, encouraged disturbing ambivalences. It moved Hughes to enjoy bourgeois pleasures while professing radicalism, to live interracially while American racism demanded militance, to enjoy seclusion even as he also wanted to join the fight.

He left the comforts of Carmel for a cheap but clean room, at $1 a day, in what would be his refuge on visits to Los Angeles for many years to come, the black-owned Clark Hotel at the corner of South Central and Washington Avenues. Here he continued work on his stories with Arna Bontemps. But their hopes soon collapsed. Late in July, Macmillan summarily rejected "The Paste Board Bandit"; an editor called the story dull and prosaic, and said that with two authors it suffered from a shifting point of view. Soon "Bon-Bon Buddy" was turned down by one magazine, then refused testily by Macmillan; eventually almost a dozen firms, including Knopf, would decline it. For Hughes, the refusals continued an ominous trend. The pieces he had translated so laboriously in Mexico were also doing poorly in the United States, although some of them appeared, through the courtesy of Ella Winter, in *Pacific Weekly,* and in *Esquire* and *Partisan Review.* To his agent, Hughes appeared to be drifting as a writer. Lieber denounced the offending Macmillan editor as a "prize dunce," but he also questioned the wisdom of Langston's continuing interest in juveniles and translations. Why not concentrate on mature stories "that can be sold for nice fat prices to some nice prosperous magazines?"

In no hurry either to join his mother in Oberlin or to begin work on his novel, Hughes hunted for a job in the film industry. For a while, his prospects seemed fine. "Things Negro are booming out there this season," he would write early in the fall, "and all the companies are planning to make colored pictures." His story "Rejuvenation Through Joy" from *The Ways of White Folks* was now being considered by Al Jolson at Paramount; Langston also heard encouraging words from a would-be collaborator, Clarence Muse, the talented black singer and actor he had met in 1932 and a black man so pliable that he was seldom out of work in Hollywood. Hughes was also impressed by the seriousness of the black composer William Grant Still, who had just capped a Guggenheim fellowship by moving permanently to California. The Mississippi-born Still, who had set Hughes's "Breath of a Rose" many years before, had trained at the Oberlin conservatory and with the avant-garde composer Edgard Varèse. He had done brilliantly both in popular and classical music. A veteran of the *Shuffle Along* orchestra and W. C. Handy's New York recording company, Still had also written several symphonic pieces and three full symphonies, including *Afro-American Symphony,* which the composer-conductor Howard Hanson had performed with his Rochester orchestra in 1931. For both Still and Hughes, the creation of a great black opera was an old ambition; in fact, Still hoped to make a career in opera. They decided to work together. Hughes's long-neglected Haitian epic would form the basis of their first collaboration.

In spite of his achievements as a writer, Langston found the film industry impossible to crack. Black performers were sometimes needed there, black writers virtually never. When one friend tried to help, the result was disastrous. On Saturday, August 10, Lillian Mae Ehrman of Carmel and Beverly Hills (whose brother-in-law was pushing "Rejuvenation Through Joy" at Paramount) gave a dinner party for Hughes at her home. Asked for suggestions for the guest list, he had offered the names of Katharine Hepburn, his favorite actress, and the writer Jim Tully, noted for his hard-bitten, socially sensitive "hobo" stories and for his hardhitting play *Black Boy.* Ehrman mistakenly invited not Hepburn but Theresa Helburn of the Theatre Guild. A noted drinker, Tully arrived ripe for a dispute. While Helburn explained to Hughes why his play with Ella Winter, "Blood on the Fields," had been turned down by the Guild, Tully seethed. Suddenly he erupted into a colorful tirade against the Theatre Guild ("the God-damndest lousy commercial lot in the world") and all the other corrupting parasites of Broadway and Hollywood. Turning to Hughes, he ended with a warning: "Stay away from Hollywood! Stay away from Beverly Hills! Stay down on Central Avenue with your poor hurt niggers!" "At the word *nigger,*" Hughes would recall vividly, "everyone gasped, except myself." The hostess, Lillian Mae Ehrman, turned red. "Langston Hughes knows what I mean," Tully insisted unsteadily but with feeling. "He's got their souls in his hands. He writes about them. He knows what I mean when I say niggers—as you people say it. As for the Theatre Guild. . . ." Hughes stayed around long after Theresa Helburn's early retreat to hear Tully explain himself further; later, he visited Tully at his home. But he also got no closer to the inside of Hollywood.

Another outing ended even more ominously, given his fresh interest in lecturing. Controversy was probably the last thing Hughes expected when he agreed to speak at a memorial service sponsored by the Los Angeles Civic League, in which his friend Loren Miller was influential, at the East 28th Street branch of the YMCA. But when the public, including some well-known Hollywood personalities, arrived for the service, they were barred. At the last minute, the YMCA officials had decided to ban the Civic League program because of the presence of the notoriously "anti-Christ" Langston Hughes. Behind the charge was his poem "Goodbye Christ," written in the Soviet Union and published in 1932 by the communist Otto Huiswood, without Hughes's consent, in the European-based *Negro Worker.* Soon after its appearance, Langston had been briskly attacked by various black clergymen; a bishop had denounced him for "making saints" of Lenin and Stalin. He had also been branded "officially a Communist" by a right-wing patriot speaking to the Cleveland Chamber of Commerce. To Hughes's defense had rushed the poet Melvin B. Tolson, who called him "a Catholic, a rebel, and a proletarian in his personal life and in his poetry and criticism [who] has always stood for the man lowest down and has sought to show his essential fineness of soul to those who were too high up—by the accident of fortune—to understand.'' Hughes himself responded to this squabble over his views on religion with a small poem, ''Personal'':

> In an envelope marked:
> *Personal*
> God addressed me a letter.
> In an envelope marked:
> *Personal*
> I have given my answer.

But he was undoubtedly dismayed by the censure in Los Angeles, and worried, not unreasonably, about its possible consequences.

The summer there yielded very little. With money from the sale of some new short stories, including ''Big Meeting'' to *Scribner's* magazine, ''Heaven to Hell'' to *American Spectator,* and ''Trouble with the Angels'' (about cowardice in the face of segregation, with clear reference to the touring company of *Green Pastures,* which had performed before Jim Crow houses), to *New Theatre,* Langston decided to leave for Oberlin. He departed knowing that Arna Bontemps, who had just accepted an offer to teach at yet another Seventh Day Adventist school, Shiloh Academy in Chicago, would also be in the Midwest soon.

On the rainy evening of August 30, Hughes reached Santa Fe, New Mexico, famous both as an old Indian and Spanish settlement and, more recently, as an artists' colony. His host was to have been Witter Bynner; in Bynner's absence, he was greeted by Bob Hunt, a close friend of the poet. Langston's visit was a flop. Although Hunt invited friends, including someone said to be Frieda Lawrence's son-in-law, to meet him, Hughes waited in vain for a spark of excitement from the artists and intellectuals of Santa Fe. He just missed meeting the queenly Mable Dodge Luhan, who had invited him to visit her on one of her trips to Carmel (after lecturing him that the house of his prize-winning poem ''A House in Taos,'' presumed by many to be about her, was nothing at all like her New Mexico home). He had even worse luck with Jean Toomer, whose ventures into mysticism had transported him to Taos, the historic little settlement at the foot of the Sangre de Cristo Mountains. Reached by telephone, Toomer was ''most evasive'' in telling Hughes where or when or, indeed, if they could meet. Perhaps Toomer had recognized himself as the charlatan Eugene Lesche in ''Rejuvenation Through Joy''; perhaps he did not wish to be associated with a Negro, now that he no longer was one. This snub left Hughes only more exasperated with Santa Fe, where the rain never ceased and the charm of turquoise and adobe was marred by squads of loud, bigoted Texan tourists. Making a quick exit, Hughes rode an evening train out of town on the last day of August.

By early September he was in Oberlin, Ohio, living with his mother and Gwyn ''Kit'' Clark in a three-room apartment at 212 South Pleasant Street. The neighborhood was quiet and shady, the historic little town charming and generally friendly. The main legacy of its famed Abolitionism was a black population of about five thousand, or roughly a quarter of the town. The street on which Hughes lived was white except for a handsome black family, Harold and Opal Gaines and their children, living two houses from his mother's apartment. Two blocks further on, the main black neighborhood began. Although Hughes had visited Oberlin only once previously, to read his poems in 1926, the town seemed familiar. As a child, he had heard much about it from his grandmother; Mary Langston and her two husbands had all lived there at one time or another, as had John Mercer Langston. Hughes, like other members of his family, believed—erroneously—that Mary Langston had not only attended Oberlin College but had been "the first Negro woman" to do so. In the college museum he found pictures of several of his ancestors, and dozens of distant cousins lived nearby.

Life at home was pleasant at first, but the pleasure did not last. Soon the old bickering with his mother flared almost every night into nasty little fights. "My mother, as a great many poor mothers do," Hughes later wrote, "seemed to have the fixed idea that a son is born for the sole purpose of taking care of his parents as soon as possible." The rent was only $9 a month, but Langston had reached Oberlin with $2 in his pocket. Although Lieber had sent $50 from the re-sale of two pieces to *Fiction Parade* and *Modern Story,* and the sale of another story, "Tragedy at the Baths," to *Esquire,* the money had not lasted long. In addition, the apartment was not large enough for three adults. Gwyn Clark, now called "Kit," was assigned to a federal relief camp, the Civilian Conservation Corps, but came home every night because the local barracks had not yet been built. At twenty-three, Kit had not yet graduated from high school. Carrie had another reason to be irritable and depressed. Fearful of what a doctor might find, she still refused to undergo a thorough examination of the lump in her breast. She had left New York impulsively, in a huff, after her long-time friend Toy Harper had criticized her for not controlling Kit and for irresponsibly "running all over the road."

Eager for relief from his situation in Oberlin, Hughes packed his bags and headed for New York as soon as word reached him that his play *Mulatto* might be produced on Broadway later that year. The news was by no means a surprise. In early June, his drama agent, the former actor John W. Rumsey of the American Play Company, had alerted him about the strong interest of one Martin Jones in a week-long staging of the play in Dobbs Ferry, New York, where Jones ran a summer theater. From the start a great admirer of *Mulatto,* John Rumsey had urged Hughes not to allow an amateur staging, but to wait instead for a professional producer brave enough to attempt a drama about miscegenation. Such a producer now seemed at hand. From Mexico, Hughes had wired his consent and waived his royalties for one week, as Rumsey had advised, because Jones promised to invite all the major Broadway producers to see the show in Dobbs Ferry. In early August, Rumsey reported to Hughes that, in spite of a very amateurish production, *Mulatto* had gripped the audience. He suggested that the third act be rewritten, perhaps by an experienced Broadway dramatist paid out of Hughes's royalties.

Anxious for a success, Hughes replied that he would be "perfectly willing" to share "up to one half" of the royalties. Rumsey then informed Jones that the author would yield one quarter of his royalty percentage; he drew up an agreement giving Jones an option on the play and the right to hire a playwright to make revisions. Hughes would receive an advance of $500 against his royalties. Serious about a production, Jones demonstrated to Rumsey's satisfaction that he had $75,000 at his immediate disposal, with an additional line of credit up to $25,000. In addition, the successful Broadway director Charles Erskine would stage the play. Everything seemed satisfactory to Hughes, who was so eager to make *Mulatto* attractive to Broadway that he was quick to suggest possible changes. There could be more topical references, more music and local color. And more violence. Perhaps the mob could lynch the "good" son as well as the "bad," he offered; "or would that be too much ironic horror?"

On Friday, September 27, when Hughes reached Manhattan on his first visit to New York in over three years, he was barely out of the subway when a friend stunned him with congratulations on his forthcoming Broadway opening. Hughes had been unaware that a date had been set for the premiere. Telephoning his drama agent John Rumsey, who had neglected to give him the news, he learned that the opening was less than a month away at the Vanderbilt Theater. Leaving his bags with Toy and Emerson Harper in their apartment at 634 St. Nicholas Avenue in Harlem, Hughes hurried down to the rehearsal hall. He found a cast hard at work; to his great satisfaction, Rose McClendon, the model for the role of the mother, Cora, when he had developed *Mulatto* at Jasper Deeter's Hedgerow Theater, was playing the part. He also discovered that the dramatist to whom he had yielded a quarter of his royalties was none other than the producer, Martin Jones, who almost tumbled out of his chair when Hughes introduced himself. Evidently Jones had not expected a colored man, or an author thought to be in California, or both. A rich young man whose family had earned a fortune in the flour business, Jones unpredictably had gone into the theater and quickly made an enormous amount of money producing a sensational play called *White Cargo*. Later he had profited more modestly from other plays, but now hoped to use *Mulatto,* another race play, to repeat his greatest success. Living the role of the producer-director to the hilt, he affected the rakish beret and polished boots of the more imperious Hollywood directors; he arrived and departed in limousines, hired and fired at will, and generally indulged a boundless confidence in his own genius.

"Over-ridden on all sides," Hughes would mourn, "I sat and watched my first play—which I had conceived as a poetic tragedy—being turned into what the producer hoped would be a commercial hit." Langston's boyish smile and yielding manner, so effective with many people, apparently meant nothing to Jones. Nor was Jones even remotely interested in *Mulatto* as art. Anxious for a killing, he would do whatever it took to get one. Hughes worked on the play until the producer's coolness turned to rudeness and contempt, which he turned equally on Langston and the few colored members of the cast (the mulatto children were played by whites). Soon the actors were reciting lines Hughes would not have written, and playing at least one scene he would not have conceived, the rape of the hero's sister by a white overseer ("Rape is for sex," Martin Jones explained. "You have to have sex in a Broadway show"). As the producer moved to maximize sex and violence in the drama, Hughes was both fascinated and appalled. "Only Shakespeare has more tragedy and death and destruction in the last act than *Mulatto* now possesses," he giggled in a letter to Marie Short about the "marvellous changes" in the play. "(But all the actors are happy because each one now either dies, kills, gets raped, or goes mad. . . .) Just so the audience doesn't rise up and kill us all!"

But the fact that his play was about to open on Broadway lifted Langston's confidence grandly. He strutted a little for the benefit of a New York *Amsterdam News* reporter, who found him looking very prosperous, about fifteen pounds heavier than since he was last in Harlem, and sporting a slick moustache. At thirty-three, Langston Hughes now seemed to be "more the successful and secure artist than the vagabond poet, in which role he formerly flourished." Smoothly Hughes revealed to the reporter his plans to lecture within a few days at the University of Minnesota, and to begin his Guggenheim novel. Would this novel be revolutionary and proletarian? "I never label my works," he responded. "Others have classified them, but I never have. I simply write what I feel. It will be a book about Negroes."

He had reason to be confident. Certainly the speech in Minneapolis, brilliantly secured by Maxim Lieber, was the major address of his life thus far. Near noon on October 10, when he rose before 4,000 students gathered in Northrup Auditorium for the opening convocation of the University of Minnesota, with a radio station poised to relay his speech to an even larger audience, Hughes faced the largest group ever gathered to hear him, and by far the largest group of whites. With its Farmer-Labor Party and its history of co-operatives and government intervention, Minnesota was an unusually socialist state by any American reckoning; Hughes's very presence as an honored guest was itself a sign of its purpose. In response, his message was about the power of interracial socialism. "As I see it," he offered, "the basic economic problem of the Negro is the same as that of his white compatriots, and it is through labor movements that some sort of solution must be reached." Economic and social injustice against whites was in many places "just as bad" as against blacks—"worse, I think, for the Negro has a certain temperamental gaiety and flexibility which makes life more bearable, by just so much, than it is for his white brother." His address, which included some of his poems, was quietly spoken but powerful. "A small, prim, negro poet," one local paper reported, "sent shivers down nearly 4,000 spines"; on the other hand, another journal emphasized a "certain boyish charm" that won the audience "almost immediately." In addition to the talk, for which he was paid $100, Hughes met classes in journalism and creative writing and read elsewhere in the area. Then, happy with his visit, he rode an express train back to New York.

But as the opening of *Mulatto* on October 24 approached, his friction with Martin Jones worsened. Jones now wanted thirty-five per cent of the royalties, plus billing as part author of the play. Feeling powerless, Hughes conceded both points but was adamant about receiving his full advance of $500. This insistence infuriated the producer. When Hughes, expecting a large party of family and friends from Ohio, requested tickets on credit, Jones refused to allow him any. Using his Minnesota fee, Langston spent $73 on tickets before discovering to his horror that Jones intended to segregate blacks in one corner of the house. Hughes countered by sending tickets for the most prominent seats to his darkest friends, including the sable Claude McKay, whom he had finally met following McKay's return from Europe. But Jones saved his most egregious insult for just before the opening, when he conspicuously handed out engraved invitations to a post-production buffet party. No black, neither the star Rose McClendon nor the author Langston Hughes, was invited. Deeply hurt, Hughes cancelled his plan to attend the premiere, showered many of his tickets on Amy Spingarn, gathered the telegrams that came from as far away as Carmel, sent one himself to the entire cast, and withdrew angrily to await the outcome.

The following morning, October 25, the New York reviewers praised the actors in general, but singled out McClendon's superb performance. Towards Langston, however, most were savage. Unaware of Jones's hand in the play, all questioned not only Hughes's treatment of the theme of miscegenation but also his basic competence as a playwright. "Whatever this fretful arrival at the Vanderbilt may be," the *World-Telegram* judged, "it is not a play"; rather it was "an attempt to dramatize an inferiority complex." The *Daily Mirror,* reporting a "feeling of revulsion" in the audience, declared miscegenation a subject unfit for parlor discussion. The *Sun* ridiculed the script as being full of the "weary, familiar stuff of scolding harangues and shrill melodrama"; the mulatto of the title was "a noisy and obnoxious little brute, richly deserving chastisement regardless of race." In the *Brooklyn Eagle,* Arthur Pollock praised Hughes's intentions but dismissed *Mulatto* as "merely a bad play"; Langston Hughes might know Georgia, but he didn't know the theater. The *Women's Wear Daily* declared that the author had missed nothing in composing the play—"neither brutality, vulgarity, bestiality nor profanity"; "if the play is at the Vanderbilt Theatre a full week," its reviewer predicted, "I shall be surprised, for it has little merit." Brooks Atkinson of the *Times,* president of the newly formed New York Drama Critics Circle, found something to praise in the play itself: "After a season dedicated chiefly to trash it is a sobering sensation to sit in the presence of a playwright who is trying his best to tell what he has on his mind." Hughes, however, had "little of the dramatic strength of mind . . . to tell a coherent, driving story in the theatre. His ideas are seldom completely expressed; his play is pretty thoroughly defeated by the grim mechanics of the stage." And yet, the presence of Rose McClendon and a playwright obviously "flaming with sincerity" made Atkinson "grateful for at least that much relief" from the shabby Broadway season.

The reception of *Mulatto* was clearly influenced not only by the excessive production mounted by Martin Jones but also by Hughes's unusual approach to a subject which many whites and blacks alike considered, in any event, to be barely legitimate as a topic of discussion. Unlike other playwrights treating the theme of the tragic mulatto, including Paul Green in his *In Abraham's Bosom,* which had won a Pulitzer Prize in 1927, Hughes approaches the drama of racial miscegenation, filial rejection, and their consequences finally from the perspective of its main victims. Attempting to be fair to Colonel Norwood, a man who himself attempted, but dishonestly, to live between two worlds, he stresses nevertheless the plight of the repudiated mulatto son, Bert, and, ultimately and most compellingly, Cora, Norwood's black mistress and the boy's loving, long-suffering mother. Bert does not tear himself apart in complete impotence or sicken and die like a stock tragic mulatto; slaying his father, he himself dies still determined to assert his right to live like a free man in the Norwood house. Cora, who has been as loyal to Norwood as if she were his legal wife, at last is stripped of her illusions about him; she recognizes him in the lynch mob hunting their son. The sight, and the events that made it possible, are too much for her. She goes mad, but speaks poetic truth in her dementia. The callow white men who enter the house near the end, however, recognize no tragic victims but only "niggers." On this grim note, the play concludes. Its final effect, one of extreme hopelessness and despair, can be overwhelming. Sometimes awkward and melodramatic, *Mulatto* is, properly approached, a harrowing orchestration of Hughes's prophetic fear that the great house of America would be brought down by racial bigotry.

Bitter and ashamed because of the reviews, Hughes met with an uncontrite Jones about restoring some of the original poetry of his *Mulatto*. Then he tried to reach his agent John Rumsey by telephone at his office, but heard Rumsey rudely shout to someone to tell Hughes that he was out. Langston slipped out of New York by train for Oberlin. These were by far the most damaging reviews of his career. They were far worse than the excoriation in 1927 of *Fine Clothes to the Jew,* which had been confined almost entirely to the black press. In less than two weeks he had tumbled from the height of his Minneapolis triumph to the depth of public ridicule in New York. Fairly sick with unspoken anger, Hughes brooded on his powerlessness before the forces that had mangled his work in the name of profit. "They completely distorted the poetic moments of the play," he would soon complain. Gloomy in the interior of the moving train, Hughes turned instinctively to verse to capture his sense of hurt; he also turned, as in his deepest moments of pain, to identify with the mass of humanity rather than with his own fate. This poem, sometimes prosaic, even banal in certain lines, nevertheless sounded a wounded nobility that made it, like the folksinger Woody Guthrie's "This Land Is Your Land," an authentic anthem of the Depression. Echoing "America the Beautiful," "My Country, 'Tis of Thee," and familiar tropes by Walt Whitman, Hughes called his poem "Let America Be America Again":

> Let America be America again.
> Let it be the dream it used to be.
> Let it be the pioneer on the plain
> Seeking a home where he himself is free.
>
> (America never was America to me.)

Let America be the dream the dreamers dreamed—
Let it be that great strong land of love
Where never kings connive nor tyrants scheme
That any man be crushed by one above.

(It never was America to me.)

O, let my land be a land where Liberty
Is crowned with no false patriotic wreath,
But opportunity is real, and life is free,
Equality is in the air we breathe.

(There's never been equality for me,
Nor freedom in this "homeland of the free.")

Say who are you that mumbles in the dark?
And who are you that draws your veil across the stars?

I am the poor white, fooled and pushed apart,
I am the Negro bearing slavery's scars.
I am the red man driven from the land,
I am the immigrant . . .

He sent the piece to Maxim Lieber, along with two other poems (that year, Hughes published fewer than a half dozen poems) "on the inequities of the system." "I'm profoundly moved," Lieber wrote about "Let America Be America Again." "It's magnificent."

The reception of the play, as distressing as it had been, was overshadowed by one other piece of news. Just before the opening of the play, Hughes received a note from a private Cleveland physician advising that his mother, who had submitted at last to an examination, was suffering from "a far advanced cancer of the breast." Immediate surgery was advised, along with expensive radium treatments on a regular basis. Hughes turned to Noël Sullivan for help. "The only thing to do, it seems," he wrote, "is to pay for the operation myself. All the clinics are overcrowded, and people must wait for their turn, if it comes at all." Soon, however, his frightened mother decided that she would not undergo an operation; she would live with the sentence of death. Almost at the same time news came of the death from cancer of Sallie Garvin, Langston's aunt in Indianapolis, whom he had visited on his way from California. At the funeral he found his uncle John Hughes, who had come from Los Angeles to be at the bedside of his last surviving sibling.

Thanksgiving Day, 1935, was snowy in Oberlin, and bleak at 212 South Pleasant Street. In the morning, as Hughes worked at his typewriter, he looked up to see hunters strolling by the house on their way to flush out rabbits for dinner. After Kit Clark came in later in the day, Langston sat down with him and Carrie, and two of their distant cousins, for a Thanksgiving meal of wild duck. Just as they finished dessert, a telegram sent by Noël Sullivan brought greetings from California and memories of balmier days in Carmel.

Just before Thanksgiving, however, Hughes had been made a provocative offer. On a visit to Cleveland for a succession of teas and dinners in his honor (his old high school English teacher Ethel Weimer arranged readings for him, including one at the prestigious Women's City Club), he had been feted, as its most famous alumnus, by the community organization started in 1915 by Russell and Rowena Jelliffe but now vastly expanded from its early days when Langston had taught art there to neighborhood children. With the reviews of *Mulatto* still burning his ears, one aspect of the organization held Hughes's attention. Its theatrical troupe, the Gilpin Players, founded in 1922 and named after the black actor Charles Gilpin, was now the most active predominantly black amateur company in the country. Since 1927 the Gilpins had performed full seasons in their own theater, Karamu House, at 3807 Central Avenue (''karamu'' being Swahili for ''the place of entertainment, or feasting, at the center of the community''). Although the players for some time had opposed performing plays about blacks, especially plays about lower-class black characters, their attitude was changing. However, finding suitable plays remained a major problem. Rowena Jelliffe, the main force behind the Gilpins, admired Hughes. In him, ''the outstanding thing to see was his wonder at the world,'' she would write. ''It shone through his deep hurt, his struggle to understand, his gaiety, his fine sense of humor, his sensitiveness to beauty and his deep liking for people.'' Would he be interested in working with the Gilpins, in effect as the resident playwright of the company? Indeed he would. Seizing the chance, Hughes was quite conscious of taking a new turn in his career when he wrote Noël Sullivan to tell about the offer and his determination to be a playwright ''in spite of it all.''

He made another decision. Unlike *Mulatto,* his first play for Karamu would not be tragic but comic. People, especially black people, wanted to laugh in these hard times. Besides, comedies about blacks were usually better at the box office than serious drama. Hughes had been struck by the affluence of many of his black boyhood friends in Cleveland, who held no important jobs (blacks were not hired as store or bank clerks even in black neighborhoods) but drove flashy cars and sported expensive clothes. Many of these friends were bankers and other executives in the ''numbers'' racket, a system of gambling that allowed small individual bets but generated huge sums of money for its organizers. ''I had lived within the law,'' Hughes later remarked wryly. ''They lived outside the law—and well.'' For Karamu, he would write a comedy about the ''numbers'' set in Harlem.

As for *Mulatto,* Martin Jones and the reviewers had so upset Hughes that for a while he ignored a crucial fact. ''That bastard of a MULATTO,'' as he called his play, was still alive on Broadway. Visiting New York over Thanksgiving, the musician William Grant Still pronounced it ''a strong dose of something! The matinee audience stumbled out—gasping—and only now, after half an hour, I'm getting my breath.'' The play was alive partly because of its raw power, but also because of Martin Jones's cleverness as a publicist. Strollers downtown were startled to see, carrying an advertisement for *Mulatto,* a majestic man in a sweeping Prince Albert coat, whose black bow tie, wide-brimmed black hat, white goatee, and black cheroot stamped him as the quintessential Southern colonel. (Hounded by a reporter as to where exactly in Dixie he was born, however, the colonel begged to be left alone: "Pleeza, mister . . . me no spika good Engleesh!") Later, an authentic Southern type staggered out of the Bowery to denounce the play as "a lewd and licentious lie" and "a vile and obscene libel on the gentlemen of the South." Early in November, another Southerner jumped up in mid-performance to harangue the cast before he was hustled away by the police. Jones's efforts became the subject of more than one amusing newspaper story—free publicity for *Mulatto*. And Jones knew how to economize. At one point, he would combine the roles of two actresses and an actor, and order the stage manager to perform the new part.

The "bastard of a MULATTO" was very much alive, but six weeks passed without a financial statement, due every Saturday, reaching Hughes. Queries by Langston to the American Play Company went unanswered. Prodded by Van Vechten, Hughes asked first Lieber, then Arthur Spingarn, to intervene. "They're about like Zora was about MULE BONE!" Langston complained to Spingarn, who had helped him in that sad affair. Examining the agreements, Lieber chastized Hughes for his concessions: "You are one of God's children all right—terribly naive." Himself the agent for Erskine Caldwell's *Tobacco Road,* the most successful play on Broadway, Lieber finally secured a lame admission from Rumsey that Martin Jones had sent him no statements. Jones, who had objected to paying Rose McClendon her full salary because he thought it too much for a Negro (or so Hughes reported), clearly saw no reason at all to pay Langston. Letters and phone calls yielded nothing but Jones's claim, impossible to prove without financial statements, that *Mulatto* was losing money—an irrelevant consideration, in any case, since royalties were paid not on profits but on receipts at the box office.

Angry at Jones's attitude when he himself needed money so badly, Hughes finished his farce about the numbers business, "Little Ham," so called after its four-foot, bootblack hero, Hamlet Hitchcock Jones. Immediately he sent copies off to Rowena Jelliffe and to Maxim Lieber, who would handle all new plays by Hughes. Accepting it at once, the Gilpins offered Hughes $50 for five performances. Lieber, whose own Broadway readers called the play mediocre, accepted the terms.

With the Gilpin money, Hughes left Oberlin on December 6 on a visit to Chicago. There he checked into the Hotel Grand, the leading black hotel in the city. He had two reasons for the visit. He wanted to work with Arna Bontemps, and he wanted to scout Chicago for the Guggenheim novel. At 731 East 50th Place, he found Alberta and Arna Bontemps barely settled into their new home. The men began work on "When the Jack Hollers," a play first outlined the previous summer in California. The title was an allusion to the folk belief that the braying of an ass arouses sexual desire. Like "Little Ham," the new play was also to be a farce, "a tragi-comedy of plantation life, voo-doo, and love among the peasants of the Mississippi Delta," as Hughes advised the skeptical Maxim Lieber. He also promised one side-splitting scene that would feature members of the Ku Klux Klan.

Having finished a first draft of the play with Bontemps, he returned to Oberlin for Christmas. To buy his train ticket and a present for his mother, he borrowed $25 from Van Vechten; Lieber also sent an ''advance'' of $20. Then, near Christmas, Martin Jones's first statement finally arrived, showing $88.50 due to the author. But Langston's elation was short-lived. Instead of sending a check, Jones threatened to close the show if Hughes did not waive all royalties until *Mulatto* showed a clear profit. Langston was enraged. He would prefer to have the show close, he insisted, rather ''than to sit here penniless to the point of notoriety,'' while everyone assumed that, with a play on Broadway, he was now rich. Finally he poured out to Lieber the nasty details of the ''discourteous and Nordic behavior'' of Jones, who showed ''distinct signs of having a Fascist soul.'' ''Since MULATTO by no stretch of the imagination, could be called a great work of art, nor a service to any cause,'' Hughes ordered Lieber, ''therefore, in the name of the proletariat, let him pay $88.50 or stop amusing himself with my MULATTO. . . . Go ahead and call his bluff!''

On this sour, demoralized note, 1935 ended. ''This has been a hard year,'' he admitted. ''If in 1936 I do not make my board and keep, I shall withdraw from the business of authoring and try to take up something less reducing to the body and racking to the soul. I'll just let ART be a sideline, like it used to be in the days when I was a bus boy and was at least sure of my meals.''

As moved as he was by her illness, Langston needed to get away from his mother. ''She was a little difficult for him,'' Arna Bontemps would recollect of Carrie Clark. ''She just kept him terribly broke, you see, so that sometimes I had to lend him money. . . . She imagined that he was rich.'' Instead of leaving her, Hughes vented his anger in a macabre little play, ''Soul Gone Home,'' in which a dead son sits up at his wake to confront his mother, evidently a cheap whore, with charges of hypocrisy and negligence. Another playlet, ''The Road,'' which he never finished, also involved arguments over politics and money between a mother and her son, who identifies with the black poor. Hughes also revised a dramatic adaptation of the short story ''Mother and Child'' from *The Ways of White Folks*. And probably about this time he wrote ''Genius Child,'' a poem he would publish the following year:

> This is a song for the genius child.
> Sing it softly, for the song is wild.
> Sing it softly as ever you can—
> Lest the song get out of hand.
>
> *Nobody loves a genius child.*
>
> Can you love an eagle,
> Tame or wild?
>
> Wild or tame,
> Can you love a monster
> Of frightening name?
>
> *Nobody loves a genius child.*
>
> *Kill him*—and let his soul run wild!

In January, Hughes announced that he would go overseas in March on his Guggenheim fellowship. He would write his novel in the Mediterranean port town of Málaga in Spain. To prepare for his leaving, he moved his mother into a room at a colored YWCA in Cleveland, near a clinic where she would receive electrotherapy treatment at the cost of $25 every two weeks.

Money remained a problem. Although Hughes promptly accepted an offer of $25 a week from the black Federal Theatre project in Harlem for work on an original play, nothing came of the plan. His struggle with Martin Jones continued. On January 3, Jones sent a wire to Oberlin threatening to close the play within the week if Langston did not waive his royalties. Backed by Maxim Lieber, Hughes replied that he would consider the waiver only if Jones settled past debts. Finally, more than two months after *Mulatto* opened, its author received a check for $88. ''I am happy as a lark,'' Hughes wrote Lieber, who had succeeded where all others had failed. ''If I'm ever a commissar in the Soviet Republic of Mississippi, I'll confer upon you a decoration. You have just about saved one member of the colored proletariat's life.'' Then, with Lieber's support, he refused to waive any royalties. At John Rumsey's office, Jones denounced Hughes in vile language, then introduced a new claim: for having brought the play to Broadway, he wanted an additional ten per cent of Hughes's money. Both sides awaited arbitration by the Dramatists Guild. Another disappointment came after Hughes won a prize of $50 from the leftist *New Theatre* magazine for a militant one-act play, *Angelo Herndon Jones*. Hughes waited and waited, but no money arrived. ''What is it about the theatre, right, left, or otherwise,'' he asked in exasperation, ''that makes it so loath to give up its cash?'' Still, when *Esquire* bought only the first fifty lines of ''Let America Be America Again,'' he balked at cutting the poem, because the last section contained ''the dialectic solution, according to the new line,'' with the people rising to reclaim their land (''And yet I swear this oath— / America will be!''). An impatient Lieber urged Hughes to forget ''the dialectic solution'' and pocket the money. He pocketed the money.

Hughes's concern with the ''dialectical solution'' was reflected in *Angelo Herndon Jones,* for which he drew on the actual history of a young black communist from Ohio, Angelo Herndon, who had been sentenced to twenty years on a Georgia chain gang for allegedly inciting insurrection, after leading a march on behalf of the relief of blacks. The play calls for a stark set consisting mainly of a poster announcing a meeting in support of Herndon. The main characters, all black, are two prostitutes, two young men, and a pregnant young woman and her mother, who face eviction. At the meeting on behalf of Herndon, one of the young men, Buddy Jones, whips up the crowd in support of the pregnant woman, who is bearing his child, and against her eviction. Buddy will throw in his lot with Herndon, whom he admires for his courage and for his rigorous, race-free Marxist ideology. The coming child will be named Angelo Herndon Jones.

Returning to Chicago to continue work on ''When the Jack Hollers,'' Hughes prowled the streets of the South Side to gather material for his novel. This was dangerous business. With almost every black on relief, the South Side had become, as Hughes would recall, a sort of Shanghai. The crime rate was appalling. "The openness of it so startled us that we could scarcely believe what we saw," Arna Bontemps would write. "In a few months my feeling ran from revulsion to despair. . . . We had to fly home each evening before darkness fell and honest people abandoned the streets to predators. Garbage was dumped in alleys around us. Police protection was regarded as a farce. Corruption was everywhere." Twice his home was burglarized; he and Alberta awoke once to find a thief in their bedroom. Hughes took notes from people who remembered the great Chicago riot of 1919, when 38 blacks and whites died, and over 500 were injured. According to his plan, this riot would feature in the young manhood of Sandy, the hero of *Not Without Laughter,* who comes to Chicago near the end of that novel.

As January deepened and the temperature sank to 20 degrees below zero, Hughes endured his coldest winter since Moscow. Encumbered by earmuffs, scarves, and layers of thin clothing, his toes freezing in cheap shoes, he pined for sunny "Ennesfree" and the beach at Carmel. "You can imagine," he wrote plaintively to Noël Sullivan, "the contrast between the bitter-cold, dirty, dangerous Chicago South Side and Carmelo Street near the clear Pacific." On the other hand, Chicago boasted a small but thriving black artistic and intellectual community that welcomed him warmly. He met and befriended the pianist and composer Margaret Bonds, a graduate in music of Northwestern University who had played the piano with the Chicago Symphony and had set several of Hughes's poems to music; the student of sociology Horace Cayton, whose sister Madge had welcomed Langston to Seattle in 1932; Ted Ward, a very promising playwright; Margaret Walker, the sixteen-year-old poet Hughes had encouraged in 1932 in New Orleans, and now a student at Northwestern University; and, already hailed as a radical poet and journalist, Richard Wright, still not thirty years old. A communist who had served as the executive director of the John Reed Club of Chicago, Wright had lectured at least once, in November 1934, on Hughes's work. Although certain blacks found him ambitious and aloof, Langston saw mostly the other side of his personality; to him, Wright was personable and ingratiating, soft and almost sweet. He was undoubtedly taking Hughes's measure. Already the author of a novel about black proletarian life in Chicago (the posthumously published *Lawd Today*), Wright was at least as interested as Hughes was in making vivid fiction out of life on Chicago's South Side. His *Native Son* was only four years away.

To get through the winter, Langston searched the near Midwest for lectures at cut-rate fees on (for him) musty topics—"Asiatic Peoples under the Soviets," for example, for a pro-Soviet group. Some invitations came, but so did rebuffs. In Gary, Indiana, he was barred as a communist atheist by the white school board, then by a group of timid black Baptist ministers.

Behind these barrings were items such as the poems "Goodbye Christ" and "Ballads of Lenin" (which was translated into Yiddish in the *Morning-Freiheit* of New York), as well as the play *Angelo Herndon Jones*. Hughes, however, had re-enlisted even more dramatically in the ranks of the far left. He was drawn out by the latest development in communist global strategy—the call for a Popular Front against fascism issued the previous year, 1935, by the Seventh World Congress of the Communist International. On January 3, he had delivered a major address, "The Negro Faces Fascism," before 10,000 persons attending the Third U.S. Congress Against War and Fascism at the Public Hall in Cleveland. The Congress had brought together more than 4,000 delegates representing peace organizations, labor unions, fraternal groups, and ethnic and youth societies, and had featured a variety of well-known speakers, including a former Marine Corps general, Smedley ("war is a racket") Butler. But the organization had been repeatedly attacked by critics, including most recently the president of the American Federation of Labor, as communist. In his own appearances, Hughes boldly endorsed key communist political approaches. On a radio broadcast, he scorned the New Deal and questioned the purpose of "the militarization of the young people of this country" through training programs in colleges and Civilian Conservation Corps camps (such as the one that employed his "brother" Kit). "Has America any real fear of invasion from abroad?" he asked. "Is not the answer rather that we have colonies and world markets to maintain which supply profits to a select [few]?" He singled out the Soviet Union for praise: "Russia can't help becoming a great country for the working man and will set new high ideals in our new civilization."

For communists, the cause of Ethiopia in its struggle with Mussolini's Italy was a major and highly successful lever in the ongoing effort to win Afro-American support. On the other hand, many black Americans took a strictly racial view of the war there. Hughes seemed sometimes to endorse the communist view, sometimes the more racial perspective. The previous September, *Opportunity* had published his poem "Call of Ethiopia" ("ETHIOPIA / Lift your night-dark face"), which made a racial appeal. The Italian invasion, on October 3, had sparked more rousing, essentially racial verses ("Ethiopia, stretch forth your mighty hand. / Let the King of Kings lead his ebony band"), soon set to marching music by one Thelma Brown. The communist view, however, was endorsed by Langston in February, 1936, when *New Theatre* published his "Air Raid Over Harlem: Scenario for a Little Black Movie," in which he linked the war in Africa to the danger posed by fascism to both black and white workers of the United States. Still later, in *American Spectator,* he would publish "Broadcast on Ethiopia":

> The little fox is still.
> The dogs of war have made their kill.
>
> Addis Ababa
> Across the headlines all year long.
> Ethiopia—
> Tragi-song for the news reels.
> Haile
> With his slaves, his dusky wiles,
> His second-hand planes like a child's.
> But he has no gas—so he cannot last.
> Poor little joker with no poison gas!
> Thus his people now may learn

How Il Duce makes butter from an empty churn
To butter the bread
(If bread there be)
Of civilization's misery . . .

The racial implications of events in Ethiopia were prominent again, however, in "White Man," a poem he would publish in December. Hughes linked its fate to that of black America, and chastized those who "enjoy Rome / and *take* Ethiopia," just as easily as they enjoy Louis Armstrong, then copyright his music as their own.

Racial or pro-communist, Hughes spoke in militant tones. And yet, in spite of these utterances, a contradiction was growing between his radicalism and his quest for commercial success, between the spirit of "The Negro Faces Fascism" and plays such as "Little Ham" and "When the Jack Hollers." About the "numbers" farce Maxim Lieber noted sarcastically that "it is rather light and presents no existing problem faced by Negroes in Harlem or elsewhere. From that point of view, the play is safe for Broadway." And "When the Jack Hollers," by Hughes and Bontemps, was an appropriate companion piece to "Little Ham." How did Langston justify such writing after his stern admonitions to black writers delivered the previous year at the first American Writers Congress? To Noël Sullivan, Hughes presented himself, in a remarkable confession, as using his influence with the left only to ease his way through the Depression. Since poverty seemed to be his lot, "the only thing I can do is to string along with the Left until maybe someday all of us poor folks will get enough to eat, including rent, gas, light, and water."

Perhaps Hughes was stringing along not with the left but with Sullivan—or even slyly rebuking his wealthy friend, who had called "Let America Be America Again" a *tour de force* written to satisfy others, and not indicative of Langston's true feeling. For a variety of reasons, however, Hughes was drifting away from the far left. At thirty-four, he was probably tired of being poor; the death of his father had perhaps removed the single most important pressure that had driven him toward radicalism and his version of bohemianism. Clearly Hughes also had realized, following the Carmel debacle, that he could not continue easily as a member of the far left when the masses of blacks viewed the left with suspicion, and when the left was almost exclusively white. In returning to his mother, to poverty, to dirty, dangerous black neighborhoods in Cleveland and Chicago, he was to some extent realigning himself racially, just as he had realigned himself following the break with Godmother. If he believed intellectually in inter-racial communism, he could not stand forever, or even for long, away from black people. His Karamu playwriting, from one angle a venture in commercialism, was perhaps also part of his renewal of racial bonding. Black audiences wanted to laugh at little Hamlet Jones, not to agonize with *Angelo Herndon Jones,* which the Gilpin Players had turned down after the actresses refused to play whores. Determined not to abandon the left, Hughes would nevertheless try to give blacks something of what they wanted, and bask in their applause.

Another mark of the decline of his political involvement came with the founding convention, between February 14 and 16 at the Eighth Regiment Armory in Chicago, of the latest radical effort to unify blacks, the National Negro Congress. The idea of the Congress had been developed the previous May at a conference at Howard University in which various groups, including the NAACP, initially had taken part, but which had quickly encountered strong communist pressure in taking shape. In charge of the new organization was an old friend, John P. Davis ("Sir John" of the *Fire!!* magazine circle in 1926), now perhaps the most powerful black intellectual in the Communist Party. Intended to supplant the moribund League of Struggle for Negro Rights, the Congress apparently had little use for the League's president; Hughes was mainly a spectator at the gathering in Chicago of nearly a thousand delegates from twenty-eight states and almost six hundred organizations. Following the Popular Front doctrine, and steering clear of inflammatory rhetoric and the chimerical call for black self-determination (on which the League had insisted), the Congress called for a unified black program on a variety of subjects—war and fascism, civil rights, agriculture, trade unions, consumerism, and political parties. Hughes made only a token appearance, on a panel to discuss "The Role of the Negro Writer and Artist on the Social Stage." Also on the panel, but a major force in organizing the conference, was Richard Wright, who had perhaps already formulated the charge he would make the following year, that "generally speaking, Negro writing in the past has been confined to humble novels, poems, and plays, prim and decorous ambassadors who went a-begging to white America."

As he awaited the start of his fellowship in March, Hughes's finances suddenly improved. Perhaps as a birthday present to Langston, Noël Sullivan countersigned a check for $196.51 he had received as a legacy into Hughes's account, then down to 33 cents at the Bank of Carmel. Next, the Chicago-based *Esquire* bought "Slice Him Down," a story of violence set in Reno. Inviting Hughes to his office, Arnold Gingrich asked him to tone down its ending (he agreed at once) and offered him $150 in immediate payment for the story. Langston's eyes grew wide "at the very mention" of the money. However, he steeled himself, as he assured his agent, and refused it: "I never handle such matters. Art is my only field. Send the check to Mr. Lieber." But he ran straight from the *Esquire* office to buy himself a shirt on Michigan Avenue. Against the still uncollected *New Theatre* magazine prize, Lieber sent Hughes $10, because he did not care "to have your death due to starvation on my conscience." Martin Jones seemed about to pay almost $500 due on *Mulatto,* which was still in full cry at the Vanderbilt. But Langston was disappointed again. William J. Clark, the long-suffering (since 1932) printer of *Scottsboro Limited,* served a lien on the *Mulatto* management for $285.50. Lieber and the American Play Company settled with Clark for a lesser sum. Then Jones further tied up Hughes's money by placing it in escrow with the Dramatists Guild, pending arbitration, while admitting freely to John Rumsey that he aimed to make it as difficult as possible for Hughes to collect any money. As for Jones's financial troubles—when the owners of the Vanderbilt tried to move *Mulatto* out at the end of its lease, he bought the theater.

On March 1, when Hughes began his Guggenheim fellowship, he still thought about going to Málaga but had not yet booked passage to Spain, where the political situation was rapidly deteriorating. Instead, he visited New York. He stayed at his ancestral home in the city—the Harlem branch of the YMCA, on 135th Street. There, a few days later, on a program shared with Carlton Beals, Horace Gregory, and Babette Deutsch, he made a vital impression (Hughes was "the soul and everything," according to an official) at a tribute to Jacques Roumain, recently freed from a jail in Haiti after a campaign by the Committee for the Release of Jacques Roumain. Feverish during the meeting, Hughes was soon down with influenza so severe that he was ordered to the Edgecombe Sanitarium in Harlem for ten days of rest. Then a telegram arrived, collect, from Kit Clark in Cleveland. Carrie Clark was dying. "SEND MAMA ANY HOSPITAL NECESSARY REGARDLESS COST," Hughes wired back. On the night of March 18, he took a train to Cleveland, where he found her in great pain and groggy from opiates. He arranged to have her moved to a hospital.

He had returned in time for the premiere at Karamu House of *Little Ham*. Advance sales had been heavy; to add to the excitement, Maxim Lieber came from New York to observe the production, as did two scouts for Columbia Pictures in Hollywood. On March 24, however, Langston could not attend the opening; a relapse of his illness sent him to bed. The next morning, he discovered that his play was a huge success. The *Cleveland News* called *Little Ham* "a hilarious comedy," and "side-splitting stuff"; the *Call and Post* praised its "hilarious lines and good clean humor." Apart from the black *Cleveland Gazette,* notoriously hostile to the Jelliffes, other newspapers were just as complimentary. And *Variety* magazine, perhaps the most important journal in the entertainment industry, speculated favorably on the chances of a Broadway run. Lieber would look into the matter, and the Paramount scouts left the impression that a motion picture was a definite possibility.

In certain of his short stories and poems, Hughes had shown a gift for comedy; *Little Ham* was his first attempt, other than in the ill-fated "Mule Bone" with Zora Hurston, to explore this gift on the stage. With this venture, and subsequent ventures in the genre, came both opportunities and problems. The major opportunity was for the expression of black culture in one of its most distinctive forms, that of laughter and the rich language and styles that make laughter culturally significant. The major problems in Hughes's comic plays—apart from the question of his basic talent as a playwright—probably derive from his instinctive choice of farce as his model, rather than some more elegant version of comedy. Set mainly in a shoe shine parlor and a beauty shop in Harlem in the nineteen twenties, *Little Ham* is about the diminutive bootblack Hamlet Hitchcock Jones, his colorful friends and acquaintances, including the many women anxious to please him, and his own passionate love affair with the "numbers" game. Although the play contains some serious moments, as when it highlights the fact that whites control and profit most from the "numbers" game while blacks sustain it, *Little Ham* seems to float lightly, buoyed mainly by stock comic stage routines. In the process, however, it also allows the black urban masses, though sometimes stereotyped, to display and express themselves with a freedom seldom seen before in American theater. Hughes was emboldened by the notion that stereotypes are most often based on some underlying if distorted truth, and that this exposure of black culture, in a play aimed mainly at a black audience, was worth the risks involved in stereotype.

For all its success, *Little Ham* brought the first of many charges that Hughes was utterly shallow in his comedies, and almost perversely unwilling as an author to aim for themes as weighty as, say, in *Mulatto*. Hughes himself insisted that "there is a serious undertone in LITTLE HAM. There is in all my plays." One reviewer indeed insisted that "underlying all that laughter you can hear at Karamu, there is the terrifying and tragic thread of life where there is no hope." Always Langston had found black laughter, like the blues, which often evoked laughter, a deeply serious matter; both were essential to the endurance of the race and its will to prevail. He himself affirmed the existence of a blues-sanctioned ambivalence between tears and laughter in his work; whether many of those who laughed also sensed the sorrow is, however, debatable. Much would depend on the skill and sensitivity of directors and actors. A seasoned Karamu actor and director later caught this unorthodox balance of spirit in discussing *Little Ham:* "To give up the hope that tomorrow may be better—that is tragedy; to live cheek-by-jowl with tragedy, but still to dream—that *can* be comedy."

Although he worked hard on his plays, Hughes would always find it difficult to adhere to strict standards of dramatic construction. Several of his dramas, loosely built to begin with, tend to trail off rather than come to a precise, well-made end. Eventually he would find virtue in necessity, and construe this tendency as an expression of artistic taste rather than a limitation. "If you entertain," he would declare of a typical play, "what does it matter . . . as long as it holds the audience, and you are able to say what you want to say." Hughes repeatedly asserted to editors and critics that he was not a "sensitive" writer—that is, he did not look upon any of his works, even his poems, as monuments to be strictly preserved, but easily accepted changes in his lines suggested by others, if the suggestions had any merit. This attitude to composition seemed too casual to some of his acquaintances. Certainly Rowena Jelliffe sometimes wished for more from him. "I think Langston should have been more responsible, and tried for more serious themes and plays, to match his gifts," she would say. "He had enormous talent, but he needed to take the whole thing more seriously than he did. It was sometimes a real struggle to get him to do what he had promised." Full of praise for *Soul Gone Home,* calling it fine, mature writing, she obviously found other plays short of its standard.

Hughes was first unable, then finally unwilling, to change his apparently wayward method of composition. Without knowing it, however, he was in fact refashioning conventional dramatic form to suit both his peculiar gift and the nature of his main subject and his ideal, if not actual, audience—the black masses. Two years after *Little Ham,* when he at last found himself in a situation that allowed him as much independence as he would ever have in the theater, Hughes almost instinctively turned to a loosely structured version of theatrical form. Not even he, however, would look past its conspicuous limitations to recognize its utility and authenticity; almost twenty years passed before Hughes began to create an entire series of plays according to this liberated model.

With the success of "Little Ham," the Gilpins immediately agreed to make the collaboration by Hughes and Arna Bontemps, "When the Jack Hollers," their next production, set to open at the end of April. While the show went into rehearsal, Langston settled into a small apartment at 2245 East 80th Street, as his mother, against the odds, slowly regained her health. In his apartment, however, Hughes faced an unpleasant fact. He did not want to write another novel. A new novel had been Blanche Knopf's idea, just as *Not Without Laughter* had been Godmother's. He had gathered raw material in Chicago, but felt no powerful urge to continue the story of Sandy into his manhood. To some extent, Hughes simply did not have the stamina for a long work of fiction, which might be the same as saying that his discomfort in revealing even in fiction an adult male psychology based on his own experience was too serious to overcome. Whatever the reason, and although he did not want to accept the Guggenheim money on false pretences, the novel of Chicago life simply refused to come. Instead, he worked on his libretto for the opera about Haiti, and even returned to his unwieldy California labor epic, "Blood on the Fields." He also attended rehearsals of "When the Jack Hollers" right up to its premiere, which was eagerly awaited in Cleveland following the success of *Little Ham.*

Certainly the Karamu Theater was full on April 28 when the new show by Hughes and Bontemps opened. Set in the Mississippi delta and in springtime, and originally titled "Careless Love" after the classic blues, "When the Jack Hollers" presents a sometimes richly folk-comic cast drawn from a sharecropping community devoted equally to the pursuit of love and a deep faith in conjuring, both of which the authors exploit for their comic potential. Along with this benign blending is a more volatile concern with the major Southern problems. In an unusual approach, the authors offer a satirical view of poverty and racism in general and the Ku Klux Klan in particular. The play, which attempted to break new ground in using humor as a political weapon in an arena, the South, where social reality was apparently too savage readily to admit it, ends with a happy reconciliation of poor blacks and whites united against the bosses, as well as a loving union of various couples hitherto at odds with one another. But the reaction of the Karamu audience to "When the Jack Hollers" was only lukewarm; the production was not so much a qualified success as a mitigated failure. To the critic of the *Cleveland News,* Hughes himself admitted that his and Bontemps's play needed tightening. Almost everyone agreed.

Early in May, when Carrie's health had improved sufficiently, they moved to a five-room apartment at 2256 East 86th Street, only four houses from their first home in Cleveland in 1916. For the first time in Cleveland, thanks to the Guggenheim Foundation, they were not living in an attic or a basement; Carrie had space for a live-in maid, which she needed. Suddenly Langston had some success with *Mulatto.* Although he lost one stage of the arbitration on a technicality, he won the case as a whole; in mid-April his theatrical agent sent him a check for $228.37. This amount was less than the full amount owed by the implacable Martin Jones (mail to Hughes at the Vanderbilt was being returned "addressee unknown" and "not here"), who was now claiming to have rewritten the play entirely. This charge did not completely displease Langston: "In fact, I'm delighted to have someone else want to take credit for the present version—which I think is pretty awful in lots of places."

He endured a flurry of rumors about a film of *Mulatto* by Paramount Pictures, just as he had been teased pointlessly by the thought of one of *Little Ham*. Another hope pulled him to New York in mid-June. A project now almost ten years old, his collaboration with the Swedish writer Kaj Gynt on a musical called "Cock o' the World," suddenly revived. Hughes signed a fresh agreement with the obsessed Gynt, who was sure once again about luring Paul Robeson to star in it, and Duke Ellington to write the music. Very much interested in a collaboration with Ellington, who was virtually his musical counterpart in black America, Hughes began writing lyrics for the musical. He also applied for membership, which would be granted a few months later, in the American Society of Composers, Arrangers, and Performers (ASCAP), and in the Song Writers' Protective Association. Although he had written lyrics before, he was now opening a new channel in his career, into which a great deal of his energy would flow for the rest of his life in return for the steady payment of royalties by ASCAP. If Hughes was becoming less of an artist, he was certainly becoming more of a professional writer.

To make way for another production by Jones, *Mulatto* had moved to the Ambassador Theater, where Hughes attended a performance and noticed that the costumes and scenery had improved. Jones, however, still refused to speak to him. He visited the seriously ill Rose McClendon, who had been forced by her doctors to leave the show. Langston sent her roses from Thorley's, where he had worked as a delivery boy in 1922. He was bitter about Jones's treatment of a gracious woman and a fine actress who had been cheated by racism of the recognition and rewards she deserved; which other Broadway star rode the subway home at night, as she had done each night after her performances in *Mulatto?* At Carl Van Vechten's home at 150 West 55th Street (he would move later that year to 101 Central Park West), Hughes attended a birthday party, complete with three cakes—one red, one white, and one blue—for Van Vechten, James Weldon Johnson, and Alfred A. Knopf, Jr., whose birthdays fell on the same day. Although Hughes had seldom been in New York in recent years, his correspondence with Van Vechten had continued, and Van Vechten was still among the first to read any major new work; moreover, Langston still owed his former mentor $200 from the Godmother disaster of 1930. On this visit to Manhattan, he also spent an evening with the composer Harvey Enders, listening with some excitement to his ballet score of Langston's urban ballad, "Death in Harlem." The only gloomy moment of the New York visit came at Yankee Stadium on June 19, when Hughes watched in heartbreaking disbelief as Max Schmeling knocked out black America's darling, Joe Louis, in the twelfth round of their heavyweight fight. "I walked down Seventh Avenue and saw grown men weeping like children, and women sitting on the curbs with their heads in their hands. All across the country that night when the news came that Joe was knocked out, people cried."

Langston was still in New York on the morning of July 6 when, in the lobby of the YMCA, he met the past and the future. The past was Alain Locke, who was about to leave for Russia, of all places, and a theater festival. As Hughes and Locke were talking politely, a young man approached.

"Dr. Locke, do you remember me?"

"Why, of course I do."

Ralph Ellison had arrived in New York the day before from Tuskegee Institute in Alabama with $75 and the hope of making enough money to return to school for his senior year; Locke, who had met him during a recent visit to Tuskegee to see Hazel Harrison, a fine pianist on the music faculty, introduced him at once to Hughes. For Ellison, this was an auspicious start to his summer adventure. Although he would remember himself as a typical "eager, young, celebrity-fascinated college junior" in their first meeting, he had read Hughes's poetry since the sixth grade in Oklahoma. In turn, Langston was immediately impressed by his conversation with a young man who obviously knew more than a little about music (his major area at Tuskegee), modern poetry, and sculpture. Hughes had with him two books, including André Malraux's *Man's Fate*. "Since you like to read so much," Ellison later recalled Hughes telling him, "maybe you'd like to read these novels and then return them to their owner." Not long after, hoping no doubt to encourage Ellison towards a Marxist perspective, he lent him a copy of John Strachey's *Literature and Dialectial Materialism* and recommended the poetry of Cecil Day Lewis. Through Langston, Ellison found friends in Louise Thompson (the owner of the books), Toy and Emerson Harper, and the black sculptor Richmond Barthé, who accepted him as his first pupil. Working as a counterman in the dining room of the YMCA, Ellison also did small chores now and then for Hughes, such as delivering manuscripts and mailing Thomas Mann's *Complete Stories* as a present to Noël Sullivan. Confident that Ellison had a bright future in spite of the hard times, Hughes jokingly shared with him a grand secret—how to make one's way in the world without money: be nice to people, and let them buy your meals. "I'm following your formula with success," Ellison soon reported earnestly; so far he had paid for only two dinners.

Returning to Cleveland, Hughes soon heard the sad news of the death in New York of Rose McClendon. Not long after, Lincoln Steffens died in California. The summer of 1936 drifted to its end without any breakthrough into Broadway or Hollywood. Duke Ellington, visiting Cleveland, played Hughes some charming music he had written for "Cock 'o the World," but nothing more was heard of the musical. At least Langston had reason to be glad that he had not left for Málaga in Spain to write his novel. On July 17, only months after the Republicans' sweeping electoral victory, the army, led by General Franco, had revolted in Morocco. The Spanish Civil War had begun.

The violence and panic sweeping Spain, with Málaga prominent in the reports, brought a double-banner headline in the black *Chicago Bee:* "LANGSTON HUGHES, NOTED AUTHOR, REPORTED MISSING." Friends of the "distinguished, fiery, eccentric colored poet and dramatist" were worried for his safety in Spain; since his name was not on the list of the evacuated foreigners, a search had begun for "the explosive, sensational, taciturn 'weaver of dreams'." Who put this hoax on the *Bee* is not known. Perhaps it was Hughes himself, or so he later hinted to a friend, although he could not have written so badly. But he had long respected the value of publicity and for years had fed items about himself to the Associated Negro Press, which seldom looked closely at gift copy before printing it. He may even have coined the phrase "Poet Laureate of the Negro Race" as applied to himself, then watched it grow into currency. Gleefully he dispatched a copy of the *Bee* to Van Vechten: "Carlo—I'm lost!"

Noël Sullivan passed through Cleveland early in September, on his way to France. Tired of his drab life in Cleveland, Hughes would have loved to stroll once again down the boulevards of Paris, "a city of charm too true to ever be untrue to those who love her," as he saluted it in a telegram to Sullivan as the *Normandie* sailed. Instead, he secluded himself for eighteen days at the Majestic Hotel in Cleveland to toil on his next play for the Karamu group. Turning aside from low-toned farces, Hughes began something loftier—a rewriting of the "singing play" about Dessalines and the Haitian revolution that he had started many years before, in September, 1928, under Mrs. Mason's patronage. At the end of his stay at the hotel, he sent off a copy of the refashioned epic, which called for 65 actors and sundry dancers, to Maxim Lieber, his agent. Lieber found the title "abominable"; subsequently, the play would have, at different times, three names—"Emperor of Haiti," "Drums of Haiti," and (for its first production, and with Hughes taking a cue from the title of his anthology of Mexican and Cuban translations) "Troubled Island."

The arrival of the touring company of *Mulatto* brought mixed feelings of triumph and chagrin for its author. Aiming to sink the visit, he craftily whispered something of his trials with Martin Jones to every journalist who would listen. But on the surface Hughes was pleasant; he threw a party for the cast, who presented him with "a swell English briar pipe" (always mechanically inept, he would never learn to keep it going). But Jones sprang a surprise on Hughes. Suddenly he was friendly, Langston wrote a friend; "And now we speak!" Not long after, when authorities in Chicago tried to ban the play as obscene, as Philadelphia had done, Hughes responded to a telegram and went there to make some cuts himself. For all the suffering it had caused him, *Mulatto* was well on its way to setting a record for the number of performances of any play written by a black American, a record not eclipsed until Lorraine Hansberry's *A Raisin in the Sun* (with its title taken from a Hughes poem) more than twenty years later.

The opening on November 18 of *Troubled Island* was the glittering highlight of the black Cleveland season. Sponsored by local chapters of the Alpha Kappa Alpha sorority as the first production of the sixteenth Gilpin Players season, the premiere brought out of the community a Depression-defying montage of tuxedos, fur coats, and gorgeous evening gowns. Prominent among the smart set was the author's mother, Carrie Clark, lately recovered from an altercation with a clerk in a local store over a fur coat she had bought and tried to return. "LANGSTON HUGHES' MOTHER BEATEN IN STORE," one paper informed the world.

The production itself was something of a success. Although the *Plain Dealer* reviewer hinted that *Troubled Island* was thin, he called the performance "an exceptionally interesting theatrical occasion, a creditable new play by a distinguished Negro dramatist." With its emphasis on black revolution and heroism, for once there was no backlash against Karamu and the Jelliffes from the hostile *Gazette*. Moderately encouraged, Hughes immediately set to reworking the play once again for its ultimate purpose now, as the libretto of the opera to be scored by William Grant Still in Los Angeles.

Emerging out of Hughes's reading but also from his visit to Haiti in 1931, *Troubled Island* is very much a study in failure, in which Hughes stresses the inability of Jean-Jacques Dessalines, the most revered figure in Haitian history, to sustain the spirit of revolution that led to the winning of political independence from France. Illiterate, headstrong, and hedonistic as Dessalines is, however, he is brought down in Hughes's play less by his ignorance and corruptibility than by the villainy of others. Eventually he commits the most deadly sin, in Hughes's eyes, by deserting the black masses from which he has come. Turning instead to the advice of treacherous mulatto counsellors, he seeks the pleasures of counterfeit European styles and customs; he also sets aside his black wife, Azelia, who goes mad (as Cora the mother goes mad in *Mulatto*) in favor of the mixed-blood Claire Hereuse. Hughes, in making Dessalines more a victim than an initiator of his own downfall—as the historical evidence strongly suggests he was—thus devalues Dessalines's worth as a tragic hero; the demented Azelia becomes a more provocative character than Dessalines himself. Along with this ambivalent characterization comes a weakening of the entire drama; the plot slackens, and injections of plebeian comic relief and the clap-trap of the Haitian royal court help little to sustain interest or raise the play above the level of mediocrity.

The premiere of *Troubled Island* in Cleveland was accounted a general success, but Langston remembered it for a more intimate reason. One person, "a lovely-looking girl, ivory-white of skin with dark eyes and raven hair like a Levantine," as Hughes would recall her, stood out among the gorgeously dressed black women. Twenty-two years old, Elsie Roxborough carried herself with the assurance of someone born to beauty, position, and money. Her sense of position came mainly from her father, a lawyer and legislator who had been the first black Michigan state senator; her uncle, his brother, had less prestige but more money, with a small fortune made in sundry business ventures. Talented and headstrong, Elsie Roxborough had grown up, despite her race, dreaming of unlimited success. "She was an extremist," one friend later judged; "she thought she was Hollywood." A gossip columnist assured the world that Elsie Roxborough "has what the girls call 'flash' "; another black newspaper dubbed her the foremost debutante in Detroit. Whatever Roxborough had, she made a mark at once on Hughes. In September, 1936, on the other hand, he was ripe for a love affair; his Moscow sweetheart, Sylvia Chen, had just stung him with the news of her marriage. Elsie Roxborough would become "the girl I was in love with then."

Passionate for the stage, Roxborough in 1936 was a talented senior at the

University of Michigan; one of her classmates was a fellow playwright, Arthur Miller, who would remember her as "the most striking girl in Ann Arbor." By this time, she had written and directed plays, handled advertising for the Little Theatre at Ann Arbor, and also written on drama for the local press. In April she had staged an adaptation of Walter White's novel *Flight,* put on by her own troupe, the Roxane Players, who had then scheduled for the coming November a drama written by Roxborough herself, "Father Forgive Them," about a young woman who wants to pass for white. As hard as she worked, Roxborough also found time to play. The previous year, she and Joe Louis, with whom she may have had a romance, and who was managed in part by Roxborough's wealthy uncle, publicly denied that they were engaged. Apparently she had one standing offer of marriage, which meant nothing once she saw Hughes.

After a mutual friend, the Chicago musician Margaret Bonds, had stirred her interest in Langston, Elsie warmed to him even before they met. First she secured from him ("and now I want to proposition you") the promise that he would send a blurb for "Father Forgive Them" and attend the premiere. They still had not met when she arrived in Cleveland for the opening of *Troubled Island.* Expecting a "staid old bore," with whom she might perhaps join forces in the theater, Roxborough instead tumbled into love. She definitely went too far at the post-premiere party that evening; she proposed marriage to Hughes, who promptly refused her. Soon she was apologizing miserably that "nothing I said or did after 1:30 is accountable. . . . I am dreadfully sorry and disgusted with myself for I had counted on our meeting so much, I had so much to say in a business-like way that wasn't said at all." Accepting her apology, Hughes gave her permission to stage the first act of *Troubled Island* at a sorority convention.

Fascinated by Langston, perhaps especially by his coolness, Elsie returned uninvited to Cleveland to see him again. Once again she raised the question of marriage; this time she promised him an unusual degree of freedom. She would keep her own name, and they could go largely in separate ways. To which he replied, as Roxborough put it, that "our affair" (a marriage of convenience) would be "tiresome." Although her boldness would have put off many men, Elsie probably upset Langston above all with her attitude to race and color. She was painfully sensitive about her extremely light skin; her drama of passing, "Father Forgive Them," was to some extent autobiography. "The tragedy of my life is being a Negro or having to be considered so, definitely, in the future," she wrote Hughes; she knew that she shouldn't "hide my head in shame because of my color," but a sense of morbid unhappiness and of helplessness haunted her.

Refused by Hughes, who probably disliked racial shame above everything else, Roxborough hung on. "I do apologize," she wrote him about her proposals. "Marriage has never appealed to me, particularly." But she made certain to tell him that she had just refused her main suitor, because "I should resent anything he did for the rest of his life because you refused me." To which Langston replied weakly that while marriage was out of the question, he would always be there for her; he pleaded poverty as his reason for bachelorhood. Elsie was not convinced. "What you need, if you would take her," she insisted, "is a wife who is game enough not to care about incomes!" For a while she switched her greeting from "querido mío" to "hermano mío," but her affection was not sisterly. Staying with a friend, she passed part of the Christmas season near him in Cleveland. On New Year's Eve, as Langston waited for her in the parlor of her friend's home to take her to a gala party to welcome 1937, Elsie listened as the radio played "You're So Easy To Love" and she thought that she "would simply die of ecstasy—or fall down the stairs."

Perhaps the two had an affair; perhaps not. But in return for her love, Roxborough met a character indomitably "level-headed and phlegmatic," as she put it, except when he teased her. Asked for a poem written to her, he sent instead a song, which she parodied into yet another plea for marriage:

> Never begged before,
> Never fallen so hard,
> But I'm set on the idea
> Of Playwright and Bard . . .

Once she signed a letter as Elizabeth Barrett, and sent it to her hoped-for Robert Browning. But finally she could not mistake the chill in his blood, at least where she was concerned: "I come to see how futile it all is, trying to get beyond your indifference." On Valentine's Day she received flowers, but not from Hughes—"so I began to think about you in a very bad way. I have decided that you are about as sentimental as a clam!" She was fed up with "all that indifferent, platonic attitude of yours." She went to the verge of anger, but love dragged her back, and a feeling of helplessness. "I do miss you terribly," she cried out once. "You are my rod and my staff as damnable as you are! I miss you so dreadfully that sometimes I could cry." And undoubtedly sometimes did.

The romance, or such as it was, ended for him less than six months after their first meeting, when the black papers began to report, to Hughes's amazement and anger, that they were about to marry. The story reached almost every major black community in the country. From Los Angeles, Nora Holt wrote to congratulate him on his coming marriage. Kit Clark, his priapic "brother," warned him not to fall for Elsie. "You are in the same category as I am," Kit reminded him. "What the boys call a coxsman." Who was responsible for the story? Langston suspected Elsie herself, working through a press agent. She denied the charge, but coyly: "Such people who would expect you to consider something so normal!" She begged him, moreover, not to refute the story publicly "and embarrass me further." But on March 27, the *Baltimore Afro-American* bannered his response: "Denies Report That He Will Marry Soon." "I'm afraid," Hughes said, "that my marital intentions have been greatly exaggerated. I am a professional poet and while poetry is so frequently associated with romance, there seems to be little compatibility between poetry and marriage, especially where one must depend on it to support a wife." As for being in love—"that much certainly could be true." Undeterred, the *Pittsburgh Courier* set his portrait next to Elsie's and asked: "Are They Bound For The Altar?"

Another story speculated that "Elsie Roxborough, Langston Hughes May Take Altar-Trek." Langston was furious. Icily he kept her in suspense about whether he would show up for her full production in Detroit of *Troubled Island,* now called *Drums of Haiti.* "For God's sake don't get temperamental," she pleaded, "—just because someone was presumptuous enough to make so great an error. I'm the only one to suffer for that, for your intentions towards anyone were made quite plain all over the world."

He attended the production, which included Joe Louis's sister, Eulalia Gaines, in a major role. And backstage Hughes jumped in fright when there suddenly emerged from the gloom a character dressed as the Haitian "Papaloi," with painted face and mask, feathers, and a grand cape. Elsie introduced the men in what the actor later called a *New Yorker* cartoon moment—he was Detroit's leading black poet, Robert Hayden. Towards Roxborough herself, Hughes was outwardly friendly, but glacial at any hint of intimacy. Their affair was definitely over.

If she was broken-hearted, at least Hughes could tell himself that he had done nothing to encourage her. He was not obliged to fall in love with everyone who wanted him. And (unlike in his dealings with Sylvia Chen) he had made his position clear almost from the start. For Hughes, Roxborough was a tragic figure, beautiful, privileged, but doomed, of the kind most poignant to him—the mulatto caught between two worlds. He sympathized with her, but could not help her or save her.

Elsie Roxborough later went off to the Pasadena Playhouse in Califonia, but dropped out of her program there. By 1938, she was in New York, passing for white as "Pat Rico" and working as a model. She did not forget Hughes. Once, writing to him to suggest a meeting, she confessed her undying admiration (and promised not to propose to him). Later, Hughes would claim to have received little gifts from her from time to time, always without a return address. He also learned that she was miserable with her life across the color line. In 1942, their original matchmaker, Margaret Bonds, wrote Hughes sorrowfully that "I wish you'd married Elsie. She would have listened to you, and might have been *quite* dependent . . . and she'd be happier now, too."

In 1949, Elsie Roxborough, then "Mona Manet," died, perhaps a suicide, of an overdose of barbiturates while living as a white among whites on the East Side of Manhattan. Some time later, Hughes began to display her framed photograph, inscribed in 1937 "To Langston, Con Amore!," high on a bookcase in his study. He answered questions about her identity only in a guarded way.

In December, 1936, the month after their first meeting, his Guggenheim fellowship expired with scarcely a word of his novel having been written. However, because of the fellowship and his *Mulatto* royalties, which totalled almost $4000 in 1936, this Christmas was much happier than the last; Hughes bought himself a new suit and a typewriter, and took out an insurance policy on his life for $300 with Metropolitan Life, paying 25 cents a week in premiums. But he was clearly at a loss how precisely to proceed in his career. His relationship with Knopf was badly wounded, if not dead. Since the spring of 1934, when

Blanche Knopf had turned down his radical poems and the Soviet book, Hughes had simmered with resentment at her. In September, 1935, after she had queried him about a new book for her list, he allowed almost six months to pass before replying. Did she wish a volume of plays? (''but you said you'd rather have a novel first''); or another book of poems? (but she had said that a book of poems now ''would also be unwise''). He did not rule out a novel or an autobiography (the latter first promised in 1925): ''Do either of these strike you as good, and which first, the novel or *ma vie?''*

He sent this slightly taunting inquiry early in March, three days after he had finished work on a book for a very different publisher. Louise Thompson, his reliable ally and companion in the Godmother and Soviet film wars, now worked for the International Workers Order, the powerful fraternal benefit society linked to the Communist Party. ''Langston told me what had happened to his poems,'' she said later. ''I liked them just as much as he did. They were powerful and uncompromising. I decided to do something about it, so I went to our education department and convinced them to bring out a pamphlet of his radical poems; I thought people needed to know them in the struggle.'' Leftist unions were becoming involved in such ventures, mainly in the theater; auto workers in Detroit had recently published a novel by Upton Sinclair. Pleased, Maxim Lieber urged Hughes to take a cut in royalties to help the project. On March 4, he completed the selection, called ''A New Song,'' the title chosen by Louise Thompson from one of the poems. The International Workers Order first planned an edition of 15,000 copies, with a preface by Joseph Freeman; eventually the edition numbered 10,000 (by far Hughes's largest to date in English), with an introduction by Mike Gold. The thirty-one-page booklet cost only 15 cents a copy.

This was Hughes's only literary success in some time. Finishing another farce for the Gilpin Players, ''Joy To My Soul,'' he decided to travel to California. There he would work with William Grant Still on their opera and visit Noël Sullivan. When an offer came for him to go even further, Hughes grabbed at it. He agreed to go that summer to Europe and the Soviet Union as a paid guide on an eight-week tour (July 3 to August 31) called ''National Minorities in Europe and the Soviet Union,'' sponsored by the progressive Inter-Racial Study Group through the firm Edutravel, Inc. In the meantime, to finance his California trip he tested the lecture circuit, working through a Fifth Avenue speakers' bureau. The main result was an appearance on March 7 at the Brooklyn Academy of Arts and Science in a symposium and poetry reading that included the progressive poet Genevieve Taggard. For young Ralph Ellison, who had not returned to Tuskegee and who had kept in touch with Hughes (''he buoyed up my hopes!''), this reading, preceded by an exciting dash in a taxi across the Brooklyn Bridge with a dressed-up Hughes and Taggard, herself perfumed and splendid in a glowing white silk dress, opened a magical window on the future: ''For the first time I felt a certain sense of possibility about a life in the arts.'' Ellison would soon complain that Hughes was ruining his favorite vices, such as hanging out at the Apollo Theatre: ''They don't seem quite the same after you and your conversation.'' In Boston, Hughes lectured in the well-known Ford

Hall Forum, where he spoke on "A Poet's Campaign Against Racial Discrimination" and stressed the economic roots of racism. "Raise the standard of living of the Negro," he asserted, "and the battle is half won." While in the area he read successfully at Harvard University, in his first visit there. In the audience was Toye Davis, a former Lincoln University man from Hughes's years there, and now a graduate student in biology at Harvard.

Returning to Cleveland, Hughes attended the premiere of his *Joy To My Soul.* The plot centers on the affairs of Buster Whitehead, an oil-rich but none too bright young man from remote Shadow Gut, Texas, come northeast to the Grand Harlem Hotel to meet and marry a sweetheart wooed through a "Lonely Hearts" newspaper column. Buster Whitehead survives a variety of adventures, including a marijuana-induced spell, to marry not his Lonely Hearts sweetheart, who is monstrous, but the pretty cigarette girl with whom he falls in love. With this play Hughes not only returned, after *Troubled Island,* to comedy but experimented with a more frantic comic style than he had ever attempted; a production note actually advised using Negro minstrelsy as a model, and directed the cast to overact. Clearly, however, Hughes also hoped to suggest something rather sinister in *Joy To My Soul.* The background, thickly populated by conmen, call girls, freaks, and a convention of the Knights of the Royal Sphinx, subtly offers a satirical, darker, at times even grotesque counterpoint to the lighter main action. Whatever Hughes's intentions, however, the contrast did not result in a sense among the audience of a deeper drama. Blacks enjoyed the references to Cleveland, and the *News* reported only that Hughes had written "a rollicking farce." The more judicious *Plain Dealer* disagreed, but saw nothing serious attempted by Hughes: "Striving too mightily to produce a commercial success, he succeeds in turning out an amusing, often crude, boisterously slangy, little potboiler. It is strange how his poetical sense of beauty is utterly missing in this mumbo-jumbo about Harlem life." Hughes reasoned that many of the whites in the audience simply did not catch the jokes, especially those about life in a black hotel, that set blacks howling. But it was clear that *Joy To My Soul* had no future on Broadway.

Disappointed once again, he shrugged off the criticism and agreed to write yet another play for Karamu, but his mood was somber on April 21 when he left Cleveland for California. The weather was no help; the Midwest was bleak with rain or dingy with piled snow. At a reading along the way in Omaha, Nebraska, a local right-wing magazine branded him a communist (but 150 persons came out to hear him). At Salt Lake City, however, where he visited the late Wallace Thurman's relatives, and purchased an ash tray as a gift for Noël Sullivan, the sun shone brightly and the wide West unfolded before him, "spacious and life-giving." In Los Angeles, William Grant Still, who now enjoyed a secure niche as a musician at Warner Brothers, seemed confident and anxious to begin work on their opera set in Haiti. Typing either in his room at the Clark Hotel or out in San Dimas Canyon with Still, Hughes pushed hard to finish the libretto. His effort was not entirely enjoyable; although he admired Still as a musician, Langston began to find him too imperious as a collaborator. Any disagreement between librettist and composer invariably had to be resolved in Still's favor, an attitude Hughes would encounter and deplore in almost all the composers with whom he worked. Nevertheless, he and Still signed a formal agreement on June 1, near the end of Langston's stay in the Los Angeles area.

In Carmel, Noël Sullivan greeted Hughes warmly at his new home, "Hollow Hills Farm," a ranch of eighteen acres sprawled on green slopes above peach orchards, about five miles within Carmel Valley (Sullivan had named the farm after a poem by "Fiona Macleod," the Celtic nom de plume of the Scots writer William Sharp). In residence was a menagerie of horses, cows, sheep, goats, and dogs, including the aging German Shepherd Greta, Hughes's loyal companion during his year at "Ennesfree." A large, comfortable house with a sunny terrace allowed Sullivan to entertain indulgently, with ample space for huge lunches and dinners and a grand piano and an organ in place for his many musicales. Eulah Pharr, as kind and accommodating as ever, supervised the running of the large household.

There was now, however, a complicating factor in Langston's relationship with Sullivan. His patron's anti-communism had grown deep. "Believing firmly as I do in your intuitive intelligence," he had written to Hughes shortly after his own visit to Cleveland en route to Paris, "I dare to hope that your loyalty to this ism is daily growing more qualified." Sullivan was now outspoken on the subject. Meeting Louise Thompson in San Francisco, he had made the mistake of attacking the party to her face; although she was being entertained in his Hyde Street home at the time, the radical Thompson had scorched him with a reply. Sullivan had then warned Hughes, "in the light of our respective feelings . . . we should not discuss the Russian experiment when you get here."

Hughes had no intention of severing all his ties with the left, but he was not as bold as Louise Thompson. Before leaving Los Angeles, he had received a telegram from Paris from André Malraux, Louis Aragon, and other organizers, secretly inviting him to attend the Second International Writers' Congress in Spain towards the end of June. He had declined the invitation, citing his Edu-travel tour, scheduled to start on the third day of July. His true attitude to being linked so openly with the left, and with the struggle in Spain, was perhaps more involved. He had allowed himself to be included on the committee for the American celebration of the new Soviet constitution in 1936. That fall, however, he flatly refused an invitation to a Communist Party banquet in New York for the vice-presidential and presidential nominees of the party, although Joseph Freeman not only guaranteed all his expenses but held up advertising until his reply. Freeman invoked Emerson, Thoreau, and Frederick Douglass as Hughes's nineteenth-century political and literary forebears, but he did so in vain. Langston would not come out publicly for or against the party. He did not vote in the presidential election; in fact, up to that time, he probably had not voted in any election whatsoever, and there had been a small but telling sign of his waning radicalism. In October, just before the election, he helped to judge an NAACP slogan contest for its youth chapters. It was the first time since 1930 and the Scottsboro dispute, when he had lashed out at the organization, that Hughes acceded to one of its requests for help. Once more he was friends with Walter White, whom he had abused almost scandalously in an essay written in

Moscow. To some extent, this was Popular Front politics; to some extent, Hughes was returning timidly to his oldest institutional ally.

The maverick English aristocrat Nancy Cunard, who had personally underwritten and edited the historic anthology *Negro* (a truly valuable and amazing book, Hughes had called her effort, which he, unlike many other black Americans, had helped eagerly), sent him an impassioned query from Madrid: "Why oh why aren't you here?" When Cunard and Pablo Neruda (who "admires and knows your work") appealed for help with an anthology of revolutionary verse on behalf of the Spanish fighters, he sent his "Song of Spain" ("I must drive the bombers out of Spain! / I must drive the bombers out of the world!") and asked Richard Wright for a contribution. In Cleveland, Hughes braved a downpour to hear the novelist Ralph Bates talk about Spain and of how Pablo de la Torriente-Brau, an attractive young radical writer Langston had met in Cuba, had been killed at the front. But in one direction Hughes made no move, except in a token way—toward the organization of the Abraham Lincoln Brigade, which was attracting hundreds of whites and blacks to fight or otherwise serve perilously in Spain. And he was making music with William Grant Still in California in May when *New Masses* listed him as a signatory, with twenty-two other writers including Erskine Caldwell, Van Wyck Brooks, Malcolm Cowley, Claude McKay, and Archibald MacLeish, to "A Manifesto and a Call" for a "National Writers' Congress," which identified Spain as the first real battlefield "in a civil and international conflict that is certain to recur elsewhere."

Still a supporter of the Soviet Union, pro-Stalinist in spite of the purges (his cheerful editor at *Izvestia,* Karl Radek, had fatally confessed to treason in the infamous "Trial of the Seventeen" that year), Hughes saw such events there as part of the price of the transformation of the state wrought by the revolution of 1917, a transformation from which the colored Soviets had greatly benefited. He made no attempt to enter the raging American debate between the supporters of Trotsky and those of Stalin; he drew the line there. The greatest American opponents of the Soviet Union were also the greatest opponents of basic rights for blacks; Hughes needed no further illumination to see Russia's or Stalin's position. Beyond quiet support for the Soviet Union, however, he would not go. He had written an occasional essay for Soviet journals, including an essay on Pushkin for *Izvestia.* But this act was hardly radical; the chairman of the American Pushkin Committee was Robert Frost.

What mattered ultimately for Hughes was not his political beliefs but their effect, impossible to measure precisely, on his art, which was his main contribution to the life about him. His art, however, was foundering. In the theater and as a poet and a short story writer, in his moribund Russian book and his aborted novel, he had declined as a writer in the previous three years. In a *Saturday Review* article, the novelist James T. Farrell listed Hughes among those writers associated with the 1935 Writers' Congress whose brave words had been followed by negligible artistic results. Only his agent Maxim Lieber knew exactly how negligible, as when the former roaring lion of the black left sent his "Sweet and Sour Animal Book," designed for near-infants, to be turned down by eight editors that year. Or was there a hidden meaning in the fact that

Lions in zoos
Shut up in a cage
Live a life of smothered rage.
Lions in the forest
Roaming free
Are happy as ever
Lions can be.

Was it merely a coincidence that Hughes's art seemed to decline in his most radical years? Probably not. From the start of his career, Hughes had shown a strong socialist conscience. But radical socialism had featured little in his greatest work: the poems leading up to and crowned by the landmark blues poems of 1927; his touching novel, *Not Without Laughter;* his harrowing play (in its original version) *Mulatto;* or his finest short stories. Radical socialist literary theory, as exhibited in his collection "A New Song," tended to short-circuit the full process of Hughes's artistic genius. Although many of his radical poems are powerful and valuable, it is almost certain that under the radical aesthetic, a poem such as "The Negro Speaks of Rivers" would have been impossible. On the other hand, Hughes could not blame radical socialist literary theory for the uneven quality of his plays. Careening from one pole to another, he sometimes had surrendered to the marketplace.

Above all, he needed a tonic, a restorative, and one came even as Hughes, conscious that he was missing the landmark Carnegie Hall conference in New York he had joined others in calling for, took his time returning to the East. He made extended stops in Salt Lake City, Denver (where he appeared at a Colorado Poetry Week Fellowship), and Chicago, where he visited Richard Wright and Arna Bontemps, now enrolled in the Graduate Library School of the University of Chicago. Back in Cleveland, however, he found another wire from Malraux and Aragon awaiting him. On behalf of the now delayed Second International Writers' Congress, they wanted him to reconsider his refusal to come. Hughes quickly sent a telegram to Edutravel, Inc., from whom he had heard nothing recently. Edutravel informed him that the tour had fallen through. On June 20, he wired Paris that he might be able to come.

Hughes also decided to go to Spain. From the *Baltimore Afro-American* and the *Cleveland Call and Post* and *Globe* he worked out an agreement to act as a foreign correspondent. When he seemed unable to secure space on a ship, he wired Paris, where Louis Aragon settled the matter with a telegram to a French line.

Leaving his mother all the money he could spare, he took the precaution of writing a check made out to her landlord for three months' rent paid in advance. He was still in Cleveland on the night of June 22, when Joe Louis knocked out the white boxer James J. Braddock in the eighth round of their fight in Chicago, to win the heavyweight championship of the world. Riding around the city for hours in a car full of carousing friends, Hughes helped celebrate the first black heavyweight champion of the world since the rebellious Jack Johnson.

Then, hoarse from shouting, he hurried the next day to New York. On Ralph Ellison's behalf, but without telling him, he wrote a letter to Richard Wright, whose poems and essays in *New Masses* Ellison admired and who was moving to New York within days. Visiting Amy Spingarn, Hughes sat for a portrait in oils. On June 29, in the law office of her brother-in-law Arthur Spingarn, he drew up a will—in the summer of 1937, Spain and death were linked words. He left all his property, such as it was, to his mother. If she died before Langston did, his estate would fall to his "brother," Gwyn Shannon Clark.

At the exclusive 21 Club, where he went to collect a large check from John Rumsey, his agent for *Mulatto,* Hughes was denied admission and humiliatingly made to loiter outside until Rumsey finished his lunch and condescended to bring the money. At last, on the morning of June 30, at the pier on the Hudson at the end of West 50th Street, Langston was given a rousing farewell by a party of friends that included Ralph Ellison, Louise Thompson, and Aaron Douglas. He sailed for Europe second class, on the *Aquitania.*

13

EARTHQUAKE WEATHER
1937 to 1939

Don Quixote! España!
Aquel rincon de la Mancha de
Cuyo nombre no quiero acordarme. . . .
That's the song of Spain . . .

"Song of Spain," 1937

AFTER two drab, often demoralizing years in the United States, Paris in the summer of 1937 was a tremendous boost to Hughes. In fact, his two weeks there before he left for Spain began a revival of idealism, especially of radical energy, that would last almost a year. Lethal as his stay in besieged Madrid might prove at any moment, his time there seemed infinitely preferable to "the dull relief W.P.A. kind of worried existence we had at home in Cleveland." Langston's spirit, dulled and blunted by poverty and disappointment in America, became honed again under the pressure of the anti-fascist struggle in wartime Spain. As the foreboding grew that an even greater war was coming to Europe, he would begin to feel life with an intensity he had not known since his first weeks in the Soviet Union. "Today," a terse poem published in *Opportunity,* captured his sense of vital urgency:

This is earthquake
  Weather:
Honor and Hunger
Walk lean together.

Paris was both a feast and a double-edged reminder to Hughes of his own past. Within a day or two of his arrival at Cherbourg on July 6, he was searching among the narrow, crowded streets on the slopes of Montmartre for memories of his months there in 1924 when, only twenty-two years old and in his ascendancy as a poet, he had passed heady, jazz-filled nights at Le Grand Duc nightclub on rue Pigalle. In 1937, Le Grand Duc was no more. But Langston was able to find Gene Bullard, its old manager, who now trained boxers at a gymnasium he ran on rue Mansart. Romeo Luppi, the Italian waiter who had kindly taken Langston home to Desenzano, was also still living in Paris. Not so one of the two singing stars of Le Grand Duc in 1924, the icily aloof Florence, or Florence Embry Jones. Soaring in popularity after quitting Le Grand Duc during Hughes's time there, Florence would nevertheless die a pauper in the United States. But her bubbling young replacement from New York, Ada "Bricktop" Smith, was now the reigning star of the Parisian night.

Although "Bricktop" was in London for the summer, Paris was full of other black entertainers. In a city that had become "the amusement center of Europe," as Hughes would recall it, Montmartre was nothing less than "the little Harlem" of the French capital. The black English chanteuse Mabel Mercer was visiting from London; a black American revue had practically taken over the Moulin Rouge; the latest Paris sensation, the American singer Adelaide Hall, was at the Big Apple; at Boeuf sur le Toit, Neeka Shaw was a great favorite; Olley Cooper performed at Melody's; and the internationally renowned Josephine Baker, from the slums of St. Louis by way of *Shuffle Along* in 1921, starred at the *Folies Bergères*. Recumbent on a chaise-longue, dressed in the diaphanous veils of the queen of a desert harem, Baker granted to a bemused Langston an interview in which she spoke only French—except for sudden, impulsive lapses into an English never learned in the slums of St. Louis: "I teel you how I fly zee plane, how I have learn zee loop-zee-loop. I love eet! I thrill eet!"

With the news from Spain mostly about Loyalist defeats, all this singing and dancing may have seemed like so much revelry by night before a Waterloo. Hughes, however, thoroughly enjoyed the freedom of his brief stay. "Paris was so alive that summer," he later wrote, "that I regretted having to leave it to fulfill my newspaper assignment in Spain." A reliable guide to its nightclubs was his former roommate in Mexico City, the photographer Henri Cartier-Bresson, now married to a beautiful Javanese dancer, Retna "Elee" Moerindiah, and living in a small apartment near the Opéra. "I was very happy to see Langston again," Cartier-Bresson would recall. "I had been to the United States in 1935 but I had missed him. He loved jazz as much as I did, so it was a pleasure to take him to all the little places—and some of the famous ones, like the Bar Boudon at the corner of rue Fontaine and rue Douai, where all the serious jazz lovers gathered." At one of Retna Moerindiah's performances, Hughes met Cartier-Bresson's mother and sisters. On Bastille Day, carrying his friend's camera case, he squeezed close to the president's reviewing stand as the armed forces marched down the Champs Elysées, "so I saw within arm's reach the President and all the dignitaries on the official reviewing stand, and was practically under the horses' feet when the famous Spahis came riding by. I was anxious to get a good view of the tall, colorful Moroccan troops in their flowing robes, and the giant Senegalese, jet-black in red fezzes." Later, at the Bastille, Hughes and Cartier-Bresson attended the workers' parade, then watched the nautical fetes on the Seine, including the fights with lances and shields in the famous Joutes Nautiques de Sete, before joining the dancers in the streets as Paris celebrated the anniversary of the French revolution.

Guiding Hughes through a different aspect of Paris was the young Afro-American scholar of French culture Mercer Cook, son of the actress and singer

Abbie Mitchell (who was now playing the key role of Cora in the touring company of *Mulatto*). Mainly through Cook, who was studying in Paris, Hughes once again met René Maran, the Prix Goncourt author of the novel *Batouala,* who had dazzled him with the speed of his talk in 1924, as well as the lesser-known but already accomplished young poets Léon Damas of French Guiana and Léopold Sédar Senghor of Senegal. "To these men, and Aimé Césaire of Martinique, the name Langston Hughes meant a tremendous amount," Mercer Cook would say, "even if he was not much older than they were. They had been reading and studying his poetry for years in the *Crisis* and in other places. His work had a lot to do with the famous concept of *Négritude,* of black soul and feeling, that they were beginning to develop." The racial spirit of the Harlem Renaissance, of which Hughes's poetry was quintessential, had suffused the journal *La Revue du Monde Noir,* founded in Paris in 1931 by a group led by Paulette Nardal of Martinique; poems by Hughes appeared in several numbers. In 1932, after the revue had died following six issues, the more radical *Légitime Défense* had appeared (to be banned after one number), in which Etienne Léro praised Hughes and Claude McKay as "les deux poètes noirs revolutionnaires" who most inspired black racial pride. In 1935, Senghor, Damas, Césaire, and other African and West Indian students in the French capital strongly endorsed the idea of radical black racial self-examination in their newspaper *L'Etudiant Noir,* which would appear irregularly over the next two years. To such people, Langston Hughes had been not only one of the prophets in his calls during the previous decade for black racial pride instead of assimilation, but the most important technical influence in his emphasis on folk and jazz rhythms as the basis of his poetry of racial pride. "They were the first to admit how important the black American writers, not least of all Hughes, had been to them," Mercer Cook would remember. "He was modest and completely unassuming, of course, but Langston was a hero to many of these young black people."

Along with such honoring must have come a sense, however muted, of superannuation. Still only thirty-five years old, Hughes had been speaking for a long time. Moreover, he did not consider himself much of a hero—least of all when he met some of the radicals gathering for the Paris meeting of the Writers' Congress, which had already convened in Madrid, Valencia, and Barcelona. One such person was Nicolás Guillén, still as impishly good-natured as when he had interviewed Hughes in Havana on a Sunday morning in 1930 and taken to heart his advice about the poetic possibilities of the Afro-Cuban *son.* But since then, with his books of verse *Motivos de Son, Sóngoro Cosongo, Sonnets for Tourists and Songs for Soldiers,* and *West Indies Ltd.*, Guillén had emerged as the most revolutionary writer in the Caribbean. Now a member of the Communist Party, he had also served time for his political beliefs in Cuba's most infamous prison, the Castillo del Principe. He had once been a droll observer of Hughes's racial radicalism, but was now far more committed as a Marxist. Another reunion brought Hughes together with the Haitian poet Jacques Roumain, also a communist, who had recently been freed from a Haitian prison after two years there and was now studying architecture at the Musée de l'Homme. A very strong personality, firmly associated with radicalism and blacks, was Nancy Cunard, tall, blonde, and amazingly thin, with "a body like sculpture in the thinnest of wire," Langston recalled (somebody else once said that she looked liked an asparagus, presumably blanched). With nerves tightly strung, Cunard vented her hatred of fascism and racism without restraint; she had long outraged her titled family and English society in general by her open friendship with black people, including an Afro-American musician who had been her lover, and her espousal of radical causes. Hughes also met her co-editor on the series of leaflets of revolutionary verse on Spain, the Chilean poet Pablo Neruda (they published Hughes's "Song of Spain" in one leaflet called *Deux Poèmes*). Among the other writers Hughes met in Paris that summer were Alejo Carpentier of Cuba; W. H. Auden, Stephen Spender, and John Strachey from Britain; Mikhail Koltzov and Ilya Ehrenburg of *Izvestia* and *Pravda;* Bertolt Brecht; and virtually every important anti-Fascist French writer, including André Malraux (the leader of a volunteer airforce in Spain), Louis Aragon, Louis Chamson, and Tristan Tzara. Hughes was particularly pleased to meet Georges Adam of *Ce Soir,* who had already translated and published short stories from *The Ways of White Folks;* Langston arranged to have Adam sent a generous collection of his other short stories for translation into French.

The restorative power for Hughes of this gathering of the engaged was immense; it overcame the growing sense of disengagement that had been overtaking him, and that he would claim much later as an evolution in his sense of self as an artist. "In the Civil War in Spain," he would assert, "I am a writer, not a fighter. But that is what I want to be, a writer, recording what I see, commenting upon it, and distilling from my own emotions a personal interpretation." Few listeners, however, would have guessed he felt this degree of distance as they listened to the actor Jean-Louis Barrault magnificently declaim some of Hughes's most radical poems at one public gathering; or on July 19, when Hughes himself addressed a session in the Théâtre de la Porte-Saint-Martin that included André Chamson, Louis Aragon, and the Spanish Catholic writer José Bergamin. As if unshackled after years of bondage, Hughes spoke out with unusual passion and singularity of purpose in an address entitled "Too Much of Race." One of four U.S. delegates (along with Malcolm Cowley, Anna Louise Strong, and Louis Fischer), he had come "most especially representing the Negro peoples of America, and poor peoples of America—because I am both a Negro and poor." He spoke for blacks, "the most oppressed group in America. . . . We are the people who have long known in actual practice the meaning of the word fascism—for the American attitude toward us has always been one of economic and social discrimination." To show the ignominy of black life in America, he touched on "the sorrows of the Scottsboro boys," now six years in jail. "In America," he said, "Negroes do not have to be told what fascism is in action. We know. Its theories of Nordic supremacy and economic suppression have long been realities to us."

> And now we view Fascism on a world scale: Hitler in Germany with the abolition of labor unions, his tyranny over the Jews, and the sterilization of the Negro chil-

> dren of Cologne; Mussolini in Italy with his banning of Negroes on the theatrical stages, and his expeditions of slaughter in Ethiopia; the Military Party in Japan with their little maps of how they'll conquer the whole world, and their savage treatment of Koreans and Chinese; Batista and Vincent, the little American-made tyrants of Cuba and Haiti; and now, Spain and Franco with his absurd cry "Viva España" in the hands of Italians, Moors and Germans invited to help him achieve "Spanish Unity." Absurd, but true!
>
> We Negroes of America are tired of a world divided superficially on the basis of blood and color, but in reality on the basis of poverty and power—the rich over the poor, no matter what their color. We Negroes of America are tired of a world in which it is possible for any one group of people to say to one another: "You have no right to happiness, or freedom, or the joy of life." We are tired of a world where forever we work for someone else and the profits are *not* ours. We are tired of a world where, when we raise our voices against oppression, we are immediately jailed, intimidated, beaten, sometimes lynched. Nicolás Guillén has been in prison in Cuba, Jacques Roumain in Haiti, Angelo Herndon in the United States. . . . I say, we darker peoples of the earth are tired of a world in which things like that can happen. And we see in the tragedy of Spain how far the world-oppressors will go to retain their power . . .

Including himself among the suppressed radical writers of color (in part because the State Department had flatly refused to grant him press credentials for Europe), Hughes closed by invoking

> the great longing that is in the hearts of the darker peoples of the world to reach out their hands in friendship and brotherhood to all the races of the earth. . . . We represent the end of race. And the Fascists know that when there is no more race, there will be no more capitalism, and no more war, and no more money for the munitions makers—because the workers of the world will have triumphed.

"He never forgets his comrades," Jacques Roumain whispered gratefully to Nancy Cunard as an ovation swelled for the most stirring speech Hughes had delivered in many years.

Now, with the Congress and the nightclubs behind him, it was time to go on to Spain. Since Guillén, who had been reporting there for the radical Cuban journal *Mediodia,* would be returning, he and Hughes decided to travel together. With credentials secured for him by Louis Aragon to cross the closed French border, Langston boarded a train late on July 24 at the Gare d'Orsay. He left Paris with some trepidation; almost a hundred people had just been killed in air raids on Barcelona, one of their major stops. Although he believed in the fight against fascism, Hughes was a pacifist at heart; certainly he had never been under fire. But death exempted no one in Spain, not even poets. Lured from safety to a place several kilometers outside Granada, Federico García Lorca had been shot the previous August by Civil Guards. García Lorca had been perhaps only the most celebrated casualty of a war civil in name but international in reality, and fought with the most modern weapons of destruction—evidence of which Hughes saw the next morning when, after stopping at a town in France to stock up on cigarettes, matches, and soap, he and Guillén crossed to the border village of Port-Bou. Windows had been shattered, walls bore the tracks of machine-gun bullets, and houses had been gutted by bombs. After idling nervously at a small café overlooking the bay, they boarded a new train for Barcelona. They passed peacefully through fields of ripe wheat, where men and women scythed the harvest, and reached Barcelona in darkness. The city itself was dark; on the previous night, twenty-three persons had died in the air raids. By murky light, a bus took Hughes and Guillén up the Ramblas, the famous tree-lined thoroughfare, to the Hotel Continental. At a café nearby, "there was a wan bulb behind the bar inside to help the barman find his bottles, but other than that no visible light save for the stars shining brightly. The buildings were great grey shadows towering in the night, with windows shuttered everywhere and curtains drawn. There must be no visible lights in any windows to guide enemy aviators." At midnight, the café echoed with news on the radio about death and destruction. To Langston, Barcelona tingled with fear. He himself was very much afraid.

During the next two days, while they waited for a car to take them down the coast, he shopped for sandals and a local woolen jacket and visited the jai-alai courts and Chinatown. On the third evening the sirens began to wail, followed by the thunder of anti-aircraft guns. Hughes retreated to the air-raid shelter. When no bombs fell, he went to bed. Soon he was deep in slumber. "The next thing I knew was that, with part of my clothes in my arms, I was running in the dark toward the stairs. A *terrific* explosion somewhere had literally lifted me out of bed. . . . I put my trousers on over my pajamas and sat down shaking like a leaf, evidently having been frightened to this dire extent while still asleep, because I had hardly realized I was afraid until I felt myself shaking. When I put one hand on the other, both hands were trembling." As a bomb exploding some distance away rattled the hotel, Langston fled to the nearest bathroom. When he returned, Guillén was in the gloomy, candle-lit lobby "sitting calmly like Buddha on a settee under a potted palm. He said, '*Ay, chico, eso es!*' Well, this is it! Which was of little comfort."

South towards Valencia, serene in the high summer sun, the countryside mocked Hughes's fear as he drove with Guillén through green groves of oranges and olives; the Mediterranean glittered calmly to the east. Finally in the distance they saw medieval towers mingled with modern high-rise buildings—Valencia, which Langston had visited briefly on his way home on the *West Cawthon* from Genoa in 1924. The third largest city in Spain, Valencia was now the seat of the Loyalist or Republican government. With the fall of Bilbao in June, the Republicans no longer controlled any territory in the north of Spain. Since the previous November, Madrid had been under siege, defended in large part by the communist-supported International Brigades. Barcelona, the major city of the autonomous province of Catalonia, and the second largest in Spain, was a historic stronghold of anarchism and socialism but was also under enormous pressure from Nationalist forces. Later, observers would look back and see the Loyalist defeat as already decided in the late summer of 1937, even if in Valencia victory still seemed a distinct possibility to the alliance of anarchists, socialists, liberals, and communists defending the electoral victory achieved early the previous year by the Popular Front. Dependent mainly for munitions on the Soviet Union, however, this imperfect coalition was steadily losing ground to the rival conservative alliance led by Francisco Franco and dominated by landed aristocrats, the traditional military leadership, the church, and the Fascist Party—the Falange—supported by Nazi Germany and, especially, by Italy.

As soon as they had secured a press card from the militia and a room at the Hotel Londres overlooking the Plaza Castelos, Hughes and Guillén visited the local house of the Alianza de Intelectuales Antifascistas, an organization of writers and artists founded in the wake of the First International Writers' Congress in 1935, with facilities in Valencia, Madrid, Barcelona, and Alicante. They lunched with Miguel Hernandez, acclaimed as the "boy poet" of the Loyalists. In no particular hurry to reach besieged Madrid, for several days they took to the beaches or dawdled in the nightclubs. For news of the war Hughes and Guillén depended not on the notoriously optimistic official dispatches but on reports by soldiers on leave in Valencia from the front. Their main interest was in news of the Fifteenth International Brigade, formed at Albacete in January and comprising separate British, Canadian, Hispanic (from Cuba, Mexico, Puerto Rico, and Latin America), and U.S. battalions. The Americans, over 3300 in number, of whom between 80 and 100 were black, formed mostly the Abraham Lincoln Battalion and a subsidiary outfit, the Washington Battalion. Although Valencia had been bombed by planes and shelled from sea, the city that summer was still largely free of the hunger and terror afflicting other parts of Republican Spain. "The cafes were full morning to night," Hughes would write, "even long after dark. . . . Valencians just didn't seem to care much. They had good wine and good food—fresh fish and melons and the sweetest of oranges and grapes." In one place in the city, however, there was mourning. At the League of Anti-Fascist Writers building, the body of a young European photographer, Gerda Taro, who had been crushed by a tank as she worked on a battlefield, lay in state. Taro had been the first foreign newspaperwoman to die in the war.

Finally, early in the second week of August, a military bus carried the two writers some thirteen hours over the battered road to Madrid. At 7, Marques del Duero, they were welcomed at the local Alianza de Intelectuales Antifascistas by the poet Rafael Alberti and his wife, María Teresa León, who acted and wrote plays and fiction. Together they ran the facility in the elegant old home of an aristocrat who evidently had left in a hurry; on the walls Langston saw valuable medieval tapestries and precious paintings by Goya and El Greco; the furniture was fine and antique, and the grandee's silver, china, and crystal were in daily use at the Alianza (Hughes himself was given a sweater from the marquis's ample wardrobe). He was pleased to discover that both Alberti and Leon knew his work, some of which had appeared the previous year with their own in *Nueva Cultura,* a monthly magazine published in Valencia. Assigned to a room on the top floor of the old mansion, with a fine old hand-carved Indian bed, Langston was grateful to have such a lovely view until he realized that his room faced enemy guns almost directly. But no building in Madrid was safe—not the Alianza, nor the repeatedly struck Hotel Florida, where Ernest

Hemingway lived, nor the Hotel Alfonso, where the facade of three floors had been blown away.

With his typewriter and box of jazz and blues records, Hughes settled in for the duration of his stay. Functioning vigorously in spite of the tightening siege, the Alianza was a major center for Loyalist propaganda. Its own press, part of a gift brought by Louis Aragon and other French writers, printed a house newspaper, posters, pamphlets, and other documents; members edited the People's Army Brigade newspaper, *El Mono Azul*; and María Teresa León, the first woman in Spain to direct a playhouse, staged plays on behalf of the Loyalist cause. Most of all, the Alianza worked hard to make a united propaganda front out of the factions warring within the Loyalist camp, notably the rival anarchist and socialist trade unions. Langston was impressed by life at the Alianza de Intelectuales and the spirit of "*no pasaron!*", the rallying cry of Madrid against the rebels. With the sound of bombs and gunfire never far away, and defeat by a vengeful enemy an imminent possibility, its staff worked cheerfully. "It is a center for every writer and artist who opposes the return to barbarism," Hughes reported. "It is a place where, now, today, art becomes life and life is art, and there is no longer any need of a bridge between the artists and the people—for the thing becomes immediately a part of those for whom, and from whom, it was created. The poem, the picture, the song is only water drawn from the well of the people and given back to them in a cup of beauty so that they may drink—and in drinking, understand themselves."

In his three months in Madrid at the Alianza de Intelectuales, which played host to the many prominent foreign artists and writers making the wartime pilgrimage to besieged Madrid to testify against fascism, he would meet more famous American writers than at any other time of his life. Whatever else they might have been in the United States, all were friendly here; Madrid in 1937 was the most dangerous foxhole in the world. Lillian Hellman, unperturbed when a shell smashed into the building in the middle of her radio speech, and Dorothy Parker, who came and went quietly, her famous wit bridled, impressed Hughes a great deal. Still another visitor was his old comrade and friend Louise Thompson, in Madrid as a member of a relief delegation; following her return home, she would receive all of Hughes's dispatches and pass them on to newspapers and magazines. Ernest Hemingway, "a big likable fellow whom the men in the Brigades adored," cordially accepted Langston as a member of the war correspondents' fraternity. Hughes admired Hemingway, who had rallied the badly divided Writers' Congress in New York in June with a blazing denunciation of fascism as "a lie told by bullies" and a political system under which no honest artist could flourish: "A writer who will not lie cannot live and work under fascism." Elsewhere, denouncing Italian actions in Ethiopia, he had ridiculed Mussolini as "the cleverest opportunist in modern history." Hughes was also impressed by the brilliant Soviet journalist Mikhail Koltzov, apparently Stalin's personal representative in Spain, and Hemingway's model for the admirable figure of Karkov in *For Whom The Bell Tolls;* Koltzov had previously solicited work from Langston for *Pravda*. When Hughes, Guillén, Hemingway, and Koltzov spent almost a day together touring encampments, the four men posed for a group photograph.

But the most compelling story of Spain, Langston knew, was on the battlefield. In the company of two experienced American correspondents, Leland Stowe and Richard Mowrer of the Chicago *Daily News,* he went to inspect the aftermath of the battle of Brunete. Here, on July 6, a thousand men of the Lincoln and Washington battalions had advanced at dawn against enemy fortifications. The Americans gained ground until the vastly superior rebel air force began to pound their lines near Villanueva de la Cañada; so many men were lost in the Brunete battle that the Washington outfit had to be absorbed completely into the larger Abraham Lincoln Battalion. In the devastated town of Quijorna, where Hughes saw the results of the first sustained aerial bombardment in the annals of warfare, the stench of flesh rotting in the summer sun almost sickened him. Bodies had been removed but arms, legs, hands, and scraps of flesh remained untouched. Walking down a country road, he came on a bizarre sight—mature trees torn apart by bombs and shells. He also heard the chirping of birds.

> ''Birds?'' Stowe said, ''There's no birds. Those are sniper bullets whistling by.''
> ''Firing at *us?*'' I exclaimed.
> ''Certainly,'' said Stowe. ''There's nobody else on this road to fire at.''
> ''I never knew bullets sounded like birds cheeping before.''
> ''Well, now you know,'' Mowrer grinned.

During a visit to a prison hospital, Langston almost collided with a very tall, very dark man, a Moor. This was one of the main reasons he had come to Spain. ''I knew that Spain once belonged to the Moors,'' he wrote from Madrid, ''a colored people ranging from light dark to dark white. Now the Moors have come again to Spain with the fascist armies as cannon fodder for Franco. But, on the loyalist side, there are many Negroes of various nationalities in the International Brigades. I want to write about both Moors and Negroes.'' The hospital was full of wounded Moors. And, on the Loyalist side, Langston had met on the boat coming over a black man from Utah, a Cuban, and two West Indian aviators, all closemouthed about their destination, but all bound for Spain. His first contact with black Americans serving in the forces was at the International Brigades Auto Park on the outskirts of Madrid, when he interviewed Thaddeus Battle, who had left Howard University at the age of 23 to come to Spain, and Bernard ''Bunny'' Rucker from Ohio. Hughes learned more about two black Americans already among the dead—Alonso Watson, the first black American to die in Spain, killed in February in the month-long battle near the Jarama River; and Oliver Law, a Texas-born organizer for the party in Chicago, who had spent six years in the segregated U.S. Army without rising above the rank of private, but had quickly risen first to command the machine gun crew, then the entire Abraham Lincoln Battalion. Apparently the first black ever to command a white American military unit, Law had died a hero in July during the Brunete offensive.

On August 27, when Hughes made the first in a series of broadcasts from Madrid to the United States on shortwave radio, he proudly told the world about the black volunteers who knew that if fascism triumphed, "there will be no place left for intelligent young Negroes at all. In fact no decent place left for any Negroes—because Fascism preaches the creed of Nordic supremacy and a world for whites alone." Here in Spain, Langston reported, there was no prejudice; in fact, many Spaniards would be considered colored in the United States. Contrary to racist American custom and expectations, blacks commanded white troops without question, and fought with skill and courage. (Years later, in his autobiographical account of the period, he would also tell of those few blacks who proved to be cowards under fire.)

Repeatedly he praised the spirit of the Spanish ("a sweeter, kinder people I've never seen," he wrote to Noël Sullivan), and he devoted one talk to the life and poetry of Federico García Lorca. There were Madrileños who fled at the sounds of war, and others who, under fierce shelling, moved calmly about their business. The Puerta del Sol, Madrid's busiest area, was as popular as ever by day. The *corrida* was shut down, with the accursed bullfighters gone to join Franco, but nineteen legitimate theaters were still playing. The gypsy dancers were still in the music halls, and the greatest flamenco artist, Pastora Pavón, or La Niña de los Peines, whose half-sung, half-spoken *soleas* "could make the hair rise on your head, could do to your insides what the moan of an air-raid siren did, could rip your soul-case with her voice," was still performing. The movie houses were crowded with fans of Shirley Temple and Bill Robinson in *The Littlest Rebel,* the Marx Brothers in *A Night at the Opera,* and Paul Robeson in *Song of Freedom.* The Madrid Symphony played on as if there were no armies hammering for admittance, no Generalissimo Franco who, on his white horse at the city gates a year before, had vowed to sip coffee and cognac soon in the Puerta del Sol.

> Midnight in Madrid, one can walk for blocks and not meet a soul save perhaps an occasional guard on the corner. When the moon is shining, the great buildings in the center of the city loom up like silver shadows against the sky. Like shadows of the city in a dream, an unearthly city where nobody lives. If it is a quiet night for the troops that hold the fascists at bay outside Madrid, then you will hear only an occasional rifle shot, or a quick run of machine gun fire from trench to trench, then silence again. But if it is a night of battle, in the city you hear quite loudly and plainly the medley of bullets, hand grenades and trench mortars. And maybe before the night is out, the boom of cannons, and the crack of shells falling, ——not on the front at all, but in the very street where you live. (You see, in quiet moments one forgets that all Madrid is the front.)

Responding to the deadly drama of Madrid, Hughes wrote with renewed brilliance. His most passionate sense of the violence and the heroism of Spain inspired three strong poems: "Air Raid: Barcelona," "Moonlight in Valencia: Civil War," and "Madrid, 1937":

Put out the lights and stop the clocks.
Let Time stand still.
Again man mocks himself
And all his human will to build and grow.
  Madrid!
The fact and symbol of man's woe.
  Madrid!
Time's end and throw-back,
Birth of darkness,
Years of light reduced:
The ever minus of the brute,
The nothingness of barren land
And stone and metal,
Emptyness of gold,
The dullness of a bill of sale:
BOUGHT AND PAID FOR! SOLD! . . .

From October 23 to early the following year, long after his departure from the city, he published almost a dozen articles in the *Baltimore Afro-American* and sent other pieces to the Associated Negro Press. At one point he expected to make a booklet of twenty-two such articles, to be entitled "Negroes in Spain," in which he would include portraits of individual blacks on both sides of the fighting, a chapter comparing Ethiopia to Spain in the light of fascism, and a concluding section on the "World Meaning of Spanish Struggle." Although the *Afro-American* displayed his dispatches sensationally ("HUGHES BOMBED IN SPAIN: TELLS OF TERROR OF FASCIST RAID"), his own tone was more disciplined. The articles viewed the war from a perspective that merged the narrowly racial *Afro-American* view with Hughes's proletarianism and anti-fascism. The result was excellent propaganda for the left, aimed directly at the black American world. One new propaganda vein was opened when he published a maudlin dialect poem in ballad-epistle form, called "Letter From Spain," addressed to "Dear Brother at home" in Alabama:

We captured a wounded Moor today.
He was just as dark as me.
I said, Boy, what you been doin' here
Fightin' against the free? . . .

Some black soldiers demurred that his use of dialect gave the wrong impression about their background; almost all the volunteers were educated. But Langston, whose aim clearly was to sway the typical readers of the Associated Negro Press, followed with more ballads about the war (and he would continue to write this kind of proletarian doggerel for years to come). The greater fight, he believed, could not be carried on without the support of the common people. From Chicago, the important black communist leader William L. Patterson wrote Langston to endorse his effort and praise his "simplicity mingled with great depth of human feeling and understanding."

Hughes also published extensively, but with a far less racial flavor, in *The Volunteer for Liberty,* the main organ of the English-speaking units of the Fifteenth International Brigade. On September 6, in the wake of Japanese incursions into Northern China, which had been met by the unexpected resistance of Kuomintang and Communist Party forces working together against a common enemy, he published "Roar, China!"

> Roar, China!
> Roar, old lion of the East!
> Snort fire, yellow dragon of the Orient,
> Tired at last of being bothered.
> Since when did you ever steal anything
> From anybody.
> Sleepy wise old beast
> Known as the porcelain-maker,
> Known as the poem-maker,
> Known as maker of fire-crackers . . .

Early in October, the magazine reprinted his old poem "October 16th," about the Abolitionist martyr John Brown (one fighting unit of the brigades was called the John Brown Field Artillery Battery). He wrote about the Spanish immigrant to Cuba, Enrique Lister, who returned to Spain to fight first as a trade unionist, then as a soldier, eventually rising to division command of the Fifth Army Corps of the Peoples Army, and leadership in the defense of Madrid. Such recognition of the gifts of an uncommon common man was possible in anti-fascist Spain, Hughes pointed out, and should inspire the masses everywhere.

Months after Langston's departure, the *Volunteer for Liberty* carried his translation, "Spanish Folk Songs Of The War":

> Frontiers that divide the people,
> Soon we'll tear apart.
> The masses speak a thousand tongues—
> But have one heart . . .

Perhaps even more than during his Mexican months, Hughes was exposed to the work of poets writing in Spanish—the Spaniards Rafael Alberti of the Alianza and León Felipe, as well as visiting Mexicans such as Maria Luisa Vera, Jorge Mansisidor, Juan de la Cabada, and Octavio Paz. In Mexico he had translated prose; now in Madrid he tried to render the poetry of García Lorca by concentrating first on two of his best-known works—the play *Bodas de Sangre,* or *Blood Wedding,* and the volume of poems *Romancero Gitano,* or *Gypsy Ballads,* that had made Lorca the most acclaimed Spanish poet of his era. A cheap, mass edition, with an introduction by Rafael Alberti, was just going on sale. Alberti, who had known García Lorca since their student days together, encouraged Hughes to proceed with his translating; with a fellow poet, Manuel Altologuirre, Alberti worked with him to capture the unrhymed but assonant, simple and direct force of Lorca's verse. In turn, the Mexican composer Silvestre Revueltas set Langston's poem "Song for a Dark Girl" to music; the score and the text, in English and Spanish, were later published in New York. And, in a tribute to a dead acquaintance, Hughes published in November in *The Champion* his translation of a story, "The Hero," by Pablo de la Torriente-Brau, the young Cuban writer killed very early in the war.

If Langston made an uneasy peace with the danger of Madrid, there were moments of terror: "I thought I might not live long." Once, going out to the University City sector in the company of some visiting women, including one American who refused to remove her big white hat, he was slightly wounded by a piece of an explosive bullet when snipers opened fire. But he was reluctant to leave Madrid, which in 1937 was the center of the world. "The longer I stayed in Madrid, the more I liked it. I might get hungry there, but I never got bored." After the dingy compromises of the recent two years, he was clearly grateful for the hard, cold glare of Spanish reality, where fascist faced anti-fascist in deadly confrontation. "It's a thrilling and poetic place to be at the moment," Hughes wrote almost gratefully; "there's surely nothing else like it in the world." Careful not to make himself out to be a hero, he allowed himself only now and then to write true trenchcoat prose. Once, early in October, he visited the theater of operations on the Ebro River, rumored to be the next great battlefield, on a night of heavy rain. In a disastrous attack on the village of Fuentes del Ebro, a section of the massive Aragon front, the Lincoln Battalion had again suffered heavy casualties; its Canadian counterpart, the Mackenzie-Papineau Battalion, was also badly mauled. Among those killed was Milton Herndon, a black American who had led a machine gun section as a sergeant with the "Mac-Paps." He was the brother of the imprisoned radical organizer Angelo Herndon, about whom Langston had written his play *Angelo Herndon Jones*. Milton Herndon was attempting to rescue a stricken comrade, a white soldier named Irving Schatz, when a bullet in the mouth, then one in the head, killed him. To open perhaps his finest dispatch from Spain, Hughes described the scene as he waited patiently for two American soldiers, one black, the other white, who would tell him exactly how Sergeant Herndon had died.

> It was quiet on the front. No action. Our attack was over. Silence in the black dark of a rainy night in a valley where perhaps twenty thousand men lie in trenches, behind barricades in ruined villages, squatting beside machine-guns, gathered around field pieces pointing toward the enemy. Rain and silence. Sporadic rifle fire. Occasionally, for a few seconds, a machine-gun spitting a row of bullets into space. Then long blanks of silence again.
>
> Afar off, the boom of cannon, steady for maybe half an hour. Perhaps the government guns trained on Zaragossa, as the enemy guns are trained on Madrid. Then silence again. And the rain coming down in a soft, steady drizzle.
>
> Where the tent sags, water drips down and spatters on the table. Two candles burn. Men come in and out with messages. . . .

The following week, Hughes visited the American Hospital Center at Villa Paz, near the village of Salices, on the combined estates of an ancient castle and a mansion owned by a Bourbon princess. There he met Salaria Kee, a graduate of the Harlem Hospital Training School, who had arrived in April in the second American volunteer medical unit; she was the only black nurse in the Fifteenth Brigade. Langston also talked to Dr. Arnold Donowa of Harlem, a West Indian-born, Harvard-trained specialist in oral surgery serving with the American Medical Bureau in Spain; Donawa would entrust a long list of badly needed medical supplies to Hughes when they parted. At Tarazona, Langston visited the major training base for the foreign volunteers. He was there on October 18 in the company of two visiting congressmen, including the only one to vote against official American non-intervention in the war, when pennants flew to celebrate the first anniversary of the formation of the International Brigades. In a ceremony inconceivable in the United States, white soldiers marched past the base commander, Lieutenant Walter Garland, a black American, twenty-four years old and twice wounded, as he relinquished his post as commander of the base. That night, when Hughes attended a farewell party for Garland, black and white soldiers sang spirituals and folksongs and made brief speeches in his honor: "And I, who did not know that soldiers cried, saw some of them cry. The next day Garland left for the United States to try to tell the folks at home what the Spanish struggle meant."

Without winter clothing of his own, Langston warmed himself in a winter coat donated by Bernard "Bunny" Rucker and the sweater borrowed from the departed marquis. He also lost weight steadily; he was fifteen pounds lighter by the time he left Madrid. But even after fulfilling his schedule of dispatches, he stayed on in the besieged city. "I didn't want to leave," he would soon confess to Marie Short in Carmel. "Nobody does that lives there." But he was in growing danger. While Hughes lived in Madrid, almost a thousand people died from the shelling, three thousand were wounded, and about three thousand buildings badly hit. At the Alianza, as elsewhere, food became pathetically scarce; sometimes a few garbanzo beans sat disconsolately on the marquis's silver plates. Rumors spread about cats being cooked; in the Madrid Zoo, the hungry lions roared in protest. "Bad cigarettes, poor wine, little bread, no soap, no sugar!" Hughes catalogued. "Madrid, dressed in bravery and laughter; knowing death and the sound of guns by day and night, but resolved to live, not die!"

Just after the middle of November, following new Loyalist reverses on the Aragon front, and with a major rebel offensive expected within a month, Hughes and Guillén left Madrid. Before his departure, Langston visited the Escorial palace and monastery in the Guadarrama Mountains outside Madrid. He would recall a farewell party in the bar of the Hotel Victoria given in his honor by the *New York Times* reporter Herbert Matthews and a few other Americans, including Ernest Hemingway, who was never at a loss for bottles of scotch. Burdened by gifts and souvenirs (but leaving his jazz records behind as a present to the Alianza) and in an alcoholic haze after the party, Langston took the bus to Valencia. On November 19, he and Guillén checked into the Hotel Bristol. Compared to Madrid, sunny, sea-blown Valencia was still relatively untouched by the war. Transportation from Valencia to Barcelona loomed as a major problem until Hughes walked casually into a Cook's Travel Agency office and, just as casually, secured a private compartment on a train otherwise crammed with passengers. On December 10, he unpacked his belongings at the Hotel Urbis in Barcelona. That evening, he attended a concert of the Catalan Symphony, lured by Gershwin's *Rhapsody in Blue,* with Maria Campmany as pianist. He also ran into Nancy Cunard, who was accompanied by the Irish painter John Banting. Slender once again and sinewy, toughened physically and mentally by Madrid, Langston seemed to John Banting "a magnificent and a magnetising man—with his sense of humanity and wide understanding of life. I wish we had seen him more often."

Five days later, leaving Guillén behind, he crossed into France. Not long afterwards Hughes was back in Paris, where he stayed in the heart of Montmartre at the Hotel Lizeux, run by the former Minister of Commerce in Selassie's Ethiopia (Selassie himself was in seclusion in England). Still reluctant to return to Cleveland, Langston passed a pleasant month in Paris. With Georges Adam, he pursued the idea of publishing a translation of the entire volume of *The Ways of White Folks* in Paris; with a leftist film group, Les Jeunes Artisans du Cinéma, he talked about the possibility of a film of *Not Without Laughter.* Finally, his money running low, Hughes sailed in chilly mid-January for New York on the Cunard liner *Berengaria.*

Back in New York, Hughes surveyed his financial situation and found it bleak. The news from Cleveland was most disheartening. Having wasted the money he had left her, his mother had persuaded her landlord to surrender the rent prepaid by Langston before his departure. Then she had solicited funds from Noël Sullivan and Amy Spingarn on the pretext of being frantically worried about Langston's safety, and had sold a precious Aztec statuette given to him by Covarrubias in Mexico. Also, Carrie had taken in a pregnant, homeless "cousin," and Gwyn "Kit" Clark had flunked out of school once again, this time at Wilberforce University in Ohio.

In Hughes's absence, Maxim Lieber had sold a few pieces of writing, but not for much. The *Nation* had bought an essay, "Death and Laughter in Madrid," for $28, and *Reader's Digest* paid $100 to reprint an excerpt. *New Masses* had carried parts of the García Lorca manuscript in a January supplement, but for nothing. For $64, *Esquire* took "Air Raid: Barcelona." Hughes's odd little play *Soul Gone Home,* about a mother and her dead son, had gone for $22.50 to *One Act Play* magazine, where it had appeared the previous July.

Needing money quickly, Langston turned to the lecture circuit. Through the International Workers Order, which had not yet brought out its promised pamphlet of his radical poems, Louise Thompson set up a series of lectures, "A Negro Poet looks at a Troubled World." The Friends of the Abraham Lincoln Brigade (the term "brigade" was used popularly to include all Americans fighting or serving in Spain), of which he was a sponsor with Lillian Hellman, James Cagney, Carl Sandburg, and others, also retained Hughes as a speaker.

After Madrid, however, money seemed far less important to Langston than a continuation of some form of radical activity. With his next major move, he boldly contradicted the commercial spirit of the plays he had written for the Karamu Theatre in Cleveland. Still in love with the stage, but recharged as a radical following his months at the Alianza, Hughes harkened back to the late summer of 1931 in New York, when, at the height of his radical zeal following his break with Mrs. Mason, he had joined Whittaker Chambers, Paul Peters, and Jacob Burck in announcing the birth of a revolutionary Suitcase Theatre within the John Reed Club of New York. Although nothing had come of that plan, Langston would now try again. "I'm afraid there's nothing left for me to do except to start a theater and produce plays," he wrote Noël Sullivan. "That will be equal to anybody's battlefront!"

"Langston was at my apartment at 530 Manhattan Avenue one rainy night just after he got back from Spain," Louise Thompson would recall. "Ralph Ellison was also there, I think. He lived around the corner and we saw a great deal of him in those days. 'I want a theater, Louise,' Langston told me. 'I'm determined to have one of my own.' He was very, very serious, and so I said I would try to help. I'd see what the I.W.O. could do." The powerful International Workers Order, which then boasted almost 145,000 members, sponsored its own schools, music and drama societies, a national cultural commission, and a magazine. Thompson also had a suggestion for Hughes. Preparatory to the appearance of his radical collection, *A New Song,* she had been reading some of the poems with great success in her travels for the I.W.O. She now urged Langston to write a play that would link the best of these pieces. By the next day, Hughes had finished the outline of a one-act play drawn from his poems, both radical and racial, with blues and spirituals added, and eventually called (as published in *One Act Play* magazine that October) *Don't You Want To Be Free?: A Poetry Play: From Slavery Through the Blues to Now—and then some!—with Singing, Music and Dancing.*

A short time later, the resourceful Thompson found both space and sponsorship for Hughes's theater. Affiliated to the leftist New Theatre League, the Harlem Suitcase Theatre was launched, according to its constitution, as "one of the cultural activities of Branch 691 of the I.W.O. and as such receives the full cooperation of the Branch toward its work." The theater was allowed the use of the second floor loft of the I.W.O. Community Center at 317 West 125th Street in Harlem. With such support came the tacit confirmation that the Harlem Suitcase Theatre would be explicitly a radical theater. Linked to the Communist Party, the I.W.O. openly appealed to the masses "to join the struggle against capitalism and for a system where all power belongs to the working class."

The first Suitcase Theatre members were all I.W.O. workers: Louise Thompson, her cousin Mary Savage, and their friends Grace Johnson and Edith Jones, who soon brought along her husband, a deep-voiced, magnificent looking youth with an electric smile named Robert Earl Jones (the father of the actor James Earl Jones). But soon a full range of personalities had been drawn to the loft on 125th Street, with forty-seven official members joining. Some, such as Dorothy Peterson, Alta Douglas, Gwen Bennett, Waring Cuney, and Dorothy Maynor, were old friends of Hughes; others were new both to Langston and the stage. Among the more experienced was Hilary Phillips, who was appointed, with an eye toward Hughes's frequent travels, co-director of the company. Toy Harper, who had once performed as a snake charmer, among various roles, in a circus, volunteered her services as a seamstress; she would keep a sharp look-out for Langston's interests when he was away. However, in spite of her presence and that of Phillips and Louise Thompson, who was elected chairman of the executive committee, the success of the Harlem Suitcase Theatre would depend almost completely on its executive director, Langston Hughes. At least at first, this power was exactly what he wanted. Langston unquestionably saw himself in the same position as that of Jasper Deeter of Hedgerow and various Moscow drama masters, who commanded their individual theaters. The Harlem Suitcase Theatre, he wrote a friend, was "my own personal playhouse. (Or workhouse, because it takes night and day to run it)." To mount its first production, he borrowed $200 from the Authors' League Fund.

His theater, Hughes vowed, would be different from all others. There would be minimal lighting, virtually no properties or sets, and no curtain; dedicated to the idea of theater-in-the-round, he saw his audience of 150 drawn intimately into the action on the stage, which would be two moveable half circles. For a true black theater, music and dance must be integral to the action; the music must be blues and spirituals, and not, as in most musicals involving Negroes, sentimental or risqué travesties of black style. The approach must always be bold and fresh. "We are doing things never done before in the Negro theater," he asserted, "—or any other for that matter." Throwing himself into the operation, when he was not dashing away on speaking engagements as far north as Canada and south to Alabama, Langston soon had its first performance planned. A two-part program would open on April 21, comprising first a brief, entertaining history of black dance from Africa to the latest Harlem step, then a production of *Don't You Want To Be Free?* In the latter's central role, that of a young black man who would act, sing, and also serve at times as a kind of chorus, he cast the promising novice Robert Earl Jones.

On February 26, he sealed his commitment to the Harlem Suitcase Theatre and New York by signing a one-year rental lease, at $50 a month, for a three-room apartment, B-53, at 66 St. Nicholas Place in Harlem. On March 1, Langston moved in. His mother followed on the last day of the month. Hughes had learned from Carrie's doctor that cancer was now destroying her lungs, and that the end was near.

Late in March, leaving Hilary Phillips to rehearse the production, he sped out of town on a twelve-city, International Workers Order tour that took him as far west as Kansas City. With the newspapers full of stories of Franco's victories, Spain was his main topic on the road. In Cincinnati, the zealot American Legion and Knights of Columbus had a court reporter take down his speech, but otherwise he was not harassed. In some places, there were encouraging developments. In Chicago, seeking links between engaged black theaters, he conferred with the director of the newly formed Chicago Negro Peoples' Theatre, Fanny McConnell (who would later marry Ralph Ellison), about a local staging of *Don't You Want To Be Free?* He also attended a reading of a brilliant new play by his friend Theodore "Ted" Ward called "Big White Fog" ("We black folks is jus los in one big white fog," a character thinks in Richard Wright's short story "Fire and Cloud," published that year). Unselfishly, Hughes immediately hailed Ward's work as the best drama ever written by a black American. Touched by such encouragement, Ward sent Langston a copy of the script and praised "the magnitude of your great soul."

Returning to Manhattan for the Easter weekend in mid-April, Langston found his mother sinking at Edgecombe Sanitarium on St. Nicholas Avenue. He also found the Suitcase Theatre production under-rehearsed. Disappointed by the dancers, he decided to cut out their part of the program and stage only *Don't You Want To Be Free?* But he was still as confident as ever about the future of his theater. "We're in the egg stage just now," he admitted to the *Daily Worker,* "but we hope to hatch out something that will live in the minds of a lot of people for a long time. We are trying to build a people's theater that will have its roots in the masses of our people."

Just in time for the premiere, his I.W.O.-sponsored booklet of poems, *A New Song,* appeared at last. The edition comprised ten thousand copies, at 15 cents a copy, with an introduction not by Joseph Freeman (who was fast falling out of favor with the Communist Party) but by the radical stalwart Mike Gold. Its seventeen poems included "Let America Be America Again," "Song of Spain," and "Ballads of Lenin"—but not "Goodbye Christ" or certain other ultra-radical pieces. Still, Mike Gold was more than satisfied with the ideological weight of *A New Song.* "This work is the fruit of a decade of experiment, of travel, and of contact with all the bewildering social and esthetic theories of our time," Gold wrote; unlike many young writers who "have lost their way in this period, mistaking some dazzling skyrocket of esthetic theory for a star," Hughes had become "a voice crying for justice for all humanity." (A generation later, not one of these pieces would appear in Hughes's *Selected Poems.*)

On Thursday, April 21, an overflow crowd of almost 200 persons turned out for the gala dollar-a-ticket premiere at the loft on 125th Street. Most were startled to see the curtainless stage; in plain sight were three chairs, a table, a screen, a slave block, and a tree stump. An American flag was hanging left of center; a heavy rope tied in a noose dangled at the center; also on stage was a carpet sweeper. Without any warning, a young black man (Robert Earl Jones) stepped forward: "Listen, folks! I'm one of the members of this group, and I want to tell you about our theater. This is it right here! . . . Now I'll tell you what the show is about. It's about me. . . ." Shortly, he launched into a recitation of Langston's "The Negro" ("I am a Negro: / Black as the night is black"). The other heroes were blacks, men and women, of various ages; the villain was a white overseer. The action moved briskly. Using techniques borrowed from musical comedy, religious pageants, and the "living newspaper," but most heavily dependent on Hughes's poems, sometimes distributed among various speakers, the play progressed without a break between episodes. One scene merged into another in an attempt to capture what Langston would call "a continuous panorama of the emotional history of the Negro from Africa to the present." Music, especially the blues and spirituals, was integral. The singers were accompanied on a rickety old piano by Carroll Tate, a veteran bandleader who served as music director of the play. Although the blues section stopped the show, the entire impact was striking. At the end, when the young black man proclaimed, "White worker, here is my hand. . . . Let's get together, folks, and fight, fight, fight!'' the packed hall rose as one. Cries for the author brought a beaming Langston to the stage to renew his promise of a Harlem repertory theater: ''We want to build a theater for you folks, a theater for which you may write and in which you may act. This is your theater.''

Thus launched, the production would run three times a week, with a ticket costing 35 cents. Attendance was very good. If the major New York papers did not notice the production uptown in Harlem, the *Amsterdam News* reported ''a significant proletarian drama and a brilliant new star''—Earl Jones, ''a majestic youth.'' His wife, Edith Jones, was also praised, as was Toy Harper as a heroic black woman enacting Hughes's ''The Negro Mother.'' In general, blacks loved the show. Claude McKay thought it decidedly better than *Mulatto,* and Owen Dodson, a cultivated, promising young poet and dramatist, called it stirring and amazing, ''a home run in sincerity.'' In early June, however, when *Don't You Want To Be Free?* played downtown as part of a national conference of the leftist Little Theatre League, the reaction was more restrained. In *New Masses,* Richard Rovere found the production exciting enough, but others thought that the play was too brief, needed original music, and fell among too many stools. In the *Crisis,* the white poet Norman Macleod compared the play somewhat less than favorably to such leftist classics as Clifford Odets's *Waiting for Lefty* and Marc Blitzstein's *The Cradle Will Rock.* Groping to locate its essential weakness, Macleod distinguished between Langston's emotional and poetic ''feeling'' of race, especially as acted out in Earl Jones's ''simple ferocity and earnestness of soul,'' and his far less successful ''thinking'' about the class struggle.

In fact, *Don't You Want To Be Free?* was a major step in the evolution of Hughes as an artist. Finally he had found his natural form as a dramatist. Unlike his previous plays, which all strained to be ''well made'' and traditional in structure, *Don't You Want To Be Free?* was loose-limbed and improvisational in effect, montage-like rather than static and monumental, always strongly lyrical and rhythmic in a linking of intense moments of Langston's own poetry, and with music—black music—blended into its tissue as an essential element. Its major criteria, in other words, came not from the white stage but from Hughes's sense of the distinguishing features of modern, urban black culture and his own poetic gifts. Apart from one play shortly to be attempted in his old style, he would never take any other approach to drama and bring the result successfully to the stage; his accomplishments twenty years later with a series of musical plays based, especially, in the gospel genre, would have their roots in *Don't You Want To Be Free?* Only one aspect of this fertile source would wither: Hughes would never again sound the explicit Marxist note on which it ends. Perhaps even he had failed to realize in 1938 that (as Norman Macleod intuited but explained condescendingly) this Marxism had been, in effect, only tacked on to the essentially racial drama, instead of forming an integral part of its action.

By the end of its first season in July, some 3500 persons, about seventy-five per cent of whom were black, would attend thirty-eight performances of the Harlem Suitcase Theatre. In the process, the group had eclipsed its main rival, the Harlem Unit of the Federal Theatre Project. Although John Houseman and Orson Welles had staged an electrifying, highly acclaimed black version of *Macbeth* in 1936, that year Houseman had been pushed out of the directorship in favor of a trio of black leaders who emphasized protest plays but failed to lure Harlemites from more irresponsible entertainments. Certainly the Federal Theatre had nothing to compare in popularity to *Don't You Want To Be Free?*, which would be used to launch new radical black theater groups in at least four additional cities, including Chicago and Los Angeles.

In the midst of this rousing start, however, the Suitcase Theatre was not without its problems. ''We never were able to hold on to a consistent core of members,'' one alumna of the group would recall. ''People came and went, there were lots of changes, and of course there were clashes of personality just as you always have in such groups. We had a lot of fun, but sometimes, especially when Langston was not there, there was too much clowning around, so much so that Hilary Phillips would sometimes have fits.'' ''Earl Jones was wonderful, with a great smile and a nice voice,'' another member would recall. ''But he had a hard time remembering his lines. You were never quite sure what he was going to do. One night he got caught up in Langston's ''Brass Spitoons'' and didn't get out of it before an hour, it seemed like. We laughed and laughed. We had a lot of fun with the Suitcase Theatre.'' An even greater problem than discipline was the search for suitable plays. In April, Langston signed an agreement to dramatize Richard Wright's ''Fire and Cloud,'' which had appeared the previous month in *Story* magazine and won a prize of $500. Wright, who was now living in New York, would soon collaborate with Hughes on a poem, ''Red Clay Blues'' (published in *New Masses* in August, 1939). But the plan to dramatize his story of the radicalization of a black preacher in the South fell through. Langston turned next to plays by the white North Carolina dramatist Paul Green—''The Man Who Died At Twelve O'Clock'' and ''The Slave.'' Later, hoping to mount a big second season starting in the fall, he considered a series that would feature Lope de Vega's *Fuente Ovejuna,* Hughes's own *Troubled Island,* his adaptation of yet another story by Wright, ''Bright and Morning Star,'' about the radicalization of an aging, devout black mother, and perhaps a comedy. None of these plays was ever staged by the Suitcase group.

In addition, the directorship was a demanding job that put no money into Langston's pocket and sometimes took money out. ''I have asked the carpenter to build the additional platform for the stage,'' one crisp memorandum to Louise Thompson read, ''and will pay for this personally, unless the theater sees fit to reimburse me.'' Nevertheless, pleased with his theater, he saw himself settling down in New York. He paid United Van Lines $53.82 to ship his belongings from Cleveland and also, after seven years of storage there, from the attic at 514 Downer Street in Westfield, New Jersey, where he had left them following his break with Mrs. Mason.

Just as Langston was settling into his home, however, his mother collapsed for the last time. Around four o'clock on the morning of Friday, June 3, Carrie died at Deaconess Hospital in Manhattan, where she had been rushed from

Hughes's apartment a few hours earlier. She was sixty-five years old. Mostly on credit, he hired an undertaker, Mamie Anderson of the Anderson-Pratt Company at 239 West 131st Street in Harlem. Unable to meet even the smaller expenses, he called Van Vechten for a loan, which came at once. The first flowers to arrive also came from Van Vechten, a bouquet of lilies so beautiful that the undertaker placed them within the casket. On Sunday, at the funeral home, Rev. Peter A. Price of Mother Zion A.M.E. Church conducted the service, during which a soprano sang Carrie's favorite hymns, "Goin' Home," "City Called Heaven," and "Beautiful Isle of Somewhere." Later, far from the grave of her parents in her native Kansas, Carolina Mercer Langston Hughes Clark was buried in the Cypress Hills Cemetery at Jamaica Avenue and Crescent Street in Brooklyn.

Not long after, at Rev. Frederick Cullen's Salem Methodist Episcopal, Langston attended another funeral. The much loved and admired James Weldon Johnson, sixty-seven years old but in fine health, had been driving with his wife, Grace Nail Johnson, through a rainstorm near Wiscasset, Maine, when a train rammed their car at a railroad crossing. Johnson was killed, his wife critically injured. On June 30, Langston joined over twenty-five hundred mourners at the service for one of the most gifted, prolific, and generous black leaders in American history. Among the most devastated mourners was Van Vechten; Johnson had long been his closest black friend. Langston, too, had moving memories not only of visits to the gracious Johnsons in Harlem and at Fisk University but also of the first great public moment of his life as a writer—the evening in May, 1925, when Johnson had recited Langston's prizewinning poem "The Weary Blues" at the landmark *Opportunity* awards dinner, which had led directly to the publication of his first book. In a sense, Hughes's career had started that evening.

By July, when the first Suitcase Theatre season ended, much of Langston's enthusiasm for the group clearly had evaporated. He not only jumped at an invitation at short notice to attend a conference of the leftist International Association of Writers in Paris, but also made plans to stay there for several weeks. The League of American Writers, prompted by the French organizers after Pearl Buck, winner of the Nobel Prize for literature that year, declined an invitation, asked Hughes to accompany Theodore Dreiser to a Congress for Peace Action and Against Bombing of Open Cities. Since Dreiser, an admirer of the Soviet Union but a staunch individualist, refused to join the League, Hughes (a vice-president) would conduct its business. Aided by perhaps his first ASCAP check, totalling $27, for song royalties, he hurried to leave New York.

Reaching Paris on July 18, he and Dreiser were met by Louis Aragon and René Blech, who would serve as Dreiser's secretary, and by Henri and Retna Cartier-Bresson. Langston stayed at a small hotel on rue Jacob, but tried to keep an eye at all times on the erratic Dreiser, whose first act on reaching Paris was to order René Blech to search the city for bottles of American gin. Decorated for a state visit by George VI and Elizabeth of Britain, Paris was, if anything, more lovely than ever. However, the excitement of the previous year, just before and just after Hughes's months in Spain, was gone. The peace congress was dull, he reported home, with too many speeches in too many languages—"to which nobody paid any attention anyway." Forty-two nations and a babel of organizations covered the political spectrum from the right wing to "as far left as the French socialists would permit it to go, which wasn't so far." He was happier when Spain's most melodramatic communist, Dolores Ibárruri, "La Pasionaria," after arriving to a tremendous ovation, started a shoving match on the stage when she was forbidden to speak. Without Dreiser, who simply failed to appear, Langston attended a luncheon for them hosted by Stephen Spender, Cecil Day Lewis, and other British writers. But babysitting Dreiser paid off at last. The novelist was "a great help to the American group, and of great interest to the French," Hughes reported. "He acquitted himself well, and I am glad he was chosen as a delegate."

Hughes's own address to the delegates on July 25 contrasted sharply with his Paris speech of the previous year. Now, in an oddly mellow, almost resigned way that no doubt reflected the Loyalist defeats in Spain, he emphasized the power of language rather than race or the danger of fascist power. "Words put together beautifully," he said, "with rhythm and meaning one as the branches and roots of a tree—if that meaning be a life meaning—such words can be of more value to humanity than food to the hungry or garments to the cold. For words big with the building of life rather than its destruction, filled with faith in life rather than doubt and distress, such words entering into the minds of men, last much longer than today's dinner in the belly or next year's overcoat on the back. And such words, even when forgotten, may still be reflected in terms of motives and actions, and so go out from the reader to many people who have never seen the original words themselves."

On August 1, after seeing Dreiser off to Spain at the Gare d'Orsay, he took a train westward towards Normandy to visit "Le Puits Carré," the country home of Nancy Cunard just outside the tiny village of La Chapelle-Réanville, near Vernon, about sixty miles from Paris. Also visiting Cunard was the novelist Norman Douglas. Other guests arrived, including a Spanish widow, whose husband had died in battle at Badajoz, and her son. Cunard was taking them in as refugees. In the shimmering summer heat, Cunard and her friends dined outdoors under a great lime tree near the stone well that gave the house its name and in which bottles of wine were cooled. In the evenings, they took long walks across the fields. Late into the night Langston told stories about his travels, the Spaniards sang flamencos and fighting songs, and Cunard played African tribal music on her gramophone. Langston browsed in Cunard's very impressive library of books about blacks and admired the volumes published by her own Hours Press, on which her former American lover, the black musician Henry Crowder, had helped her. The visit was brief but memorable. Twenty-five years later, unhappy and ill, Cunard would remind Hughes of his droll stories and her music and "our hot evenings" at La Chapelle-Réanville, and a sweeter life "alas ALL gone now in the beastly war." And, thinking also of this visit, after her death, Langston would commend Cunard's "infinite capacity to love peasants and children and great but simple causes across the board and a grace in giving that was itself gratitude.''

Returning to Paris, Hughes took a room at the Ethiopian-run hotel in Montmartre where he had stayed the previous year. Seeing a great deal of the Cartier-Bressons, he passed the rest of the month working on a one-act musical play, radical in tone, called ''The Organizer,'' intended as a collaboration with James P. Johnson, the great jazz pianist and composer. Relaxing one day, he enjoyed an aperitif on the Boulevard St. Michel with Countee Cullen; the poets were friends once again. The only excitement was provided by Theodore Dreiser, who was supposed to rejoin Hughes in Paris and go with him to London but instead had vanished. Telegrams flew between the League office in New York and the Spanish embassy, the International Association of Writers, and Hughes in Paris until word came that Dreiser had simply gone from Spain to London without his baggage and without informing anyone, and was even now on the high seas bound for New York.

On September 7, on his first visit to Great Britain, Langston reached London. Talking about Spain and reading his poems before the Left Book Club at the Group Theatre Rooms at 9 Great Newport Street, he also took notes on British leftist writers' groups for the League of American Writers in New York. He dined at the fashionable Simpson's restaurant with Essie Robeson, who evinced a sincere interest in *Don't You Want To Be Free?* as a vehicle for her husband. Hughes, who had heard much talk over the years about vehicles for Paul Robeson, paid at least as much attention to his roast beef and Yorkshire pudding, although he knew that Robeson had been performing his ballad poems from Spain, and had also recorded one of his songs. At the Café Royale on Regent Street, Langston dined with the British jazz enthusiast Leonard Feather, then left England after a thoroughly enjoyable visit.

He returned to New York third-class on the French liner *De Grasse* on September 18, in time to fulfill a round of speeches in Connecticut secured by Louise Thompson. Hughes's most pressing problem, however, was the Harlem Suitcase Theatre. With almost no progress made in his absence, the fall season—and the future of the group—was in jeopardy. He had also committed himself to a new play for the Gilpin Players at Karamu Theatre in Cleveland. Hoping to satisfy both the Suitcase and Karamu groups, Hughes quickly sketched ''Young Man Of Harlem,'' a politically aggressive pastiche not unlike *Don't You Want To Be Free?* But when ''Young Man Of Harlem'' reached Cleveland, the Gilpins rejected it outright; Rowena Jelliffe wrote Langston that they found it ''awfully naive and obvious and with pretty bad writing.'' Hughes took the news calmly. Nor was he apparently either surprised or offended when the bourgeois group also rejected *Don't You Want To Be Free?* Rashly, as it turned out, he then promised Karamu a three-act play called ''Front Porch,'' which the Gilpins promptly announced would open in mid-November.

Evidently the Harlem Suitcase Theatre also found ''Young Man of Harlem'' unsatisfactory. To open the new season in early October, as planned, was now out of the question. In a clear sign of stagnation, if not desperation, Langston decided to revive *Don't You Want To Be Free?* To balance the program, he hammered out five linked satirical skits called "Limitations of Life," of which three were finally used—"Little Eva's End," a parody of *Uncle Tom's Cabin;* "Limitations of Life," which satirized a sensational film about passing, *Imitation of Life,* by reversing the roles of the white star Claudette Colbert and the black actress Louise Beavers; and "Em-Fuehrer Jones," in which a crazed Hitler runs about the Black Forest. This skit gave Hughes a chance at last to express some of his distinct reservations about Eugene O'Neill's primitivistic vision of black psychology.

The need to revive *Don't You Want To Be Free?* lowered the morale of the Suitcase Theatre. Ominously, some of the members began to question its basic direction. Dorothy Peterson, who served for some time as technical director of the theater, voiced the main objection: "We can't go on forever doing sceneryless plays." "Two or three of the members came to me," Louise Thompson later remembered, "with the suggestion that we should do plays by Noel Coward. I was horrified. I can't imagine what they were thinking." And even plays without scenery called for money. A fund-raising scheme was drawn up, with monthly teas, dances on Sunday afternoons, four big parties spread over the year, and special lectures and presentations. Sylvia Chen Leyda, now in New York, offered to donate an evening of dance, and a blues concert was scheduled for the fall. Once, when cash was needed immediately, Van Vechten sent $10.

Langston had put himself on an impossible schedule. In addition to writing "Front Porch" for the Gilpins and travelling to Long Island to work with the composer James P. Johnson on "The Organizer," he found himself managing the new Suitcase season almost singlehandedly, including going at least once to New Jersey to buy props at a country auction. Then, while he more or less ignored "Front Porch," he started sketching yet another play, "Sold Away," a love story about a slave couple. On October 19, less than a month before its scheduled opening, he finally sent the first scene of "Front Porch" to Cleveland, along with the promise of fresh portions every day by air mail, special delivery. But Hughes soon fell behind. On November 2, a telegram from Rowena Jelliffe demanded the rest of the play; she reported that the cast was seriously demoralized. There is some question as to whether the last act arrived from Langston on time. Two different versions of the third act exist, along with a note from Hughes to the effect that the actors balked at his original ending, which was very unhappy, and requested something more pleasant. Rowena Jelliffe remembered the situation differently. "Finally there was nothing for me to do," she would recall, "but to write the last act myself. Which I did, with hours to spare before the opening. Langston was more than a little ashamed of his performance, but he was very sweet about it. He told me the next time I saw him that he had committed himself to too many people, but let *me* down because I had known him longer than anyone else and he was sure that *I* would forgive him." In any event, *Front Porch* was not a great success. The *Plain Dealer* saw in it more restraint than usual for Hughes, and a "unique hewing to Caucasian standards.'' To another reviewer, it seemed ''tepid and non-descript'' compared to other Karamu plays.

Front Porch concerns the problems not of the folk or the masses but of the black bourgeoisie, represented by a family, the Harpers, who live in an allegedly integrated neighborhood where they are scarcely tolerated; blacks like them often own large, comfortable houses with front porches and scorn their poorer brethren. The mother, Mrs. Harper, professing to be delighted by her years in the mainly white district, is determined not to backslide into the masses. Her son, on the other hand, wants to return to the black community, and her daughter becomes involved with a working-class black man who is involved in a labor strike, instead of with a pompous graduate student who is courting her. In Hughes's version of the play, the rebellious moves by the children end disastrously. When the daughter attempts an abortion, which is demanded by her mother, she falls seriously ill. Her lover, whose labor strike also collapses in failure, is barred from the Harper house. The daughter dies. In the produced version, which Rowena Jelliffe apparently wrote, the daughter refuses to have an abortion after learning of the victory of the strikers, then goes home to be united joyfully with her lover. Neither ending, however, is enough to counter the general ineffectiveness and inconsistency of the rest of the play, in spite of its force as a vehicle for Hughes's hostility to the black middle-class and the bourgeoisie in general.

Meanwhile, belying its true condition, the Suitcase Theatre launched its second season in November with a gala evening that drew dozens of celebrities, black and white, to the International Workers Order community center in Harlem. The satirical skits were a great hit—at least with the blacks in the audience, who howled throughout ''Limitations of Life'' at seeing a black society lady return home from the opera to the ministrations of her shuffling white maid. Most whites were less amused. Reviewing *Don't You Want To Be Free?*, the communist *Daily Worker* applauded a change that brought in a white worker early ''who identifies his class interests with that of the Negro''; but the reviewer still found the production falling short both technically and as propaganda.

On November 16, Hughes and Muriel Rukeyser read their poetry at the New York Public Library on 42nd Street. The next evening, the fund-raising blues program came off well. Langston read some poems, his old companion in Haiti, Zell Ingram, now living in New York, showed art work, and the talented dancer Felicia Sorel performed to Herbert Kingsley's piano accompaniment. By this time, too, the little musical with James P. Johnson, ''The Organizer,'' was ready. When it went into rehearsal, Hughes had more or less put his Suitcase house in order. But he had done so mainly to abandon it; he quietly applied for work with the New York Federal Theatre, but also made plans to go to California. Ostensibly he was going on a brief trip; in truth, he intended to stay much longer if he could.

He needed money badly. Although *Mulatto* was still on tour in New York and New England, the American Play Company, his agent for the play, had gone out of business the previous summer owing him several hundred dollars. Martin Jones was so far behind in payments that Hughes once again requested arbitration by the Dramatists' Guild. Unquestionably he was beleaguered by debts. By far the most embarrassing of these was the sum of $125 still owed to the undertaker Mamie Anderson for his mother's burial.

Passing through Cleveland, he apologized insouciantly to Rowena Jelliffe and attended a performance of *Front Porch*. In Chicago, where he saw a production by Fanny McConnell and Ted Ward of *Don't You Want To Be Free?*, he joined Arna Bontemps for a lecture tour, beginning with readings at Northwestern University in Evanston, Illinois, before heading through the Southwest to California. In Denver to speak before a film and book club, they were refused rooms by thirteen hotels, as the Associated Negro Press reported in a dispatch almost certainly sent by Langston himself. On December 4, they reached Los Angeles, where the Philippa Pollia Foundation, which sponsored educational projects for children, had retained them for a series of lectures. Following a similar tour by the guitar-playing Carl Sandburg, they spoke to school audiences on "How Stories and Poems Are Born" and "Making Words Sing, Talk and Dance." The tour was so enjoyable that not long after Bontemps left Los Angeles to return to his home in Chicago, he proposed a two-month jaunt with Langston, provided that they could each net $500, on the subject, "We Write For The Young!" On his own, Hughes would continue to speak to children for the foundation after the departure of Bontemps.

But he had stayed on for a more promising reason. Clarence Muse, the black Hollywood actor and musician with whom Langston had long talked about collaborating on movie scripts, and who now ran the Negro unit of the Federal Theatre Project in Los Angeles, needed his help. Muse's staging of Hall Johnson's *Run Little Chillun,* sold out for twenty-one weeks, had so impressed the powerful young Hollywood producer Sol Lesser that Lesser had asked Muse to develop the next film vehicle for the popular white boy singer Bobby Breen. The movie, they agreed, would be set in the South and include spirituals and blues. For the talented Muse, who hoped to produce and direct films one day, this was an extraordinary chance; in any event, Lesser's decision, as one newspaper caustically put it, created "something of a precedent in Hollywood, where it has not been considered necessary for a writer to have a knowledge of his subject matter." Muse needed an outline of a story quickly. On December 6, Langston finished "Pirates Unawares," about a rich little boy from Park Avenue and Bar Harbor (perhaps Hughes was thinking of one of Godmother's little Biddles), who falls happily among Negroes on a vacation in the Sea Islands off South Carolina, then enjoys adventures not unlike those of Tom Sawyer and Huck Finn with a little black Jim.

Langston and Muse also began work on another major project. Before Arna Bontemps had left town, he had discussed with Langston and Muse the possibility of Muse's Federal Theatre group producing "St. Louis Woman," a play adapted by Bontemps and Countee Cullen from Arna's novel *God Sends Sunday*. Muse agreed to take the play if Hughes would revise it completely on a retainer from the Federal Theatre Project. Langston accepted the offer. But nobody was interested ultimately in a Federal Theatre staging. They all hoped for a Broadway hit or, even better, a Hollywood film. The moment seemed right for such a film; even before a formal agreement beyond the Federal staging was worked out by the various agents, Langston and Muse went to work. When negotiations among the various parties, including agents for Bontemps and for Cullen and the publisher of the novel, became intense, the infinitely flexible Langston smoothed over all disagreements. "Make the agreement solely with me, if you wish," he once informed Bontemps after the latter had questioned Muse's role as a part-author of the revised play. "Or not at all, as you *darkies* choose. . . . But as co-writer with me, he has been a great help on the script, being an old theatre man and knowing all the angles of building sure-fire theatre reactions. Should you cut me in on it, of course I would divide with Clarence. (Which you need not mention further.)" Early in May, Langston and Arna would complete a confidential agreement guaranteeing Hughes a share of the property.

With money coming in at last, Hughes settled into an indefinite stay in Los Angeles. "I am hell bent on paying my debts before too many more weeks roll around," he explained to Louise Thompson. "So pray for me that I may grow strong in the power of the Almighty Dollar, for nothing else in this capitalistic world seems to possess the same strength and vigor." Obviously he would not be in New York soon. "We shall do all we can to keep things going here," a dismayed Thompson replied patiently, "though I must say it will be a happy day for us when you return."

Around New Year's Day, 1939, following a post-Christmas visit to Carmel, Langston signed a contract to work with the Federal Theatre on "St. Louis Woman." For three months, he would be paid just under $1200, which was much more money than he had earned in a long time. Meanwhile, Sol Lesser rejected Langston's first outline, "Pirates Unawares," but kept his offer open. On January 13, the three men met to discuss fresh ideas—or Lesser and Muse talked, and Hughes demurely took notes. Should Bobby be the son of a plantation owner? Or a poor white boy with a cruel, Bible-thumping father and a faithful old black "Uncle"? Langston appeared to have no difficulty creating stereotypes; in "Pirates Unawares," he had already given Breen an old black mammy. Lesser decided on a nineteenth-century setting, rather than one in the contemporary South, as in "Pirates Unawares," where it would be harder to skirt the painful issue of segregation. After several long days in the Los Angeles Public Library researching the South of the 1840's, Hughes delivered on January 26 an outline he called "Dixie, or The Little Master." Eventually the title was changed to "Way Down South."

Langston and Muse wanted this chance so badly that neither dared to question either the financial or the political aspects of their involvement. For their work on the synopsis they were each paid $150, a deal that Maxim Lieber called "nothing short of raw" when, belatedly, he heard about it. Although Lieber had sent Hughes the name of his agency's Hollywood representative, Sidney Biddell of the Feldman, Blum Corporation, apparently neither Langston nor Muse dared to have an agent face Sol Lesser on his behalf. But the arrangement began to trouble Hughes. Finally, pushed by Lieber, he sheepishly turned to Sidney Biddell, who immediately secured a contract, essentially open-ended as long as they worked, at $125 a week for each man. In addition, the contract ensured at last that they would receive formal credit for their efforts.

Working for just over two months under this arrangement, Hughes finally accumulated some money. With the Federal Theatre stipend, a *Mulatto* royalty check for almost $90 (after Martin Jones's first check was incorrectly dated and returned, and the second bounced), his ongoing lectures for the Philippa Pollia Foundation and on his own, and the $250 he expected from an unauthorized publication of "Let America Be America Again," he was within sight of earning at least $3000 during the first three months of 1939. Never before had he gained money at such a rate. Hughes sent $100 to Van Vechten (owed on a loan of $200 when Langston had broken with Mrs. Mason), $150 to Amy Spingarn (almost certainly he still owed her money for his Lincoln education), and the balance owed on his mother's funeral to the undertaker. But he kept most of the money for a special task: to fund the writing, at long last, of his autobiography.

For his Hollywood contract, however, Hughes paid dearly in a loss of pride. He and Muse were made to feel lucky to be employed, as indeed they were, given the industry's attitude to blacks. No branch of the America media had been so consistently humiliating to blacks, or excluded them so thoroughly except in demeaning ways. Twelve years later, Langston would vent some of his anger at what Hollywood had done to black Americans: "Hollywood is our *bête noire*. It is America (and the world's) most popular art. . . . Yet, shamelessly and to all the world since its inception, Hollywood has spread in exaggerated form every ugly and ridiculous stereotype of the deep South's conception of Negro character." In 1939, working on *Way Down South,* the insults were also personal. In the most humiliating episode, Langston was forced to eat a sandwich under the broiling California sun because a white executive refused to enter any restaurant with him. "Even allowing for a certain amount of Langstonian exaggeration," Lieber wrote, "I am positively scandalized," especially since he had made it clear "who and what you were, assuming that this would eliminate any misunderstanding." Hughes was deeply hurt to find many Jewish Americans, who were prominent in the film industry, in the vanguard of discrimination. Two years later, he mockingly threatened in a letter to Arna Bontemps to disrupt a lecture and "tell all about how much prejudice there is in Hollywood under Israel." But he would never protest directly, if at all. Instead, he moaned about the often gruelling pace of Hollywood work. "Never take a Hollywood job," he warned Bontemps. "They vary from nothing to do at all, to rush! rush! rush! No nice in-between kinds." "I find it amusing and not unprofitable working for Hollywood," he wrote to Van Vechten, who often complained that blacks were not aggressive enough in challenging racism. Langston characterized Hollywood leaders as being anxious not to offend blacks, but almost totally ignorant of the offensiveness of most of the movies. Interviewed on KMPC radio in Beverly Hills, he raised the general question of the exclusion of black writers and actors from the industry. The interviewer swiftly changed the topic. And yet the chance for big Hollywood money kept Hughes enthralled. He was about to sign a contract linking his talents to those of Clarence Muse, with whom he had just written a song, "Refugee Road," when Lieber stopped him: "No, Langston, please don't let the enthusiasm of the moment run away with you again."

In New York, meanwhile, the Harlem Suitcase Theatre was dying. From various members came various opinions on what ailed the group: Louise Thompson was too domineering, Hilary Phillips too weak, Dorothy Peterson too cosmopolitan, Toy Harper too defensive about Langston's interests. To Langston, Mary Savage lamented that "personality problems" were killing the group. All agreed that he alone could save it, but he would not promise to return. Instead, he founded in Los Angeles the New Negro Theatre, a counterpart to the Harlem group. On March 19, his new organization made its debut in a theater at 41st Street and Central Avenue with *Don't You Want To Be Free?,* directed by Clarence Muse, and the three parody sketches.

Completing a flurry of work on *Way Down South,* Hughes also finished his revision of Bontemps and Cullen's "St. Louis Woman" for Muse's black Federal Theatre unit. Adding a variety of devices inspired by grandiose Hollywood musicals, including the entry of the hero in a hack drawn by two milk-white bobtailed horses, he estimated that the play was "75% a new show now." From W. C. Handy, long an admirer, came permission to use his celebrated "St. Louis Blues" freely throughout the show. Muse liked the revised version and accepted it for his Federal Theatre unit—although some of the members suspected that it was beneath them and proposed staging an opera instead.

Early in April, Langston reached Kansas City in time for a series of programs with Bontemps at churches and high schools and at least one socialist group, the Workers' Institute of Greater Kansas. His Lincoln schoolmate Thomas Webster of the Urban League welcomed Hughes to the city and promised to sponsor him on future visits. With Franco's total triumph the previous month, Spain was a pressing topic for Hughes, who defended the lost Loyalist cause and especially its attempts to break with tradition and educate the masses in a country one-half illiterate. Franco would kill these efforts, he predicted, because fascism had a vested interest in mass ignorance. To the largely white Workers' Institute, and with growing reports of persecution in Germany, he declared the unswerving black American sympathy for the Jews there. But he also urged the white workers to see the plight of black Americans caught in the vise between employers who denied them jobs and trade unions that forbade them membership. And deploring the continued existence of the slums he had seen in Kansas as a boy, Hughes also suggested that blacks must share at least some of the blame for their condition.

His tour with Bontemps took him into Detroit and through Missouri, Kentucky, Illinois, Indiana, Ohio, and northern New York into Pennsylvania. At last they returned to Chicago, where Langston took a room at the black Hotel Grand on South Parkway and settled down to six weeks of work on his autobiography. Cutting back on his social engagements, Hughes also postponed one literary task he was anxious to begin—a translation, recently authorized by the poet's estate, of García Lorca's *El Poeta en Nueva York.* Unwilling to set foot in New York, where he could retrieve his extensive files of correspondence and other documents, he worked from memory in his hotel room; in front of him he set a Van Vechten photograph of Ethel Waters, who was even then making history in *Mamba's Daughters* as the first black woman to be billed as a star on Broadway. His project was known to only a few people, including Blanche Knopf and Van Vechten, who had first urged Hughes to write an autobiography in 1925, and Arna Bontemps.

Aware of his friend's smiling reticence, Bontemps begged Langston now, as he faced the task of autobiography, to "leave nothing out, make it a real titan's book." Certainly Hughes responded to the challenge with a grand display of energy. By May 27, when his hard-earned savings were almost exhausted, he had written 250 pages. No end was in sight, however; he joked nervously that his book would be another *Gone With The Wind.* Determined to finish the task, he asked Noël Sullivan to allow him to stay at Hollow Hills Farm in Carmel Valley until the task was done. Sullivan invited him to stay as long as he wished.

Before heading west, Hughes travelled to New York for an important public appearance and to settle certain personal matters. On June 2, against the background of a world lurching towards war, the Third American Writers' Congress opened at the New School for Social Research on West 12th Street in Greenwich Village. In addition to the fall of the Spanish Republic announced by Franco on April 1, the Nazis had invaded Czechoslovakia in March, Italy had seized Albania in April, and the long-expected Hitler-Mussolini pact had been concluded in May. The brutal reality of the Fascist threat was underscored by the presence of several leading German writers among the five hundred delegates at the Congress, who had been forced into exile by the Nazis after the tragic *Kristallnacht* of the previous November.

On June 2, however, when Hughes addressed a special public session at Carnegie Hall, he did not emphasize world politics, or even the general place of blacks in America. Absorbed in his weeks of work recounting his own life, he chose to speak on the problems facing blacks who try to live as writers in the United States. The speech, "Democracy and Me," was perhaps an apologia for the compromises forced on him in the years since his return from the Soviet Union in 1933, especially the compromise soon to be broadcast through the nation, the film *Way Down South.* Comparing the life of blacks to that of Jews in Germany, Hughes saw the only important difference to be the American principle of freedom of speech in contrast to Nazi intolerance. Otherwise, Hughes saw no difference between the two nations. "Magazine offices, daily newspapers, publishers' offices are as tightly closed to us in America as if we were pure non-Aryans in Berlin," he declared. As for Hollywood, no studio "in all the history of motion pictures had yet dared make one single picture using any of the fundamental dramatic value of Negro life. . . . On the screen we are servants, clowns, or fools. Comedy relief. Droll and very funny." Hughes recounted incidents in which distinguished blacks had been publicly humiliated by racism (he might have mentioned the barring of Marian Anderson from Constitution Hall by the Daughters of the American Revolution earlier that year). In perhaps his most pointed remark, given the Jewish presence there, he declared that Hollywood, "in so far as Negroes are concerned, might just as well be controlled by Hitler."

At the end of the conference, however, his appeal was more general. In "one of the most dramatic incidents of the session," according to a *New York Times* reporter, the audience stood with bowed heads as Hughes slowly read the names of forty-five writers from nine countries who had been killed under fascism.

At the Harlem Suitcase Theatre, Langston watched Dorothy Peterson rehearse her role as the mother for her debut in *Don't You Want To Be Free?* He was frequently at the theater, but offered no promise about his future involvement. He did not give up his apartment at 66 St. Nicholas Place, which his "brother" Gwyn "Kit" Clark was sharing with two other men. According to all reports, Kit was drinking heavily and was generally indigent.

As soon as Hughes was back in Los Angeles, he took the final step with the Harlem Suitcase Theatre. On July 14, in a letter to Louise Thompson, he resigned from the post of director. Offering as the main reason his enforced absences from New York, he wished the theater well. Four days later he attended the first showing of *Way Down South,* which Clarence Muse had not only acted in but also co-directed with Bernard Vorhaus and generally managed the corps of some 300 black and 50 whites actors, singers (including the Hall Johnson group), and dancers. On the screen Hughes read his credits for the original screenplay and lyrics for two songs, "Good Ground" and "Louisiana." *Way Down South* was no better than he had expected, nor much worse than he had feared. He hoped only that his radical friends would understand the reasons for his involvement in such a film; he probably also hoped that it would succeed and lead to other lucrative offers from Hollywood. In some quarters, *Way Down South* was indeed a success. The influential *Variety* judged "the Muse-Hughes yarn" to be "by far the best" of Breen's pictures; the *Los Angeles Times* called it "picture-perfect" as well as "box-office perfect." But another journal, voicing the judgment Hughes feared, disagreed; *Way Down South* was *Uncle Tom's Cabin* with Bobby Breen as Little Eva—and Hughes and Muse, no doubt, as a composite Uncle Tom; the film had set back the cause of honestly portraying blacks by about a hundred years. The adverse criticism began to roll in. Louise Thompson, writing from New York, did not spare Hughes: "Everybody says they cannot understand how you could have written such a scenario or if they changed it why you permitted it to come out under your name. I tell you this unpleasant news because I know you want to know what they are saying about the picture here."

Stung by the criticism, especially from blacks, and resentful of what he saw as ingratitude and insensitivity after his years of sacrifice for the leftist and the black causes, Hughes drew up a sarcastic "STATEMENT IN ROUND NUMBERS CONCERNING THE RELATIVE MERITS OF 'WAY DOWN SOUTH' AND 'DON'T YOU WANT TO BE FREE' AS COMPILED BY THE AUTHOR MR. LANGSTON HUGHES." The Hollywood film had paid his mother's funeral bill, her doctors' bills in Cleveland and New York, "several hundred dollars back debts that worried my soul,'' secured for him ''two new suits, two pairs of shoes, and a half dozen shirts—the first in three years,'' and had left enough money over to allow him to start his autobiography. The radical play, on the other hand, although performed over two hundred times in six large cities, had netted him exactly $40 to date in royalties, and cost him much more in typing and other expenses. ''Of course, I am delighted to present DON'T YOU WANT TO BE FREE as a gift to the Negro people. But, on the other hand, starvation is not always amusing. For public consumption: *Take your choice.*''

Criticism from the left served only to push Hughes further away from radicalism. In desperate need of money, he had made a pragmatic choice as he pursued the overriding ambition of his life: to earn a living exclusively by his pen while living in the United States. In reality, he saw no choice. Trying to keep the radical faith, he had done the best he could do and still survive as a writer with the broad, racially encompassing influence that he wanted to preserve. Yes, he had helped perpetuate certain of the very stereotypes he hated; but the alternative was raw protest or grey poverty, and exclusion into a silence and a void in which he saw nothing noble. He had compromised, but he had also survived to write again.

Now, for example, as he looked for the means to complete his autobiography, he had no choice but to turn for help to a patron warm to him personally but bitterly hostile to the left. Who else was ready to help him? Determined to press on with his life in spite of all criticism, in the middle of August Hughes boarded a train out of Los Angeles for Carmel Valley and Noël Sullivan's Hollow Hills Farm.

14

THE FALL OF A TITAN
1939 to 1941

But if you was to ask me
How de blues they come to be,
Says if you was to ask me
How de blues they come to be—
You wouldn't need to ask me:
Just look at me and see!

"Evenin' Air Blues," 1941

BENT on making the most of this chance to hammer and forge his autobiography into "a real titan's book," in Arna Bontemps's challenging words, Langston settled down near mid-August at Noël Sullivan's sunny Hollow Hills Farm in Carmel Valley. "I have a charming little Mexican style house on a hillside above a pear orchard," he wrote to Bontemps, "quiet and all my own for work—one little room, a fireplace, and a patio." There he revised the drafts completed during his weeks at a hotel on the South Side of Chicago. Although Chicago might have been more stimulating, Hughes definitely preferred the comforts of Carmel Valley. After seven years of friendship, he had very little trouble accepting Noël Sullivan's largesse. Grown more politically conservative and reflective, almost sacerdotal in his manner, Sullivan apparently had lost little of his affection for blacks in general or Langston in particular, and nothing of his passion for hospitality. "In those days," his housekeeper Eulah Pharr would recall, "forty-five guests for dinner was not at all uncommon, or seventy-five for a picnic lunch at the farm. Mr. Sullivan couldn't live without lots of people coming out, and everyone more or less expected that he would entertain them in a grand style." Along with the local gentry at Sullivan's table were stars from among the thousands of visitors to Carmel and Monterey; Hughes was soon meeting famous singers and musicians, painters and sculptors, actors and actresses, as well as the inevitable sprinkle of itinerant aristocrats from Europe. He did his part so well that within a month or so he gained twelve pounds. "I don't think anyone enjoyed the eating and the drinking and the talking more than Langston did," Eulah Pharr would remember. "He was the perfect guest."

Sullivan's help had come at a crucial time; Hughes's finances were once again in a sad way. His only summer suit, recently purchased with his Hollywood windfall, was hanging not in his own closet but on the rack at a pawnbroker's shop in Los Angeles. To hire a typist for the final copy of his autobiography, and also to pay for his apartment on St. Nicholas Place in New York after his sub-tenants fell behind on the rent, he borrowed $200 from Knopf. He then accepted with alacrity an advance of $200 from the Rosenwald Fund in Chicago for a children's book to be written with Arna Bontemps. For "Boy of the Border," about a little Mexican driving a herd of broncos to California, he consulted with Robinson and Una Jeffers's son Garth, who had been a horse herder in New Mexico. Still, his autobiography remained Langston's major concern. Clearly he remembered the terms of Arna Bontemps's challenge even when the task began to seem like drudgery. "The old book drags along just like the old novel did, and just won't seem to get itself done," Langston would write to Louise Thompson. "I've got to make a Titanic effort."

But even as he tried to see his life, and what he had accomplished as a black writer, on an epic scale, certain political events rattled and confused Hughes. He had been only days in Carmel Valley when the news exploded about the appalling non-aggression pact signed in mid-August by Nazi Germany and the U.S.S.R. From almost any point of view on the radical American left, the timing was very bad; for Hughes and some others, it was deeply humiliating. A week before the announcement, he had joined more than four hundred signers of an open letter calling for close cooperation between the Soviet Union and the United States against fascism, and denying "the fantastic falsehood that the U.S.S.R. and the totalitarian states are basically alike." An admirer of the strongman Joseph Stalin, throughout the decade Hughes had never questioned publicly either the internal practices or the international strategies of the Soviet Union. In fact, in April, 1938, following Stalin's campaign to quell dissent by ruthlessly prosecuting his opponents, Hughes had signed a statement by American progressives in support of what the *Daily Worker* called "the recent Moscow trials of the Trotskyite-Bucharinite traitors." That statement had ended with a call to liberals "to support the efforts of the Soviet Union to free itself from insidious internal dangers, principal menace to peace and democracy."

Both gestures left Langston shabbily positioned after the Nazi-Soviet pact. He could hardly disagree with one former colleague in Spain who denounced the agreement as "a betrayal of people like the men we knew in the Lincoln battalion, especially those who gave a leg or a life to fight Fascism." When the German invasion of Poland on September 1 brought an almost immediate declaration of war by Britain, the Soviet position became almost impossible to defend. The Russian entry into Poland further embarrassed the American left, especially its pacifists; for some, their last vestige of sympathy for the U.S.S.R. vanished in November when the Soviets invaded Finland—"the Russian barbarism in Finland," as it was condemned in *The New Republic* by Ralph Bates, whose speeches in America on Spain, where Hughes later met him, had inspired Langston and other Americans artists and intellectuals.

Under Noël Sullivan's roof, the merits of communism became a forbidden topic of conversation. Its demerits were taken for granted; over lunch one day, an Italian baroness spoke wistfully to an incredulous Langston about her dear friend Benito Mussolini, then denounced Eleanor Roosevelt as a flaming radical. Hughes, meanwhile, clearly represented himself to Noël Sullivan as alienated in general from communism, if still pro-socialist. Sullivan expected no less; in a scorching letter to the still-radical Ella Winter, who had left Carmel following the death of Lincoln Steffens, Sullivan declared himself "humiliated and mortified" to think that he had once been an ally of pro-Soviet personalities like herself. Only Langston Hughes and one other friend, Sullivan wrote, still made him pause in his anti-communism. Sullivan believed so deeply "in their purity of heart and goodness and kindness and intelligence that I could never dismiss as unworthy or unimportant the tenets of their faith." As a black, Langston had suffered from every evil of the capitalist system: "he was bait for the Communist trap and, at least partially, fell into it." Taking the long view, however, Langston "now recognizes himself as a creative artist and is no longer willing to be a propagandist."

Although Hughes almost certainly did not see the situation in such melodramatic terms, he knew that the time was wrong for radical socialist propaganda. "I am laying off of political poetry for a while," he wrote to a radical friend in the new year, "since the world situation, methinks, is too complicated for so simple an art. So I am going back (indeed have gone) to nature, Negroes, and love." Langston had retreated from radical socialism, but not for the reasons proposed by Noël Sullivan—although Langston perhaps encouraged Sullivan in thinking that these were his reasons. To a large extent, he gave up on radicalism not on ideological grounds, but as an impractical involvement that endangered his career as a writer. Radicalism paid very poorly in America; it also tended to estrange him from the black masses. Accordingly, he had been returning the needle of his conscience to its oldest and deepest groove, that of race. But instead of attempting to explain or justify this realignment, Hughes had done everything he could to conceal it. Leftists still thought of him as being of the left; to friends like Sullivan, however, he could point to his renewed emphasis on race as proof of his distance from communism, and pass off as deep alienation what was in fact more of a pragmatic withdrawal. Hughes might have done better to follow the lead of, say, Diego Rivera and André Breton in their "Manifesto: Towards a Free Revolutionary Art" published in the fall of 1938 in *Partisan Review,* where the writers boldly staked out a revolutionary position equally contemptuous of Germany and the "reactionary police-patrol spirit represented by Joseph Stalin." But Langston had no intention of attacking the Soviet Union; his belief in radical socialism had become clandestine but remained strong. The basis of his faith remained the progressive treatment of the darker minorites in the Soviet Union, as compared to overt racism and segregation in the U.S.A.

Finishing his autobiography, on October 30 he sent off to Blanche Knopf and Carl Van Vechten the first draft, about 450 pages, of "The Big Sea: The Saga of A Negro Poet." Within a few days he had their responses. Although Van Vechten found the quoted "Advertisement for Opening of the Waldorf-Astoria" to be "bad economics and bad poetry," and meticulously offered a laundry list of factual errors, his enthusiasm stood "at white heat." Blanche Knopf, however, had serious objections. She called the manuscript "a fine peformance," but one that "badly needs certain revisions." The Harlem Renaissance section was tedious—"too full of Carl, Thurman, Toomer" and so on, until it became "a sort of Harlem Cavalcade, which is wrong." But her objections were met head-on by Van Vechten, even if he made no effort to hide one of his motives from Langston. "Please assure me," he begged, "you aren't taking any of ME out of *The Big Sea.*"

Blanche Knopf's reaction dismayed Hughes. The differences between a black author and his white publisher never seemed more grating. Now more than ever, on the other hand, he was unwilling to fight her; he wanted this book badly. Eager to please her and also Van Vechten (without fully understanding the latter's position) he quickly pulled fifty pages from the Harlem section, including the Waldorf-Astoria poem. But first he defended the details of Harlem life as "the background against which I moved and developed as a writer, and from which much of the material of my stories and poems came." He had cut most of the abstractions, he said, but perhaps "some small portions of it may have vital meaning only to my own people." These portions would add "to the final integrity and truth of the work" as a whole: "I am trying to write a truthful and honest book." Black readers "would certainly give me a razzing if I wrote only about sailors, Paris night clubs, etc., and didn't put something 'cultural' in the book—although the material is not there for that reason, but because it was a part of my life. And the fault lies in the writing if I have not made it live."

Largely at the insistence of Van Vechten, who had read the material as a Harlem Renaissance intimate, almost all the cuts were restored. Langston was "the last historian of that period who knows anything about it," Van Vechten declared in urging Langston to expand rather than trim the section. And much as he detested the Waldorf-Astoria poem, Van Vechten was all for publishing it, since it was part of Hughes's essential history. The diminishing subtitle, "The Saga of a Negro Poet," was abandoned.

The cause of Hughes's misunderstanding with Blanche Knopf may indeed have been in his writing. *The Big Sea* is a study in formal sleight of hand, in which deeper meaning is deliberately concealed within a seemingly disingenuous, apparently transparent or even shallow narrative. Its very opening, when Langston heaves his books overboard at the start of his voyage to Africa in 1923, is ambiguous; this radical gesture implies either adolescent, anti-intellectual rebellion or the achievement of a Prospero-like wisdom. Or perhaps something else. As with so many of his important utterances, the opening had been conceived on a train, in the hour or so after Arna Bontemps, his travelling companion, had stressed to Langston the urgent need to write this book. In his demand for "a real titan's book," Bontemps had inspired a truly Titanic opening. The event itself was not contrived for *The Big Sea;* Hughes first wrote about his act of book-drowning only two years after his departure for Africa. But Langston deleted one highly significant fact from the final draft of *The Big Sea.* He threw all his books overboard—except for his copy of Whitman's *Leaves of Grass.*

In a genre defined by confession, Hughes appears to give nothing away of a personal nature. But if only three admissions in *The Big Sea* qualify as personal exposé, these are fundamental to the integrity of the text. The first concerns the religious revival in his childhood, when Langston had waited and waited—in vain—for Jesus; Hughes takes pains to dramatize not only his disappointment and despair but also his grievous lie (that Jesus *had* come) to the congregation and his subsequent guilty weeping into the night. The second admission concerns Langston's feeling for his father: "I hated my father." The third, near the end of the book, tells of his wrenching break with Godmother, Charlotte Mason. Deliberately, Hughes links the last episode to the others: "The light went out with a sudden crash in the dark, and everything became like that night in Kansas when I had failed to see Jesus and had lied about it afterwards. Or that morning in Mexico when I suddenly hated my father." All three moments unveil Hughes's extreme fear of abandonment; the second, however, involving his father, allows us to see his courageous response to this terror. Langston's hatred of his father is juxtaposed against two other cardinal hatreds: "My father hated Negroes. I think he hated himself, too, for being a Negro."

The crucial related episode of the book, apparently casually but very carefully set by Hughes, establishes these hatreds in a consequential way, when father and son collide at the dramatic intersection of race, art, power, and the future. "We were in a wild and lonesome-looking country as the shadows grew long in the late afternoon, and the mountains hid the sun," Langston wrote about a ride during his year-long stay in Mexico, where James Hughes lived. "My father's ranch seemed to take in a whole mountain side and on over the rim beyond that. Little fires were glowing on his mountain, as we rode upward in the dusk toward a cluster of peasant huts." Previously, at a high pass called Las Cruces, or the crosses, they had passed the swaying bodies of three barefooted Indian men hanged for banditry. Against this chiaroscuro of isolation and poverty, of mountain chill and death and Indian peonage, the two men rest. The next day, they debate Langston's future. " 'A writer?' my father said. 'A writer? Do they make any money?' " Langston must go away to Europe or South America, and not "stay in the States, where you have to live like a nigger with niggers." With his father's expansive promise of untold wealth, but in a future far from other blacks and the United States, comes a strange, almost magical brightening of the natural world. Father and son ride together slowly, "with the distant rim of the mountains all around and the sky very blue." But young Langston is neither intimidated nor seduced. Dreaming of poetry and the black race, he declines his father's plan. "My father shut up. I shut up. Our horses went on down the mountain into the blue shadows."

Having led the young black Christ-poet up to a high place, Satan has threatened him with poverty and ignominy, then offered him the world in exchange for his soul. But Langston remains true to his God, which is poetry and the black folk, two elements seen by him as virtually one and the same. His admission that he hated his father establishes for the black reader the depth of his will to be a writer and his undying love of the black race. What his father detests, the son loves; thus Langston Hughes almost subliminally whispers to his black readers, his ultimate audience in spite of the realities of the marketplace and the intentions of Knopf, that his life is completely devoted to their own. For them, he had "killed" his own father. And killed his mother, too, who wished him to work and support her, instead of singing the black race at the price of poverty. In *The Big Sea,* Carrie Clark is praised a little but more often (although gently) ridiculed and dismissed.

This extraordinary devotion to the race is quietly juxtaposed by Hughes against the central paradox of his life, announced after his discussion, in the first chapter, of his rather cool reception by blacks in Africa in 1923. "You see, unfortunately," he confesses in the first line of his next chapter, "I am not black." On the surface, this statement refers simply to a biological fact and to racist custom in the United States, where "the word 'Negro' is used to mean anyone who has *any* Negro blood at all in his veins. In Africa, the word is more pure." Hughes probably also means to underscore, however, two other points: the extraordinary nature of his devotion to the African race, given the variety of blood in his veins; and, more importantly, the lesson that, for a Negro in America, racial pride cannot be a cloistered virtue but one that constantly must be tested and intensified by sacrifice and effort. Whatever his hopes in making these points, however, Hughes unintentionally also allows sensitive readers a clear glimpse of his vulnerability.

The Big Sea then unfolds as if written by a man completely at ease with himself. The book covers Hughes's life until 1931, or his departure for Haiti following his break with his wealthy patron Godmother (who is never identified, and who was still alive) and his controversy with Zora Neale Hurston over their play "Mule Bone." But apart from his somber words about his break with Godmother, no controversy, whether over race, class, sex, or personality, is allowed to ruffle the smiling surface of Langston's tale. His leftist opinions and involvements have vanished without a trace. The poem about the Waldorf-Astoria is presented in excerpts which criticize the rich without actually asserting any radical sentiment; Hughes carefully omits its last, most radical section ("Hail Mary, Mother of God! / The new Christ child of the Revolution's about to be born"). He recalls the poem as his innocent reaction to the oncoming Depression ("The thought of it made me feel bad, so I wrote this poem"), and—erroneously—as a cause of his break with Godmother. While one love affair, with Anne Coussey in Paris in 1924, is orchestrated into a much more dramatic episode than it was in reality, it is also finally sacrificed to Langston's life mission and his greatest love, poetry and the black race.

To a newspaper editor, Hughes described the book as written with the "pace and incident value of a series of short stories. It is not 'literary' in the high-brow sense." And yet the text remains curiously poised. The balancing act is forced on Hughes by the double audience that almost any black writer bearing a racial message but dependent on a white publisher has to keep in mind. The smiling poise of *The Big Sea* is, in fact, the poise of the blues, where laughter, art, and the will to survive triumph at last over personal suffering. *The Big Sea* replicates the classic mode of black heroism as exemplified in the blues; it spurns the more violent, cynical, embittered mode, one which Langston, for all his racial radicalism, had never embraced. He wrote very gently, though not unmovingly, about his devastating break with Godmother, and only humorously about Zora Neale Hurston ("a very gay and lively girl") and their controversy. Again, Hughes depends on his black readers to understand at once how the style and substance of his book, and his life, cohere with the style and substance of their lives, as they laughed to keep from crying and, in so doing, endured and even prevailed. The variety of audience and the unusual model of heroism make the depth of *The Big Sea* truly puzzling to both black and white readers. Because it is firmly grounded, however, the narrative eventually succeeds. The powerful ability of the text to convince its readers derives most from its astonishingly simple, water-clear prose, which certifies the integrity of Hughes's narrative. His sentences are utterly devoid of linguistic affectation, as pure and compelling as an innocent child's might be; Hughes wrote a quintessentially American prose. As for the extended section on the Harlem Renaissance, "when the Negro was in vogue," Van Vechten was correct; it would never be surpassed as an original source of insight and information on the age.

With the autobiography virtually set, Langston turned to other projects. He sold two poems to *The New Yorker,* "Hey-Hey Blues" (which appeared in late November) and "Sunday Morning Prophecy." He also sent a revised libretto of *Troubled Island* to William Grant Still, with whom he had worked intermittently on his various trips to Los Angeles. Hughes also pressed on with his García Lorca translations. Typing and binding three private booklets of poems from "Gypsy Ballads," he dedicated one to his old Carmel friend Marie Short, whose mixture of personal sadness and obvious affection for him touched Langston, and who had recently moved to Carmel following a divorce. A fund-raising program also kept him busy. In Cleveland, fire had gutted the Karamu theater after a long-envious black newspaper had suggested that the Jelliffes' playhouse needed to be burned down. Appalled, Hughes wrote to a number of his wealthier friends soliciting money on behalf of Karamu, and attacking "the intolerance of many whites and the bigotry and ignorance of many Negroes in the Cleveland community."

In need of cash himself, Langston also began to plan a reading tour for the new year, 1940. But more immediately he needed a vacation, and spent several days in San Francisco. He visited the Golden Gate International Exposition, where Count Basie's orchestra was a sensation; at the city opera house, Hughes heard Kirsten Flagstad in *Tristan und Isolde.* Inside the landmark Coit Tower he proudly inspected his name and Van Vechten's inscribed among others on a mural. He visited Matt and Evelyn "Nebby" Crawford and their little daughter "Nebby Lou," "the prettiest child in the world," nicknamed after her mother and Louise Thompson; and Roy Blackburn, his former secretary at ''Ennesfree'' in Carmel, now a married man. Then Hughes returned for the Christmas holidays to peaceful Hollow Hills Farm, where the residents hailed the birth of four white lambs and twin black goats on Christmas Eve as a sign from heaven. Under a huge tree in the main house were many gifts for him, including one from Noël Sullivan that was perfect for a restless traveller—an overcoat with a zippered lining. Almost all of Langston's gifts were pieces of clothing; his needs were obvious to his friends.

With *The Big Sea* set for publication, Hughes determined that his next book would be as a lyric and racial poet, an identity generally submerged in the previous decade in favor of fiction and the theater, or altered in the direction of Marxist verse. ''I am about to go back to one of my earlier art forms,'' he wrote Lieber, ''having flirted so passionately (but to no purpose) with the theatre. That hussy!'' He took yet another step into his past with his choice of poems. Picking through his uncollected pieces, he set aside his radical socialist verse, which Knopf would not have published in any case. Culling about two dozen blues pieces, he added almost a dozen more dialect or ballad poems, including his brilliant ''Death in Harlem'' about Arabella Johnson and the Texas Kid—''one of the best poems I ever wrote.'' He then capped them with a sequence of seven poems called ''Seven Moments of Love: An Un-Sonnet Sequence in Blues,'' in which he had tried to merge the blues with a vaguely sonnet-like form. He foresaw ''a light and amusing book'' just right for an America haunted by fears of war. With the right illustrations, it could make ''a marvellous and grand HEY HEY cullud and colorful book'' that would be ''lots of fun. And ought to sell, too!''

Hughes's judgment of America and the publishing industry seemed on the mark. *Esquire* bought ''Seven Moments of Love'' for its May, 1940, number for just over $100 (the most ever paid for one of his poems), and *Poetry* accepted two blues poems for its April issue. Robinson Jeffers admired the collection, and Van Vechten was ecstatic. But once again Blanche Knopf was cool to his efforts; she postponed a decision on whether to publish the book. She further disappointed Langston by declining to bring out a cheap edition of *Not Without Laughter,* which he wanted for his coming tour, or to provide advertising leaflets to give away on the tour. To these rejections he could say little. After twenty years in the magazines, and fourteen years with the firm of Alfred A. Knopf, he had still not brought himself or his publishers anything like financial success. He was a poor man's literary Titan, without a best-seller.

As in his return to poetry in general and the blues in particular, and his avoidance of radical writing, Hughes signalled in other ways his backward turn in life. Once again he published in the *Crisis,* thus ending the estrangement that had begun with the Scottsboro controversy, and also in *Opportunity,* the journal of the Urban League. In these magazines he published mostly ballads. Hughes showed he had not lost the ability to tap the reservoir of hurt, resentment, and determination in the race, as in his quickly popular ''Note on Commercial Theatre,'' about the often shameless plundering of black music and culture by whites:

You've taken my blues and gone—
You sing 'em on Broadway
And you sing 'em in Hollywood Bowl,
And you mixed 'em up with symphonies
And you fixed 'em
So they don't sound like me.
Yep, you done taken my blues and gone.
. . . .
But someday somebody'll
Stand up and talk about me,
And write about me—
Black and beautiful—
And sing about me,
And put on plays about me!
I reckon it'll be
Me myself!

Yes, it'll be me.

Reluctantly on February 10, just after his thirty-eighth birthday, Langston set out for the East. Ironically, his poetry tour started in Downingtown, Pennsylvania, exactly where his epic 1931 to 1932 journey through the South had begun. Then, backed by the Rosenwald Fund but fired by radical zeal, he had taken poetry heroically to the people. This time, although he was sometimes sponsored by progressive groups such as the International Workers Order, Hughes's main interest was in eking out a living. He read his way south from Downingtown, then turned eastward and north to New York, where he appeared on February 28 at the influential Young Men's Hebrew Association on Lexington Avenue at 92nd Street in Manhattan, brought there by his friend Norman Macleod, the founder of its new Poetry Center. (The previous year, 1939, Macleod had published a novel, *You Get What You Ask For,* in which Langston briefly appeared, thinly disguised as a radical writer named "Larry," just back from the Soviet Union and Japan around 1935, and living first in Carmel, then in New York.) Near mid-March, Hughes spoke to excellent crowds in Lexington and Frankfort, Kentucky; on March 17, over 600 persons turned out for his reading sponsored by the NAACP in Columbus, Ohio. In Dayton two days later to speak to the leftist League for Progressive Action, he eulogized Otto Reeves, a local black youth who had been killed in Spain. He lectured at black Wilberforce University, and in nearby Yellow Springs at the liberal white Antioch College, then went down to Nashville and Chattanooga, Tennessee, the most southern points of the tour. The tour proceeded as planned, except for a quick trip to New York when lawyers at Knopf warned that his portrait of Jean Toomer, especially its mockery of Toomer's flight from the race, was possibly libelous.

Hating the Jim Crow bus above all, he endured the more spacious Jim Crow railroad car, and longed for the luxury of touring in his own automobile as he had in 1931-32. Most of all, he longed to be back in Carmel Valley: "Your farm, Noel, is a little heaven,'' he had written Sullivan. Hardly an improvement compared to his previous tour, his average fee was around $35, his highest probably $95; his profit would be small. In most places, Hughes simply spun the old scratchy Gramophone disc of his program; he read his poems (more than once Hughes declared that he did not like them enough to memorize them) interspersed with uplifting, gently humorous, only occasionally barbed commentary. He was well aware that this atmosphere of mildness and decorum contrasted sharply with the news from Europe, especially France, where Nazi stormtroopers and the Luftwaffe were inflicting terrible losses on opposing forces and hapless civilian communities. ''If Paris is taken I don't know how to stand it,'' Langston wrote Sullivan in May. Doubtless he had inflated his tone to match his patron's solemnity, but Hughes's feeling was also genuine. When France fell, he agonized about the black writers trapped by the Nazis. ''What has become, what will become of them?'' he wondered. ''The barbarism of the whole thing is more then I can ascribe to human intelligence.''

By March 29, he was sick of the road when the toughest part of his schedule began. An eleven-day tour of West Virginia institutions sponsored by the state supervisor of Negro schools led Hughes dizzily through steel and coal towns in ''a kind of continual going round the mountain, mighty interesting but tiring, dusty, and dirty.'' In the Louisville area, dozens of long-lost relatives on his father's side of the family vied with each other to entertain him, with hot toddies for breakfast and so much food that he feared he would soon resemble ''Fats'' Waller. Near the end of April he reached Cleveland. Then, at long last, the tour ended with a lecture at the University of Chicago on May 8, and another at the inter-racial forum of the Plymouth Congregational Church in Detroit four days later, after which he returned to Chicago and the restful company of Arna and Alberta Bontemps.

As he toiled from town to town, Langston could reflect sorrowfully on how little his financial state had improved since 1931, and how much he still struggled merely to meet his basic expenses. A Titan should not have to go on tour. He had a particular reason now to dwell on this subject of money. On a train north to Cornell University in Ithaca, New York, he finally finished reading a novel he had carried around for some time, Richard Wright's *Native Son.* The previous November, Arna Bontemps had sent the extraordinary news, almost impossible to believe, that the Book of the Month Club had selected the yet unpublished *Native Son* as its offering for January, 1940. As Bontemps wanly observed, Wright's financial problems were solved forever. Langston must have felt at least a twinge of pure envy. After twenty years of publication, he was still poor; with only his second book, Wright now bathed in a shower of gold.

Hughes liked Richard Wright. Their professional relationship was excellent, as when Langston proposed to dramatize a Wright short story for the Suitcase Theatre, or Wright approached Hughes about the magazine *New Challenge,* a revival of Dorothy West's *Challenge,* which Wright had hoped to develop into a significant journal; the men had even collaborated on a song, ''Red Clay Blues,'' published in *New Masses* the previous August. Although some other blacks found Wright ruthlessly self-serving, to Langston the young writer from Chicago out of Mississippi was both charming and highly talented. In any event, the power of *Native Son* could not be denied. "It is a tremendous performance!" Hughes congratulated Wright. "A really great book which sets a new standard for Negro writers from now on. Congratulations and my very best wishes for a great critical and sales success." Langston was doing whatever he could to help Wright: "I've been ballyhooing the book at all my lectures."

Calling *Native Son* a "tremendous performance" did not mean, however, that Hughes liked the book. He did not. And the fact that the novel set new standards for black writers must have brought almost as much pain as pleasure to the most prolific of Negro authors. The publication of *Native Son* was, moreover, a disheartening challenge to Hughes in that he found the black world it described both familiar and utterly repugnant. The raw, phallic realism and naturalism of Wright's novel, the unrelieved sordidness of his depiction of black life, repelled Langston. Probably the greatest wonder for him, however, was that *Native Son* had succeeded financially and critically in spite of the fact that it bared Wright's almost unrelieved distaste for blacks, on one hand, and his evident love-hatred of whites, on the other. Hughes found almost beyond comprehension the notion that this distaste and love-hatred would yield so compellingly to fictive treatment, and that white America would then embrace such a book and so generously reward its aggressive author. And Langston could hardly forget that he himself had scouted the same South Side of Chicago for his own still-unwritten second novel, intended as a continuation of *Not Without Laughter*.

In three weeks, the story of Bigger Thomas, black rapist and murderer, sold over 200,000 copies, a number far greater than all of Hughes's sales in almost two decades of publishing. Unflattering to Langston, furthermore, Knopf decided to attach *The Big Sea* to the coattails of *Native Son,* hustling for an August publication "to get some benefit out of all that publicity." But whatever Langston's reservations, the brilliant novel opened his eyes. "I keep looking for Bigger running over the roof tops," he wrote to Van Vechten from Chicago. "See plenty of his brothers in the streets." His chance to speak publicly on the novel came on April 28, at a tea sponsored by the Chicago Public Library for hundreds of its employees. Declining, according to one newspaper, to appraise the book as literature, he instead stressed its social import. Bigger Thomas was an exception rather than the rule, he argued, in the reaction by blacks to their suffering; blacks as a whole disapproved of the book because they feared to be identified with a murderer. If Hughes's remarks were accurately reported, they were limp, narrow, and really deprecating, and unworthy of the man who had once excoriated the taste of the black bourgeoisie. Powerfully prophetic, *Native Son* was too divisive, too harsh in its judgment of black and white alike for him to accept Wright's triumph easily.

"The hosannas are all for Dick Wright," Arna Bontemps had comforted him, "from now till *The Big Sea* comes out." Hearing about Hughes's speech, Bontemps sent further encouragement: "Folks think about same concerning Dick's book up here." While he was accurate in his reporting and also loyal to Hughes, Bontemps himself, a far more liberal judge of writing than Langston, had the highest regard for Wright and *Native Son.* Building on their common Seventh Day Adventist background, and as serious readers in a community hardly devoted to books, Wright and Bontemps had developed a strong affection and respect for one another in Chicago. Disdainful of envy and somewhat more objective about their race than Langston, Bontemps had no difficulty facing up to the fact, as a black novelist, that Wright had revolutionized the fictional presentation of Afro-American life with *Native Son.* At the same time, however, Bontemps lost none of his respect for Hughes, especially as a poet. ''According to my father more than once,'' one of Bontemps's children later recalled, ''almost from the start of their relationship he had recognized in Langston an original, authentically American artist, a grand creative figure like Mark Twain or Walt Whitman, in spite of his flaws as a writer, and thus essentially beyond criticism. No new rival, no matter how accomplished, could diminish Langston's standing in the national literature. My father admired other black writers, but he never lost that lofty view of Langston.'' For one very simple reason, Bontemps had wanted *The Big Sea* to be ''a real titan's book''; he saw Hughes, with all his shortcomings as a man of literature, as nevertheless a figure of epic proportion.

Hughes's situation, however, left him little opportunity for epic gestures. While he nervously awaited his share of the hosannas, he snapped at any offer of work that came his way. Jobs with the radio networks were so rare for blacks that he jumped at a request from the Columbia Broadcasting System for scripts for a dramatic series, ''The Pursuit of Happiness.'' But CBS rejected the first two scripts he offered. His musical play with James P. Johnson, *The Organizer,* was ''too controversial for us to give it an emotional treatment on an essentially dramatic show.'' (The play had been performed successfully at a convention of the progressive International Ladies Garment Workers Union the previous year.) But a script on the accommodationist educator Booker T. Washington was just right for airing on April 7, the day the U.S. Post Office released a stamp bearing Washington's likeness—the first time any black American had been so recognized.

In this crucial period of assessment and reassessment of his life, Langston drew more and more on his relationship with Arna Bontemps, whose career as a writer had itself progressed steadily. With a children's book written with Hughes and two novels, *God Sends Sunday* and *Black Thunder,* behind him, in 1938 Bontemps had published another ''juvenile,'' *Sad-Faced Boy,* then brought out his third novel, *Drums at Dusk*, the following year. After starting graduate work in English at the University of Chicago, he had won a Rosenwald Fellowship to write and travel in the Caribbean. Along with other writers such as Jack Conroy and Nelson Algren, he had become a recognized part of the younger literary set in Chicago; Bontemps terminated his job with Shiloh Academy and joined Conroy in writing a WPA volume on blacks in Chicago. By this time, his friendship with Langston had become tightly woven by a shuttle of letters, generally circumspect and to the point, but enlivened by an effervescent good humor, that flitted back and forth between them with scarcely a pause. For all their levity, the letters between the men show the high value they both placed on restraint, order, and discipline.

Unquestionably the friendship deepened as Hughes's sense of self as a writer became more materialistic and pragmatic, as his original myth-making power and his radical socialist force became less exuberant. At the base of the relationship was probably Langston's need for a confidant, a shelter on which he could depend in his wanderings, and Bontemps's willingness to fulfill that need in return for a link to someone he saw as both good and great—a spiritual brother and, at the same time, the best of black writers. The cement was their careful integrity in dealing with each other and the world. They trusted one another implicitly; until they died, nothing strained their fraternal relationship. Slowly through the thirties, then accelerating in importance as the decade ended, Arna Bontemps's voice became the sound most trusted by Langston. In some senses, Bontemps was an island Hughes had passed in his wandering, then circled, and finally come back to in a gathering storm.

By 1940, the two men had become, as Alberta Bontemps would remember them, "close, closer than blood brothers." Blood brothers fought or drifted apart; these two men stayed with one another not physically but in their thoughts, expressed in letters. "He was always wherever we were," Alberta Bontemps would recall of Hughes in spite of the infrequency of his visits. "I liked him. I really liked him. I was a little jealous at first because he consumed so much of Arna's time, but that feeling passed. They were different in some ways. You couldn't put either on the best dressed list, but while Langston was only careless, Arna dressed like a gangster; he wore black shirts and white or yellow ties. You see, they were handsome and knew it; they didn't need anything. Langston drank; Arna hardly touched liquor. Langston read very little. Arna read everything, except detective stories. He would drive me crazy—every time he passed a newsstand he picked up a different paper. He always had to have what was current in books and magazines. I used to jump on him about that, when we needed so many things. As for Langston, I think he picked up most of his ideas out of the gutter, if you know what I mean. He had the touch. He could dine with kings and with common people. Arna was not so flexible. But otherwise they were very much alike. And with all the hard work, Langston was considerate. He was thoughtful about other people. He would always laugh and tell me: 'Alberta, when we come into the money, we are going to buy you a tiara!' "

In Bontemps's children, Hughes clearly vested much of his paternal feeling; to Joan, Paul, Poppy, Camille, and Constance, he was a warm, bubbling presence easily drawn into comradeship. "I have no early memory that doesn't include Langston," Paul Bontemps (Arna's second child) would recall. "His manner with children was wonderful. We loved him. He never patronized us, never descended to baby talk or pats on the head. And he always brought gifts, no matter how small, whenever he visited." Hughes was like Paul's father in "the way he was calm and in control of his emotions. Langston's largest emotion was laughter, or humor. I do not remember him ever being irritable, much less shouting."

Early in 1940, Bontemps accepted the post of cultural director of the state-sponsored American Negro Exposition in Chicago, set to run later that year, from July 4 to Labor Day, in conjunction with the Diamond Jubilee of black emancipation. At once he looked for a place for Hughes, who was paid $140 to confer in New York with Exposition officials about entertainment to be staged in conjunction with the fair. Langston then agreed to write the book for "Jubilee: A Cavalvade of the Negro Theatre," a compendium of scenes from celebrated musicals such as "Shuffle Along." Hughes also was to prepare a smaller, original revue called "The Tropics After Dark." These assigments would keep him in Chicago for some time.

While the now famous Richard Wright (the "sepia Steinbeck") leisurely purchased a house in Chicago for his mother, then prepared to sail for Mexico with his white wife of several months, an elegant ballet dancer named Dhima Meadman (their best man had been young Ralph Ellison), Langston unpacked his belongings in a drab hotel room in Chicago. "The ladies of the race, I presume, are raising hell!" he speculated about the marriage. "Equality, where is thy sting?" As for the Exposition, he began work without a contract. Charmed by the smooth young black men running the fair, he had the promise of generous terms once the shows were launched. Among his co-workers, in addition to Bontemps, was the musician Margaret Bonds; Horace Cayton was director of research, assisted by the scholarly young sociologist St. Clair Drake; the poet Frank Marshall Davis took charge of publicity. The absence of a contract seemed quite unimportant, since Hughes's two projects were prominently featured in advertisements for the fair. Besides, the cause was noble: here were talented young blacks organizing to celebrate the achievement of the race in America. Near the end of March, his book for "The Tropics After Dark" was ready, with music by Margaret Bonds and Zilner Randolph. Then, with the help of W. C. Handy and other veteran black entertainers, the historical "Cavalcade" script was complete. Langston had written two songs for the shows, one set to music by the gospel composer Thomas Dorsey, author of "Precious Lord, Take My Hand," the other by Duke Ellington.

Neither show, however, went into rehearsal as planned. The state grant of $75,000, plus the promise of a matching sum by the federal government, attracted the more brazen elements of Chicago's political and criminal establishments, which were sometimes indistinguishable. One white politician called in his secretary, dictated the History of the Negro Race, and demanded that it be purchased. The carpenters' union insisted that sets built by non-union labor be torn down. The musicians' union set an extortionate scale of payment. For a while, independent financing for "The Tropics After Dark" proved elusive; then, after a skirmish over concession rights, a beer company offered $5000 to launch the show. But the grant was delayed while a fake corporation was established to launder the money. As tempers flared in the summer heat, Hughes began to wonder exactly when he would be paid. Merely to raise the subject of money at the Exposition, he wrote Maxim Lieber, one needed a gun in one hand, a sledgehammer in the other, and a knife between one's teeth.

On July 4, when the Exposition opened at the Chicago Coliseum, neither the

"Cavalcade" nor "The Tropics After Dark" was ready. The glories of black achievement proved inadequate as an attraction; without music and dance, the crowds quickly thinned. Hughes, who had warned against expecting people to come simply for information, was fatalistic. "Just how much culture can the masses take," he joked, "unless Little Egypt is around to shake?"

On July 12, the alarmed Exposition leaders stripped "The Tropics After Dark" of Hughes's dialogue and pushed it onto a stage at the beer hall. Rehearsals were a formality: the orchestra itself showed up only four hours before the first performance. The epic "Cavalcade" script remained unused as the Exposition daily fell deeper into debt. When Hughes made a final appeal for payment, the young black lawyer in charge of the Exposition, Truman K. Gibson, meticulously itemized the losses of the fair, then respectfully declined to sign the contracts for Langston's work. When Hughes threatened to sue, Arthur Spingarn sent the names of two eminent black Chicago lawyers, but advised against pressing the case. Disgusted, Hughes saw the entire affair as typical "cullud" mismanagement. The Diamond Jubilee of freedom? "We need twice 75 more years to be ready to celebrate this present period." After visiting the black lawyers named by Spingarn, he gave up all plans to sue. "I think a gangster would have more success collecting our money than a lawyer," he decided, "and would probably ask no higher per centage."

Chicago, he judged sourly, was "certainly one of the chief abodes of the Devil in the Western Hemisphere." His spring and summer, the ill-paying, exhausting tour and the fiasco of an Exposition, had been almost a total loss. One respite had come with a visit to Cleveland to mark both the twentieth anniversary of his high school class and the removal of Central High, now ninety-five percent black, to a new building on 40th Street. Another event of mixed appeal, however, was the appearance of the first book devoted exclusively to him and his work, *Langston Hughes: Un Chant Nouveau,* by the Haitian physician, social scientist, and amateur historian René Piquion, with an introduction by Arna Bontemps. Although Langston was flattered by the attention, Piquion's work was assailed by critics as an unhappy mixture of weak biography and Stalinist propaganda. Hughes had fared better the previous year when the excellent black scholar and critic J. Saunders Redding had praised him as "the most prolific and the most representative of the New Negroes," as well as the finest interpreter of the black masses in their distinctive racial feeling, in Redding's landmark study of black American verse, *To Make A Poet Black.*

In July, Richard Wright was back in town, estranged from his bride but hardly daunted, apparently; within a few months he would marry another bright woman, Ellen Poplar, also white, whom he had known in the party, and whom Bontemps would report as being "perhaps the best bet—if one is to judge by looks." Wright and Hughes were honored at a reception given by Jack Conroy and Nelson Algren for the launching of *New Anvil* (in which the first editorial promised to reveal the Richard Wrights of tomorrow; Langston Hughes, presumably, was passé). Acting more and more with the assurance of a celebrity, like a man who would never again live on the South Side or in any black community, Wright joined Hughes on a tour of the Exposition. Although Langston gave Wright no sign about his reservations concerning the book, he was not happy about the turn of events; a little too gleefully he reported to Noël Sullivan about *Native Son* that "most Negroes hate the book! And lots of left-wingers, too!"

Anxiously awaiting the first copies of *The Big Sea,* he was eager to match Wright's success with a best-seller of his own. At the end of July, when his ten free copies came ("my most handsome jacket so far!"), Langston threw himself into what even for him was a gigantic publicity effort. He stalked the booksellers, checking on their orders; to Knopf in New York went a long list of prospective buyers, including every person mentioned in the book who was still alive. Concerned with sales overseas, he pressed for a Spanish translation. Portions were sold to *Town and Country* and *Saturday Review of Literature,* and to the *Afro-American* group of newspapers. Blurbs were secured from Erskine Caldwell, Fannie Hurst, Granville Hicks, Janet Flanner, Paul Green, and Louis Adamic; he nudged Mrs. Knopf to try for one from the saintly Eleanor Roosevelt. Appearing on every radio show that would have him, Hughes cocked an ear for the hosannas Bontemps had predicted. The owner of the popular Argus Book Shop in Chicago anticipated "a tremendous sale because it's a terrific book." On August 22, days before the official publication of the book, Blanche Knopf sent him the "really fine" *New York Times* review set to appear the following Sunday. And *Newsweek* magazine called *The Big Sea* perhaps the most readable book of the year.

Still, *Native Son* loomed over Hughes and his autobiography, blocking the sun. "That boy is sure kicking up the dust," Bontemps had written about Wright's latest triumph; *Native Son* was to be made into a play, with Wright collaborating with Paul Green, and Orson Welles, hailed as a genius after *Citizen Kane* the previous year, directing. On September 6, at a dinner in New York, Langston and five thousand others heard Wright tell in great detail exactly how he had come to focus on the extraordinary character of Bigger Thomas. The occasion was a benefit for the Negro Playwrights Company, a group formed earlier in the year to further black drama (it would soon collapse, with a loss of $12,000) and including Hughes in an honorary position, Wright, Theodore "Ted" Ward, and Powell Lindsay.

Given its blend of disingenuousness and complexity, *The Big Sea* not surprisingly drew sometimes contradictory reviews. Alert, the *New York Times Book Review* saw it as "both sensitive and poised, candid and reticent, realistic and embittered." Bitterness, or the lack of bitterness, was a favorite topic with white reviewers. The New York *Sun* noted that Hughes "does not allow much bitterness to creep in"; a syndicated reviewer urged whites to read it "because it is true and honest and not bitter." Where some reviewers found honesty, others saw undue reticence. In the *Nation* Oswald Garrison Villard spoke of initial disappointment, but finally of Hughes's "absolute intellectual honesty and frankness." Alain Locke, however, regretted that Hughes had glossed over important matters (his thanks to Langston, no doubt, for passing over Locke's sexual aggression and his duplicity with Mrs. Mason). Ella Winter, who once had begged Hughes to omit her name from the title page of their radical play, found him "too gentle with us 'white folks.' " Ralph Ellison, by then a Marxist, complaining that "too much attention is apt to be given to the aesthetic aspects of experience at the expense of its deeper meanings," questioned whether the style was appropriate to "the autobiography of a Negro writer of Hughes's importance." Walt Carmon, with whom Hughes had lived for a while in Moscow, was shocked to notice "not a single mention of a radical publication you've written for or a single radical you have met or has meant anything to you." Several readers linked *Native Son* to *The Big Sea;* one called publicly for Wright to read Hughes's book before it was too late. In *The New Republic,* Wright himself, now suddenly the most prestigious voice in black writing, praised Hughes as "a cultural ambassador" and for having carried on "a manly tradition" in literature when other writers "have gone to sleep at their posts." But Wright liked the book much less than he let on. In calling Hughes "a cultural ambassador," he invited recollection of a passage in his most important essay, "Blueprint for Negro Writing." There, he had scorned past black writers in general as the "prim and decorous ambassadors," the "artistic ambassadors" of the race, "who went a-begging to white America . . . in the knee-pants of servility."

A few days after a reception given for him and Wright by the *New Anvil* group in Chicago, Langston left for Carmel Valley. Spending one day in Los Angeles, he conferred with officials of the progressive Hollywood Theatre Alliance about its plan for a socially conscious "Negro Revue," to include members of the New Negro Theatre, which Hughes had started there. The Alliance found Hughes, after his Chicago experience, a tough bargainer. Refusing to begin work without a signed contract, he declined to manage, direct, or raise funds for the project. He also declined to wait while the Alliance discussed his terms. Retaining Loren Miller as his local lawyer, he took the noon train to Monterey and Hollow Hills Farm.

To Langston's surprise and delight, he now had a mansion of his own in Noël Sullivan's "little heaven." On August 1, the Feast of the Transfiguration, a day considered significant for artists, the devout Sullivan had started construction of a one-room cottage for Hughes; a month later, Sullivan and a group of friends, including Robinson Jeffers, had christened the cottage with champagne. Langston was touched. Seven years before, he had fled from Carmel in confusion; now he felt almost at home in Carmel Valley. The autumn rains had not yet come, the summer hills were brown, the orchards in the valley dark green and heavy with peaches. Comfortable in his little house, he played with Greta, the aging German Shepherd who had kept him company in Carmel in 1933, and whose framed portrait he had carried about with him on his recent travels.

He revised his collection of poems, which Blanche Knopf had finally approved for publication after Van Vechten had sweetly ("Dear Grand Duchess") petitioned her. On November 2, Hughes sent her the manuscript, now called "Shakespeare in Harlem." He also requested a loan of $50 to pay a dentist to repair his teeth, which were in very bad shape. The money came, but Mrs.

Knopf turned down his request for illustrations by Covarrubias of *The Weary Blues* (another glance backwards for Langston), because Covarrubias was never on time with his work. She vetoed Hughes's suggestion of the black *Esquire* cartoonist E. Simms Campbell, the illustrator of *Popo and Fifina,* as inappropriate (perhaps too crude) for Langston's style; she dismissed samples of work by Zell Ingram, Hughes's old friend, as not good enough. She also balked at the title "Shakespeare in Harlem," about which Hughes himself had serious reservations, but Van Vechten rushed to praise it and the entire volume, which had been revised largely according to his advice. "The whole book sings," he assured Hughes, "with that kind of wistful loneliness you have made peculiarly your own."

By this time Langston was back in Los Angeles, staying at the Clark Hotel at the corner of South Central Avenue and Washington Boulevard, lured by an advance of $25 a week against royalties to help Donald Ogden Stewart, the president of the League of American Writers and the husband of Ella Winter, to prepare the "Negro Revue" for the Hollywood Theatre Alliance. Hughes's royalty as sketch "director" would be half of one per cent of the gross, with separate payment for original material.

In Los Angeles, he seized every opportunity that came his way to promote *The Big Sea.* His major chance came on November 15 at the expensive Vista del Arroyo Hotel in the wealthy suburb of Pasadena, when five hundred guests were expected for a "Book and Author" luncheon to be chaired by George Palmer Putnam, the retired publisher and the widower of the aviator Amelia Earhart. Hughes was at the hotel, no doubt anticipating gales of applause and the sale of at least one hundred copies, when a motley caravan of motor vehicles, including a sound-truck blaring a recording of "God Bless America," rumbled up to the elegant entrance. A hundred truculent members and supporters of the Four-Square Gospel Church and the Angelus Bible College had come to protest his appearance by picketing the luncheon. Behind this act was one of the more notorious of American evangelists, Aimee Semple McPherson, who had survived sundry lawsuits, public fights with her mother and her daughter, and a sex scandal involving a cottage in Carmel, to prevail as head of her influential fundamentalist religious group. Just before noon, she telephoned George Palmer Putnam to confirm her leadership of the picketers. McPherson had a specific reason to harass Hughes. She was one of the allegedly fraudulent ministers of religion mentioned by name in his "Goodbye Christ." The entire poem was reprinted on leaflets that denounced Hughes as a communist and challenged the guests: "ATTEND THE LUNCHEON CHRISTIANS . . . and eat, *if you can.*" Hughes and Putnam wanted to press on with the reading, but the manager of the hotel preferred to cancel the lunch rather than have his hotel picketed. The police tried to help by citing two protestors for traffic offences; the group, however, refused to be intimidated. Badly embarrassed, Hughes then offered to withdraw. The Chief of Police of Pasadena and George Palmer Putnam escorted him to an automobile for the ride back to Los Angeles.

That night, he wrote a careful account of the incident to Noël Sullivan, who may not have known about the poem. The next morning, to his relief, the *Los*

Angeles Times treated his mishap farcically; Langston Hughes was "Old Satan," wrestled to defeat in a "modern mat encounter" with some local evangelicals. Before a group of black clubwomen, Langston disparaged McPherson's act as a mere publicity stunt, and breezily announced that he would not press a lawsuit against her. But he was frightened by the most formidable personal opposition he had ever encountered, and by the exposure of "Goodbye Christ" after years of silence or inconsequential reprintings. McPherson's followers included some blacks, and certain black ministers had denounced the poem soon after its publication.

Hughes continued work on the revue but did so reluctantly; he longed to be back in his little cottage at Hollow Hills Farm. He was the only black writer on the project, his salary was small, and a long daily commute from the Clark Hotel was necessary because, as he wrote his agent, "in this charming democracy of ours there seems to be no place for Negroes to live in Hollywood even if they do work out there occasionally." And much as he admired fellow writer Charles Leonard, "a swell fellow" who had worked with Wallace Thurman on his Broadway play *Harlem,* and Donald Ogden Stewart, he found the Rodeo Drive radicals in the leftist Hollywood Theatre Alliance exasperating. One day the committee was militant, the next day frivolous, "so the love songs go back in and the lynching numbers come out." He also had trouble convincing "good and clever white writers" that simply transcribing material into black dialect did not necessarily make it genuine black material, or material appealing to blacks. To Maxim Lieber and his wife, Minna, he summed up his trials in dealing with the various Hollywood attitudes and factions: "What would you say is the ratio of *two plus* multiplied by *x-y* over *black z* divided by *liberal q* plus *left a* over the *necessary b* of the unified *zip* desired for a revue? I await your answer for classification."

He stayed in Los Angeles because he needed the money and because he hoped that two or three songs might "click." He hated the thought of going back on tour soon. In addition, songwriting was steadily becoming something of an addiction to him, like gambling, perhaps like the "numbers," where a small investment of time might bring a windfall of reward. Lieber had scolded him the previous year, when he had jumped at Clarence Muse's offer of a long-term contract, that he was more than a lyricist. But if his agent would help him to be a song writer just "one MORE time I will never bother you again," Langston later begged. He sent along an explanatory lyric: "The sweetest words a letter knows / Is *check enclosed.*" Two songs, each part of an elaborate skit, with music by the French-trained composer Elliot Carpenter, were promising. One was "Mad Scene From Woolworth's" or "Going Mad With A Dime," in which a woman with only ten cents cavorts against a typical dime store set. The other, even more important to Hughes, was a parody of "Old Black Joe" called "America's Young Black Joe," the climax of a skit about Joe Louis based on an idea by Charles Leonard:

> I'm America's YOUNG BLACK JOE.
> Most times good natured, smiling and gay

My sky is sometimes cloudy
But it won't stay that way.
I'm comin', I'm comin'—
But my head *ain't* bending low!
I'm walking proud! I'm speaking out loud!——
I'm America's YOUNG BLACK JOE!

On December 15, when the skit was staged at an NAACP gala, with film clips of black athletes and heroes projected behind the singer, "America's Young Black Joe" created such a stir that Hughes sped it off to Van Vechten, John Hammond of Columbia Records, and Lieber—the song had "a good chance of becoming a kind of Negro GOD BLESS AMERICA, expressing their patriotic and democratic sentiments." Thus Langston began a campaign to counter the charges of un-Americanism made against him, charges especially telling with the surge of patriotism that accompanied the growing expectation that the United States would soon be at war.

One skit that failed to mature showed Hughes's continuing fascination, perhaps obsession, with Richard Wright's success:

My name's Richard Wright and I try to write
Like Bigger Thomas talks.
As a matter of fact for the sake of this act,
I try to walk like Bigger walks,
Cause:
  I'm a boogie woogie man.
  I'm a native son with a fountain pen gun . . .

The singer is a "great big hairy . . . / Cash and carry, best-selling literary / Boogie, boogie woogie man."

Just after the NAACP gala, and in spite of the success of his patriotic song, Langston slipped out of Los Angeles. He went so quietly that two weeks passed before the Alliance leaders knew that he was gone. Although he was in effect breaking a contract, he was sick of the revue and the factions in the Alliance, tired of the long commute to Hollywood, and anxious to rest at the Farm. But he had barely unpacked in his *retiro* when the *Saturday Evening Post* of December 21 arrived to jolt him. Inside was a reprint of the Aimee McPherson handbill, including the dreaded poem. Like McPherson, the magazine was exacting revenge for having been named in "Goodbye Christ": the *Post* had earned its place in 1932 by curtly refusing to publish "Good Morning Revolution." Langston wrote Blanche Knopf to suggest a lawsuit against the *Post,* which he hoped might deter others from reprinting the poem. But the damage continued. Stanford University, the most fashionable school among the California rich, cancelled a planned appearance there.

The Christmas season, 1940, was haunted by the ghost of "Goodbye Christ" and Hughes's desire to repudiate this extreme evidence of his radical past. In a fateful step, he spent several days laboring on a statement about the poem which he sent to everyone who mattered, including Knopf, the Associated Negro Press, and the Rosenwald Fund, from whom he hoped to win a fellowship soon. Hughes capitulated to his critics. Characterizing himself as "having left the terrain of the 'radical at twenty' to approach the 'conservative at forty,' " he affirmed that he "would not and could not write" such a poem now, "desiring no longer to *épater le bourgeois.*" He asserted that the poem had been withdrawn from circulation (but he did not say how or when he had done so). In any case, it had been simply a satire on the exploiters of religion, ironically cast from a Communist point of view that had not been authentically his own. "Lord help me!" Langston sighed to Arna Bontemps about his statement. "Better show it to the Rosenwald folks so they'll be clarified, too."

The end of the year found him, he confessed, "broke and remorseful as usual," "broke and ruint." Langston was also sick. He told everyone who inquired that he felt vaguely arthritic and influenza-ridden, although he had no clear signs of the flu. Alone in his little cottage, he nursed a deepening sense of depression, of financial and perhaps even artistic failure in his life. He had cleared less than $1200 for the year, or so he reported for his income tax. His long, tiring spring tour had yielded little profit. His trust in the leadership of the Negro Exposition celebrating black emancipation had ended bitterly. "Boy of the Border," the Mexican children's book with Arna Bontemps, had been rejected repeatedly by publishers. And the story of his life, *The Big Sea,* had done poorly, given his high hopes and the grand success of *Native Son.* In spite of decent reviews and Bigger Thomas's bloody coattails, only 2845 copies had been sold so far in the United States. Apparently he was not as popular as he had believed, Hughes might well have thought. From Arna Bontemps he heard that sales of *Native Son* had declined abruptly, once people had discovered that it was not a detective story. But Langston was not much consoled by the news.

Although the Hollywood Theatre Alliance chided him about his sudden, unannounced departure ("a great loss to the project") and pressed for his return, Hughes made no plans to leave the farm. He roused himself on January 10 only to attend a dinner given by Noël Sullivan for Robinson Jeffers on the poet's fifty-fourth birthday. Poet to poet, Hughes praised Jeffers to the skies: "All of human terror and frailty, and all the mud of mankind's earth-rooted feet, and the sea-wind and sky-wind in men's hair, and their hands that would find a star, and the shock of rock here and star there . . . these things come to rest in Jeffers." Robinson Jeffers was so many things that Hughes was not. Settled in a house built virtually with his own hands, protected by a stalwart wife (to whom Langston had just addressed a poem) and handsome sons, Jeffers was a man consistent in his saturnine vision, tenaciously a tragic poet. In Langston's depression, the contrast must have seemed severe; he had described Jeffers in his tribute exactly as one might describe a real Titan.

His illness worsened. Once again, as in the greatest moments of anxiety in his life, he felt his body poisoned, his limbs locking, the bones, especially in his left knee, hurting terribly. To those who sympathized he talked still of sudden arthritis and of a coming cold that never finally came. On January 14, however, he asked Eulah Pharr to drive him to see a local physician, Dr. Russell

Williams. The doctor urged his immediate removal to the Peninsula Community Hospital in Carmel.

At half past four that afternoon, Hughes entered the institution as a patient. He was quickly diagnosed as suffering not from influenza or any similar ailment, as he knew well, but from gonorrhea.

By this time, the infection was in an advanced stage. He had picked up the venereal infection about five or six weeks previously, during the first week of December, toward the end of his stay in Los Angeles—under what conditions, it is impossible to say. The length of time before seeking expert treatment suggests strongly that it was his first venereal infection. A Wasserman test for syphilis proved negative. Langston was treated with sulfathiazole, the standard drug in the age before penicillin, but showed only slight response. A course of sulfapyridine brought no response and was stopped. A painful catheter through his penis, applied "with tenderness & caution!!" (as someone teased in his medical record) drained the infection and allowed a gradual clearing of the discharge. His diet of sulfa led to constant grogginess and even delirium, as well as a dramatic loss of appetite and severe constipation. Hughes's temperature remained high for several days, and he was also treated for sinusitis. His bones continued to hurt him, especially in his left leg—a condition he attributed to running down a hill a few weeks before, although the pain was thought to be a side-effect of his infection.

As he lay in the hospital suffering from this unspeakable illness (sciatica, he wrote one friend; arthritis, he assured another), with his medical bills piling up and with no idea of how he might pay them, a succession of lesser blows struck at Langston's crumbling self-esteem. Blanche Knopf advised him that her lawyers could do nothing about "Goodbye Christ," because the poem, published outside the country apparently with his consent, since he had not immediately repudiated it, was in the public domain. His grave doubts about the wisdom of the circulated statement deepened when the tough-minded Carl Van Vechten, who detested the poem, nevertheless questioned whether Hughes should have apologized at all. Perhaps worst of all for Langston, as a symbol of his headlong fall from grace, came the news that he had just been evicted as the tenant of the three-room apartment at 66 St. Nicholas Place in Harlem, after his most recent sub-tenants, including his "brother" Kit, had fallen $98 behind in the rent. Langston's possessions would be removed by Toy Harper.

"Last straw N.Y. eviction," he wrote Arna Bontemps briefly, gasping and in obvious pain. "Guys let their rent get behind—way behind—without my knowing it. And put out. Well, anyhow, here I am moaning and groaning. Fever is down, so can write with effort. . . . More news when able to write. Love to Alberta and kids. I'm weak and groggy."

A newspaper item that must have come to his attention sooner or later brought the chill of public contempt, and the accusation of cowardice and dishonor, to Langston's sick bed. On January 15, the communist *People's World* of San Francisco, long one of his supporters, heaped scorn on his statement about "Goodbye Christ" and his other recent efforts, such as the marching song "America's Young Black Joe," to put as much distance as possible between himself and his old views. ''Hughes has been bitten with the war bug,'' a popular book columnist named Ben Burns judged. ''Not only is he primping for ye imperialism but now has renounced all the sentiments expressed in his 'Goodbye Christ.' . . . Instead of defending the poem which appropriately said goodbye to Christ because of all the sins against Negroes committed in His name,'' Langston Hughes had timidly apologized for having written it. ''This Hughes is a long way from the Hughes who not so many years ago wrote:

> 'Put one more 'S' in the USA
> and make it Soviet
>
> 'Put one more 'S' in the USA,
> we'll live to see it yet.'

The time had come for the faithful to turn their backs on this traitor. ''So goodbye Huges [*sic*],'' the columnist jibed in the *People's World,* mocking the poet with his own reckless phrase from his days of radical glory in 1932 in Moscow. ''This is where you get off.''

ABBREVIATIONS

ABSP	Arthur B. Spingarn Papers
ALLP	Alain Leroy Locke Papers
APC	American Play Company Papers, Berg Collection
ARC	Amistad Research Center, New Orleans
Bancroft	Bancroft Library, University of California, Berkeley
Berg	Berg Collection, New York Public Library
CCP	Countee Porter Cullen Papers
CMP	Claude McKay Papers
CVVP, Yale	Carl Van Vechten Papers
CVVP, NYPL	Carl Van Vechten Papers, Manuscripts Division
FPGP, UNC	Frank Porter Graham Papers, Southern Historical Collection, University of North Carolina
JESP, NYPL	Joel Elias Spingarn Papers, Manuscripts Division
JTP, Fisk	Jean Toomer Papers, Fisk University
JWJ	James Weldon Johnson Collection, Beinecke Rare Book and Manuscript Library, Yale University
LHP	Langston Hughes Papers, James Weldon Johnson Collection, Beinecke. Comprising mainly Correspondence (General, Family, Fan Mail, Miscellaneous, etc.) and Manuscripts (numbered).
MSP	Marie Short Papers
MSRC	Moorland-Spingarn Research Center, Howard University
NSP	Noël Sullivan Papers
NYPL	New York Public Library, Astor, Lenox and Tilden Foundations
WEBDP, UM(A)	W.E.B. Du Bois Papers, University of Massachusetts at Amherst
WTP	Wallace Thurman Papers
WWP, LC	Walter White Papers, Library of Congress

NOTES

1. A Kansas Boyhood

For the years until Langston Hughes left Lawrence, Kansas, in 1915, the best source outside of his own writing is Mark Scott, "Langston Hughes of Kansas," *Kansas History: A Journal of the Central Plains* 3 (Spring, 1980): 3–25. An invaluable unpublished source is Paulette D. Sutton, "Langston in Lawrence," presented to Prof. William Tuttle for History 96 at Kansas University, April 27, 1972. This essay is based on taped interviews with childhood acquaintances of Hughes, as well as original research into Lawrence history.

In Hughes's work, see *The Big Sea* (New York: Alfred A. Knopf, 1940), pp. 11–26. For earlier drafts of *The Big Sea,* see the Vivian G. Harsh collection, Carter G. Woodson Regional Library, Chicago. Among very brief mss. and published essays, see LH, "L'histoire de ma vie" (1925), LHP 505; LH, "The Fascination of Cities," *Crisis* 31 (Jan. 1926): 138–140; LH, "The Childhood of Jimmy: Six Pictures in the Head of a Negro Boy," *Crisis* 34 (May 1927): 84; LH, "Just Travelling," Carmel *Pine Cone,* May 2, 1941 (original in the Marie Short Papers, Bancroft Library, University of California, Berkeley); LH, "A Sketch of My Life" (1943), LHP 3598; LH, "My Grandmother," in "Autobiographical Notes For Edwin R. Embree" (1943), LHP 82; and LH, "Memories of Christmas" (1946), LHP 671.

For general information on black life in Lawrence, especially at the turn of the century, see the archives of the Douglas County Historical Society in the Elizabeth M. Watkins Community Museum, Lawrence, Kansas. Particularly important is Kaethe Schick's extensive notes toward a dissertation study of blacks in Lawrence for the University of Kansas. See also Donald B. Zavelo, "The Black Entrepreneur in Lawrence, Kansas, 1900–1915," submitted to the Department of American Studies, University of Kansas, May 1, 1975. See also a folder of transcripts of taped interviews with several older black Lawrence residents, conducted in 1977 by Curtis Nether.

For an introduction to much of this material and to Lawrence history in general, I am indebted to Katie Armitage of Lawrence, Kansas.

page 3

"by the youngest good writer": LH to Marianne Moore, Dec. 26, 1966, LHP.

"Inimitable, irresistible Langston": Marianne Moore to LH, Dec. 30, 1966, LHP.

page 4

"I was pretty conscious": LH to Arthur P. Davis, Dec. 12, 1953, LHP.

"My grandmother raised me": LH, *The Big Sea,* pp. 13–14.

"I hated my father": LH, *The Big Sea,* p. 49.

"Some time I feel": Carrie M. Clark to LH, May 13 [1928?], LHP.

"When my grandmother died": LH, *The Big Sea,* p. 17.

"My theory is . . . children": LH to Richard Rive, Mar. 27, 1962, LHP.

page 5

"a small woman, brown": LH, "My Grandmother," in "Autobiographical Notes for Edwin R. Embree," July 7, 1943, LHP 82.

"She never shouted or got happy": LH, "My Grandmother."

"Through my grandmother's stories": LH, *The Big Sea,* p. 17.

"She had been away": LH, "My Grandmother."

"she is of most respectable": Deposition in Fayetteville, North Carolina, May 2, 1855 by B. W. Robinson and Joshua Carman concerning Mary S. Patterson; endorsed by Clerk of the Court, Cumberland County. Courtesy of George Houston Bass. Contrary to many published reports, Mary Sampson Patterson [Leary Langston] never attended Oberlin College.

page 6

"His soul is marching": Mary Leary Langston to Richard J. Hinton, Sept. 7, 189[9?]; Richard Hinton Collection, Kansas State Historical Society.

"I was tried by a jury": For the entire text, see Charles Howard Langston, *"SHOULD COLORED MEN BE SUBJECT TO THE PAINS AND PENALTIES OF THE FUGITIVE SLAVE LAW!" Speech before U.S. District Court for the Northern District of Ohio* (Cleveland, 1859).

page 7

"Capt. Brown was engaged": *Cleveland Plain Dealer,* Nov. 18, 1859.

"Gen. Nat Turner, the hero": *Ibid.* In "My Grandmother," LH asserts that Nathaniel Turner was the "only son" of Charles Langston; Desalines was "a foster child of my grand-mother's."

scorned him as a "saddle colored idiot": *Lawrence Standard,* Sept. 1, 1880.

page 8

"As the mountain eagle": *Report of the 22nd Annual Meeting of the American Anti-Slavery Society, May 9, 1855, Metropolitan Theatre, New York* (New York, 1855).

"My earliest thought of riches": LH, "My Grandmother."

page 9

"We are a childish race": Mary J. Dillard to LH, Nov. 3, 1929, LHP.

"The colored children have": *Lawrence Daily Journal,* Aug. 17, 1910.

she was elected "Critic": *The Historic Times* (Lawrence), Sept. 26, 1891.

"Deputy District Clerk" in the city courthouse: Census of Lawrence, Douglas County, Kansas, Mar. 1, 1895, Ward 1, Schedule 1, p. 16.

considered novels "cheap": LH, "My Grandmother."

hailed once as a "black Mecca": *New York Times,* Apr. 9, 1891.

page 10

"My father hated Negroes": LH, *The Big Sea,* p. 40.

the "Nigger Diggins": Dolph Shaner, *The Story of Joplin* (New York: Stratford House, 1948), p. 2.

"EXCEPTIONAL Opportunities for Prospecting": *Joplin Daily Globe,* Jan. 7, 1900.

page 11

"some of them never came back": Joel T. Livingston, *A History of Jasper County Missouri and Its People* (New York: Lewis Publishing Co., 1912), Vol. I, p. 501.

page 12

"long-branch kindling": LH, *The Big Sea,* p. 15.

the spitting image of "that devil": LH "Childhood Memories," *Chicago Defender,* Dec. 18, 1948.

"You're just like Jim Hughes": LH, "Notes for the 'Big Sea' about 1940," LHP 79.

"I am well": LH to Carrie F. Battle, n.d., LHP.

page 13

"You don't want to eat": Fragment in LH, "I Wonder as I Wander," LHP 519.

a "deep thinker": Mary J. Dillard to LH, Nov. 3, 1929, LHP.

page 14

"Your people can't come": LH, "I Wonder As I Wander," LHP 519.

kind but "old, old": LH, "My Grandmother."

began to believe "in nothing but books": LH, *The Big Sea,* p. 16.

"Catch it, Melvin. Catch it": Personal Papers, n.d., LHP 863.

"that taketh all things under wing": LH, *Dear Lovely Death* (Amenia, N.Y.: Troutbeck Press, 1931), n.p.

page 15

"Wilder and wilder I mugged": LH, *The Big Sea,* p. 25.

"She kept me neatly": LH, "My Grandmother."

"washed his overalls every Sunday morning": LH, *The Big Sea,* p. 18.

page 16

"For me, there have never been": LH, *The Big Sea,* p. 18.

"I got de weary blues": LH, *The Big Sea,* p. 215.

page 17

"Rock Chalk! Jay Hawk!": Cited incorrectly in LH, *The Big Sea,* p. 22.

"He was a very gentle": May Hampton to Curtis Nether, interview, June 6, 1977. Courtesy of the Douglas County Historical Society, Lawrence.

"a quiet, calm and collected person": Cited in Paulette D. Sutton, "Langston in Lawrence." Courtesy of Paulette D. Sutton.

"She kept him kind of under": Sutton, "Langston in Lawrence."

"Of course . . . that stirred up": Sutton, "Langston in Lawrence."

page 18

because Langston was a "nigger": LH, "The Childhood of Jimmy," *Crisis* 34 (May 1927): 84.

"THE BOTTOMS now I sing": *Lawrence Daily Journal,* Apr. 25, 1908.

page 19

"My earliest memories of written words": LH, Tribute to W. E. B. Du Bois, *Freedomways* (1st Quarter 1965): 11.

"fecal impaction": State of Kansas Standard Certificate of Death No. 223598: Mary S. Langston. Date of death: April 8, 1915. State Board of Health, Division of Vital Statistics.

page 20

"the mortgage man who always came": LH, "My Grandmother."

"a kind of strong, bronze cowboy": LH, *The Big Sea,* p. 36.

"It was dusk-dark": In the section "The Body" in "The Childhood of Jimmy," Hughes describes a repeated warning from "my aunt" to avoid the example of "Clarence," who lived next door and was arrested and forced to marry his pregnant girlfriend. Hughes also mentions a proposed tryst with a girl whose mother worked all day in "Mrs. Ronnermann's kitchen." "Jimmy" was Hughes's legal first name; according to the census records of the 4th ward, city of Lawrence, March 1, 1915, a 19-year-old boy named Clarence lived next door to the Reeds. The quotation is from a handwritten piece called "Sex Silly Season" (draft and notes for "a book about sex"), Mar. 25, 1963, LHP 3486.

page 21

"when you were saved": LH, *The Big Sea,* p. 19.

"The whole congregation prayed for me": LH, *The Big Sea,* p. 20.

"And the preacher said": LH, "The Childhood of Jimmy," p. 84.

"I cried, in bed alone": LH, *The Big Sea,* p. 21.

"I had waited for Him": LH, "The Big Sea," 1st draft, leaves 11–12; Vivian G. Harsh collection, Carter G. Woodson Regional Library, Chicago.

page 22

"Nobody loves a genius child": LH, "Genius Child," *Selected Poems* (New York: Alfred A. Knopf, 1959), p. 83. See also *Opportunity* 15 (Aug. 1937): 239.

2. Outsetting Bard

page 23

"when his only reputation": Raymond Dooley, ed., *The Namesake Town: A Centennial History of Lincoln, Illinois* (Lincoln, Illinois: Centennial Booklet Committee, 1953), p. 3.

he "never knew anything named Lincoln": Dooley, *The Namesake Town,* p. 11.

"Negro Confesses He Is Guilty": *Lincoln Evening Star,* Aug. 30, 1915.

page 24

"my classmates, knowing that a poem": LH, *The Big Sea,* p. 24.

"a coloured boy, yellow rather than black": Laura M. Armstrong to Frances Dyer, Jan. 13, 1952; copy in LHP.

"In the first half of the poem": LH, *The Big Sea,* p. 24.

"My life has been filled with a great envy": LH, "I Wonder As I Wander," LHP 519.

page 25

"a great dark tide": LH, *The Big Sea,* p. 27.

"As always, the white neighborhoods": LH, *The Big Sea,* p. 27.

"the gayest and bravest people possible": LH, *The Big Sea,* pp. 54–55.

page 26

"about law and order in life and art": LH, *The Big Sea,* p. 28.

"I was continually amazed": LH, *The Big Sea,* p. 32.

"I couldn't afford to eat": LH, *The Big Sea,* p. 33.

The family "had about them a quaint": LH, *The Big Sea,* p. 30.

page 27

"We don't 'low no niggers": LH, "The Fascination of Cities," LHP 363; published in *Crisis* 31 (Jan. 1926): 138–140.

He found Chicago "vast, ugly, brutal, monotonous": LH, "The Fascination of Cities."

"Angels of Mercy": LH, "The Red Cross Nurse," Central High School *Monthly* 19 (Feb. 1918): 13.

"Central's heart has a memory": LH, "Our Service Flag," *Monthly* 19 (Mar. 1918): 18.

"I loves to see de big white [moon]": LH, "My Loves," *Monthly* 19 (Apr. 1918): 13.

"When you have gone your way thru life": LH, "To Youth," *Monthly* 19 (Apr. 1918): 10. For a copy of the April 1918 issue of the Central High School *Monthly,"* I am indebted to Thomas H. Wirth of The *Fire!!* Press, Metuchen, N.J. Many numbers of the *Monthly* may be found in the archives of the Western Reserve Historical Society, Cleveland. I was unable to locate any copies of the *Monthly* for the school year 1919 to 1920, when Hughes was a senior.

page 28

"Well, ma . . . I never knew": LH, "Those Who Have No Turkey," *Monthly* 20 (Dec. 1918): 36.

"Very lonesome" Joe wants desperately: LH, "Seventy-Five Dollars," *Monthly* 20 (Jan. 1919): 22.

"Come with me to Little-Boy Land": LH, "Play-Toy Town," *Monthly* 20 (Jan. 1919): 12.

"Oh Central, . . . Great, grey Alma Mater": LH, "A Song of the Soul of Central," *Monthly* 20 (Jan. 1919): 9–10.

page 29

Sandburg became his "guiding star": LH, *The Big Sea,* p. 29.

"Carl Sandburg's poems fall": LH, *The Big Sea,* p. 29.

page 30

"Many of the metropolitan daily papers": LH, "Mary Winosky," LHP 666. This story was apparently never published, or perhaps was published in the school year 1919 to 1920, for which I have found no copies of the *Monthly*. Hughes evidently forgot about his two short stories published in the *Monthly* when he directly led James A. Emanuel to publish an article called "Langston Hughes' First Short Story: 'Mary Winosky,' " *Phylon* 22 (Fall 1961): 267–272. The essay in part attempts to refute a statement published in 1927—thereafter accepted as fact, until Emanuel's correction—that Hughes's first short story was written that year. Manuscripts of the two *Monthly* stories apparently do not exist, which perhaps led Hughes to forget that he had written them. Of

Hughes's three extant high school short stories, it seems likely that "Those Who Have No Turkey" was written first. It also appeared in *Brownies' Book* 2 (Nov. 1921): 324–326.

They "were almost all interested in more": LH, *The Big Sea,* p. 31.

"I learned from it the revolutionary attitude": LH, "Claude McKay: The Best" [1933], LHP 29.

"but none moves us deeply": Floyd Dell to LH, n.d., LHP.

page 31

"her face growing more and more belligerent": LH, "My Most Humiliating Jim Crow Experience," LHP 734; published in *Negro Digest* 4 (May 1945), pp. 33–34.

"You are to accompany me": LH, *The Big Sea,* p. 35.

page 32

"Are you Langston?": LH, *The Big Sea,* pp. 37–38.

"Look at the niggers": LH, *The Big Sea,* p. 41.

page 33

"Seventeen and you can't add": LH, *The Big Sea,* p. 45.

"Langston Hughes is crazy about 'eats' ": LH, ed., *Central High School Annual: 1920* (Cleveland, Ohio), n.p.

page 34

"I began to wish": LH, *The Big Sea,* p. 46.

"One day, when there was no one": LH, *The Big Sea,* p. 47.

"Hurry up!" he barked: LH, *The Big Sea,* p. 48. For subsequent quotations, see pp. 48–49.

page 35

"You're a nigger, ain't you?": LH, *The Big Sea,* p. 50.

"I knew I was home": LH, *The Big Sea,* p. 51.

page 36

"Just Because I Loves You": See LH, *The Big Sea,* pp. 28–29, for the texts of "Just Because I Loves You" and "The Mills."

page 37

"When Susanna Jones wears red": *The Big Sea,* pp. 52–53. "When Sue Wears Red" first appeared in *Crisis* 25 (Feb. 1923): 174, but in edited form—all references to Jesus were omitted.

"Langston was absolutely ecstatic": Rowena Woodham Jelliffe to author, interview, Dec. 7, 1980.

page 38

"You have a fine appearance": Charles E. Ozanne to LH, Aug. 16, 1921, LHP.

"In certain ways she had become": Rowena Woodham Jelliffe to author, interview.

page 39

making him "really want to be a writer": LH, *The Big Sea,* p. 34.

"I've known rivers": LH, "The Negro Speaks of Rivers," *Crisis* 22 (June 1921): 71.

pages 40 to 41

"nothing except stares of mild disapproval": LH, "A Diary of Mexican Adventures (If there be any)," July 20–23, 1920, LHP.

page 42

a plan "I had never dreamed of": LH, *The Big Sea,* p. 62.

page 43

"Aunt Sue has a head full of stories": LH, "Aunt Sue's Stories," *Crisis* 22 (July 1921): 121.

"a jingle in a broken tongue": Paul Laurence Dunbar, "The Poet," *Complete Poems of Paul Laurence Dunbar* (New York: Dodd, Mead, 1913), p. 191.

"Well, son, I'll tell you": LH, "Mother to Son," *Crisis* 25 (Dec. 1922): 87.

page 44

"I am a Negro": LH, "The Negro," *Crisis* 23 (Jan. 1922): 113.

"I am the nigger": Carl Sandburg, "Nigger," *Complete Poems of Carl Sandburg* (New York: Harcourt, Brace, Jovanovich, 1970), p. 23.

"the dust of dreams": LH, "Fairies," *Brownies' Book* 2 (Jan. 1921): 32.

"Flowers are happy in summer": LH, "Autumn Thought," *Brownies' Book* 2 (Nov. 1921): 307.

page 45

she found one poem "very charming": Jessie Redmon Fauset to LH, Oct. 10, 1920, LHP.

"The little house is sugar": LH, "Winter Sweetness," *Brownies' Book* 2 (Jan. 1921): 27.

"I think it has a very fine touch": Jessie Fauset to LH, Jan. 18, 1921, LHP.

page 46

"the moon had gone down": See LH, "Up to the Crater of an Old Volcano," *Brownies' Book* 2 (Dec. 1921): 334–338.

"In Cleveland . . . the boys I knew": "The Big Sea," 1st draft, leaf 65; Vivian G. Harsh collection, Carter G. Woodson Regional Library, Chicago. Hughes later deleted this observation.

"Langston . . . was like the legendary Virgin": Richard Bruce Nugent to author, interview, Aug. 16, 1981.

"dust and tobacco and animals": LH, *The Big Sea,* p. 71.

page 47

"they were beautiful things": LH to Carl Van Vechten, May 17, 1925, CVVP, Yale.

"is not a game": LH, *The Big Sea,* p. 71.

a "precious" poem: E. J. Mullen, "A Conversation with Carlos Pellicer," June 2, 1972; courtesy of Prof. Edward J. Mullen.

page 48

"I began to wish": LH, *The Big Sea,* p. 58.

"completely charmed" by it: Jessie Fauset to LH, Feb. 25, 1921, LHP.

"Ah!" she sighs, "I would sell all": LH, "The Gold Piece: A Play for Children," LHP 456; published in *Brownies' Book* 2 (July 1921): 191–194.

"very much pleased": Jessie Fauset to LH, July 11, 1921.

"How long did it take you": LH, *The Big Sea,* p. 72.

page 49

"Beast-strong / Idiot-brained": LH, "The South," *Crisis* 24 (June 1922): 72.

"Gracias a dios!": LH, *The Big Sea,* p. 79.

3. My People

page 50

"There is no thrill in all the world": LH, "The Fascination of Cities" (1925), LHP 363; in *Crisis* 31 (Jan. 1926): 138–140.

"I stood there, dropped my bags": LH, *The Big Sea,* p. 81.

"the greatest Negro city": *New York Age,* Jan. 10, 1920.

page 51

"better, cleaner, more modern": James Weldon Johnson, *Black Manhattan* (New York: Atheneum, 1977), p. 148.

"holding the handle of Manhattan": Claude McKay, *Harlem: Negro Metropolis* (New York: Harcourt Brace Jovanovich, 1968), p. 21.

"I was in love with Harlem": LH, "My Early Days In Harlem," *Freedomways* 3 (Summer 1963): 312.

"I had come to New York": LH, "My Early Days In Harlem."

page 52

"There were maybe a dozen of us": George N. Redd to author, telephone conversation, May 12, 1981.

"It hurt me to the soul": LH, "I Wonder As I Wander," n.d., LHP 520.

page 53

"See you tomorrow, Lang": LH, *The Big Sea,* p. 83.

"they assigned me to gather": LH, *The Big Sea,* p. 83.

"you have been so generous": Jessie Fauset to LH, n.d., LHP.

"What would I say?": *The Big Sea,* p. 93.

"a cold, acid hauteur": Claude McKay, *A Long Way From Home* (New York: Harcourt Brace Jovanovich, 1970), p. 110.

"a gracious tan-brown lady": *The Big Sea,* p. 94.

"too prim school-marmish and stilted": David Levering Lewis, *When Harlem Was In Vogue* (New York: Knopf, 1981), p. 124.

"I mean this," she assured him: Jessie Fauset to LH, Jan. 22, 1922, LHP.

page 54

"from that day until this": Augustus Granville Dill to LH, Jan. 21, 1951, LHP.

study without any "prescribed limits": James N. Hughes to LH, Dec. 26, 1921, LHP.

chosen to be one of "a picked group": LH to James N. Hughes, Feb. 14, 1922, LHP.

"stay up half the night": *Ibid.*

"I want this statement": James N. Hughes to LH, Feb. 21, 1922, LHP.

page 55

Hughes was performing "admirably": Herbert E. Hawkes to James N. Hughes, Mar. 7, 1922, LHP.

"if, in the future, you attend to these things": LH to James N. Hughes, Mar. 2, 1922, LHP.

The failure . . . was "sufficient proof": James N. Hughes to LH, Mar. 16, 1922, LHP.

page 56

"That winter I spent as little time": LH, "My Early Days In Harlem," p. 312.

mourned the "little dark girls": Claude McKay, "Harlem Shadows," *Pearson's Magazine* 39 (Sept. 1918): 276.

"a white man's house of assignation": Eric Walrond, "The Black City," *Messenger* 6 (Jan. 1924): 14.

"the chief past-time of Harlem": George Schuyler, "New York—Utopia Deferred," *Messenger* 7 (Oct.–Nov. 1925): 344–349, 370.

"Dream-singers, / Story-tellers": LH, "My People," *Crisis* 24 (June 1922): 72.

"the old junk man death": LH, "Question," *Crisis* 23 (Mar. 1922): 210.

"A half-shy young moon veiling her face": LH, "The New Moon," *ibid.*

"This ancient hag": LH, "Mexican Market Woman," *ibid.*

page 57

Under the name "LANG-HU" or "LANGHU": The four poems are "Passionate Love" and "Reasons Why" (Apr.) and "Utopia" and "Time's Prohibition Bar" (May).

"The moon still sends its mellow light": LH, "To A Dead Friend," *Crisis* 24 (May 1922): 21.

"Beautiful, like a woman": LH, "The South," *Crisis* 24 (June 1922): 72.

"Loud-mouthed laughers in the hands / of fate": *Ibid.*

"Nor do I wish to return": LH to R. J. M. Danley, May 14, 1922, LHP.

page 58

"But I didn't advertise": LH, *The Big Sea,* p. 86.

"Now, / in June": LH, "After Many Springs," *Crisis* 24 (Aug. 1922): 92.

"I liked it": LH, "L'histoire de ma vie" (1925), LHP 505.

page 59

"the low beating of the tom-toms": LH, "Danse Africaine," *Crisis* 24 (Aug. 1922): 167.

"They seem to take them seriously": LH to James N. Hughes, Aug. 20, 1922, LHP.

"I don't want to go back": *Ibid.*

"everywhere I roomed": LH, "My Early Days In Harlem," pp. 312–313.

"sometimes you would catch a glimpse": LH, *The Big Sea,* p. 88.

page 60

"It seemed to me now": LH, *The Big Sea,* p. 89.

"Where are we going?": LH, *The Big Sea,* p. 90.

"a most picturesque place": LH to James N. Hughes, Oct. 23, 1922, LHP.

page 61

"the finest fellows I've ever met": LH to Alain Leroy Locke, Feb. 6, 1923; ALLP, MSRC.

"Does a jazz-band ever sob?": LH, "Three Poems of Harlem," "Gods," and "Our Land: Poem for a Decorative Panel," in LH to Countee Cullen, Feb. 12, 1923, CCP, ARC.

page 62

"driving the butcher's cart": LH, "Joy," *ibid.*

"I loved my friend": LH, "To F.S.," *ibid.* There is no evidence to support the published conjecture that "F.S." refers to Ferdinand Smith, a prominent West Indian-American leftist maritime union official of the late thirties and forties.

"I am waiting for my mother": LH, "Poem," *ibid.*

"I certainly do not think": Jessie Fauset to LH, May 17, 1922, LHP.

page 63

"That brown girl's swagger gives a twitch": Countee Cullen, "To a Brown Boy," *Color* (New York: Harper & Brothers, 1925), p. 8.

"I don't know what to say": LH to Countee Cullen, Mar. 4, 1923, CCP, ARC.

page 64

"Ay ya! / Ay ya!": LH to Countee Cullen, Mar. 7, 1923, CCP, ARC.

"very beastly of me": LH to Countee Cullen, Mar. 4, 1923, CCP, ARC.

"your beloved" A. E. Housman: LH to Countee Cullen, Apr. 7, 1923, CCP, ARC.

page 65

"Droning a drowsy syncopated tune": LH, *The Weary Blues* (New York: Alfred A. Knopf, 1926), pp. 23–24.

page 66

"beauty of a cabaret poem": LH to Countee Cullen, Apr. 7, 1923, CCP, ARC.

page 67

"Write to him": Countee Cullen to Alain Locke, Jan. 12, 1923, ALLP, MSRC.

"insists on my knowing you": Alain Locke to LH, Jan. 17, 1923, LHP.

"I am chasing dreams": LH to Alain Locke, Feb. 6, 1923, ALLP, MSRC.

page 68

"First impressions count with me": Alain Locke to LH, Feb. 10, 1923, LHP.

"and then, at the end of your journey": LH to Alain Locke, Feb. 19, 1923, ALLP, MSRC.

"He doesn't need": Alain Locke to Countee Cullen, Mar. 15, 1923, CCP, ARC.

"Is Mr. Locke married?": LH to Countee Cullen, Apr. 7, 1923, CCP, ARC.

"but I love them": LH to Alain Locke, Apr. 6, 1923, ALLP, MSRC.

"pagan to the core": Alain Locke to LH, n.d., LHP.

page 69

"some little Greenwich Village of our own": LH to Alain Locke, n.d., ALLP, MSRC.

"a sort of shock-absorber": Alain Locke to LH, n.d., LHP.

page 70

"Don't you ever get that feeling": LH to Countee Cullen, Mar. 4, 1923, CCP, ARC.

"I can't go with you": LH to Alain Locke, n.d., ALLP, MSRC.

"Be sure to get this to him": Jessie Fauset to LH, Apr. 26, 1923, LHP.

"you assuredly have the true poetic touch": Jessie Fauset to LH, May 28, 1923, LHP.

"thoroughly efficient, honest, and reliable": Letters from Chief Steward H. J. Hicks, s.s. *Bellbuckle,* Jan. 29, 1923, and Fleet Steward A. W. Ingram, Caldwell's Fleet, June 4, 1923, LHP 863.

"No more skeins have been tangled": Countee Cullen to Alain Locke, May 31, 1923, ALLP, MSRC.

page 71

"He was sympathetic": Countee Cullen to Alain Locke, June 8, 1923, ALLP, MSRC.

"I think you are probably doing": Countee Cullen to Alain Locke, June 21, 1923, ALLP, MSRC.

"The spring is not so beautiful there—": LH, "The Water-front Streets," *The Weary Blues,* p. 71.

page 72

"It was like throwing a million bricks": LH, *The Big Sea,* p. 98.

"I had no intention": LH, "The Big Sea," draft ms., n.d.; Vivian Harsh Collection, Carter G. Woodson Regional Library, Chicago. This admission was left out of *The Big Sea.*

4. On the Big Sea

page 73

"a picture-book town": LH, "Ships, Sea and Africa: Random Impressions of a Sailor on His First Trip Down the West Coast of the Motherland," *Crisis* 27 (Dec. 1923): 70.

"fairy islands": LH, *The Big Sea,* p. 10.

"My Africa, Motherland": LH, *The Big Sea,* p. 10.

"You should see the clothes": LH to Carrie M. Clark, July 3, 1923, courtesy of George Houston Bass.

page 75

"Is it true," the boy asked: LH, *The Big Sea,* p. 105.

"a million fallen stars": LH, "Ships, Sea and Africa," p. 71.

"the dirtiest saddest lot": LH, "Ships, Sea and Africa," p. 71.

page 76

"the worst trip in the world": LH, "L'histoire de ma vie" (1925), LHP 505.

"like the father of the Katzenjammer Kids": LH, *The Big Sea,* p. 113.

"The poor missionaries": LH, *The Big Sea,* p. 111.

"James L. Hughes": Ledger of Shipping Articles, 1923, No. 3672, s.s. *West Hesseltine,* Maritime Division, National Archives, Washington, D.C. See also the Port Deck Department Log Books 7656–7659, ss. *West Hesseltine,* Merchant Marine Ships' Logs, 1918–1941; Division of Operations, U.S. Shipping Board Emergency Fleet Corp., National Records Center, Suitland, Maryland.

"He is really squeezing life": Countee Cullen to Harold Jackman, July 20, 1923, CCP, JWJ.

page 77

"Mercedes is a jungle-lily": LH, *The Weary Blues,* p. 90.

"tous aux tavernes": LH, "Ships, Sea and Africa", p. 71.

"vile houses of rotting women": LH, "L'histoire de ma vie."

"Tenerife. All those houses": LH, "Sex Silly Season," Mar. 25, 1963, LHP 3486.

"the bare, pointed breasts": LH, *The Big Sea,* p. 102.

"evidently not realizing it": LH, *The Big Sea,* p. 110.

"save for a whisp of loin-cloth": LH to Countee Cullen, July 21, 1923, CCP, ARC.

"ten-year-old wharf rats": LH, *The Big Sea,* p. 106.

page 78

"The white man dominates Africa": LH, *The Big Sea,* p. 102.

"I do not hate you": LH, "The White Ones," *Opportunity* 2 (Mar. 1924): 68.

"The Africans looked at me": LH, *The Big Sea,* p. 11.

"They looked at my copper-brown skin": LH, *The Big Sea,* p. 103.

"We are related—you and I": LH, "Brothers," *Crisis* 27 (Feb. 1924): 160.

"The night is beautiful": LH, *The Weary Blues,* p. 58.

"To fling my arms wide": LH, "Dream Variation," *Crisis* 28 (July 1924): 19.

page 79

"Oh, the little black girls": LH, "L'histoire de ma vie."

"tall, black, sinister ships": LH, *The Big Sea,* p. 120.

"Singing black boatmen": LH, "African Fog," *New York Herald-Tribune,* Feb. 15, 1926.

page 80

"the millions of whiskey bottles": LH, "L'histoire de ma vie."

"Each time a man would rise": LH, *The Big Sea,* p. 108.

"we took away riches": LH, *The Big Sea,* p. 102.

"We are homeward bound": LH to Countee Cullen, Oct. 1, 1923, CCP, ARC.

page 81

"Langston is back": Countee Cullen to Alain Locke, Nov. 24, 1923, ALLP, MSRC.

"She seemed just a tiny": LH, *The Big Sea,* p. 131.

"I could hear my mother": LH, *The Big Sea,* pp. 132–134.

"Oh, singing tree!": LH, "Jazzonia," *Crisis* 26 (Aug. 1923): 162.

page 82

"As for Langston—": Alain Locke to Countee Cullen, n.d., CCP, ARC.

in spite of Hughes's "irresponsible streak": Countee Cullen to Alain Locke, May 12, 1924, ALLP, MSRC.

"I am humbly and desperately in need": Alain Locke to LH, n.d., LHP.

"James L. Hughes": Ledger of Shipping Articles, 1924, Nos. 254, 869, s.s. *McKeesport,* Maritime Division, National Archives, Washington, D.C.

page 83

"MAY I COME NOW": LH to Alain Locke, Feb. 2, 1924, ALLP, MSRC. An undated letter by Locke in reply is also in ALLP, MSRC. It was either never sent or returned to Locke, probably unopened.

"Forgive me for the sudden": LH to Alain Locke, Feb. 4, 1924, ALLP, MSRC.

page 84

"Sale étranger!": LH, *The Big Sea,* p. 154.

"this rather disreputable young Negro": Rayford W. Logan to author, interview, Aug. 6, 1979.

page 85

"Kid, stay in Harlem!": LH to Countee Cullen, Mar. 11, 1924, CCP, ARC.

page 86

"that beast of a place": Anne Marie Coussey to LH, July 26, 1924, LHP.

"Too green the springing April grass": LH and Arna Bontemps, eds., *The Poetry of the Negro: 1746–1949* (New York: Doubleday, 1949), p. 329.

page 87

"I wonder how you have been able": Anne Marie Coussey to LH, July 26, 1924, LHP.

"So you are in love": Countee Cullen to LH, May 14, 1924, LHP.

"blind goddess to which we black are wise": *New York Amsterdam News,* Apr. 25, 1923.

"The haunting face of poverty": LH, "Grant Park," *Messenger* 6 (Mar. 1924): 75.

page 88

"O, Great God of Cold and Winter": LH, "Prayer for a Winter Night," *Messenger* 6 (May 1924): 153.

"the circus of civilization": LH, "Lament for Dark Peoples," *Crisis* 28 (June 1924): 60.

"in the hope that you will revise": Jessie Fauset to LH, June 24, 1924, LHP.

"In the Johannesburg mines": LH, "Johannesburg Mines," *Messenger* 7 (Feb. 1925): 93.

"my brown goddess": LH to Harold Jackman, Aug. 1, 1924, LHP.

"Me an' ma baby's": LH, "Negro Dancers," *Crisis* 29 (March 1925): 221.

page 89

"Do you like it?": LH to Countee Cullen, July 24, 1924, CCP, ARC.

"Oh, the sea is deep": LH, "Song for a Suicide," *Crisis* 28 (May 1924): 23.

"people are taking it all wrong": LH to Countee Cullen, June 27, 1924, CCP, ARC.

"Her teeth are as white as the meat of an apple": LH, "Fascination," *Crisis* 28 (June 1924): 86. A published assertion that the poem is about Anne Coussey is incorrect; it was in the collection sent to Cullen by LH in Feb. 1923 from Jones Point (CCP, ARC).

"And we sat in my room": LH, *The Big Sea,* p. 168.

page 90

"I am afraid you are getting into a groove": Anne Marie Coussey to LH, July 26, 1924, LHP.

"I wonder if you really did": Anne Marie Coussey to LH, June 3, 1926, LHP.

"Since I left you, Anne": LH, "A Letter to Anne," *Lincoln News* (Lincoln University, Pa.), Feb. 27, 1927.

"In the mist of the moon I saw you": LH, "In the Mist of the Moon," *ibid.*

Swept by "a sudden notion": LH to Countee Cullen, June 27, 1924, CCP, ARC.

"that you may have whatever happiness": Alain Locke to LH, May 5, 1924, LHP.

page 91

"a darned nice fellow": LH to Countee Cullen, July 4, 1924, CCP, ARC.

"Do come home": Countee Cullen to LH, June 13, 1924, LHP.

"Qui est il?": LH, *The Big Sea,* p. 184.

"as deadly sociological as most": Alain Locke to LH, May 22, 1924, LHP.

page 92

"Locke is here": LH to Countee Cullen, n.d., CCP, ARC.

"I do not love your Montmartre": Alain Locke to LH, May 5, 1924, LHP.

"See Paris and die": Alain Locke to Countee Cullen, July 26, 1924, CCP, ARC.

"As to the joy of having met you": Alain Locke to LH, n.d., LHP.

"Many of them clung to me": LH, *The Big Sea,* p. 188.

page 93

"And the sun!—Oh!": LH to Countee Cullen, Aug. 13, 1924, CCP, ARC.

a "trance-like walk": Alain Locke to LH, n.d., LHP.

"I like you immensely": LH to Alain Locke, Aug. 12, 1924, ALLP, MSRC.

"Lovingly yours, Alain": Alain Locke to LH, n.d., LHP.

"I would love to see Venice": LH to Alain Locke, Aug. 16, 1924, ALLP, MSRC.

page 94

"I began to wonder": LH, *The Big Sea,* pp. 189–190.

"certainly as pleasant as anything": LH to Alain Locke, n.d., ALLP, MSRC.

"If you have need of a man": U.S. Vice Consul Leo J. Callahan to Master, s.s. *West Mahomet,* Sept. 22, 1944, LHP.

page 95

"I, too, sing America": LH, "Epilogue," *The Weary Blues,* p. 109.

page 96

"fix your eye on some professional goal": Jessie Fauset to LH, Oct. 8, 1924, LHP.

an "ugly, childish, little review": LH to Countee Cullen, n.d., CCP, ARC.

page 97

Van Vechten would record . . . "Kingston Hughes": Diary entry, Nov. 10, 1924, Box F, CVVP, NYPL.

"The atmosphere was already heady": Arna Bontemps, "The Awakening: A Memoir," in Bontemps, ed., *The Harlem Renaissance Remembered* (New York: Dodd, Mead, 1972), p. 19.

"Countee! Countee! . . . Mr. Hughes is here": Bontemps, "The Awakening," p. 18.

page 98

"we forgot the cold": Bontemps, "The Awakening," p. 20.

"one week in New York": LH, "L'histoire de ma vie."

"Any news of Langston?": Countee Cullen to Alain Locke, Nov. 23, 1924, ALLP, MSRC.

5. We Have Tomorrow

page 99

"It must be rich": LH, "Our Wonderful Society: Washington," *Opportunity* 5 (Aug. 1927): 226.

"My cousins introduced me": LH, *The Big Sea,* p. 203.

"The people themselves . . . assured me": LH, "Our Wonderful Society: Washington," p. 226.

page 100

"They had all the manners": LH, *The Big Sea,* p. 207.

"just so it yields the bucks": LH to Harold Jackman, Dec. 14, 1924, LHP.

"Oh, wash-woman": LH, "A Song to a Negro Wash-Woman," *Crisis* 29 (Jan. 1925): 115.

page 101

"the most striking piece of scientific work": Rayford W. Logan and Michael R. Winston, eds., *Dictionary of American Negro Biography* (New York: Norton, 1982), p. 666.

"Although I realized what a fine contribution": LH, *The Big Sea,* p. 211.

page 102

"10 p.m.—you are out": LH to Alain Locke, n.d., ALLP, MSRC.

"crude-boned soft-skinned wedge": Jean Toomer, *Cane* (New York: Boni and Liveright, 1923), p. 71.

"looked at the dome of the Capitol": LH, *The Big Sea,* p. 209.

"Son, what's the matter?": LH, *The Big Sea,* p. 217.

"Like the waves of the sea": LH, *The Big Sea,* p. 209.

"I tried to write poems like the songs": LH, *The Big Sea,* p. 209.

"I always put them away": LH, *The Big Sea,* p. 217.

page 103

"Play that thing": LH, "To A Negro Jazz Band in a Parisian Cabaret," *Crisis* 31 (Dec. 1925): 67.

"You are no friend of mine": LH, "To Certain Intellectuals," *Messenger* 7 (Feb. 1925): 103.

"Ice-cold passion": LH, "Poem to a Dead Soldier," *Workers Monthly* 4 (Apr. 1925): 261.

page 104

"Death comes like a mother": LH, "Ways," *Buccaneer* 1 (May 1925): 20.

"I ask you this": LH, "Prayer," *ibid.*

"She stands / In the quiet darkness": LH, "Troubled Women [*sic*]," *Opportunity* 3 (Feb. 1925): 56. Later called "Troubled Woman."

"My soul / Empty as the silence": LH, "Summer Night," *Crisis* 31 (Dec. 1925): 66.

"My old man's a white old man": LH, "Cross," *ibid.*

page 105

"a common condition": Alain Locke, "Harlem," *Survey Graphic* 6 (Mar. 1925): 630.

"vibrant with a new psychology": Alain Locke, "Enter the New Negro," *Survey Graphic* 6 (Mar. 1925): 631.

"We have tomorrow": *Ibid.* First published as "Poem," *Crisis* 28 (Aug. 1924): 163. Later called "Youth."

"tricked and misused me": Darwin T. Turner, ed., *The Wayward and the Seeking* (Washington, D.C.: Howard University Press, 1980), p. 133.

Locke destroyed "every vestige": David L. Lewis, *When Harlem Was in Vogue,* p. 154.

page 106

"a soft young fellow with a purr": Paul Peters to LH, May 29, 1932, LHP.

"We struck up a . . . friendship": Richard Bruce Nugent to James Hatch, transcript of interview, 1982; courtesy of the Hatch-Billops Collection and Richard Bruce Nugent.

"He was a made-to-order Hero": Richard Bruce Nugent in Thomas H. Wirth, ed., "Lighting FIRE!!," broadside (unpaginated) on publication of facsimile of *Fire!!* (Metuchen, N.J.: The *Fire!!* Press, 1982).

page 107

"a rather unique contrast": Charles S. Johnson to LH, Dec. 20, 1924, LHP.

"worse than the coasts of Africa": LH to Alain Locke, n.d., ALLP, MSRC.

"Your ship . . . will be coming in": Jessie Fauset to LH, n.d., LHP.

She "is a clever girl": LH to Carl Van Vechten, June 4, 1925, CVVP, Yale.

page 109

"With him . . . 'amusing' things were essential": Mabel Dodge Luhan, *Intimate Memoirs: Movers and Shakers* (New York: Harcourt, Brace, 1936), p. 16.

page 110

"I shall do my best": Carl Van Vechten to LH, n.d., LHP.

"Your work has such a subtle": Carl Van Vechten to LH, n.d., [May 13, 1925], LHP.

"It is amusing": LH to Carl Van Vechten, n.d. [May 12, 1925], CVVP, Yale.

"I would be very, very much pleased": LH to Carl Van Vechten, May 15, 1925, CVVP, Yale.

"LITTLE DAVID PLAY ON": Arna Bontemps, "Langston Hughes," *Ebony* 1 (Oct. 1946): 20.

"THANKS IMMENSELY": LH to Carl Van Vechten, May 18, 1925, CVVP, Yale. Quoted erroneously in Bontemps, "Langston Hughes."

"Mr. Van Vechten has sent": Blanche Knopf to LH, May 18, 1925, LHP.

"How quickly it's all been done": LH to Carl Van Vechten, n.d. [May 18, 1925], CVVP, Yale.

page 111

"Never in my wildest dreams": *Ibid.*

"Gypsy done told me": LH to Carl Van Vechten, n.d. [May 17, 1925], CVVP, Yale. See also Carl Van Vechten, "The Black Blues," *Vanity Fair* 24 (Aug. 1925): 86.

"What you write me": Carl Van Vechten to LH, n.d., LHP.

"In six of them": LH to Carl Van Vechten, n.d. [May 25, 1925], CVVP, Yale.

"Undoubtedly, the greatest kind": Jessie Fauset to LH, n.d., LHP.

"Please don't think this an attempt": Countee Cullen to LH, May 18, 1925, LHP.

page 112

"You have grown to be": M. J. Reed Campbell to LH, June 25, 1925, LHP.

Langston was "so *sincere*": Hall Johnson to LH, June 9, 1925, LHP.

"We want to make it": Carl Van Vechten to LH, n.d., LHP.

"Treat it romantically if you will": Carl Van Vechten to LH, n.d., LHP.

"I hate to think backwards": LH to Carl Van Vechten, n.d., CVVP, Yale.

"you like best the lighter ones": LH to Carl Van Vechten, n.d. [June 4, 1925], CVVP, Yale.

page 113

"Humming-birds and bright flowers": LH to Carl Van Vechten, Aug. 9, 1925, CVVP, Yale.

"Cullen looks like one": Richard Bruce Nugent to James Hatch, transcript of interview.

"Strut and wiggle": See LH, "Cabaret," "To Midnight Nan at Leroy's," "Fantasy in Purple," and "Suicide," *Vanity Fair* 24 (Sept. 1925): 62.

"I know Carl is coining": Countee Cullen to Harold Jackman, Oct. 7, 1925, CCP, JWJ.

"who want us to do only Negro things": Countee Cullen to Alain Locke, May 15, 1925, ALLP, MSRC.

page 114

"a real African type": LH to Carl Van Vechten, Aug. 23, 1925, CVVP, Yale.

"like the petals of a bright flower": LH to Carl Van Vechten, Oct. 29, 1925, CVVP, Yale.

"Because you want me to: LH to Carl Van Vechten, n.d. [Aug. 23, 1925], CVVP, Yale.

page 115

"rather too deliberate eccentricities": Jessie Fauset to LH, n.d., LHP.

"but he wore their clothes": LH to Carl Van Vechten, n.d. [Sept. 7, 1925], CVVP, Yale.

"All your work has the flavor": George S. Schuyler to LH, Sept. 24, 1925, LHP.

"You'd think I was famous": LH to Carl Van Vechten, Oct. 9, 1925, CVVP, Yale.

"You'd get just the mere formal": Jessie Fauset to LH, Oct. 23, 1925, LHP.

"Do you happen to know": LH to Carl Van Vechten, Oct. 29, 1925, CVVP, Yale.

"there may be money in *that": Carl Van Vechten to LH, n.d., LHP.

"Some big hearted person": LH to Walter White, Oct. 29, 1925, WWP, LC.

page 116

"I want to come to Lincoln": LH to Dean of the College, Lincoln University, Oct. 20, 1925; Office of the Registrar, Lincoln University, Pa.

"we shall be very glad": Rev. George Johnson to LH, Oct. 22, 1925, LHP.

"almost as lovely as the poems": LH to Countee Cullen, Nov. 7, 1925, CCP, ARC.

the man who "has made things possible": Jessie Fauset to LH, Oct. 14, 1925, LHP.

"I should be honored": W. E. B. Du Bois to LH, Oct. 20, 1925, LHP.

"And if they who have a chance": Gwendolyn Bennett to LH, Dec. 2, 1925, LHP.

page 117

"It is fine to have such letters": Carl Sandburg to Charles S. Johnson in Johnson to LH, Oct. 15, 1925, LHP.

"I never in life thought": LH to Carl Van Vechten, n.d. [Dec. 8, 1925], CVVP, Yale.

page 118

people kept up "social barriers": LH to Countee Cullen, Nov. 24, 1925, CCP, ARC.

"has been playing with Negroes": Bruce Kellner, *Carl Van Vechten and the Irreverent Decades* (Norman: Univ. of Oklahoma, 1968), p. 195.

"perhaps I might be able": Amy Spingarn to LH, Oct. 7, 1925, LHP.

page 119

"I love to give pleasure": Wallace Thurman to LH, Feb. 6, 1926, LHP.

"the most closed-mouth and cagey": Wallace Thurman, *Infants of the Spring* (New York: Macaulay, 1932), p. 232.

"He seems to be a kind, shy sort": LH to Harry C. Block, Dec. 7, 1925, Alfred A. Knopf Inc. files, N.Y.

"My dear Langston Hughes": Nicholas Vachel Lindsay to LH, Dec. 6, 1925; presentation copy in the Special Collections Department, Langston Hughes Memorial Library, Lincoln University.

"Something of what you tell me": LH to Vachel Lindsay, draft, Dec. 7, 1925, LHP.

page 120

"I must insist that you never": Jean Toomer to Horace Liveright, Sept. 5, 1923, JTP, Fisk.

"Thunder of the Rain God": LH, "A House in Taos," *The Big Sea,* pp. 260–261.

page 122

''One must do something'': LH to Sartur Andre [Andrzejewski], n.d., LHP.

''You are delightfully generous'': Charles S. Johnson to LH, Dec. 15, 1925, LHP.

''indubitably now is the psychological moment'': Carl Van Vechten to LH, n.d., LHP.

''although culturally Harvard would offer'': Amy Spingarn to LH, Dec. 23, 1925, LHP.

''I have been thinking a long time'': LH to Amy Spingarn, draft, n.d., LHP.

page 123

''I like my book'': LH to Harry C. Block, Jan. 16, 1926, Alfred A. Knopf Inc. files, N.Y.

''nice music, but nothing grotesque'': *Ibid.*

''a few more thousands'': LH to Carl Van Vechten, Jan. 20, 1926, CVVP, Yale.

''or else, their souls have grown deep'': *Ibid.*

page 124

''If I ever get in the school books'': *Ibid.*

''He was so black'': Richard Bruce Nugent to James Hatch, transcript of interview.

''extraordinarily anxious'' to have him read: V. F. Calverton to LH, Feb. 6, 1926, LHP.

Hughes's ''entirely original'' treatment: W. C. Handy to LH, Feb. 8, 1926, LHP.

''You see, . . . I'm going'': LH to Carl Van Vechten, Jan. 20, 1926, CVVP, Yale.

6. A Lion at Lincoln

page 125

''Lincoln is wonderful'': LH to Countee Cullen, Feb. 21, 1926, CCP, ARC.

''Out here with the trees'': LH to Alain Locke, Mar. 8, 1926, ALLP, MSRC.

''I like Lincoln so well'': *Ibid.*

page 126

''the black Princeton'': Toye G. Davis to author, interview, Oct. 13, 1980.

''a most unfortunate effect'': Horace Mann Bond, *Education For Freedom: A History of Lincoln University, Pennsylvania* (Lincoln University, 1976), p. 374.

''in very very poor condition'': Therman B. O'Daniel to author, interview, Oct. 14, 1980.

''adulterers and fornicators'': Bond, *Education For Freedom,* p. 386.

''To set foot on dozens'': LH, ''Cowards from the Colleges,'' n.d., LHP. Published in *Crisis* 41 (Aug. 1934): 226–228.

''their morals were higher'': Toye G. Davis to author, interview.

page 127

''because then one is always likely'': LH to Alain Locke, Mar. 8, 1926, ALLP, MSRC.

''One can use, wear, or borrow'': LH to Alain Locke, Mar. 8, 1926, ALLP, MSRC.

''beaten till I couldn't sit'': LH to Countee Cullen, Feb. 21, 1926, CCP, ARC.

''Lincoln is more like what home'': LH to Carl Van Vechten, Mar. 26, [1926], CVVP, Yale.

''a very quiet, very nice guy'': Leroy D. Johnson to author, interview, Oct. 13, 1980.

acclaimed the ''most popular'': *Lincoln News,* ''Commencement Number,'' 1929, p. 3.

''Langston was a remarkable person'': Therman O'Daniel to author, interview.

page 128

''rather dull'': LH to Alain Locke, Mar. 8, 1926, ALLP, MSRC.

''besides, I like Jewish people'': LH to Alain Locke, Mar. 29, 1926, ALLP, MSRC.

''I liked that you on the platform'': Sartur Andrzejewski to LH, May 9, 1926, LHP.

''immense inspiration to us'': Harold Williams to LH, Apr. 18, 1926, LHP.

page 129

''well worth watching'': *New York Herald-Tribune,* Aug. 1, 1926.

''sensitive and intelligent'': *New Orleans Times-Picayune,* Apr. 4, 1926.

''just the right thing'': Claude McKay to LH, Apr. 24, 1926, LHP.

''I am your son'': LH, ''Mulatto,'' *Saturday Review of Literature* 3 (Jan. 29, 1927): 547.

page 130

''simply marvellous'' new black male dancer: LH to Alain Locke, June 4 [1926], ALLP, MSRC.

''I've been home too long'': LH to Alain Locke, May 4, 1926, ALLP, MSRC.

''Rather flippant in tone'': *Nation* (Freda Kirchwey) to LH, June 3, 1926, LHP.

''a poet—not a Negro poet'': See LH, ''The Negro Artist and the Racial Mountain,'' *Nation* 122 (June 23, 1926): 692–694.

page 131

''Do not read any more blues'': LH, ''My Adventures as a Social Poet,'' *Phylon* (3rd Quarter, 1947): 206.

page 132

''the nicest town in the world'': LH to Alain Locke, Aug. 12, 1926, ALLP, MSRC.

''very precious, autocratic, commanding'': Richard Bruce Nugent to author, interview, June 10, 1984.

''the pen that set tongues to wagging'': W. C. Handy, *Father of the Blues: An Autobiography,* ed. Arna Bontemps (New York: Macmillan, 1970), p. 236.

page 133

''You are in the final analysis'': Wallace Thurman to LH, July 29, 1929, LHP.

''who seemed to thrive without'': Bruce Kellner to author, May 22, 1984.

a ''dramatico-musical composition'': LH, ''Leaves: A novelty song number for singer and chorus,'' n.d. (''copyrighted 1926 by Jimmy Hughes''), LHP.

page 134

''He suggested that maybe someone'': Richard Bruce Nugent, ''Lighting FIRE!!,'' essay on publication of facsimile, Thomas H. Wirth, ed., *Fire!!,* (Metuchen, N.J.: The *Fire!!* Press, 1982). Courtesy of Richard Bruce Nugent.

''Colored people can't help but'': LH to Alain Locke, Aug. 12, [1926], ALLP, MSRC.

''in spite of its pro-African propaganda'': Alain Locke to LH, Sept. 2, 1926, LHP.

''our people to a 'T' '': Carolyn Clark to LH, Nov. 3 [1926], LHP.

page 135

''that hideous black 'nigger' '': George M. McClellan, July 14, 1926, LHP.

''Fire gonna burn ma soul!'': LH, ''Fire,'' *Fine Clothes to the Jew* (New York: Alfred A. Knopf, 1927), p. 50.

page 136

''When hard luck overtakes you'': LH, *Fine Clothes to the Jew,* p. 18.

page 137

''ho hum'': Wallace Thurman to LH, n.d., LHP.

would not burn ''without Nordic fuel'': Wallace Thurman to LH, n.d., LHP.

''*FIRE* weaving, vivid hot designs'': Foreword, *Fire!!: A Quarterly Devoted To The Younger Negro Artists* 1 (Nov. 1926): 1.

''a beautiful piece of printing'': *Crisis* 33 (Jan. 1927): 158.

''*Physically,* if not mentally'': Wallace Thurman, ''Cordelia the Crude,'' *Fire!!* 1 (Nov. 1926): 5.

''I have just tossed the first issue'': LH, *The Big Sea,* p. 237.

page 138

''Under such boredom . . . no wonder'': LH to Carl Van Vechten, Nov. 26, 1926, CVVP, Yale.

page 139

''Writing of original narratives'': *Lincoln University Herald* 31 (Nov. 1926): 46.

page 140

''A woman can have two lovers'': LH, ''Luani of the Jungles,'' in Nathan Irvin Huggins, *Voices from the Harlem Renaissance* (New York: Oxford Univ. Press, 1976), p. 152.

''More out of the core'': Amy Spingarn to LH, n.d., LHP.

''Jacob and the Negro'': Arthur B. Spingarn to LH, Feb. 3, 1927, LHP.

''the poet of the modern Negro proletariat'': George Schuyler to LH, Jan. 27, 1927, LHP.

''It's harder and more cynical'': LH to Dewey Jones, Feb. 5, 1927; cited in James A. Emanuel, *Langston Hughes* (New York: Twayne Publishers, 1967), pp. 31–32.

proclaiming Hughes a ''SEWER DWELLER'': *New York Amsterdam News,* Feb. 5, 1927.

Gay confessed ''[it] disgusts me'': *Philadelphia Tribune,* Feb. 5, 1927.

Rogers called it ''piffling trash'': *Pittsburgh Courier,* Feb. 5, 1927.

"a literary gutter-rat": *Chicago Whip,* Feb. 26, 1927.

"obsession for the more degenerate": *Philadelphia Tribune,* Feb. 12, 1927.

pages 141 to 143

"Put on yo' red silk stockings": LH, "Red Silk Stockings," *Fine Clothes to the Jew,* p. 73. Also cited: "Beale Street Love," p. 57; "Laughers," pp. 77–78; "Mulatto," p. 71; "Sport," p. 40; "Hey!" p. 17; "Homesick Blues," p. 24; "Bad Man," p. 21; "Po' Boy Blues," p. 23; "Listen Here Blues," p. 85; "Lament Over Love," p. 81; "Gypsy Man," p. 22; "Hard Daddy," p. 86; "Gal's Cry For A Dying Lover," p. 82; "Ma Man," p. 88.

page 144

"even I myself belong": *Pittsburgh Courier,* Feb. 26, 1927.

"we should display our higher selves": Cleveland *Plain Dealer,* Mar. 27, 1927.

page 145

cultural training "from which to view": LH, "These Bad New Negroes: A Critique on Critics," Mar. 22, 1927, LHP. Published in *Pittsburgh Courier,* April 16, 1927.

"you and I . . . are the only colored": Carl Van Vechten to LH, Mar. 25 [1927], LHP.

"a rare poet": *Washington Eagle,* Mar. 11, 1927.

"Its open frankness will be a shock": *Saturday Review Of Literature* 3 (Apr. 9, 1927): 712.

"In a sense . . . he has contributed": *Chicago Daily News,* June 29, 1927.

"tawdry": *Boston Transcript,* Mar. 2, 1927.

"uneven and flawed": *New York Times Book Review,* Mar. 27, 1927.

"a great honor for me": LH to Walter White, n.d. [Feb. 1927], WWP, LC.

page 146

"I believe . . . that poetry": *Camden Evening Courier,* Mar. 3, 1927.

"All seems beautiful to me": LH to Carl Van Vechten, Mar. 1, 1927, CVVP, Yale.

"It's a— . . . you will": Charles S. Johnson to LH, May 6, 1927, LHP. See LH, "Our Wonderful Society: Washington."

"They really stopped discussing": LH to Alain Locke, n.d., ALLP, MSRC.

"Honey, don't let nobody tell you": LH, *The Big Sea,* p. 251.

page 147

she felt "tremendous rapport": Mrs. R. O. Mason, notebook, "A. L. February & March 1927," ALLP, MSRC. Letters to Mrs. Mason from Locke and from Zora Neale Hurston may be found in the Alain Locke Papers, Moorland-Spingarn Research Center, Howard University. Hughes's letters to Mrs. Mason are apparently lost, although a few drafts or copies are in his papers at Yale.

"I believed I could find leaders": Mrs. R. O. Mason to Alain Locke, May 1, 1932; draft in her notebook "Alain Locke Sept. 1927," ALLP, MSRC.

page 148

"a gift for a young poet": LH, *The Big Sea,* p. 313.

page 149

"first real hours together": Mrs. R. O. Mason to LH, n.d., LHP.

"my winged poet Child": Mrs. R. O. Mason to LH, June 5, 1927, LHP.

"they do have an air": LH to Alain Locke, June 11, 1927, ALLP, MSRC.

"May your flight through the South": Mrs. R. O. Mason to LH, June 19, 1927, LHP.

page 150

I'm having too grand a time": LH to Alain Locke, June 24, 1927, ALLP, MSRC.

"I bought some wishing powder": LH, *The Big Sea,* p. 291.

"to stand in a ring": LH, *The Big Sea,* p. 293.

"I knew it would be fun": LH, *The Big Sea,* p. 296.

page 151

"jump at de sun": Robert E. Hemenway, *Zora Neale Hurston: A Literary Biography* (Urbana, Ill.: Univ. of Illinois, 1977), p. 14.

"certainly the most amusing": LH, *The Big Sea,* pp. 238–239.

"de heavenly depot": LH to Carl Van Vechten, July 15, 1927, CVVP, Yale. For notes on this Southern journey, see LH, "Journals: 1920–1937," LHP 586.

page 152

"with so much love": George Washington Carver to LH, Oct. 23, 1927, LHP.

"Look a yonder!": see LHP 586.

''with greater emphasis on Great Negroes'': *Ibid.*
''his pleasing voice'': *Tuskegee Messenger,* Sept. 10, 1927.
''As I watched the stars'': Mrs. R. O. Mason to LH, July 22, 1927, LHP.
''later when you are ready'': Mrs. R. O. Mason to LH, July 26, 1927, LHP.

page 153

''Our Father who art in Heaven'': LH to Carl Van Vechten, Aug. 15, 1927, CVVP, Yale.
''The trouble with white folks'': LH, *The Big Sea,* p. 296.
''It seemed rather shameless'': LH to Alain Locke, Oct. 8, 1927, ALLP, MSRC.
''sting'' her errant husband: LHP 586.
''Can you guess, dear Langston'': Mrs. R. O. Mason to LH, n.d., LHP.

page 154

''How splendidly . . . you are conducting'': Mrs. R. O. Mason to LH, Sept. 28, 1927, LHP.
''I think we got on famously'': Zora Neale Hurston to LH, Sept. 21, 1927, LHP.
''seven hours that went like one'': LH [''Personal Papers: Memoranda and Address Lists''], LHP 863.

7. Godmother and Langston

page 156

''The corn ears are ripe'': LH to Mrs. R. O. Mason, draft, n.d., LHP.
''Of course, Langston, there is no where'': Mrs. R. O. Mason to LH, Dec. 1, 1927, LHP.
''to your storehouse of delightful memories'': Mrs. R. O. Mason to LH, Sept. 10, 1929, LHP.

page 157

''the horrible menu took my breath away'': Mrs. R. O. Mason to LH, May 6, 1928, LHP.
''On the way, I saw Langston'': Harold Jackman to Countee Cullen, Jan. 3, 1929, CCP, ARC.
''one good hearted kid'': Carrie Clark to LH, March 27, 1926, LHP.

page 158

''I no longer read'': LH to Claude McKay, Sept. 13, 1928, CMP, JWJ.
''May the river of your life'': Mrs. R. O. Mason to LH, Feb. 1, 1928, LHP.

page 159

''You are a golden star'': Mrs. R. O. Mason to LH, Sept. 9, 1928, LHP.
''You may feel, Langston'': *Ibid.*
''that yapping crowd'': Mrs. R. O. Mason to LH, May 6, 1928, LHP.
''so I sat there unmoved'': LH to Carl Van Vechten, Feb. 27, 1928, CVVP, Yale.
''There was always some small talk'': Thomas A. Webster to author, interview, Dec. 14, 1980.
''this 'hit and miss' condition'': Mrs. R. O. Mason to LH, May 6, 1928, LHP.
''the vagabond of Negro poets'': Mary White Ovington, *Portraits in Color* (New York: Viking, 1927), pp. 194–195.

page 160

''Undoubtedly, it is the finest'': LH to Claude McKay, Mar. 5, 1928, CMP, JWJ.
''that Miss Reagan'': Mrs. R. O. Mason to LH, Sept. 9, 1928, LHP.
a *''terrible''* new Miller and Lyles black musical'': LH to Carl Van Vechten, n.d. [Feb. 1928], CVVP, Yale.
''I discover therein that one had'': LH to Alain Locke, n.d., ALLP, MSRC.
''There is a barrel house on the avenue'': LH, ''Barrel House: Chicago,'' *Lincoln University News* (October 1928): 7.
''I've never tried a standard form'': LH to Carl Van Vechten, Dec. 10, 1926, CVVP, Yale.

page 161

''the faith that keeps a homeless race'': *Opportunity* 6 (May 1928): 149.
''The sun, / Like the red yolk of a rotten egg'': LH, ''Sunset—Coney Island,'' *New Masses* 3 (Feb. 1928): 13.

page 162

thought several of Hughes's poems ''somewhat ridiculous'': *Daily Princetonian,* Mar. 23, 1928.
''pardon the impertinence'': Countee Cullen to LH, Apr. 2, 1928, LHP.
Cullen's ''parade'': LH to Carl Van Vechten, Feb. 27, 1928, CVVP, Yale.
''GROOM SAILS WITH BEST MAN'': Cited in LH to Arna Bontemps, Apr. 13, 1963, LHP.

page 163

"these tips of budding life": LH to Mrs. R. O. Mason, draft, May 18, 1928, LHP.

"wonderfully emblematic": Mrs. R. O. Mason to LH, May 20, 1928, LHP.

page 164

"Aunt Hagar Williams stood": LH, *Not Without Laughter* (New York: Alfred A. Knopf, 1930), p. 3.

"I realize that the old hurts": Mrs. R. O. Mason to LH, n.d., LHP.

"most sacred to your godmother": Mrs. R. O. Mason to LH, July 26, 1928, LHP.

"It is marvellous the way": Mrs. R. O. Mason to LH, June 29, 1928, LHP.

page 165

McKay "has used every art": W. E. B. Du Bois, "The Browsing Reader," *Crisis* 35 (June 1928): 202.

"If I'm ruined": LH to Claude McKay, Sept. 13, 1928, CMP, JWJ.

"The leaders and their followers": See LH, "Emperor of Haiti," Notes, Sept. 19, 1928, LHP 318.

page 166

"tearing off the veils": Mrs. R. O. Mason to LH, Sept. 23, 1928, LHP.

"I was so tired of it": LH to Alain Locke, Oct. 1, 1928, ALLP, MSRC.

Describing her . . . as "my girl friend": LH to Alain Locke, Nov. 20, 1928, ALLP, MSRC.

"pitiful, dirty, sloven boisterous": Laudee Williams to LH, Mar. 14, 1929, LHP.

"Prince of Fairies": Laudee Williams to LH, Apr. 28, 1929, LHP.

"No word—not even a single word": Laudee Williams to LH, Dec. 5, 1928, LHP.

"Most Henpecked": *Lincoln News* (June 1929): 3.

"To be approved and admired": Carrie Clark to LH, Feb. 1, 1929, LHP.

page 167

"Mama Clark is some chippie": Harold Jackman to Countee Cullen, n.d., CCP, ARC.

"I am hungry to see you": Carrie Clark to LH, Mar. 27, 1926, LHP.

"Dear, . . . why don't you love": Carrie Clark to LH, Oct. 29, 1928, LHP.

"Some time I feel": Carrie Clark to LH, May 13 [1928?], LHP.

"an amazing piece of writing": Mrs. R. O. Mason to LH, Nov. 11, 1928, LHP.

page 168

"in answer to your question": Amy Spingarn to LH, n.d., LHP.

"All the week I have been thinking": LH to Mrs. R. O. Mason, draft, Feb. 23, 1929, LHP.

page 169

"some little, less important things": *Ibid.*

page 170

"All deductions, conclusions, opinions": LH *et al.*, "Three Students Look At Lincoln," ms., n.d., LHP 3782. Another copy may be found in Special Collections, Langston Hughes Memorial Library, Lincoln University.

"staid contented Presbyterians": LH to Carl Van Vechten [May 8, 1929], CVVP, Yale.

"obvious incompetency" in the teaching: LH, "Three Students Look At Lincoln."

"In the primitive world": LH, *The Big Sea,* p. 311.

page 171

"LINCOLN VOTED 81–46 AGAINST": *Baltimore Afro-American,* Apr. 27, 1929.

"they are scared stiff": Alain Locke to Mrs. R. O. Mason, Apr. 16, 1929, ALLP, MSRC.

"the only way of creating self-respect": Claude McKay to LH, May 14, 1929, LHP.

"you're person No. 1": LH to Carl Van Vechten [May 8, 1929], CVVP, Yale.

"Oh! Langston, . . . I was so proud": Carrie Clark to LH, n.d., LHP.

"Now dat yo done gat": Waring Cuney to LH, Aug. 21, 1929, LHP.

page 172

"I *never* understood Wallace": Louise Thompson Patterson to author, interview, May 23, 1984.

"Awfully bad colored shows": LH to Claude McKay, June 27, 1929, CMP, JWJ.

"the oldest, largest, darkest": LH to Wallace Thurman, July 29, 1929, WTP, JWJ.

"The whole book . . . needs literary welding": Mrs. R. O. Mason to LH, May 29, 1929, LHP.

"What I object to": Mrs. R. O. Mason to LH, July 5, 1929, LHP.

page 173

"Well, . . . who cares": LH, "Journals: 1929", LHP 586.

"my spirit moving with you": Mrs. R. O. Mason to LH, Aug. 2, 1929, LHP.

"every colored lady has at least": LH to Wallace Thurman, July 29, 1929, WTP, JWJ.

she was "first and foremost": Hall Johnson to Mrs. R. O. Mason, n.d., ALLP, MSRC.

page 174

"a cane-clumping ancient": Richard Bruce Nugent to author, interview, June 10, 1984.

"be treated with every tenderness": Alfred A. Knopf to Carl Van Vechten, copy, May 28, 1929; Alfred A. Knopf files, New York City.

page 175

"Have you consulted your individual": Mrs. R. O. Mason, Dec. 15, 1929, in notebook "Alain Locke Sept. 1927 [*sic*]", ALLP, MSRC.

"inner emotional conflict": Alain Locke to LH, Dec. 30, 1929, LHP 801.

"controversial detail": Mrs. R. O. Mason to LH, Sept. 24, 1929, LHP.

"People were sleeping in subways": LH, *The Big Sea,* p. 319.

"indoor aviation": Melvin Ross to LH, July 18, 1930, LHP.

"Langston Hughes would probably do it": *Nation* (H. R. Mussey) to LH, May 7, 1930, LHP.

"Now . . . I am free": LH, "Journals: 1930", LHP 586.

page 176

"She gives me her gift": LH, "Journals: 1930."

"distinguished American Negro poet": Walter White to Harry F. Guggenheim, Feb. 19, 1930, WWP, LC.

"Make them prove that I am": LH, "Journals: 1930". 586.

"How many are going?" LH, "Journals: 1930."

page 177

"It seems years since I've felt": LH to Mrs. R. O. Mason, draft, Feb. 25, 1930, *ibid.*

"person extraordinary of this or any other world": LH, *The Big Sea,* p. 324.

page 178

"Yo también honro a América": José Antonio Fernández de Castro, "Yo, también," *Social* [Cuba] (September 1928): 28.

"Presentación de Langston Hughes": José Antonio Fernández de Castro, "Presentación de Langston Hughes," *Revista de la Habana* 1 (Mar. 1930): 367–368.

page 179

"Drums like fury; like anger": LH, "Journals: 1930."

page 180

"a little Cuban mulatto": Nicolás Guillén, "Conversación con Langston Hughes," in Edward J. Mullen, ed., *Langston Hughes in the Hispanic World and Haiti* (Hamden, Ct.: Archon Books, 1977), p. 172.

"The Americans . . . seem to clot": LH, "Journals: 1930."

page 181

"un verdadero escándolo": Nicolás Guillén to LH, Apr. 21, 1930, LHP.

"*eight formidable negro* poems": Gustavo E. Urrutia to LH, Apr. 20, 1930, LHP.

"the best kind of negro poetry": Gustavo E. Urrutia to LH, May 1, 1930, LHP.

8. Flight and Fall

page 182

"She used to talk about Zora": Louise Thompson Patterson to author, interview, May 23, 1984.

"Langston, Langston . . . this is going to be big": Zora Neale Hurston to LH, Mar. 8, 1928, LHP.

page 183

"so much more practical": Zora Neale Hurston to LH, May 1, 1928, LHP.

"Without flattery . . . you are the brains": Zora Neale Hurston to LH, July 10, 1928, LHP.

"You are always helpful": Zora Neale Hurston to LH, Apr. 30, 1929, LHP.

"NO niggers," she vowed: Zora Neale Hurston to LH, May 31, 1929, LHP.
"with no furniture at all": LH, *The Big Sea,* p. 239.
"I need money worse than you": LH, *The Big Sea,* p. 240.
"Studies from Obeah Land": In presentation copy, Aug. 12, 1929; courtesy of Thomas H. Wirth.
"The trouble with Locke": Zora Neale Hurston to LH, n.d., LHP.
"terribly worried because Zora and I": LH to Mrs. R. O. Mason, draft, n.d., LHP.

page 184

"So far in this world": LH to Mrs. R. O. Mason, draft, June 6 [1930], LHP.
"the plot, construction, and guiding the dialog": LH to Carl Van Vechten, Jan. 16, 1931, CVVP, Yale.

page 185

"restless and moody, working": LH, *The Big Sea,* p. 320.
"The way she talked to Langston": Louise Thompson Patterson to author, interview.
"I cannot write here": LH, *The Big Sea,* p. 325.

page 186

"or rather to release yourself": LH to Mrs. R. O. Mason, draft, n.d., LHP.
"His patron has failed him": Carl Van Vechten, Diary (May 26, 1930), CVVP, NYPL.
"Do you think the idea": LH to Carl Van Vechten, June 6, 1930, CVVP, Yale.
"Dear child, what a hideous spectre": Mrs. R. O. Mason to LH, June 6, 1930, LHP.

page 187

"I was afraid it was happening": LH to Mrs. R. O. Mason, draft, n.d., LHP.
"Under present conditions it is useless": Mrs. R. O. Mason to LH, June 17, 1930, LHP.
"Congratulations on your redemption": Alain Locke to LH, n.d., LHP.
"For the beauty of your eyes": LH to Mrs. R. O. Mason, draft, n.d., LHP.
"Oh, Alamari, as the sun sets": Mrs. R. O. Mason to LH, July 10, 1930, LHP.

page 188

"I ask you to help the gods": LH to Mrs. R. O. Mason, draft, Aug. 15, 1930, LHP.
"the first steps" toward intellectual rapprochement: Elmer A. Carter to LH, July 2, 1930, LHP.
"Dear lovely Death": LH, "Dear Lovely Death," *Opportunity* 8 (June 1930): 182.
"Strange, / That in this nigger place": LH, "Aesthete in Harlem," *ibid.*

page 189

"So long, / So far away": LH, "Afro-American Fragment," *Crisis* 37 (July 1930): 235.
"GLAD THEY CHOSE YOU": LH to Joel E. Spingarn, Dec. 10, 1930, JESP, NYPL.
"with my deep appreciation": LH to Joel E. Spingarn, May 28, 1930, JESP, NYPL.

page 190

"without bitterness or apology": Arthur B. Spingarn to LH, July 15, 1930, LHP.
"an artistic and most gripping human document": Robert M. Labaree to LH, Dec. 12, 1929, LHP.
"Bring out the laurel wreath": *Pittsburgh Courier,* July 26, 1930.
"the simplicity of great art": Sterling Brown, "Not Without Laughter," *Opportunity* 8 (Sept. 1930): 279.
"where I used to climb": Alain Locke to LH, Aug. 6, 1930, LHP.
"not up to your usual": Anne Marie Coussey [Wooding] to LH, n.d. [Aug. 1930], LHP.
"had gotten awfully bored": LH to Claude McKay, Sept. 30, 1930, CMP, JWJ.

page 191

"It's a pretty spot": *Ibid.*
"the lost loveliness that was": Robert Lewis, "Rose McClendon," in Nancy Cunard, ed., *Negro: An Anthology* [abridged, ed. Hugh Ford] (New York: Frederick Ungar, 1970), p. 199.
"My old man's a white old man": LH, "Cross," *Crisis* 31 (Dec. 1925): 66.

page 192

"He was not large": John Mercer Langston, *From the Virginia Plantation to the National Capitol* (Hartford, 1894), p. 21.
"short but excruciating session": Louise Thompson to LH, Oct. 4, 1930, LHP.
"I had thought she was wonderful": Louise Thompson Patterson to author, interview.

page 193

"You love me. You have proved": Zora Neale Hurston to Mrs. R. O. Mason, Nov. 25, 1930, ALLP, MSRC.

"I am helping myself forget": Mrs. R. O. Mason to Alain Locke, Aug. 8, 1930, ALLP, MSRC.

"the winter sunshine on Park Avenue": LH, *The Big Sea,* p. 326.

"Merry Christmas, China": LH, "Merry Christmas," *New Masses* 6 (Dec. 1930): 4.

page 194

"your two letters telling me of": Mrs. R. O. Mason to LH, Nov. 30, 1930, LHP.

"Violent anger makes me physically ill": LH, *The Big Sea,* p. 327.

"by mere accident it came to me": Mrs. R. O. Mason to LH, Jan. 10, 1931, LHP.

page 195

"Would you do anything": LH to Carl Van Vechten, Jan. 16, 1931, CVVP, Yale.

"cried and carried on": Carl Van Vechten to LH, Jan. 20, 1931, LHP.

in an absolute "tantrum": Carl Van Vechten to LH, Aug. 17, 1942, LHP.

"I am not using one single solitary": Zora Neale Hurston to Mrs. R. O. Mason, Jan. 20, 1931, ALLP, MSRC.

"By this time . . . I had come to feel": Zora Neale Hurston to LH, n.d., LHP.

page 196

"25 percent original research": Robert E. Hemenway, *Zora Neale Hurston: A Literary Biography* (Urbana, Ill.: Univ. of Illinois, 1977), p. 96.

"went off to his rooms": Entry, Jan. 22, 1931, in Mrs. Mason's notebook, "1928 *[sic]* Data as we close 399 [Park Avenue]"; ALLP, MSRC.

"It is just as we know": Zora Neale Hurston to Mrs. R. O. Mason, Jan. 20, 1931, ALLP, MSRC.

"well, she's just lying": Louise Thompson Patterson to author, interview.

page 197

"the split is over": LH to Carl Van Vechten, Jan. 22, 1931, CVVP, Yale.

"Thirty days in jail": *Ibid.*

"I shouldn't wonder if you are": Carl Van Vechten to LH, Jan. 19, 1931, LHP.

An editor . . . was "rather disappointed": Alfred A. Knopf Inc. (Bernard Smith) to LH, Mar. 2, 1931, LHP.

page 198

"The tragedy is the credit will go": Alain Locke to Mrs. R. O. Mason, Jan. 29, 1931, ALLP, MSRC.

"She made such a scene": LH to Carl Van Vechten, Feb. 4, 1931, CVVP, Yale.

"Carrie was absolutely magnificent": Rowena Woodham Jelliffe to author, interview, Dec. 7, 1980.

"DARLING GODMOTHER ARRIVED SAFELY": Zora Neale Hurston to Mrs. R. O. Mason, Feb. 3, 1931, ALLP, MSRC.

"Do you think she is crazy": LH to Carl Van Vechten, Feb. 4, 1931, CVVP, Yale.

"What a sorrowful misguided way": Mrs. R. O. Mason to LH, Feb. 12, 1931, LHP.

"When the sun shines again": LH to Mrs. R. O. Mason, draft, Feb. 22, 1931, LHP.

page 199

"Langston . . . do hurry to recognize": Mrs. R. O. Mason to LH, Mar. 1, 1931, LHP.

"a signed author's portrait racket": Alain Locke to Mrs. R. O. Mason, Mar. 5, 1931, ALLP, MSRC.

"a mad careening": *Ibid.*

"isn't that shameful": Alain Locke to Mrs. R. O. Mason, Mar. 20, 1931, ALLP, MSRC.

Hughes's "real mean-ness": Alain Locke to Mrs. R. O. Mason, Apr. 16, 1931, ALLP, MSRC.

"Why can't he die!": Alain Locke to Mrs. R. O. Mason, Mar. 29, 1931, ALLP, MSRC.

"Dear Godmother, the guard-mother": Zora Neale Hurston to Mrs. R. O. Mason, Mar. 10, 1931, ALLP, MSRC.

"Knowing you is like Sir Percival's glimpse": Zora Neale Hurston to Mrs. R. O. Mason, Sept. 28, 1932, ALLP, MSRC.

"I can't conceive of such lying": Zora Neale Hurston to Mrs. R. O. Mason, Mar. 25, 1931, ALLP, MSRC.

"Honest Godmother it requires all my self-restraint": Zora Neale Hurston to Mrs. R. O. Mason, May 17, 1932, ALLP, MSRC.

"Or perhaps a nice box of apples": Zora Neale Hurston to Arthur B. Spingarn, Mar. 27, 1931, ABSP, MSRC.

she was "a changed woman": Arna Bontemps to LH, Nov. 24, 1939, LHP. In Charles H. Nichols, ed., *Arna Bontemps-Langston Hughes Letters: 1925–1967* (New York: Dodd, Mead, 1980), p. 44

page 200

"He got sick": Louise Thompson Patterson to author, interview.

"I hear almost no news": Alain Locke to Mrs. R. O. Mason, Mar. 29, 1931, ALLP, MSRC.

"Well . . . I guess my play": LH to Carl Van Vechten, Mar. 13, 1931, CVVP, Yale.

page 201

"Zell was built like a stevedore": Blyden Jackson to author, interview, Oct. 15, 1980.

"appropriate libation" on Dixie soil: LH [and Zell Ingram], "Journals: 1931. The official daily log book—Jersey to the West Indies", LHP 586.

"Sunrise in a new land": *Ibid.*

page 202

"El poeta Langston Hughes": *Diario de la Marina,* April 8, 1931. Cited in Edward J. Mullen, *Langston Hughes in the Hispanic World and Haiti* (Hamden, Ct.: Archon Books, 1977), p. 31, 44n.

"Presentación de Langston Hughes": José Antonio Fernández de Castro, "Presentación de Langston Hughes," *Revista de la Habana* (March 1930): 367–368. Reprinted in Mullen, *Hughes in the Hispanic World,* pp. 169–171.

"¡Hombre! ¡Que formidable": LH to Nicolás Guillén, July 17, 1930, in Angel Augier, *Nicolás Guillén* (Havana: Contemporáneos, 1971), p. 95; also in Mullen, *Hughes in the Hispanic World,* p. 30.

page 203

policemen arrived "with drawn sabres": LH, *I Wonder As I Wander* (New York: Rinehart, 1956), p. 13.

"This is outrageous": LH, *I Wonder As I Wander,* p. 14.

"But now / Against a pirate called": LH, "To the Little Fort of San Lázaro, On the Ocean Front, Havana," *New Masses* 6 (May 1931): 11.

page 204

"my most verbatim story": LH to James A. Emanuel, Sept. 19, 1961, LHP.

"The dream is a cocktail at Sloppy Joe's—": LH, "Havana Dreams," *Opportunity* 11 (June 1933): 181. See also LH, *Selected Poems,* p. 49.

"the palace band plays immortally": LH, "People Without Shoes," *New Masses* 7 (Oct. 1931): 12.

page 205

"groups of marines in the little cafés": LH, "White Shadows in a Black Land," *Crisis* 39 (May 1932): 157.

"about all for which one can give the Marines credit": LH, "People Without Shoes," p. 12.

"has its hair caught in the white fingers": LH, "White Shadows in a Black Land," p. 157.

"all the work that keeps Haiti alive": LH, "People Without Shoes," p. 12.

page 206

"Stronger, vaster, more beautiful": LH, "A Letter from Haiti," *New Masses* 7 (July 1931): 9.

page 207

objected "violently to our carrying groceries": LH, *I Wonder As I Wander,* p. 24.

"a sex dance undisguised": LH, *I Wonder As I Wander,* p. 22.

"But the black Haitians of the soil": LH, *I Wonder As I Wander,* p. 22.

"It was in Haiti that I first realized": LH, *I Wonder As I Wander,* p. 28.

page 208

"I haven't done any work": LH to Carl Van Vechten, May 27, 1931, CVVP, Yale.

which he found "very amusing": James N. Hughes to LH, Jan. 13, 1931, LHP.
"In one way or another": LH to James N. Hughes, June 30, 1931, LHP.
"This is a hell of a country": LH, "Journals: 1931" (July 6), LHP 586.
"Farewell to Haiti": LH, "Journals: 1931" (July 10), LHP 586.
"the deep fiery eyes": LH, *I Wonder As I Wander,* p. 29.

page 209

"I descended the hill": LH, *I Wonder As I Wander,* p. 30.
"I saw approaching a long line": LH, *I Wonder As I Wander,* p. 31.
"one of the finest people": LH, "Journals: 1931" (July 11), LHP 586.
"inspiré par votre existence": Jacques Roumain to LH, Nov. 13, 1931, LHP.
"At Lagos you knew sad faced girls." Jacques Roumain, "Langston Hughes," in Edna Worthley Underwood, trans., *The Poets of Haiti, 1782–1934* (Portland, Maine: The Mosher Press, 1934), p. 66.

9. Starting Over

page 211

"Get in my car": LH, *I Wonder As I Wander,* p. 33.
"America's leading Negro woman": LH, *I Wonder As I Wander,* p. 40.
"What luck for us!": LH, *I Wonder As I Wander,* p. 40.

page 212

"She was a wonderful sport": LH, *I Wonder As I Wander,* p. 40.
Levin of the Julius Rosenwald fund was . . . "almost certain": Walter White to LH, May 5, 1931, LHP.
Moe of the Guggenheim Foundation had "practically assured": Charles S. Johnson to LH, May 15 [1931], LHP.
"yes, I'll be very glad": Amy Spingarn to LH, July 15, 1931, LHP.
"made a warmhearted little talk": LH, *I Wonder As I Wander,* p. 41.

page 213

"to make a loafing tour": Zora Neale Hurston to LH, Nov. 22, 1928, LHP.
"very affable and extrovert": Blyden Jackson to author, interview, Oct. 15, 1980.
"There is not in the entire country": *New York Amsterdam News,* Aug. 5, 1931.
"the end of the gay times": LH, *The Big Sea,* p. 247.
"All he got for his trouble": Zora Neale Hurston to Mrs. R. O. Mason, Aug. 14, 1931, ALLP, MSRC.
"YOU HAVE WRITTEN A SWELL BOOK": LH to Wallace Thurman, Mar. 2, 1932, WTP.
"smiling and self-effacing": Wallace Thurman, *Infants of the Spring* (New York: Macaulay, 1932), pp. 231–232.

page 214

"dilapidated and garbage-strewn": Claude McKay, *Harlem: Negro Metropolis* (New York: Harcourt Brace Jovanovich, 1968), p. 23.
"still in the process of making": James Weldon Johnson, *Black Manhattan* (New York: Atheneum, 1977), p. 281.
"all of Harlem a vast slum": McKay, *Harlem,* p. 28.
"I'd like to make a reading tour": LH to Walter White, Aug. 5, 1931; Rosenwald Papers (microfilm), ARC. The Rosenwald Papers are at Fisk University, but the Langston Hughes folder (Box 422, no. 13) was missing when I visited there.
"a rare personality": Frank Olmstead to Walter White, Aug. 6, 1931, WWP, LC.
"Lots of New Yorkers tried to tell me": LH to Julius Rosenwald Fund (Edwin R. Embree), Mar. 17, 1932, LHP.

page 215

"I am Africa": *Lincoln University News* (Thanksgiving Number, 1927): 8.
"It beats being a Red Cap": LH, *I Wonder As I Wander,* p. 42.
"Fight against white chauvinism": "Draft Manifesto of John Reed Clubs," *New Masses* 7 (June 1932): 4.

"a group of proficient actors": *New Masses* 6 (Sept. 1931): 21.

page 216

the "days when you and Whittaker Chambers and I": Paul Peters to LH, Nov. 22, 1965, LHP.

"with Jesuitical zeal and cleverness": Walter White, "The Negro and the Communists," *Harper's Monthly* 164 (Dec. 1931): 70.

"There is a tenseness": *Washington Tribune,* June 29, 1940; quoted in Robert M. Farnsworth, "What Can A Poet Do? Langston Hughes and M. B. Tolson," *New Letters* (Fall 1981): 19–29.

page 217

"Voices crying in the wilderness": LH, "To Certain Negro Leaders," *New Masses* 6 (Feb. 1931): 4.

"I am so tired of waiting": LH, "Tired," *ibid.*

page 218

'Give up beauty for a moment.": LH, "Call to Creation," *ibid.*

"God slumbers in a back alley": LH, "A Christian Country," *ibid.*

"The CRISIS must live": LH to W. E. B. Du Bois, Apr. 28, 1932, WEBDP, UM(A).

"Perhaps today / You will remember John Brown": LH, "October the Sixteenth," *Opportunity* 9 (Oct. 1931): 299.

"8 BLACK BOYS IN A SOUTHERN JAIL": LH, "SCOTTSBORO," *Opportunity* 9 (Dec. 1931): 379.

page 219

"Look! See What **Vanity Fair** says about the new": LH ["Advertisement for Opening of the Waldorf-Astoria"], *New Masses* 7 (Dec. 1931): 16–17.

page 220

"Eight Black Boys, A White Man": LH, "Scottsboro, Limited: A One Act Play," *New Masses* 7 (Nov. 1931): 18–21.

"went out in rags to conquer": LH and Kaj Gynt, "Cock o' the World" (outline), LHP 254.

"that we were planning to stop": LH, *I Wonder as I Wander,* p. 55.

"a very valiant and forthright servant": Prentiss Taylor to author, interview, Aug. 29, 1980.

page 221

"the modern Negro Art Movement": LH to Prentiss Taylor, Oct. 13, 1931, LHP.

"What the black masses need": George S. Schuyler to LH, Oct. 14, 1931, LHP.

"I did not know how": LH, "My Life and Times (My Lively Times)", LHP 733.

page 222

"Children, I come back today": LH, "The Negro Mother," *The Negro Mother and Other Dramatic Recitations* (New York: Golden Stair Press, 1931), p. 16.

"He never gave away much": Prentiss Taylor to author, interview.

page 223

"the most beautiful of Negro campuses": LH, *I Wonder As I Wander,* p. 43.

"I was deeply touched": LH, "Cowards from the Colleges," *Crisis* 41 (Aug. 1934): 227.

"That is not Hampton's way": LH, *I Wonder As I Wander,"* p. 44.

"This young artist, swaying the emotions": *Hampton Script,* Nov. 11, 1931.

his words . . . "burned" him: John W. Cooper to LH, Nov. 7, 1931, LHP.

page 224

"most of us white folks": Guy B. Johnson to LH, Oct. 27, 1931, LHP.

"Christ is a Nigger": LH, "Christ in Alabama," *Contempo* 1 (Dec. 1931): 1.

"so they won't need to be prostitutes": *Ibid.*

page 225

"half-baked, uneducated, and wholly reprehensible": R. B. House to Kemp D. Battle, Dec. 17, 1931, FPGP, UNC.

"We had Langston Hughes for dinner": Anthony Buttitta, "A Note on *Contempo* and Langston Hughes," in Nancy Cunard, ed., *Negro* (New York: Negro Universities Press, 1969), p. 142.

"a million bottles of home brew": LH to Walter White, Dec. 8, 1931, WWP, LC.

"the insulting and blasphemous articles": *Southern Textile Bulletin,* Dec. 10, 1931.

"the angels of darkness": *Daily Tar Heel,* May 8, 1932.

"nothing but a corrupt, distorted brain": J. B. Whittington to Frank Porter Graham, Sept. 16, 1932, FPGP, UNC.

"It's bad enough to call Christ a bastard": LH, *I Wonder As I Wander,* p. 46.

"I have been a little fearful": Elmer A. Carter to LH, Jan. 8, 1932, LHP.

"Please Please *don't go":* Carrie Clark to LH, Dec. 11, 1931, LHP.

"a swell time": LH to Walter White, Dec. 8, 1931, WWP, LC.

page 226

"furore in North Carolina": Thomas L. Dabney to LH, Dec. 24, 1931, LHP.

"I believe that anything which makes people think": *Atlanta World,* Dec. 18, 1931.

acknowledged receipt of an "enormous check": Prentiss Taylor to LH, Dec. 1, 1931, LHP.

"exactly my idea of what a *true* Southern gentleman": LH, *I Wonder As I Wander,* p. 49.

"What has been said of Hughes": *Nashville Banner,* Jan. 30, 1932.

page 227

"quiet and scholarly, looking like a young edition": LH, *The Big Sea,* p. 248.

"I thought I saw an angel flying low": Arna Bontemps, "Nocturne at Bethesda," in LH and Arna Bontemps, eds., *Poetry of the Negro: 1746–1970* (New York: Doubleday, 1970), p. 211.

page 228

"We lived in a decaying plantation mansion": Arna Bontemps, "Why I Returned," *The Old South* (New York: Dodd, Mead, 1973), p. 14.

"the world's worst school": Arna Bontemps to Walter White, Nov. 13, 1931, WWP, LC.

"I find him a steady and consistent writer": LH to Rosenwald Fund (George R. Arthur), Dec. 31, 1931, LHP.

page 229

"My wife did not favor this": Arna Bontemps, *The Old South,* p. 15.

"Sh-SSS-SS-S!" he begged Hughes: LH, "Sh-SSS-S-S!" *Masses* (Toronto), (May–June 1933): 9.

"if I dared remain that long": LH to Carl Van Vechten, Jan. 2, 1932, CVVP, Yale.

"Scottsboro's just a little place": LH to Carl Van Vechten, Jan. 2, 1932, CVVP, Yale. Also in *Contempo* 1 (Feb. 15, 1932): 2.

"Mrs. Bethune knew how": LH, *I Wonder as I Wander,* p. 51.

"My son, my son!": LH, *I Wonder As I Wander,* p. 50.

"His work was a living encouragement": LH, "Vachel Lindsay: Incident," LHP 3809.

page 230

"A great many teachers": LH, "Cowards from the Colleges," p. 227.

"It sho is pityful": LH, ["Autobiographical and biographical notes"], LHP 78.

"And all the human world is vast and strange": LH, "Ph.D.," *Opportunity* 10 (Aug. 1932): 249.

"Dear Negro Leaders" LH, "Dear Negro Leaders" [1932], LHP 279.

page 231

"For a moment the fear came": LH, "Brown America in Jail: Kilby," *Opportunity* 10 (June 1932): 174.

"vacuous and insulting essay": Thomas Mabry to Allen Tate, Jan. 22, 1932, photocopy, WWP, LC.

"both very interesting writers": Allen Tate to Thomas Mabry, Jan. 20, 1932, WWP, LC.

"The south as an institution": Eugene Levy, *James Weldon Johnson: Black Leader, Black Voice* (Chicago: Univ. of Chicago, 1973), p. 328.

"introduced me most graciously": LH, *I Wonder As I Wander,* p. 52.

"my fellow poet": Recollection of Ruth Britten to Roberta Miller, in Roberta Miller to LeRoy Percy, n.d. [1983]; courtesy of LeRoy Percy.

page 232

"and over the years that followed": LH, *I Wonder As I Wander,* p. 52. No such letters have been found.

"such regimentation as practiced in this college": LH, "Cowards from the Colleges," pp. 226–227.

"Fear has silenced their mouths": LH, "*Negro Songs of Protest,* by Lawrence Gellert: Introduction" [1932], LHP 765.

"American Negroes in the future": LH, "Cowards from the Colleges," p. 228.

"a grand high yellow audience": LH to Carl Van Vechten, Feb. 12, 1932, CVVP, Yale.

''I was so very nervous'': Margaret Walker Alexander to author, interview, Dec. 20, 1981.

page 233

''in gratitude for his encouragement'': LH, *I Wonder As I Wander,* p. 51.

''maintains a guest house on its campus'': LH, ''Cowards from the Colleges,'' p. 227.

''the nice Negroes living like parasites'': LH to Claude McKay, Sept. 13, 1928, CMP, JWJ.

''Deep in Alabama earth'': LH, ''Alabama Earth: At Booker Washington's Grave,'' *Tuskegee Messenger* 6 (June 28, 1930): n.p.

''with such great depth and sincerity'': LH to Effie Lee Power, Feb. 15, 1932, copy, LHP.

''sold like reefers on 131st Street'': LH to Carl Van Vechten, Feb. 17, 1932, CVVP, Yale.

''I feel that for the first time'': LH to Rosenwald Fund (Edwin R. Embree), Mar. 5, 1932, LHP.

''I've never had a finer response'': LH to Rosenwald Fund (Edwin R. Embree), Mar. 17, 1932, LHP.

page 234

''the belle of the community'': *Kansas City Call,* March 18, 1932.

''our needs are many'': *Ibid.*

''This tour m'en fiche'': LH to Prentiss Taylor, Mar. 12, 1932, LHP.

''It was essentially a jazz version'': Alain Locke, ''Sterling Brown: The New Negro Folk Poet,'' in Nancy Cunard, ed., *Negro: An Anthology,* p. 115.

''burnt out and trading on the past'': Alain Locke to Mrs. R. O. Mason, Mar. 26, 1932, ALLP, MSRC.

page 235

''I am sure the whole plan'': Louise Thompson to LH, Mar. 10, 1932, LHP.

''the usual kinky headed caricatures'': LH to Helen Sewell, Feb. 15, 1932, copy, LHP.

''too red to be included'': LH to Prentiss Taylor, Feb. 23, 1932, LHP.

''I'm more excited about this Scottsboro booklet'': LH to Prentiss Taylor, Apr. 12, 1932, LHP.

page 236

''Where women shoot pool'': LH to Carl Van Vechten, Apr. 9, 1932, CVVP, Yale.

''miles of yards'': LH to Toy Harper, Apr. 26, 1932, LHP.

''my representative on the coast'': LH to Noël Sullivan, Apr. 26, 1932, NSP, Bancroft.

''the swellest apartment'': LH to Prentiss Taylor, Apr. 19, 1932, LHP.

''like a big kid'': *California Eagle* [n.d., misc. poems and clippings], LHP Uncat.

''That day is past'': LH, *A New Song* (New York: International Workers Order, 1938), pp. 24–25.

page 237

''That the land might be ours'': LH, ''An Open Letter to the South,'' *New Masses* 7 (June 1932): 10.

''Many of your poems insist'': LH to Ezra Pound, Apr. 22, 1932, LHP.

page 238

''America's Foremost Negro Poet'': Program for appearance by LH, May 14, 1932, Berkeley, California; courtesy of Matt Crawford.

''That poem had a tremendous impact'': Matt Crawford to author, interview, May 28, 1980.

''the only Noel Sullivan'': In presentation copy to Noël Sullivan of LH, *The Weary Blues,* May 22, 1932; courtesy of Gladys Sullivan Mahoney.

page 239

''Deep down in my heart'': Noël Sullivan, *Forty Years Remembered: A Letter in the Form of a Memoir to the Children of my Sister Gladys S. Doyle.* Privately printed, 1954; courtesy of Gladys Sullivan Mahoney.

''would feel very much honored'': Noël Sullivan to LH, Mar. 3, 1932, LHP.

a ''very amusing boy'': LH to Carl Van Vechten, May 16, 1932, CVVP, Yale.

''an evening of spirit'': San Francisco *Daily News,* May 25, 1932.

''He seemed amazed about everything'': Eulah Pharr to author, interview, June 20, 1980.

''Without any possible disloyalty'': Noël Sullivan to LH, May 18, 1932, LHP.

page 240

''One evening at dinner'': Matt Crawford to author, interview.

''Be sure to take soap'': LH, *I Wonder as I Wander,* pp. 65–66.

page 241
"DE LUXE EDITION MOST BEAUTIFUL BOOK": LH to Prentiss Taylor, June 6, 1932, LHP.
"YOU HOLD THAT BOAT": LH to Louise Thompson, June 6, 1932, courtesy of Louise Thompson Patterson.

10. Good Morning, Revolution

page 242
"in a mock jaunty tone": Alain Locke to Mrs. R. O. Mason, June 16, 1932, ALLP, MSRC.
grown "stuffy" and "a frightful bore": Louise Thompson to Louise ("Mother") Thompson, n.d. [June 20, 1932]. Letters to her mother are quoted by courtesy of Louise Thompson Patterson.
page 243
"but most of the twenty-two": LH, *I Wonder As I Wander,* p. 70.
"a magnificent person": Paul Peters to LH, May 29, 1932, LHP.
"We have some giddy people": Louise Thompson to Louise ("Mother") Thompson [June 20, 1932].
page 244
"the pathos and poverty of Berlin's low-priced market": LH, *I Wonder As I Wander,* p. 71.
an almost "fairy-tale journey": LH, *I Wonder As I Wander,* p. 72.
"a plainly pleasant town": LH, *I Wonder As I Wander,* p. 73.
page 245
"Bow down my soul in worship very low": Claude McKay, "St. Isaac's Church, Petrograd," in Wayne F. Cooper, ed., *The Passion of Claude McKay: Selected Poetry and Prose, 1912–1948* (New York: Schocken, 1973), p. 127.
"in terms of Russian buying power": LH, *I Wonder As I Wander,* p. 75.
"Don't mention Hollywood": LH, *I Wonder As I Wander,* p. 75.
"Of all the big cities": LH, *I Wonder As I Wander,* p. 74.
page 246
"like royalty used to be entertained": Louise Thompson to Louise "Mother" Thompson, July 14, 1932.
"famous Negro novelist and poet": Moscow *Daily News,* June 28, 1932.
"Quite truthfully, there was no": LH, "Moscow and Me: A Noted American Writer Relates his Experiences," *International Literature* (No. 3, July 1933): 60–66.
Lewis . . . "has disappointed everybody": Louise Thompson to Louise "Mother" Thompson, July 14, 1932.
page 247
"We have had to argue": *Ibid.*
"Vot ist matter?": LH, *I Wonder As I Wander,* p. 89.
"At first I was astonished": LH, *I Wonder As I Wander,* p. 76.
"It would have looked wonderful": LH, *I Wonder As I Wander,* p. 79.
page 248
"nearly became a nervous wreck": LH, *I Wonder As I Wander,* p. 80.
nicknames "Madame Moscow" and "Glupie": Louise Thompson Patterson to author, interview, May 23, 1984.
"amazed at the näiveté": LH, *I Wonder As I Wander,* pp. 75–76.
page 249
"naked as birds and as frolicksome": LH, *I Wonder As I Wander,* p. 94.
Lewis's "disgraceful conduct": Louise Thompson to Louise ("Mother") Thompson, Aug. 24, 1932.
"Comrades! We've been screwed!": Henry Lee Moon to author, interview, Nov. 27, 1981; also in LH, *I Wonder As I Wander,* p. 95.
"SOVIET CALLS OFF FILM": Paris *Herald-Tribune,* Aug. 12, 1932.
"No Negroes went bathing": LH, *I Wonder As I Wander,* p. 95.
page 250
"Rejecting as unsound, insufficient and insulting": Text of statement by McNairy Lewis, Theo-

dore R. Poston, Henry Lee Moon, and Laurence Alberga, Aug. 22, 1932; courtesy of Louise Thompson Patterson.

"gross administrative inefficiency": "Summary," statement unsigned, n.d.; courtesy of Louise Thompson Patterson.

"a shameful, foolish experience": Louise Thompson to Louise ("Mother") Thompson, Aug. 24, 1932.

denounced Poston as . . . another "son-of-a-bitch": Matt Crawford to author, interview, May 20, 1980.

page 251

"NEGROES ADRIFT IN 'UNCLE TOM'S' ": New York *Herald-Tribune,* Aug. 12, 1932.

"It was the first time I realized": LH, *I Wonder As I Wander,* p. 97.

"The two assailed Joseph Stalin's": *New York Amsterdam News,* Oct. 19, 1932.

"a strange concoction of lies": Loren Miller to *Pittsburgh Courier* (Floyd Calvin), Sept. 17, 1932; courtesy of Louise Thompson Patterson.

"a few of us had to go act": Louise Thompson to Louise ("Mother") Thompson, Sept. 7, 1932.

"O, Movies. Temperaments": "Moscow and Me," p. 63.

"The fact of the case": Matt Crawford to Evelyn Crawford, Sept. 21, 1932; courtesy of Matt Crawford.

"I realize now how much": *Ibid.*

page 252

"Good-morning, Revolution": LH, "Good Morning Revolution," *New Masses* 7 (Sept. 1932): 5.

"Listen, Christ": LH, "Goodbye Christ," *The Negro Worker* 2 (Nov.–Dec. 1932): 32. Hughes sometimes later referred to the poem as "Goodbye, Christ."

page 253

"It is the same everywhere for me": LH, "The Same," *The Negro Worker* 2 (Sept.–Oct. 1932): 31–32.

page 254

"He is certainly doing very well": Louise Thompson to Louise "Mother" Thompson, Sept. 18, 1932.

"a model of its kind": *New York Times Book Review,* Oct. 23, 1932.

"To an American Negro": LH, *A Negro Looks at Soviet Central Asia* (Moscow: Co-operative Publishing Society of Foreign Workers in the U.S.S.R., 1934), p. 5.

"mandolins and guitars, balalaikas and accordions": LH, *I Wonder As I Wander,* p. 103.

page 255

"He is the mayor of Bokhara": LH, "Going South in Russia," *Crisis* 41 (June 1934): 162.

"We sent a telegram to Comrade Gorky": LH, "Going South in Russia," p. 163.

"the colour line is hard and fast": LH, *A Negro Looks at Soviet Central Asia,* p. 5.

page 256

"If our [tour] continues": Matt Crawford to Evelyn Crawford, Sept. 30, 1932; courtesy of Matt Crawford.

"There is an exact parallel": *Ibid.*

page 257

"the most typically Asiatic": Matt Crawford to Evelyn Crawford, Oct. 4, 1932; courtesy of Matt Crawford.

page 258

"We have been able to see": LH *et al.,* "To the Workers and Peasants of Uzbekistan Socialist Soviet Republic," Oct. 5, 1932; courtesy of Louise Thompson Patterson.

"The past ten days have been the greatest": Louise Thompson to Louise ("Mother") Thompson, Oct. 7, 1932.

"I seldom get angry": LH, *I Wonder As I Wander,* p. 107.

page 259

"It was difficult not to say": Arthur Koestler, *The Invisible Writing: An Autobiography* (London: Collins, 1954), p. 111.

"A writer must write": LH, *I Wonder As I Wander,* p. 113.

"a timid little mouse": Arthur Koestler, *Invisible Writing,* p. 112.
"was so fascinated by this sleepy-eyed trial": LH, *I Wonder As I Wander,* p. 116.
"Atta Kurdov looks guilty": LH, *I Wonder As I Wander,* p. 117.

page 260

"The Poet listens like a child": Robert M. Farnsworth, ed., *Caviar and Cabbage: Selected Columns by Melvin B. Tolson from the Washington Tribune, 1937–1944* (Columbia, Missouri: University of Missouri Press, 1982), p. 255.
"a poet with a purely humanitarian approach": Arthur Koestler, *Invisible Writing,* p. 112.
one of the "emotional hypochondriacs": LH, *I Wonder As I Wander,* p. 120.
"behind the warm smile": Arthur Koestler, *Invisible Writing,* p. 111.
"as drab and ugly a city": LH, *I Wonder As I Wander,* p. 119.

page 261

"paler lads who looked like Persian figures": LH, *I Wonder As I Wander,* p. 123.
"slobbering in each other's bowls": LH, *I Wonder As I Wander,* p. 115.
"Koestler almost keeled over": LH, *I Wonder As I Wander,* p. 130.
"always, if he had any notes": LH, *I Wonder As I Wander,* p. 133.
"I saw other former synagogues": LH, *I Wonder As I Wander,* p. 136.

page 262

"had been lured under false pretences": Arthur Koestler, *Invisible Writing,* p. 114.
"Something hard and young in me": LH, *I Wonder As I Wander,* pp. 147–148.

page 263

"I am afraid I did not": LH, "Dixie Christmas USSR," in "A Christmas Sampler" (1958), LHP 247.
"The gates were locked": LH, *I Wonder As I Wander,* p. 187.

page 264

"I know what imperialist exploitation": Moscow *Daily News,* Feb. 6, 1933.
"Langston was charming, listening": Si-lan Chen Leyda (Sylvia Chen), *Footnote to History,* ed. Sally Barnes (New York: Dance Horizons, 1984), p. 162.
"My admiration for his integrity": Leyda, *Footnote to History,* p. 162.
"A delicate, flowerlike girl": LH, *I Wonder As I Wander,* p. 256.

page 265

"I am so sad": Leyda, *Footnote to History,* p. 163.
"the girl I was in love with": LH, *I Wonder As I Wander,* p. 256.
"Langston was the first man": Si-lan Chen Leyda to author, interview, June 6, 1980.
"People are just committing suicide": Carrie Clark to LH, Sept. 19, 1932, LHP.
"There is so much to do": Louise Thompson to LH, Feb. 26, 1933, LHP.
"Hey! Hey! . . . I would like": LH to Prentiss Taylor, Mar. 5, 1933, LHP.
"even in places where there is almost nothing": LH, *I Wonder As I Wander,* p. 182.
"Now I know why the near-by Indian Empire": LH, *A Negro Looks at Soviet Central Asia,* p. 52.

page 266

"bourgeois aestheticism" of *The Weary Blues:* See Lydia Filatova, "Langston Hughes American Writer," *International Literature* (No. 1, 1933): 99–104.
"Never must mysticism or beauty": LH to Prentiss Taylor, Mar. 5, 1933, LHP.
"The revolutionary poems seem very weak": Carl Van Vechten to LH, Apr. 4, 1933, LHP.
"Blanche bases her note to me": LH to Carl Van Vechten, May 23, 1933, CVVP, Yale.
"Bring us with our hands bound": LH, "Our Spring," *International Literature* (No. 2, 1933): 4.
"Columbia, / My dear girl": LH, "Columbia," *International Literature* (No. 2, 1933): 54.

page 267

"The gentlemen who have got to be classics": LH, "Letter to the Academy," *International Literature* (No. 5, 1933): 112.

page 268

"Black world / Against the wall": LH, "A New Song," *Crisis* 40 (Mar. 1933): 59.
"The bees work.": LH, "Black Workers," *Crisis* 40 (Apr. 1933): 80.

seemed "to have a new ring to them": Amy Spingarn to LH, Apr. 13, 1933, LHP.
"a buxom body": LH, *I Wonder As I Wander,* p. 201.

page 269

"I'm writing you as to a ghost": Sylvia Chen to LH, Aug. 2, 1935, LHP.
"made my hair stand on end": LH, *I Wonder As I Wander,* p. 213.
"They were people who went in for Negroes": LH, "Slave on the Block," *Scribner's Magazine* 94 (Sept. 1933): 141–144.
"someone had offered him something": LH, "Poor Little Black Fellow," *American Mercury* 30 (Nov. 1933): 326–335. For "Cora Unashamed," see *American Mercury* 30 (Sept. 1933): 19–24.

page 270

"the world's worst travel bureau": LH to Noël Sullivan, June 12, 1933, NSP, Bancroft.
showed him "a swell time": LH to Carl Van Vechten, Mar. 12, 1933, CVVP, Yale.
"a suitcase crammed with tin goods": Leyda, *Footnote to History,* p. 163.
"like very pungent kennels": LH, *I Wonder As I Wander,* p. 201.
"This has even the hundreds of tanks": LH to Carl Van Vechten, May 23, 1933, CVVP, Yale.
"the outstanding living Negro writer": LH, "American Negro Writers. IX. Claude McKay: The Best"; n.d. [1933], LHP 29.

page 271

"OUR KEATS AND SHELLEY": LH, "Countee Cullen: OUR KEATS AND SHELLEY," and "Walter White: A WHITE NEGRO"; copies enclosed in Walter White to Amy and Joel Spingarn, Aug. 3, 1933; WWP, LC.
"could recite like small bronze Ciceros": Countee Cullen, *One Way To Heaven* (New York: Harper and Brothers, 1932), p. 154.
"Niggers have a hell of a time": Countee Cullen, *One Way To Heaven,* p. 154.
"Yes your mother is an actress": Carrie Clark to LH, Feb. 15, 1933, LHP.

page 272

"All the peace and happiness": LH, *I Wonder As I Wander,* p. 231.
"a gray fortress of a town": LH, *I Wonder As I Wander,* p. 234.
"as impersonal as a Technicolor movie": LH, *I Wonder As I Wander,* p. 240.
"The Japanese militarists are quite open": LH, "Swords Over Asia," *Fight* (June 1934) [misc. poems and clippings], LHP Uncat.
"but they were tanks": *Ibid.*

page 273

"the grandest time imaginable": LH to Noël Sullivan, June 30, 1933, NSP, Bancroft.
"the only large group of dark people": *Japan Advertiser,* July 1, 1933.
"predicted that there would one day": (Special Agent) R. P. Hood to Director, F.B.I. (J. Edgar Hoover), Feb. 7, 1942; LH files, F.B.I.
"Arms bristle everywhere": LH, "Swords Over Asia."
"as lovely to look at": LH, *I Wonder As I Wander,* p. 255.

11. Waiting on Roosevelt

page 276

"saying little, but radiating kindness": LH, *I Wonder As I Wander,* p. 281.
"Our acquaintanceship was for such an hurried moment": LH to Noël Sullivan, Jan. 31, 1933, NSP, Bancroft.

page 277

"The country faces the gravest crisis": "Mr. Roosevelt Must Lead," *Nation* 136 (Feb. 8, 1933): 137.

page 278

"Colored Poet Here": *Village Daily* (Carmel), Sept. 6, 1933.

page 279

"just one more California mecca": From the San Francisco *News,* reported in *The Carmelite* (Carmel) Dec. 1, 1932.

"a good old-fashioned tar-and-feather": *The Carmelite,* Apr. 28, 1932.

"gifts of snakes and poisoned candy": *Pine Cone* (Carmel), July 28, 1933.

page 280

"a grand place to live": LH to Harold Jackman, Oct. 17, 1933, HJP, JWJ.

"I am living so much like white folks": LH to Carl Van Vechten, Nov. 1, 1933, CVVP, Yale.

"The remoteness of this part of the world": LH to Countee Cullen, Nov. 19, 1933, CCP, ARC.

"rich young kids who were bored": LH to Carl Van Vechten, Nov. 1, 1933, CVVP, Yale.

"noble work!": LH to Carl Van Vechten, Nov. 1, 1933, CVVP, Yale.

page 281

"We could not stay": Ella Winter, *And Not To Yield: An Autobiography* (New York: Harcourt, Brace & World, 1963), p. 193.

"a very thrilling struggle": LH to Carl Van Vechten, Nov. 1, 1933.

"Bear in mind / that death is a drum": LH, "Drum," *Dear Lovely Death* (Amenia, N.Y.: Troutbeck Press, 1931), unpaginated.

"I was never personally interested": Maxim Lieber to author, interview, Aug. 3, 1979.

"By all means send me all": Maxim Lieber to LH, July 13, 1933, LHP.

page 282

"I have tried to assure him": Maxim Lieber to LH, Oct. 15, 1933, LHP.

"the kind of story no commercial magazine": "Three Characters In Search Of A Magazine That Is Unhampered By The Old Taboos," *Esquire* 1 (Jan. 1934): 15.

"the violence of both the Yeas and the Nays": *Esquire* 1 (Feb. 1934): 11.

"BEING ENTHUSIASTIC ADMIRER": *Esquire* 1 (Mar. 1934): 12.

"too sanguine as to their sales possibilities": Maxim Lieber to LH, Oct. 4, 1933, LHP.

"Nothing I can do will stop you": Maxim Lieber to LH, Oct. 20, 1933, LHP.

"Why is it that authors": *Atlantic Monthly* to Maxim Lieber, Jan. 18, 1934, LHP.

page 283

"and was THRILLED": Carl Van Vechten to LH, Dec. 15, 1933, LHP.

"absolutely top notch": Blanche Knopf to LH, Dec. 18, 1933, LHP.

"These stories are for you": LH, "The Ways of White Folks," ms., Dec. 25, 1933, NSP, Bancroft.

"who have based their inspiration": LH to Eugene O'Neill, Oct. 31, 1933, copy, LHP.

"the beauty you have given them": *Ibid.*

page 284

"I didn't have an invitation": Roy Blackburn to author, interview, May 29, 1980.

"At least . . . I shall hire two lawyers": LH to Carl Van Vechten, Dec. 26, 1933, CVVP, Yale.

"We often worked until two": Roy Blackburn to author, interview.

page 285

"Whoopee!!" Lieber exulted: Maxim Lieber to LH, Apr. 19, 1934, LHP.

"I silently": LH, "Wait," *The Partisan: Journal of Art, Literature & Opinion* 1 (Dec. 1933): 3.

"Great mob that knows no fear": LH, "Revolution," *New Masses* 10 (Feb. 20, 1934): 28.

"Alive in a marble tomb": LH, "Ballads of Lenin," *A New Song* (New York: International Workers Order, 1938), p. 20.

page 286

"Put one more 'S' in the U.S.A.," LH, "One More 'S' in the U.S.A.," *Daily Worker,* Apr. 2, 1934.

"the confiscation without compensation": *Equality Land And Freedom: A Program For Negro Liberation* (New York: League of Struggle for Negro Rights, 1934), pp. 10–11; courtesy of Thomas H. Wirth.

"Dear Sylvia . . . —Think about you": LH to Sylvia Chen [Si-lan Chen Leyda], Jan. 1, 1934; letters from LH may be found in the Chen-Leyda Papers, Special Collections, Bobst Library, New York University.

"Langston—Langston—However did you happen": Sylvia Chen to LH, n.d., LHP.

"COME OVER HERE": LH to Sylvia Chen, Apr. 1, 1934, copy, LHP.

page 287

never "to write you anymore": Sylvia Chen to LH, June 26, 1934, LHP.

"Darling kid, You know": LH to Sylvia Chen, July 7, 1934.

"the sweetest little girl": LH to Sylvia Chen, Oct. 18, 1934.

"What nationality would our baby be?": *Ibid.*

page 288

"or is it habit": Sylvia Chen to LH, n.d., LHP.

"as to a ghost": Sylvia Chen to LH, Aug. 2, 1935, LHP.

"Yes I got tired of waiting": Sylvia Chen to LH, Aug. 16, 1936, LHP.

"Langston wrote me a very angry letter": Si-lan Chen Leyda to author, interview, June 16, 1979.

"Seemingly there are too many people": LH to Carl Van Vechten, Apr. 18, 1934, CVVP, Yale.

page 289

"Why not thirty?": Justin Kaplan, *Lincoln Steffens: A Biography* (New York: Simon and Schuster, 1974), p. 318.

"Langston never brought it up": Roy Blackburn to author, interview.

"always the center of lots of laughter": John D. Short, Jr. to author, interview, June 9, 1980.

"was totally uninterested in women": Benjamin H. Lehman, "Recollections and Reminiscences of Life in the Bay Area from 1920 Onward: An Interview Conducted by Suzanne B. Reiss," Univ. of California Regional Oral History Office; ms., Bancroft, leaves 161–162.

page 290

"some of the best stories": "The Literary Landscape," *North American Review* 238 (Sept. 1934): 286.

"spiritual prose style": "Genius of Langston Hughes," *New York Herald Tribune Books* (July 1, 1934): 4.

"greater artistry, deeper sympathy": "Negro Angle," *Survey Graphic* 23 (Nov. 1934): 565.

"My hat is off to you": "Paying for Old Sins," *Nation* 139 (July 11, 1934): 49–50.

"either sordid and cruel": "White Folk Are Silly," *The New Republic* 80 (Sept. 5, 1934): 108–109.

page 291

"The play follows the strike": LH to Maxim Lieber, Aug. 8, 1934, LHP.

"The result . . . ought to be a sound": LH to Theatre Union (Michael Blanfort), June 15, 1934, LHP.

page 292

"radical capitalist" Noël Sullivan: LH to Carl Van Vechten, Aug. 12, 1934, CVVP, Yale.

"one's laundry man and one's druggist": LH, "The Vigilantes Knock At My Door", LHP 3811.

"There is trouble in store": *Carmel Sun,* July 26, 1934.

page 293

"I, as a Negro member": LH, "The Vigilantes Knock."

"not wishing to be tarred and feathered": LH to Sylvia Chen, Oct. 18, 1934.

"charming and pleasant, but": Blanche Knopf to LH, Aug. 3, 1934, LHP.

"I know you *don't* like them": LH to Carl Van Vechten, Mar. 5, 1934, CVVP, Yale.

"lacking in any of the elementary requisites": Carl Van Vechten to LH, Mar. 20, 1934, LHP.

page 294

"Langston Hughes has been a very 'distinguished' guest": *Sun,* Aug. 23, 1934.

"a tireless and violent advocator": *Sun,* Aug. 30, 1934.

page 295

"I would hate to *have* to marry": LH, "Journals: 1934–37", LHP 586.

"You got to be hard inside": *Ibid.*

page 296

"a very Armenian looking young man": LH to Carl Van Vechten, Oct. 31, 1934, CVVP, Yale.

"I'm waitin' on Roosevelt, son": LH, "Ballad of Roosevelt," *The New Republic* (Nov. 14, 1934): 9.

"political opportunists hoping for a rake-off": LH, "The Vigilantes Knock."

page 297

"Every Negro receiving a regular salary": *Daily Worker,* Oct. 9, 1934.
"As a fellow writer": LH to Walter White, Nov. 15, 1934, WWP, LC.
one of the twenty-five "most interesting": *Sunday Mirror* (New York), Aug. 26, 1934.
"urban Negro life in America": LH to John Simon Guggenheim Foundation, n.d., draft, LHP.
"the world owes a lot more": Noël Sullivan to LH, Oct. 16, 1934, LHP.

page 298

"no reason . . . why you should": LH to Noël Sullivan, Oct. 24, 1934, NSP, Bancroft.
"strangely enough insists on *not* going": LH to Maxim Lieber, Nov. 3, 1934, LHP.
"the Negroes exploded on the fact": Carrie Clark to LH, Jan. 15, 1934 [1935?], LHP.
"I think my *métier*": LH to Noël Sullivan, Oct. 24, 1934, NSP, Bancroft.
to "anyone for 7 or 8 bucks": LH to Maxim Lieber, Nov. 3, 1934, LHP.
"a forlorn little mountain cemetery": LH, *I Wonder As I Wander,* p. 287.
"I have no faith in him": James N. Hughes to LH, Aug. 31, 1934, LHP.

page 299

"They might think I was trying": LH to Noël Sullivan, Oct. 24, 1934, NSP, Bancroft.
"I have hit upon an idea": LH to Maxim Lieber, Nov. 3, 1934, LHP.
"Then Philip's father died": LH, "Postal Box: Love" ["about 1934"], LHP 2967.
"Of course, if my father is still living": LH to Noël Sullivan, Nov. 5, 1934, NSP, Bancroft.
"All of the misunderstandings": Carrie Clark to LH, Nov. 8, 1934, LHP.

page 300

"es *necesario":* Dolores Villaseñor y Patiño to LH, Nov. 8, 1934, LHP.
"It may be that I am an heir": LH to Blanche Knopf, Nov. 9, 1934, copy; LHP.
"I was happy myself": LH, *I Wonder As I Wander,* p. 289.

page 301

"I am afraid of those Mexicans": Sallie J. Garvin to LH, Dec. 29, 1934, LHP.
"perfect darlings": LH to Carl Van Vechten, Feb. 2, 1935, CVVP, Yale.

page 302

"most amusing . . . so simply and charmingly written": LH to Noël Sullivan, Jan. 11, 1935, NSP, Bancroft.
"he must have been colored": *Ibid.*
"poeta militante negro": José Antonio Fernández de Castro, "Langston Hughes, poeta militante negro," *El Nacional,* Mar. 3, 1935.
"el poeta de los negros": Luis Cardoza y Aragon, "Langston Hughes, el poeta de los negros," *El Nacional,* Mar. 17, 1935.

page 303

"I never lived in Greenwich Village": LH, *I Wonder As I Wander,* p. 295.
"It was no apartment": Henri Cartier-Bresson to author, interview, Sept. 29, 1981.

page 304

"Langston was a fine man": *Ibid.*
"A wall can be merely a wall": LH, "Pictures More Than Pictures: The Work of Manuel Bravo and Cartier-Bresson," Mar. 6, 1935, LHP 871.
he could "go on staying here": LH to Douglas and Marie Short, May 20, 1934, MSP, Bancroft.
"There are certain practical things": LH, "To Negro Writers," in Henry Hart, ed., *American Writers' Congress* (New York: International Publishers, 1935), pp. 139–140.

page 305

"I wish I could write you": Carrie Clark to LH, Apr. 15, 1935, LHP.

12. Still Waiting on Roosevelt

page 306

"We each put what we had": LH, *I Wonder As I Wander,* p. 300.
"that an able-bodied young Negro": Arna Bontemps, "Why I Returned," *The Old South* (New York: Dodd, Mead, 1973), p. 17.

page 307

"He seems to preserve a core": Ione Morrison Rider, "Arna Bontemps," *The Horn Book* 15 (Jan.–Feb. 1939): 15.

"extremely good looking, a beautiful brown": Horace R. Clayton, *Long Old Road* (New York: Trident Press, 1965), p. 247.

"I don't have that fury": Arna Bontemps to Ann Allen Shockley, July 14, 1972; Black Oral History Interview, Fisk University Library.

"a fantasy with lots of humor": LH to Noël Sullivan, June 7, 1935, NSP, Bancroft.

"In their way they are as selfish": Noël Sullivan to LH, Jan. 29, 1936, LHP.

page 308

"You really should just settle": Marie Short to LH, Aug. 6, 1935, LHP.

denounced the . . . editor as a "prize dunce": Maxim Lieber to LH, Aug. 2, 1935, LHP.

"Things Negro are booming": LH to Paul P. Boswell, Oct. 6, 1935, LHP.

page 309

"the God-damndest lousy commercial lot": LH, ["Personal Papers"], LHP 863.

"Stay away from Hollywood:"LH, *I Wonder As I Wander*, p. 305.

"anti-Christ" Langston Hughes: *California News* (Los Angeles), Aug. 10, 1935.

"making saints" of Lenin and Stalin: Beverly C. Ransom, "All Hail to Christ—A La Langston Hughes," *Voice of Missions* 39 (April 1933): 6.

branded "officially a Communist": *Cleveland Press*, Oct. 10, 1934.

"a Catholic, a rebel, and a proletarian": Robert M. Farnsworth, "What Can A Poet Do? Langston Hughes and Melvin B. Tolson," *New Letters* 48 (Fall 1981): 21.

page 310

"In an envelope marked: / *Personal*": LH, "Personal," *Selected Poems*, p. 88.

Toomer was "most evasive": LH to Noël Sullivan, Aug. 31, 1935, NSP, Bancroft.

page 311

"the first Negro woman": LH, *I Wonder As I Wander*, p. 309.

"My mother, as a great many poor mothers": LH, *I Wonder As I Wander*, p. 308.

"running all over the road": Toy Harper to LH, n.d., LHP.

page 312

"perfectly willing" to share: LH to American Play Company (John W. Rumsey), Aug. 13, 1935; APC archives, Berg Collection, NYPL. For the history of the *Mulatto* controversy see Folders 94–97 ("Agreements, correspondence, forms, statements from various theatrical agencies concerning Langston Hughes," comprising over 600 leaves), plus an additional folder and a typescript of the play.

"Over-ridden on all sides": LH, *I Wonder As I Wander*, p. 311.

page 313

"Rape is for sex": LH, *I Wonder As I Wander*, p. 311.

"Only Shakespeare has more tragedy": LH to Marie Short, Oct. 16, 1935, MSP, Bancroft.

"more the successful and secure artist": *New York Amsterdam News*, Oct. 5, 1935.

"As I see it, the basic economic problem": *St. Paul Pioneer Press*, Oct. 11, 1935.

"a small, prim, negro poet": *Minnesota Daily*, Oct. 11, 1935.

a "certain boyish charm": *St. Paul Recorder*, Oct. 11, 1935.

page 314

"Whatever this fretful arrival": *New York World-Telegram*, Oct. 25, 1935.

reporting a "feeling of revulsion": *New York Daily Mirror*, Oct. 25, 1935.

"weary, familiar stuff": *New York Sun*, Oct. 25, 1935.

"merely a bad play": *Brooklyn Eagle*, Oct. 25, 1935.

"neither brutality, vulgarity, bestiality": *Women's Wear Daily*, Oct. 25, 1935.

"After a season dedicated chiefly to trash": *New York Times*, Oct. 25, 1935.

page 315

"They completely distorted the poetic moments": LH to Noël Sullivan, Oct. 28, 1935, NSP, Bancroft.

"Let America be America again": LH, "Let America Be America Again," *A New Song* (New York: International Workers Order, 1938), p. 9.

page 316

"on the inequities of the system": LH to Carl Van Vechten, Nov. 29, 1935, CVVP, Yale.

"I'm profoundly moved": Maxim Lieber to LH, Nov. 19, 1935, LHP.

"a far advanced cancer": Dr. Stanley E. Brown to LH, Oct. 23, 1935, LHP.

"The only thing to do": LH to Noël Sullivan, Jan. 29, 1936, NSP, Bancroft.

page 317

"the place of entertainment": Reuben Silver, *A History of the Karamu Theatre of Karamu House;* doctoral dissertation, Ohio State University, 1961, p. 174.

"the outstanding thing to see": *Cleveland News,* Mar. 29, 1947.

"in spite of it all": LH to Noël Sullivan, Jan. 29, 1936, NSP, Bancroft.

"I had lived within the law": LH, "The Wages of Sin Are Not Always Death," May 26, 1951, LHP 3819.

"That bastard of a MULATTO": LH to Carl Van Vechten, May 11, 1936, CVVP, Yale.

"a strong dose of something!": William Grant Still to LH, Nov. 28 [1935], LHP.

page 318

"Pleeza, mister . . . me no spika": *New York World-Telegram,* Oct. 13, 1935.

"a lewd and licentious lie": *New York Telegraph,* Oct. 17, 1935.

"They're about like Zora": LH to Arthur B. Spingarn, Dec. 4, 1935, LHP.

"You are one of God's children": Maxim Lieber to LH, Dec. 9, 1935, LHP.

"a tragi-comedy of plantation life": LH to Maxim Lieber, Dec. 26, 1935, LHP.

page 319

"than to sit here penniless": LH to Maxim Lieber, Dec. 30, 1935, LHP.

"This has been a hard year": *Ibid.*

"She was a little difficult": Arna Bontemps to Ann Allen Shockley, interview, July 14, 1972; Black Oral History Interview, Fisk University Library.

"This is a song for the genius child": LH, Genius Child," *Selected Poems,* p. 83.

page 320

"I am happy as a lark": LH to Maxim Lieber, Jan 8, 1936, LHP.

"What is it about the theatre": LH to Maxim Lieber, Jan. 20, 1936, LHP.

Lieber urged Hughes to forget "the dialectic solution": *Ibid.*

page 321

"The openness of it so startled us": Arna Bontemps, *The Old South,* pp. 17–18.

"You can imagine . . . the contrast": LH to Noël Sullivan, Jan. 29, 1936, NSP, Bancroft.

page 322

"war is a racket": quoted in LH, ["Radio": broadcast re Third United States Congress Against War and Fascism], Jan. 2, 1935, LHP 2994.

"Russia can't help becoming": *Cleveland Press,* Jan. 2, 1936.

"ETHIOPIA / Lift your night-dark face": LH, "Call of Ethiopia," *Opportunity* 13 (Sept. 1935): 276.

"Ethiopia, stretch forth your mighty hand": LH and Thelma Brown, "Ethiopian Marching Song", LHP 2229.

"The little fox is still": LH, "Broadcast on Ethiopia," *American Spectator* (July–Aug. 1936): 16–17.

page 323

"enjoy Rome / and *take* Ethiopia": LH, "White Man," *New Masses* 21 (Dec. 15, 1936): 34.

"it is rather light": Maxim Lieber to LH, Dec. 18, 1935, LHP.

"the only thing I can do": LH to Noël Sullivan, Jan. 29, 1936, NSP, Bancroft.

page 324

"generally speaking, Negro writing": Richard Wright, "Blueprint for Negro Writing," *New Challenge* 2 (Fall 1937): 53.

"at the very mention" of the money: LH to Maxim Lieber, Feb. 8, 1936, LHP.

"to have your death due to starvation": Maxim Lieber to LH, Feb. 3, 1936, LHP.

page 325

"the soul and everything": Francine Bradley to LH, Mar. 8, 1936, LHP.

"SEND MAMA ANY HOSPITAL": LH to Gwyn ("Kit") Clark, n.d., LHP.

"a hilarious comedy": *Cleveland News,* Mar. 25, 1936.

"hilarious lines and good clean humor": *Cleveland Call and Post,* Apr. 2, 1936.

page 326

"there is a serious undertone": LH to Reuben Silver, May 6, 1961, in Silver, *History of the Karamu Theatre,* p. 233.

"underlying all that laughter": *Cleveland News,* March 26, 1936.

"To give up the hope": Silver, *History of the Karamu Theatre,* p. 232.

"If you entertain": LH to Reuben Silver, May 7, 1961, in Silver, *History of the Karamu Theatre,* p. 236.

"I think Langston should have been": Rowena Jelliffe to author, interview, Dec. 7, 1980.

page 328

"In fact, I'm delighted": LH to Noël Sullivan, May 27, 1936, NSP, Bancroft.

"I walked down Seventh Avenue": LH, *I Wonder As I Wander,* p. 315.

page 329

"Dr. Locke, do you remember me?": Ralph Ellison, "Remembering Richard Wright" (Paris) *Delta* (April 1984): pp. 2–3.

"I'm following your formula": Ralph Ellison to LH, July 17, 1936, LHP.

"LANGSTON HUGHES, NOTED AUTHOR": *Chicago Bee,* Aug. 2, 1936,

page 330

"Carlo—I'm lost": LH to Carl Van Vechten, n.d., CVVP, Yale.

"a city of charm": LH to Noël Sullivan, Sept. 15, 1936, NSP, Bancroft.

Lieber found the title "abominable": Maxim Lieber to LH, Oct. 6, 1936, LHP.

"a swell English briar pipe": LH to American Play Company (Beatrice Rumsey), Oct. 29, 1936, APC, NYPL.

"LANGSTON HUGHES' MOTHER BEATEN": *Cleveland Call and Post,* Nov. 19, 1936.

page 331

"an exceptionally interesting theatrical occasion": *Cleveland Plain Dealer,* Nov. 19, 1936.

"a lovely-looking girl": LH, *I Wonder As I Wander,* p. 328.

"She was an extremist": Kathleen A. Hauke, "The 'Passing' of Elsie Roxborough," *Michigan Review Quarterly* 23 (Spring 1984): 158.

"has what the girls call 'flash' ": *Baltimore Afro-American,* Apr. 24, 1937.

"the girl I was in love with": LH, *I Wonder as I Wander,* p. 328.

page 332

"the most striking girl": Hauke, "Elsie Roxborough," p. 155.

"and now I want to proposition": Elsie Roxborough to LH, Sept. 29, 1936, LHP.

"a staid old bore": Elsie Roxborough to LH, n.d. (Letter begins: "Apparently I was born . . ."), LHP.

"our affair" would be "tiresome": Elsie Roxborough to LH, n.d. ("As usual, Thursday morning . . ."), LHP.

"The tragedy of my life": Elsie Roxborough to LH, n.d. ("I am terribly disappointed! . . ."), LHP.

"hide my head in shame": Elsie Roxborough to LH, n.d. ("With my usual bravado . . ."), LHP.

"I do apologize": *Ibid.*

page 333

"What you need, if you would": Elsie Roxborough to LH, n.d. ("Delighted to receive you . . ."), LHP.

"would simply die of ecstasy": Elsie Roxborough to LH, Sept. 22, 1940, LHP.

"level-headed and phlegmatic": *Ibid.*

"Never begged before": Elsie Roxborough to LH, n.d. ("Your adorable letter came . . ."), LHP.

"I come to see how futile": Elsie Roxborough to LH, n.d. ("I am terribly disappointed! . . ."), LHP.

"so I began to think": Elsie Roxborough to LH, n.d. ("I have presented two . . ."), LHP.

"all that indifferent, platonic attitude": *Ibid.*

"I do miss you terribly": *Ibid.*

"You are in the same category": Gwyn "Kit" Clarke to LH, [Apr.] 12, 1937, LHP.

"Such people who would expect": Elsie Roxborough to LH, n.d. ("My sister has been . . ."), LHP.

"and embarrass me further": Elsie Roxborough to LH, Apr. 9, 1937, LHP.

"Denies Report": *Baltimore Afro-American,* Mar. 27, 1937.

"Are They Bound for The Altar?": *Pittsburgh Courier,* Apr. 10, 1937.

page 334

"Elsie Roxborough, Langston Hughes": *Ibid.*

"For God's sake don't": Elsie Roxborough to LH, n.d. ("Why haven't I heard . . ."), LHP.

"I wish you'd married Elsie": Margaret Bonds to LH, Oct. 19, 1942, LHP.

page 335

"but you said you'd rather": LH to Blanche Knopf, Mar. 1, 1937, LHP.

"Langston told me what had happened": Louise Thompson Patterson to author, interview, May 23, 1984.

"He buoyed up my hopes": Ralph Ellison ["Address at Langston Hughes Festival," City College of New York], Apr. 11, 1984.

"They don't seem quite the same": Ralph Ellison to LH, Apr. 27, 1937, LHP.

page 336

"Raise the standard of living": *Boston Traveller,* Mar. 15, 1937.

"a rollicking farce": *Cleveland News,* Apr. 1, 1937.

"Striving too mightily": *Cleveland Plain Dealer,* Apr. 3, 1937.

"Spacious and life-giving": LH to Noël Sullivan,, May 5, 1937, NSP, Bancroft.

page 337

"Believing firmly as I do": Noël Sullivan to LH, Nov. 5, 1936, LHP.

"In the light of our respective feelings": Noël Sullivan to LH, Mar. 3, 1937, LHP.

page 338

"Why oh why aren't you here?": Nancy Cunard to LH, Aug. 20 [1936], LHP.

"admires and knows your work": Nancy Cunard to LH, Mar. 5, 1937, LHP.

"I must drive the bombers out of Spain": LH, "Song of Spain," *A New Song,* pp. 21–23.

"in a civil and international conflict": *New Masses* 23 (May 4, 1937): 25.

page 339

"Lions in zoos": LH, "Sweet and Sour Animal Book," n.d., LHP 3707.

13. Earthquake Weather

page 341

"This is earthquake / Weather": LH, "Today," *Opportunity* 15 (Oct. 1937): 310.

page 342

"the amusement center of Europe": LH, "Colored Artists Musical Ambassador to Europe" (1937), LHP 259.

"I teel you how I fly zee plane": *Chicago Defender,* Dec. 5, 1942.

"Paris was so alive": LH, *I Wonder As I Wander,"* p. 320.

"I was very happy to see Langston": Henri Cartier-Bresson to author, interview, Sept. 30, 1981.

"so I saw within arm's reach": LH, *I Wonder As I Wander,"* p. 318.

page 343

"To these men, and Aimé Césaire": Mercer Cook to author, interview, Aug. 5, 1981.

"les deux poètes noirs": *Légitime Défense* 1 (June 1, 1932): 12.

"They were the first to admit": Mercer Cook to author, interview.

page 344

"a body like sculpture": LH, "Nancy: A Piñata in Memoriam, If One Could Break It In Her Honor," prologue to *Nancy Cunard: Brave Poet, Indomitable Rebel,* ed. Hugh Ford (N.Y.: Chilton Book Co., 1968), p. xxi.

"In the Civil War in Spain": LH, *I Wonder As I Wander,"* pp. 400–401.

"most especially representing the Negro peoples": LH, "Too Much Of Race,"*The Volunteer For Liberty* 1 (Aug. 23, 1937): 3.

page 345

"He never forgets his comrades": Nancy Cunard, "3 Negro Poets," *Left Review* 3 (Oct. 1937): 531.

page 346

"there was a wan bulb": LH, *I Wonder As I Wander,* p. 324.

"The next thing I knew": LH, *I Wonder As I Wander,* p. 325.

page 347

"The cafés were full": LH, *I Wonder As I Wander,* p. 327.

page 348

"It is a center for every writer": LH, "Madrid's House Of Culture," *The Volunteer For Liberty* 1 (Oct. 18, 1937): 6.

"a big likable fellow": LH, *I Wonder As I Wander,* p. 363.

"a lie told by bullies": *New Masses* 23 (June 22, 1937): 4. See also *The Writer in a Changing World,* ed. Henry Hart (New York: Equinox Cooperative Press, 1937), pp. 69–73.

"the cleverest opportunist in modern history": Ernest Hemingway, "Wings Always Over Africa," *Esquire* 5 (Jan. 1936): 31.

page 349

"Birds? . . . There's no birds.": LH, *I Wonder As I Wander,* p. 347.

"I knew that Spain once belonged": *Baltimore Afro-American,* Oct. 30, 1937.

page 350

"there will be no place": LH, "Negroes in Spain," *The Volunteer For Liberty* 1 (Sept. 13, 1937): 4.

"a sweeter kinder people": LH to Noël Sullivan, Jan 17, 1938, NSP, Bancroft.

"could make the hair rise on your head": LH, *I Wonder As I Wander,* p. 333.

"Midnight in Madrid": LH, "Around the Clock in Madrid: Daily Life in a Besieged City," Sept. 23, 1937, LHP 45.

page 351

"Put out the lights and stop the clocks": LH, "Madrid, 1937"; courtesy of Louise Thompson Patterson.

"HUGHES BOMBED IN SPAIN": *Baltimore Afro-American,* Oct. 23, 1937.

"We captured a wounded Moor Today": LH, "Letter from Spain: Addressed To Alabama," *The Volunteer for Liberty* 1 (Nov. 15. 1937): 3.

"simplicity mingled with great depth": William L. Patterson to LH, Jan. 5, 1938, LHP.

page 352

"Roar, China!": LH, "Roar, China!" *The Volunteer For Liberty* 1 (Sept. 6, 1937): 3.

"Frontiers that divide the people": LH, "Spanish Folk Songs Of The War," *El Voluntario de la Libertad* 2 (June 15, 1938): 15.

page 353

"I thought I might not live": LH, *I Wonder As I Wander,"* p. 357.

"The longer I stayed": LH, *I Wonder As I Wander,* p. 385.

"It's a thrilling and poetic place": LH to Arthur B. Spingarn, Sept. 18, 1937, ABSP, MSRC.

"It was quiet on the front": LH, "Milt Herndon Died Trying to Rescue Wounded Pal," *Baltimore Afro-American,* Jan. 1, 1938.

page 354

"And I, who did not know": LH, *I Wonder As I Wander,* p. 367.

"I didn't want to leave": LH to Marie Short, Jan. 15, 1938, MSP, Bancroft.

"Bad cigarettes, poor wine": LH, "Laughter in Madrid," *Nation* 146 (Jan. 29, 1938): 124.

page 355

"a magnificent and a magnetising man": John Banting, "Nancy Cunard," in *Nancy Cunard,* ed. Hugh Ford, p. 182.

page 356

"I'm afraid there's nothing left": LH to Noël Sullivan, Jan. 15, 1938, NSP, Bancroft.

"Langston was at my apartment": Louise Thompson Patterson to author, Nov. 30, 1981.

"one of the cultural activities": Constitution of the Harlem Suitcase Theatre; courtesy of Louise Thompson Patterson.

"to join the struggle": *A New Workers' Stronghold* (New York: International Workers Order, 1930), p. 6.

page 357

"my own personal playhouse": LH to Mrs. William Grant Still (Verna Arvey), n.d., LHP.

"We are doing things": *Ibid.*

page 358

"the magnitude of your great soul": Theodore Ward to Harlem Suitcase Theatre (LH), Apr. 3, 1938, LHP.

"We're in the egg stage": *Daily Worker,* Apr. 20, 1938.

"This work is the fruit": Michael Gold, Introduction, in LH, *A New Song* (New York: International Workers Order, 1938), pp. 7–8.

"Listen, folks!": LH, *Don't You Want To Be Free?, One-Act Play Magazine* (Oct. 1938): 360.

"a continuous panorama": LH to "Mr. Copeland," Feb. 16, 1963, LHP.

page 359

"We want to build a theater": *New York Amsterdam News,* Apr. 30, 1938.

"a significant proletarian drama": *New York Amsterdam News,* Apr. 30, 1938.

"a home run in sincerity": Owen Dodson to LH, June 11, 1938, LHP.

"simple ferocity and earnestness": Norman Macleod, "The Poetry and Argument of Langston Hughes," *Crisis* 45 (Nov. 1938): 359.

page 360

"We never were able to hold on": Mary Savage to author, interview, June 7, 1984.

"Earl Jones was wonderful": Grace Johnson to author, interview, June 7, 1984.

"I have asked the carpenter": LH to Harlem Suitcase Theatre (Louise Thompson), May 5, 1938, LHP.

page 362

"to which nobody paid any attention": LH to League of American Writers (Franklin Folsom), Aug. 16, 1938, LHP.

"Words put together beautifully": LH, "Words, Writers, and the World," July 25, 1938, LHP 2995.

"our hot evenings": Nancy Cunard to LH, May 25, 1962, LHP.

"infinite capacity to love": LH, prologue to *Nancy Cunard,* ed. Hugh Ford, p. xxi.

page 363

"awfully naive and obvious": Rowena Jelliffe to LH, Oct. 20, 1938, LHP.

page 364

"We can't go on": Dorothy Peterson to LH, n.d. [Aug. 1938], LHP.

"Two or three of the members": Louise Thompson Patterson to author, interview, Nov. 30, 1981.

"Finally there was nothing": Rowena Jelliffe to author, interview, Dec. 7, 1980.

"unique hewing to Caucasian standards": Cleveland *Plain Dealer,* Nov. 17, 1938.

page 365

"tepid and non-descript": *Cleveland Press,* Nov. 17, 1938.

"who identifies his class interests": *Daily Worker,* Nov. 12, 1938.

page 366

"We Write For The Young!": Arna Bontemps to LH, Dec. 19, 1938, LHP.

"something of a precedent in Hollywood": *Los Angeles Daily News,* Feb. 6, 1939.

page 367

"Make the agreement solely with me": LH to Arna Bontemps, n.d. [Mar. 1939]; in Charles Nichols, ed., *Arna Bontemps–Langston Hughes: Letters 1925–1967* (New York: Dodd, Mead, 1980), p. 29.

"I am hell bent on paying": LH to Louise Thompson, Dec. 14 [1938]; courtesy of Louise Thompson Patterson.

"We shall do all we can": Louise Thompson to LH, Dec. 17, 1938, LHP.

"nothing short of raw": Maxim Lieber to LH, Jan. 20, 1939, LHP.

page 368

"Hollywood is our bête noire": LH, ["The Negro in American Entertainment"], June 25, 1951, LHP 757.

"Even allowing for a certain amount": Maxim Lieber to LH, n.d., LHP.

"tell all about how much prejudice": LH to Arna Bontemps, Oct. 30, 1941, LHP.

"Never take a Hollywood job": LH to Arna Bontemps, n.d. [Mar. 1939], in Nichols, *Letters,* p. 29.

"I find it amusing": LH to Carl Van Vechten, Feb. 25, 1939, CVVP, Yale.

page 369

"No, Langston, please don't": Maxim Lieber to LH, Feb. 22, 1939, LHP.

"personality problems": Mary Savage to LH, Mar. 22, 1939, LHP.

"75% a new show": LH to Arna Bontemps, n.d. [Mar. 1939], LHP.

page 370

"leave nothing out": Arna Bontemps to LH, n.d. [Sept. 1939], LHP.

"Magazine offices, daily newspapers": *Daily Worker* (New York), June 5, 1939.

page 371

"one of the most dramatic incidents": *New York Times,* June 5, 1939.

"the Muse-Hughes yarn": *Variety,* July 19, 1939.

"picture-perfect" as well as "box-office perfect": *Los Angeles Times,* July 19, 1939.

"Everybody says they cannot understand": Louise Thompson to LH, Oct. 13, 1939, LHP.

"STATEMENT IN ROUND NUMBERS": LH, "To My Public," Nov. 6, 1939, LHP +3899.

14. The Fall of A Titan

page 373

"I have a charming little Mexican style house": LH to Arna Bontemps, Sept. 9, 1939, in Nichols, *Letters,* p. 36.

"In those days . . . forty-five guests for dinner": Eulah Pharr to author, interview, June 20, 1980.

page 374

"The old book drags along": LH to Louise Thompson, Jan. 19, 1940; courtesy of Louise Thompson Patterson.

"the fantastic falsehood that the U.S.S.R.": *Daily Worker,* August 14, 1939.

"the recent Moscow trials": *Daily Worker,* Aug. 28, 1938.

"a betrayal of people": Harry Dunham to LH, Sept. 10, 1939, LHP.

"the Russian barbarism in Finland": Ralph Bates, "Disaster in Finland," *The New Republic* 101 (Dec. 13, 1939): 221.

page 375

Sullivan declared himself "humiliated and mortified": Noël Sullivan to "Peter" [Ella Winter], copy in Sullivan to LH, Feb. 21, 1940, LHP.

"I am laying off of political poetry": LH to Louise Thompson, Jan. 19, 1940; courtesy of Louise Thompson Patterson.

"reactionary police-patrol spirit": André Breton and Diego Rivera, "Manifesto: Towards a Free Revolutionary Art," trans. Dwight Macdonald, *Partisan Review* 6 (Fall, 1938): 52.

page 376

"bad economics and bad poetry": Carl Van Vechten to LH, Nov. 12, 1939, LHP.

"at white heat": Carl Van Vechten to LH, n.d. [Dec. 1939], LHP.

"a fine performance": Blanche Knopf to LH, Nov. 28, 1939, LHP.

"Please assure me you aren't": Carl Van Vechten to LH, n.d. [Dec. 1939], LHP.

"the background against which I moved": LH to Blanche Knopf, Feb. 8, 1940, LHP.

"the last historian of that period": Carl Van Vechten to LH, n.d. [Dec. 1939], LHP.

page 377

"I hated my father": LH, *The Big Sea* (New York: Alfred A. Knopf, 1940), p. 49.

"the light went out": LH, *The Big Sea,* p. 325.

"My father hated Negroes": LH, *The Big Sea,* p. 40.

"We were in a wild and lonesome-looking country": LH, *The Big Sea,* pp. 60–62.

page 378

"You see, unfortunately, . . . I": LH, *The Big Sea,* p. 11.

"The thought of it": LH, *The Big Sea,* p. 321.
"pace and incident value": LH to *Afro-American* Newspapers (Carl Murphy), Jan. 25, 1940, LHP.

page 379

"a very gay and lively girl": LH, *The Big Sea,* p. 320.
"when the Negro was in vogue": LH, *The Big Sea,* p. 228.
"the intolerance of many whites": LH to Arthur B. Spingarn, Jan. 20, 1940, ABSP, MSRC.
"the prettiest child in the world": LH to Louise Thompson, Nov. 3, 1940; courtesy of Louise Thompson Patterson.

page 380

"I am about to go back": LH to Maxim Lieber, Jan. 4, 1939 [1940?], LHP.
"one of the best poems": LH to Carl Van Vechten, Dec. 15, 1939, CVVP, Yale.
"a light and amusing book": *Ibid.*

page 381

"You've taken my blues and gone": LH, "Note on Commercial Art" [later, "Note on Commercial Theatre"] *Crisis* 47 (March 1940): 79.

page 382

"Your farm, Noel, is a little heaven": LH to Noël Sullivan, May 16, 1940, NSP, Bancroft.
"If Paris is taken": LH to Noël Sullivan, May 20, 1940, NSP, Bancroft.
"What has become, what will become": LH to Noël Sullivan, July 19, 1940, NSP, Bancroft.
"a kind of continual going round": LH to Noël Sullivan, May 16, 1940, NSP, Bancroft.

page 383

"It is a tremendous performance": LH to Richard Wright, Feb. 29, 1940; courtesy of Professor Michel Fabre.
"To get some benefit out of": Blanche Knopf to LH, Apr. 19, 1940, LHP.
"I keep looking for Bigger": LH to Carl Van Vechten, Apr. 23, 1940, CVVP, Yale.
"The hosannas are all for Dick Wright": Arna Bontemps to LH, Nov. 24, 1939, LHP.
"Folks think about same": Arna Bontemps to LH, n.d., LHP.

page 384

"According to my father": Arna Alexander Bontemps to author, interview, June 16, 1984.
"too controversial for us": Columbia Broadcasting System (Davidson Taylor) to Maxim Lieber, Mar. 22, 1940, LHP.

page 385

"close, closer than blood brothers": Alberta Bontemps to author, interview (with Arna Alexander Bontemps), Dec. 23, 1984.
"I have no early memory": Paul Bontemps to author, interview (with Arna Alexander Bontemps), Dec. 23, 1984.

page 386

the "sepia Steinbeck": *Baltimore Afro-American,* Mar. 9, 1940.
"The ladies of the race": LH to Arna Bontemps, Sept. 9, 1939, in Nichols, *Letters,* p. 37.

page 387

"Just how much culture": LH to Maxim Lieber, July 9, 1940, LHP.
"We need twice 75 more years": LH to Margaret Bonds, July 14, 1940, LHP.
"I think a gangster would": LH to Maxim Lieber, Aug. 2, 1940, LHP.
"certainly one of the chief abodes": LH to Noël Sullivan, July 19, 1940, NSP, Bancroft.
"the most prolific and the most representative": J. Saunders Redding, *To Make A Poet Black* (Chapel Hill: Univ. of North Carolina Press, 1939), p. 115.
"perhaps the best bet": Arna Bontemps to LH, Apr. 21, 1941, in Nichols, *Letters,* p. 80.

page 388

"most Negroes hate the book": *Ibid.*
"my most handsome jacket": LH to Blanche Knopf, July 30, 1940, LHP.
"a tremendous sale": Ben Abrahamson to LH, Aug. 13, 1940, LHP.
"really fine" *New York Times* review: Blanche Knopf to LH, Aug. 22, 1940, LHP.
"That boy is sure kicking up dust": Arna Bontemps to LH, Jan. 21, 1940, LHP.
"both sensitive and poised": *New York Times Book Review,* Aug. 25, 1940.

"does not allow much bitterness": *New York Sun,* Aug. 28, 1940.

"because it is true": *Knoxville Journal,* Aug. 25, 1940.

"absolute intellectual honesty": *Saturday Review of Literature* 22 (Aug. 31, 1940): 12.

page 389

"too gentle with us": Ella Winter to LH, Oct. 15 [1940], LHP.

"too much attention is apt to be given": Ralph Ellison, "Stormy Weather," *New Masses* 36 (Sept. 24, 1940): 20–21.

"not a single mention of a radical publication": Walt Carmon to LH, Sept. 18, 1940, LHP.

"a manly tradition": Richard Wright, "Forerunner and Ambassador," *The New Republic* 103 (Oct. 24, 1940): 600.

"prim and decorous ambassadors": Richard Wright, "Blueprint for Negro Writing," *New Challenge* 2 (Fall 1937): 53.

"Dear Grand Duchess": Carl Van Vechten to Blanche Knopf, Dec. 28, 1939, Alfred A. Knopf Inc. files, New York, N.Y.

page 390

"The whole book sings": Carl Van Vechten to LH, Nov. 22, 1940, LHP.

"ATTEND THE LUNCHEON CHRISTIANS": Copy in LH files, Federal Bureau of Investigation: File 100–15139 (section 1).

page 391

Hughes was "Old Satan": *Los Angeles Times,* Nov. 16, 1940.

"in this charming democracy": LH to Maxim Lieber, Dec. 17, 1940, LHP.

just "one MORE time": LH to Maxim Lieber, June 16, 1940, LHP.

"I'm America's YOUNG BLACK JOE": LH, "America's Young Black Joe," n.d., LHP 2037.

page 392

"a good chance of becoming": LH to Maxim Lieber, Dec. 28, 1940, LHP.

"My name's Richard Wright and I try to write": LH, "Native Son: The Boogie Woogie Man," n.d., LHP 749.

page 393

"having left the terrain": LH, "Statement Concerning 'Goodbye, Christ'," Jan. 1, 1941, LHP 262.

"Lord help me!": LH to Arna Bontemps, Dec. 30, 1940, in Nichols, *Letters,* p. 64.

"broke and remorseful as usual": *Ibid.*

"a great loss to the project": Hollywood Theatre Alliance (Elliott Sullivan) to LH, Dec. 31, 1940, LHP.

"All of human terror": LH, "Jeffers: Man, Sea, and Poetry," LHP 559; in *Carmel Pine Cone,* Jan. 10, 1941.

page 394

"with tenderness & caution!!": Files of the Community Hospital of the Monterey Peninsula (Admission Sheet, Personal History, Physical Examination, Progress Record, Doctor's Orders, and Laboratory Report); courtesy of the Medical Records Department of the hospital and by permission of George Houston Bass. I am indebted to Dr. Howard M. Spiro, Head of Gastro-Enterology at the Yale University School of Medicine, for his assistance in securing the full report, and his analysis of its contents.

"Last straw N.Y. eviction": LH to Arna Bontemps, n.d. [Jan. 1941], in Nichols, *Letters,* p. 72.

page 395

"Hughes has been bitten": *People's World* (San Francisco), Jan. 15, 1941.

ACKNOWLEDGMENTS

I BEGAN THE RESEARCH for this book sometime in the summer of 1979, a few months after being introduced at a concert in Cambridge, Massachusetts, to George Houston Bass, a professor of English at Brown University and executor-trustee of the Langston Hughes estate. Hughes had expected that his closest friend, Arna Wendell Bontemps, would write his biography; unfortunately, Mr. Bontemps died in 1973, before he could begin the task. Since that time, the Hughes papers at Yale had been closed to almost all researchers. A few weeks after our meeting, Mr. Bass asked me to think about accepting the task. I am grateful to him for his quick and generous responses to my appeals from time to time for help, and also for scrupulously allowing me to write the book I wanted to write, without interference.

Much misinformation concerning Langston Hughes has passed into print—some, perhaps most, from his own pen, and especially in his two volumes of autobiography. I have tried to correct the record without pockmarking my text with reprimands. However, I would like to touch on one point in particular here. Although the widely used "James Mercer Langston Hughes" has a certain ring to it, and although Hughes joked once, late in his life, that this was indeed his name, no other document represents his full name as being other than James Langston Hughes. As for the increasingly fashionable tendency to assert, without convincing evidence, that Hughes was a homosexual, I will say at this point only that such a conclusion seems unfounded, and that the evidence suggests a more complicated sexual nature.

I am grateful to a number of institutions for their support, especially the National Endowment for the Humanities, the Rockefeller Foundation, and the Center for Advanced Study in the Behavioral Sciences. A grant-in-aid from the American Council of Learned Societies made possible my first summer of research. While I was on the faculty of Stanford University, I was always able to draw on its generous support; I thank in particular Bliss Carnochan, William Chace, Diane Middlebrook, Albert Gelpi, Robert Polhemus, Anne Mellor, and Jay Fliegelman, who read most of the manuscript at an important point. At Rutgers University, I am especially indebted to George Levine, Richard Poirier, Thomas Van Laan, Donald Gibson, and Cheryl Wall of the Department of English, as well as to Tilden Edelstein, Dean of the Faculty of Arts and Sciences, and Kenneth Wheeler, Provost of the New Brunswick campus.

Among librarians, I would like to thank above all the administrators and staff of the Beinecke Rare Book and Manuscript Library at Yale University, where the Langston Hughes Papers, donated by Hughes himself, may be found in the James Weldon Johnson Memorial Collection founded by Carl Van Vechten. I am indebted to Donald Gallup and David Schoonover, respectively the former and the present curator of the Collection in American Literature, as well as to Lisa Browar, Joan Hofmann, Steve Jones, and many others on the staff (past and present) for their unfailing skill and courtesy. I am also grateful to George P. Cunningham who, with Patricia Gaskins and Martha Schall, organized the Langston Hughes correspondence at the Beinecke.

I must also thank members of the staff at the Amistad Research Center (New Orleans), the Archives Department of the Atlanta University Center Woodruff Library, the Bancroft Library of the University of California, the Bibliothèque Nationale, the Harrison Memorial Library of Carmel, California, the Library of Congress, the Columbiana Collection at Columbia University, the Butler Library at Columbia University, the archives of the Federal Bureau of Investigation, the Fisk University Library, the Civica Biblioteca Berio of Genoa, Italy, the Hatch-Billops Collection, the Hoover Institution, the Huntington Library, the Carl Sandburg Collection of the University of Illinois, the Lenin Library in Moscow, the Lincoln, Illinois, Public Library, Lincoln University (Pa.) Library, the Kenneth Spencer Research Library of Kansas University, the University of Massachusetts, Amherst, the Moorland-Spingarn Research Center at Howard University, the National Archives, the New York Historical Society, the Bobst Library of New York University, the New York Public Library, Tilden, Lenox, and Astor Foundations, including the Schomburg Center for Research in Black Culture, the Berg Collection in English and American Literature, and the Manuscripts and Archives Division, the Southern Historical Collection of the University of North Carolina, the Oklahoma Department of Libraries, the Oberlin College Archives, Rutgers University Library, Stanford University Library, the George Arents Research Library of Syracuse University, the library of the University of Texas, Austin, the Western Reserve Historical Society, and the Vivian Harsh Collection at the Carter G. Woodson Regional Library Center of the Chicago Public Library. For collections of material still in private hands, I would like to thank especially Louise Thompson Patterson and Gladys Sullivan Mahoney.

I was granted interviews by the following people, all of whom had known Langston Hughes and received me courteously: Raoul Abdul, Gonzales Austen, Louis Aschille, Margaret Walker Alexander, Alberta Bontemps, Arna Alexander Bontemps, Paul Bontemps, Amiri Baraka, Roy Blackburn, James Baldwin,

Richmond Barthé, Lebert "Sandy" Bethune, the late Marguerite Cartwright, Matt N. Crawford, John Henrik Clarke, Maxwell T. Cohen, Mercer Cook, Leander James Crowe, Sims Copans, Henri Cartier-Bresson, William R. Cole, St. Clair Drake, Robert Dudley, Arthur P. Davis, Ann Dufaux-Rhodes, Jimmy "Lover Man" Davis, Roy De Carava, Toye Davis, Mr. and Mrs. Robert D. Ericson, Ralph Ellison, James A. Emanuel, Albert A. Edwards, Michel Fabre, Elton Fax, W. Edward Farrison, Blyden Jackson, Adele Glasgow, the late Paul Green, Eugene and Thomasena Grigsby, Harold L. Gaines, Jean Blackwell Hutson, Grace Johnson, Rowena Woodham Jelliffe, Ted Joans, Leroy D. Johnson, John Henry Jones, Bruce Kellner, the late George Kent, Si-lan Chen Leyda, Maxim and Minna Lieber, the late Rayford W. Logan and Henry Lee Moon, Milton Meltzer, William and Alice Mahoney, Jan Meyerowitz, Richard Bruce Nugent, Therman B. O'Daniel, Louise Thompson Patterson, Lindsay Patterson, Eulah Pharr, George N. Redd, Jay Saunders Redding, Mary Savage, the late John D. Short, Jr., Elie Siegmeister, Sue Bailey Thurman, Prentiss Taylor, Nate V. White, Thomas A. Webster, the late Jean Wagner, Zeppelin Wong, and Bruce M. Wright.

Many people contributed in a variety of ways also indispensable to my research—by responding to requests for information, aiding other aspects of my research, reading portions of the manuscript, or generously opening their homes to me as I travelled in the United States, Europe, and the Soviet Union. I thank Katie Armitage of Lawrence, Kansas (who did all of the above), Michel Auffray of the Institut Francophone de Paris, M. Margaret Anderson, Sam Allen, Ivan von Auw, Florence E. Borders, Nancy Abrams, Jervis Anderson, Ursula Berg-Lunk, Faith Berry, Jean-Claude Boffard, Esme Bhan, Richard Brophy, Phyllis Bischof, E. Randolph Biddle, William E. Bigglestone, Thomas H. Bresson, William C. Beyer, Richard Barksdale, Marcellus Blount, Rae Linda Brown, Rudolph Byrd, Dottie Berryman, Harold W. Billings, Schuyler Chapin, Brigitte Carnochan, Jack Chen, Leslie Collins, Wayne Cooper, Michael Clegg, Charles Cooney, Jackie Coogan, Sophy Cornwell, Catherina Caldwell, Marina Castañeda, James Casey, the late Ida Cullen Cooper, Dean Crawford, Walter Daniel, Gail Cohen, the late Charles T. Davis, Thomas Donaldson, Carolyn Davis, Martin Duberman, Ruth Donaldson, Mari Evans, Katherine Emerson, Dolores Elliott, Marcia Ellis, Lillie Johnson Edwards, Grace Frankowsky, Wendell Foster, Robert Farnsworth, Geneviève Fabre, Darra Goldstein, Mary Green-Cohen, Peter Gibian, Sandra Govan, Christine Gilson, J. Lee Greene, Selma George, Chryss Galassi, James Geiwetz, Richard Green, Henry Harder, Nathan Irvin Huggins, Ken Hall, Louis R. Harlan, Kathleen Hauke, Paul Horgan, Melvin Hinton, Akiba Sullivan Harper, Michael Wesley Harris, James Hatch, Gary Imhoff, Donald Joyce, Eloise McKinney Johnson, Millicent Dobbs Jordan, Steve Jansen, Philip L. Jones, Howard M. Jason, Ivy Jackman, Charles H. Johnson, the late Arthur Koestler, William Koshland, Roberta Klatzsky, Allen Klots, Phyllis R. Klotman, Jamie Katz, William Kuntsler, the late Alfred A. Knopf, Vera Kubitskaya, Hilja Kukk, Jay Leyda, Werner Lange, Gardner Lowell, Diana Lachatanere, Vaughn C. Love, Kenneth Lohf, Susan Lardner, Cliff Lashley, David Levin, David Levering Lewis, Edward Lyon, Richard A.

Long, Paule Marshall, Marlene Merrill, Robert W. McDonnell, John H. Mackey, Harry Murphy, Alice Clarke McClanahan, Henry G. Morgan, James Murray, Penelope Niven McJunkin, Michael Miller, Daphne Muse, Edward J. Mullen, Mark Naison, Minda Novek, Olivia Martin, Kenneth P. Neilson, Richard Powell, Paul R. Palmer, Zella J. Black Patterson, Bernard Perry, LeRoy Percy, Youra Qualls, Virginia Rust, Natasha Russell, Sylvia Lyons Render, Daniel and Florence Ryan, Leslie Sanders, Everett Sims, Richard Schrader, Mark Scott, Jessie Carney Smith, Ann Allen Shockley, Deborah Willis-Ryan, Paulette Sutton, Reuben Silver, Eloise Spicer, Sandra Solimano, Kaethe Schick, Murray Sperber, John Stinson, Marian and Howard M. Spiro, James Spady, G. Thomas Tanselle, Arthur L. Tolson, John Edgar Tidwell, A. Morgan Tabb, Ann Tanneyhill, Joy Thompson, Roy Thomas, Robert Thomas, William M. Tuttle, Jr., Honor Tranum, Lenora Vance-Robinson, Kraig S. Weston, Beatrice White, Grace Walker, Emery Wimbish, Jr., Emily Woudenberg, H. A. Selby Wooding, William H. Waddell, Kenneth Warren, R. B. Wellington, Sheryl K. Williams, Allen Wright, Thomas Weir, Ella Wolfe, Richmond B. Williams, Mary Jane Welch, Rhoda Wynn, Michael R. Winston, Thomas H. Wirth, Irving and Marilyn Yalom, and Andrea Lerner Young.

Bruce Kellner, Arna Alexander Bontemps, and my wife Marvina White read the entire manuscript and provided advice, criticism, and encouragement. At Oxford University Press, I thank in particular William Sisler for his confidence in the book, and Susan Meigs for her unstinting efforts to make it better. Peggy Humphreys and Quitman Marshall read the galleys expertly. My thanks, also, to Patricia Powell of Harold Ober Associates. No doubt, I have failed to acknowledge some of the persons who helped me along the way; but I am grateful to them.

For permission to quote excerpts from previously unpublished material written by Langston Hughes, I thank George Houston Bass, executor-trustee of the Langston Hughes Estate. For permission to quote material and reproduce photographs from the Langston Hughes Papers, Claude McKay Papers, Countee Cullen Papers, and Wallace Thurman Papers, I thank the Collection of American Literature, Beinecke Rare Book and Manuscript Library, Yale University. I also gratefully acknowledge permission from Amistad Research Center (Clifton H. Johnson, curator), for quotations from the Countee Cullen Papers; Atlanta University Center Woodruff Library, Archives Department, for quotations from the Countee Cullen-Harold Jackman Collection; Bancroft Library, University of California, Berkeley, for letters from Noël Sullivan and Marie Short to LH, and transcript of interview with Ben Lehman, Regional Oral History Office; Edmund R. Biddle, for letters from Mrs. Rufus Osgood Mason; Carl Cowl and the Claude McKay Estate, for letters from Claude McKay; Ida M. Cullen-Cooper, for unpublished letters by Countee Porter Cullen; Clive E. Driver and the Marianne C. Moore Estate for a letter from Marianne Moore; Ralph Ellison, for letters to LH: Hill and Wang, for excerpts from *The Big Sea;* Dr. C. J. Hurston, for letters from Zora Neale Hurston; Ivie Jackman, for letters from Harold Jackman; Rowena Woodham Jelliffe, for a letter to the Rosenwald Foundation concerning LH; Alfred A. Knopf, Inc. for letters from Blanche Knopf, Alfred A. Knopf, and other Knopf officials to LH, and for quotations from *The Weary Blues, Fine Clothes to the Jew,* and *Not Without Laughter;* Si-lan Chen Leyda, for her letters to LH; Maxim Lieber, for letters to LH; Nicholas C. Lindsay, for an inscription by Vachel Lindsay; Moorland-Spingarn Research Center (Thomas C. Battle, curator), for letters from Alain Locke; New York Public Library, Astor, Lenox and Tilden Foundations, Henry W. and Albert A. Berg Collection, for excerpts from the American Play Company Papers, and Rare Book and Manuscripts Division, for excerpts from the Joel E. Spingarn Papers; Louise Thompson Patterson, for letters to LH and to her mother, as well as for various photographs and other documents; Charles Roxborough, for letters of Elsie Roxborough; Mary A. Savage, for a letter to LH; Schomburg Center for Research in Black Culture, for photograph of Alain Locke; Madge S. Short, for letters from Marie Short to LH; Judith Anne Still, for a letter from William Grant Still; Prentiss Taylor, for letters to LH; Carl Van Vechten Estate (Donald Gallup, literary trustee), for excerpts from letters from Van Vechten; H. A. Selby Wooding, for letters from Anne Marie Coussey. For permission to quote Arna Bontemps' "Nocturne at Bethesda," I thank Mrs. Alberta Bontemps and Harold Ober Associates.

INDEX